Solid State Physics

Vimal Kumar Jain

Solid State Physics

Third Edition

Ane Books
Pvt. Ltd.

Vimal Kumar Jain
Department of Physics
Maharshi Dayanand University
Rohtak, India

ISBN 978-3-030-96019-3   ISBN 978-3-030-96017-9 (eBook)
https://doi.org/10.1007/978-3-030-96017-9

Jointly published with ANE Books India
ISBN of the Co-Publisher's edition: 978-9-384-72601-0

This Springer imprint is published by the registered company Springer Nature Switzerland AG
The registered company address is: Gewerbestrasse 11, 6330 Cham, Switzerland

To my grandkids
Suvyam, Siona, Sijjans and Seeyal

Preface to the Third Edition

This book Solid State Physics is an introductory text for graduate and post-graduate students in physics, material science and engineering. The first edition of this book was published in the year 2013, while the second edition appeared in 2015.

In the present edition, some new topics are introduced, while some of the existing topics like phonons, Drude–Lorentz model, Fermi levels, electrons, holes, etc. are modified. The segment on band theory has been revised and expanded by the addition of a nearly free electron model and tight-binding approximation. The chapter on semiconductors has been revised, and subjects of scattering, doping and defects in semiconductors are included. Some interesting features of high-temperature superconductors are also incorporated in Chap. 14. A text dealing with the density of states, Normal and Umklapp processes, Boltzmann equation and electrical conductivity also forms a part of this edition.

Meerut, India
Vimal Kumar Jain

Preface to the Second Edition

I am thankful to the readers for the interest for this book. The second edition has been enlarged by including the reflection, absorption, photoconductivity, luminescence, colour centres, interband transitions, and excitons. Semiconductor devices like tunnel diode, Gunn diode, photodiode, photoconductive diode, varactor diode, solar cell, LED, semiconductor lasers and semiconductor detectors are also described in the 2nd edition of the book. I hope this will provide a basis for understanding the various concepts.

Meerut, India   Vimal Kumar Jain

Preface to the First Edition

The book 'Solid State Physics' has been written as a textbook for graduate and postgraduate students of physics as well as for the course of B.Tech. students. This textbook has grown out of many years of 'course notes' that I developed for M.Sc. students covering many aspects of solids including its structure, bonding, defects, electrical, thermal, dielectric, magnetic properties, NMR, ESR, Mössbauer effect, and superconductivity. A part on nanomaterials is also included.

A large number of solved, unsolved problems based on the articles are given at the end of each chapter to supplement the text. Objective questions along with their answers help students to evaluate their skills. The students will find a fair amount of detail in the examples and derivations given in the chapter. The phrase *it can be shown that* has been avoided wherever possible. The vectors are denoted with bold face in the book.

I am grateful to the authors whose work I have consulted in the process of developing this book. A list of some of these is given at the end. I must also acknowledge the unreserved cooperation and support of my dear parents and my children Ira, Manu and Manish. Finally I must thank Mr. Sunil Saxena and Mr. J. R. Kapoor of Ane Books Pvt. Ltd. for readily agreeing to undertake this project.

Meerut, India   Vimal Kumar Jain

Contents

Contents

About the Author

Dr. Vimal Kumar Jain Alexander von Humboldt Fellow, received his Ph.D. from Indian Institute of Technology Kanpur in 1977. He is a former Professor and Head of Physics Department at Maharshi Dayanand University, Rohtak. He was also a Professor of Applied Science in P. D. M. College of Engineering, Bahadurgarh, Professor and Head of ECE Department, Vidya College of Engineering, Meerut and Professor of Physics, IIMT Engineering College Meerut. Dr. Jain was also a Visiting Professor at Instituto de Fisica, Universidad Federal do Rio de Janeiro, Brazil. He has been working in the field of Electron Paramagnetic Resonance and published four review articles and more than hundred research papers in National and International Journals. Dr. Jain has supervised the research work of several M.Phil. and Ph.D. students and worked on various sponsored research projects. He has also authored books on *Atomic and Molecular Spectroscopy, Classical Mechanics and Quantum Mechanics.*

Crystal Structure

1

1.1   Amorphous and Crystalline Materials

When atoms are brought together, there are forces of interaction acting between them. At comparatively large distances the forces of attraction increase rapidly as the distance between them decreases. At small distances the force of repulsion comes into picture which increases with further decrease in distance. At a particular distance, the force of attraction and repulsion balance each other, and the resultant force of interaction vanishes. This is a state of stable equilibrium, and a state of minimum energy is achieved. Therefore, the free atoms arrange themselves in a strict order at equilibrium distance from one another forming a three dimensional periodic arrangement known as crystalline state since the energy of the ordered atomic arrangement is lower than that of an irregular packing of atoms. The atoms forming the solids may be identical or different. The counterpart of the crystalline state of solids is the amorphous state. Amorphous materials are condensed phase which do not possess the long range translational (or orientational) order (or periodicity) characteristic of a crystalline state. For a crystalline solid, when the periodic and repeated arrangement of atom is perfect or extends throughout the entirety of the specimen without interruption, the result is a single crystal. The term amorphous and non-crystalline are synonymous under this condition. The term glassy has the same structural meaning, but in addition it also usually implies that the material exhibit a glass transition. The fact that an amorphous solid has no structural periodicity means that the structure can be described in terms of a periodically continued, finite sized unit cell. The structure of an amorphous solid could be described in terms of an infinite unit cell. Examples of amorphous solids are (i) glasses, (ii) ceramics, (iii) gels, (iv) polymers, and (v) thin film systems deposited on a substrate at low temperatures.

© The Author(s) 2022
V. K. Jain, *Solid State Physics*, https://doi.org/10.1007/978-3-030-96017-9_1

1.2   Lattice Points, Basis and Crystal Structure

Figure 1.1 shows a pattern which repeat itself periodically in one dimension.

Suppose that an object (for example number 7) is translated to a distance t_1, in one dimension. It is further moved repeatedly by the same distance then a collection of 7's results as shown in Fig. 1.1. The single translation in Fig. 1.1 produces an infinite linear array of the repeated object. If such a translation, t_1, is combined with another non-collinear translation t_2, then a two dimensional array obtained as follows. The entire linear array due to translation t_1 is repeated an infinite number of times by the second translation t_2 (Fig. 1.2). Another way of looking at this is to say that the linear array due to t_2 is repeated by t_1.

Since the nature of repeated objects in Fig. 1.2 does not affect the translational periodicity, it is conventional to represent this periodicity by replacing the object in the array with a point. Any collection of points shown in Fig. 1.3 is called a lattice. A point is an imaginary infinitesimal spot in space, and consequently a lattice of point is imaginary also. On the other hand the array of 7's shown in Fig. 1.2 is real. It is not a lattice of 7's

7   7   7   7   7   7   7

$\longleftrightarrow$
t_1

Fig. 1.1 One dimensional array of objects

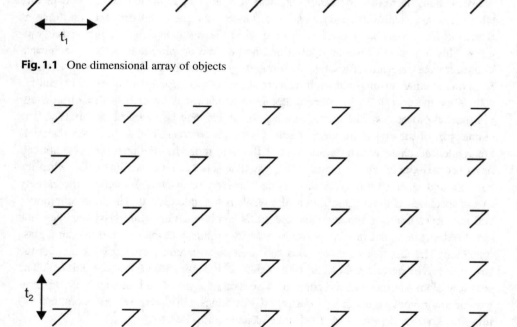

Fig. 1.2 Two dimensional array of objects

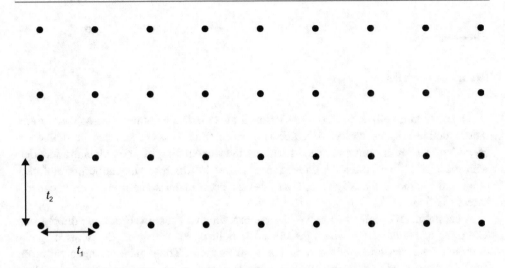

Fig. 1.3 Two dimensional array of points

because a lattice is an imaginary concept; instead it is correctly called a lattice array of 7's.

It is possible to add a third translation to the plane lattice array. The third translation repeats the entire plane at equal interval, t_3. This third translation thus produces a space lattice or a three dimensional lattice, array (Fig. 1.4).

Fig. 1.4 Three dimensional array of objects

Fig. 1.5 A linear lattice

Thus two non-collinear translations define a plane lattice and three non-coplanar translations define a space lattice. A lattice is a geometrical concept and can be defined as 'A lattice is a set of points in space such that the surrounding of one points are identical with those of all the others'. A collection of points in which the neighbourhood of each point is the same as the neighbourhood of every point under some translation is called a Bravais lattice.

A crystal is an excellent example of a pattern which repeats itself in three dimensions. The units of pattern from which the crystal is built up are either single atoms or group of atoms. Each repeating unit is known as basis or motif. The regular arrangement of the motif in the crystal is described by assigning to each a lattice point and forming a three dimensional or space lattice. A crystal structure is formed only when a basis of atoms is attached identically to each lattice point. This can be written as

$$\textbf{Lattice} + \textbf{Basis} = \textbf{Crystal Structure}.$$

1.2.1 The Linear Lattice

A linear lattice can be described by just one translation vector **a** (called fundamental or conjugate or primitive vector) drawn between the adjacent points (Fig. 1.5).

A vector drawn between any pair of points is called a lattice translation vector **T**. It is always an exact multiple of the primitive translation vector which is itself a special case of lattice translation vector. Therefore,

$$T = ua \tag{1.1}$$

where u is any integer, positive or negative.

1.2.2 The Plane Lattice

The planar lattice is described by a pair of primitive translation vectors, **a** and **b**. They must be selected so that a suitable combination of them can define the relative position of two lattice points. A pair of such vectors is normally drawn from the same origin.

The operation of linear combinations of the vector **a** and **b** of the form

$$ua + vb$$

(where u and v are positive or negative integers) can give translation from one lattice points to any other. The vectors **a** and **b** therefore constitute a primitive vector pair. These combinations are then equivalent to any lattice translation vector **T** given by

$$T = ua + vb \qquad (1.2)$$

1.2.3 The Space Lattice

A three dimensional space lattice can be described by the three vectors forming a primitive vector group. They are drawn from the same origin. Any translation vector **T** can then be described by a suitable combination of the three vectors **a**, **b** and **c** forming the primitive group

$$T = ua + vb + wc \qquad (1.3)$$

where u, v and w are positive or negative integers. Any two lattice points are connected by a vector of the form (1.3).

1.3   Unit Cell and Primitive Cell

It is convenient to subdivide the space occupied by a lattice into small parallelepipeds whose edges are three non-coplanar translations. Each of these cells is related to any of the others through a translations parallel to its edges. A lattice can be blocked out in cells in numerous ways (Fig. 1.6).

The parallelepiped which is formed by the translation vector a_1, b_1; a_2, b_2; a_3, b_3; a_4, b_4 is called a unit cell. If the lattice points are only at the corners of a unit cell, then it is called a primitive cell; otherwise it is non-primitive. The translation vectors a_i, b_i which forms a primitive unit cell are called primitive translation vector.

Properties of Primitive Cells

1. The cell will completely cover the lattice area/volume when it is repeated with its origin at each lattice point.
2. A primitive cell is a minimum volume cell.
3. The area or volume of different primitive cells is equal, irrespective of the choice of primitive translation vectors.

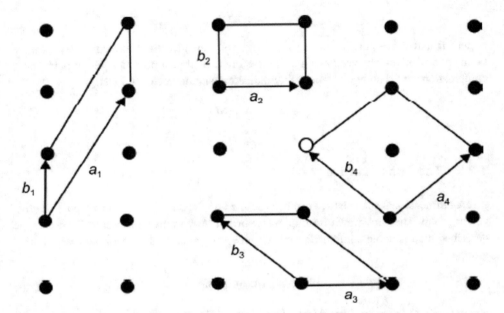

Fig. 1.6 Lattice points, of a space lattice in two dimensions. Lattice points are represented by filled circles. All pairs of vectors are translation vectors of the lattice

4. Each cell contains the equivalent of one lattice point. Take for example, cell formed by a_1 and b_1. There is a lattice point at each of the four corners, total four lattice points, but each point is shared between a total of four cells and thus has an equivalent value of 1/4. The cell formed by a_4 and b_4 is non-primitive unit cell because it is outlined by a pair of lattice translation vectors which do not constitute a primitive vector pair.

1.3.1 Wigner–Seitz Cell

A Wigner–Seitz cell is an alternative way of selecting a primitive unit cell of area equal to the other conventional primitive cells. To create a Wigner–Seitz cell, the following steps are taken

1. Choose any lattice site as the origin.
2. Starting at the origin draw vectors to all neighbouring lattice points.
3. Construct a plane perpendicular to and passing through the midpoint of each vector.
4. The area enclosed by these planes is the Wigner–Seitz cell (Fig. 1.7).

Fig. 1.7   Wigner–Seitz unit
cell in two dimensions

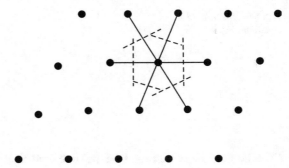

The Wigner–Seitz cell has the following characteristic

1. Wigner–Seitz cells associated with all lattice points are identical in size, shape and orientation as follows with the translational symmetry of the lattice.
2. When stacked the Wigner–Seitz cell fill all the space.
3. The Wigner–Seitz cell is polyhedron.
4. The Wigner–Seitz cell has the full point symmetry of its lattice point.

1.3.2   Areas and Volume of Unit Cells

(1) The Area of Planer Unit Cell

The area of a parallelogram (Fig. 1.8) is given by.

$$\text{Area} = \text{base} \times \text{height}$$
$$\text{Height} = h = |b|\cos(90 - \theta) = |b|\sin\theta$$
$$\text{Base} = |a|$$
$$\text{Area} = |a||b|\sin\theta = |a \times b|$$

(1.4)

Fig. 1.8   Area of a
parallelogram

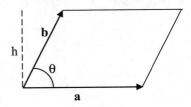

Fig. 1.9 Volume of a
parallelepiped

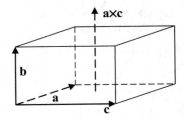

(2) **The Volume of a Three Dimensional Unit Cell**

The volume V of a parallelepiped (Fig. 1.9) is given by.

$$V = (\text{height}) \times (\text{areaofbase}) \tag{1.5}$$

The base area is the magnitude of the vector $a \times c$. The height is the vector $\mathbf{b}$ resolved
in the direction perpendicular to the base, that is, parallel to the vector $a \times c$

$$V = \left(\mathbf{b} \cdot \frac{a \times c}{|a \times c|} \cdot (a \times c) \right) = |\mathbf{b} \cdot a \times c| \tag{1.6}$$

1.4   Symmetry Operations

A symmetry operation is an operation that can be performed either physically or imagi-
natively that result in no change in the appearance of an object. In crystals, the symmetry
is internal, that is an ordered geometrical arrangements of atoms and molecules on the
crystal lattice. The internal symmetry is reflected in the external form of perfect crystal. Symmetry operations are used to describe the crystal's outward symmetry. Symmetry
operations help to define the manner in which a crystal can repeat the facets or faces on
their crystal surface. There are three main types of symmetry operations:

(i)   Symmetry with respect to a line (rotational symmetry);
(ii) Symmetry with respect to a plane (mirror symmetry); and
(iii) Symmetry with respect to a point (inversion symmetry).

1.4.1 Rotational Symmetry

If an object can be rotated about an axis and repeats itself every $2\pi/n$ rotation, then it is said to have an axis of n-fold rotational symmetry. The axis along which the rotation is performed is an element of symmetry referred to as a rotation axis. The following types of rotational symmetry axes are possible in crystals

> **1-fold rotation axis**: An object that requires rotation of 2π $(n = 1)$ in order to restore to its original appearance has no rotational symmetry (Fig. 1.10). Since it repeat itself one time every 2π, it is said to have a 1-fold axis of rotational symmetry.
> **2-fold rotation axis**: If an object appears identical after a rotation of π (Fig. 1.11), that is twice in a 2π rotation, then it is said to have a 2-fold rotation axis $(n = 2)$. The rotation axis is an imaginary line that extends outwards perpendicular to the page. A filled oval shape represents the point where the 2-fold rotation axis intersects the page.
> **3-fold rotation axis**: If an object appears identical after a rotation of $2\pi/3$ (Fig. 1.12) that is thrice in 2π rotation, then it is said to have a 3-fold rotation axis $(n = 3)$. A filled triangle shape represents the point where the 3-fold rotation axis exists.
> **4-fold rotation axis**: If an object appears identical after a rotation of $\pi/2$ (Fig. 1.13) that is four times in 2π rotation, then it is said to have a 4-fold rotation axis $(n = 4)$. A filled square shape locates the point where the 4-fold rotation axis exists.

Fig. 1.10 Onefold rotation

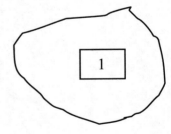

Fig. 1.11 Twofold rotation

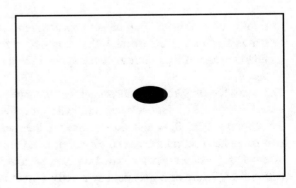

Fig. 1.12 Threefold rotation

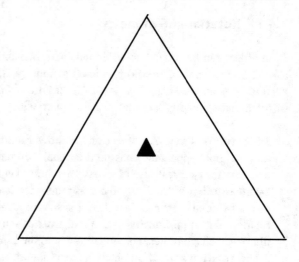

Fig. 1.13 Fourfold rotation

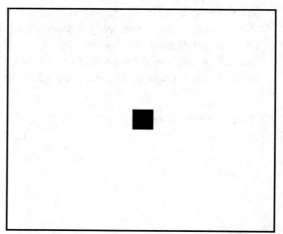

6-fold rotation axis: If an object appears identical after a rotation of $\pi/3$ (Fig. 1.14) that is six times in 2π rotation, then it is said to have a 6-fold rotation axis ($n = 6$). A filled hexagon shape represents the symbol for the 6-fold rotation axis.

These are the only rotational symmetries that can exist in crystals; all others are disallowed. These five rotational axes are called the five proper axes and are described by the symbols 1, 2, 3, 4 and 6. The objects themselves may appear to have 5-fold, 7-fold or higher-fold rotation axes. However, these are not possible in crystals because the external shape of a crystal is based on the geometric arrangement of atoms. The objects with 5-fold, 7-fold or higher fold symmetry cannot be combined in such a way that they completely fill the space. A single molecule may well have a fivefold symmetry.

Fig. 1.14 Sixfold rotation axis

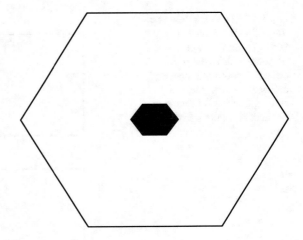

1.4.2 Mirror Symmetry

A mirror symmetry operation is an imaginary operation that can be performed to reproduce an object. The operation is done by imagining that is cut into half, and then mirror is placed next to one of the halves of the object along the cut. If the reflection in the mirror reproduces the other half of the object, then the object is said to have mirror symmetry. The plane of the mirror is an element of symmetry referred to as a mirror plane and is symbolized with the letter *m*.

The rectangle shown in Fig. 1.15 has two planes of mirror symmetry. The rectangle has a mirror plane that runs vertically on the page and perpendicular to the page. The rectangle of Fig. 1.16 has a mirror plane that runs horizontally and is perpendicular to the page. The broken lines of the rectangle show the part that would be seen as a reflection in the mirror. The rectangle has two planes of mirror symmetry. Three dimensional and more complex objects could have more.

1.4.3 Inversion Symmetry

The centre of symmetry operation refers to an operation which would invert the original face of a crystal through the centre of the crystal. If an object has only a centre of symmetry, we say that it has a 1-fold rotoinversion axis. Such an axis has the symbol $\bar{1}$. Let $f(x, y, z)$ is any function of Cartesian coordinates (x, y, z), then inversion I through the origin is

$$\text{If}(x, y, z) = f(-x, -y, -z)$$

If $(x, y, z) = f(-x, -y, -z)$, then the origin $(0,0,0)$ is said to be the inversion centre or centre of symmetry. It is a single point. Under inversion operation each point of the objects moved through inversion centre to other side, that is, coordinate (x,y,z) goes to (-x,-y,-z).

Fig. 1.15 The rectangle at the top has a mirror plane that runs vertically on the page and is perpendicular to the page. The dashed part of the rectangle below shows the part the rectangle that would be seen as a reflection in the mirror

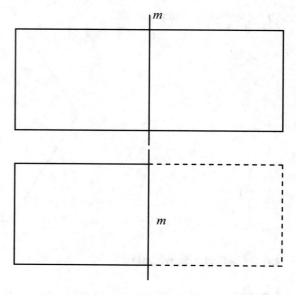

Fig. 1.16 The rectangle at the top has a mirror plane that runs horizontally on the page and is perpendicular to the page. The dashed part of the rectangle below shows the part the rectangle that would be seen as a reflection in the mirror

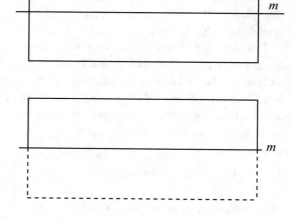

1.4.4 Rotoinversion

Combination of rotation with inversion symmetry performs the symmetry operation of rotoinversion. Objects that have rotoinversion symmetry have an element of symmetry called a rotoinversion axis (improper axes of rotation). A 1-rotoinversion axis is the same as a centre of symmetry. Other possible rotoinversion are as follows:

(i) **2-fold rotoinversion**: The operation of 2-fold rotoinversion involves first rotating the object by π then inverting it through an inversion centre. This operation is equivalent to having a mirror plane perpendicular to the 2-fold rotoinversion axis. A twofold rotoinversion axis is symbolized as a 2 with a bar over the top and would be pronounced as bar two $(\bar{2})$.

(ii) **3-fold rotoinversion**: The operation of 3-fold rotoinversion involves first rotating the object by $2\pi/3$ then inverting it through an inversion centre A threefold rotoinversion axis is symbolized as $\bar{3}$.

(iii) **4-fold rotoinversion**: The operation of 4-fold rotoinversion involves first rotating the object by $\pi/2$ then inverting it through an inversion centre. A fourfold rotoinversion axis is symbolized as $\bar{4}$.

(iv) **6-fold rotoinversion**: The operation of 6-fold rotoinversion involves first rotating the object by $\pi/3$ then inverting it through an inversion centre. A sixfold rotoinversion axis is symbolized as $\bar{6}$.

A combination of five proper and five improper rotational axes in combination with other rotational axes and mirror planes produce a total of 32 possible combination, referred to as 32 point groups. Not only does the crystal possess external symmetry but they also have internal symmetry as well. Atoms are usually symmetrically arranged within the crystal lattice. Their symmetry may also be due to rotation, inversion or mirror plane, but in addition may be due to glide planes or screw axes.

(i) **Glide planes**: Glide planes are similar to mirror planes except that the reflected atomic positions are moved 1/2 of the distance between equivalent positions.

(ii) **Screw axes**: These are similar to glide planes except that the atomic positions are rotated along with each translation, hence producing a screw like positioning of the atomic positions.

1.5   Two Dimensional Lattices

A two dimensional lattice is spanned by two vectors a and b. Every point on the lattice can be reached by a translational vector of the form (1.2). Depending on the ratio of the lengths of the vector a and b and the angle γ between them, lattices of various geometries can be constructed. The most general lattice, with no additional symmetry, is obtained when $a \neq b$, $\gamma \neq 90°$ (Fig. 1.17).

Fig. 1.17 Plane oblique lattice
(parallelogram lattice)

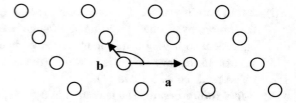

Other planer lattices of higher symmetries are obtained when a, b and γ take on certain special values. Various two dimensional lattices (Fig. 1.18) are mentioned below:

1. **Parallelogram Lattice**

$$a \neq b, \gamma \neq 90°$$

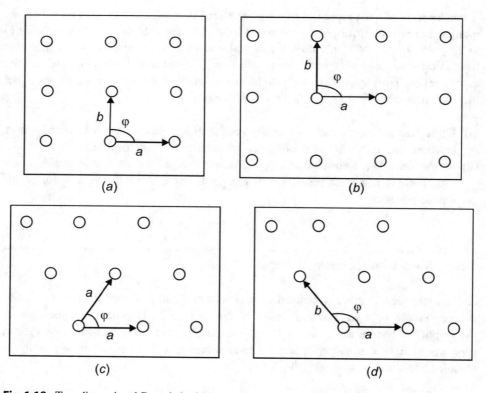

Fig. 1.18 Two dimensional Bravais lattices **a** rectangular, **b** square, **c** hexagonal, **d** diamond

2. **Rectangular Lattice**

$$a \neq b, \gamma = 90°$$

3. **Diamond or Centred Rectangular Lattices**

$$a = b, \gamma \neq 90°$$

4. **Square Lattice**

$$a = b, \gamma = 90°$$

5. **Hexagonal Lattice**

$$a = b, \gamma = 120°$$

1.6   Three Dimensional Lattices

The combination of all available symmetry operations (point group plus glides and screw) with Bravais translations leads to exactly 230 combinations, the 230 space groups. In three dimensions, there are fourteen different Bravais lattices. The fourteen lattices types are conveniently grouped into seven systems according to seven types of conventional cells. The division into system is summarized in Table 1.1 in terms of special axial relation for the conventional unit cells. The axes a, b, and c and angles α, β and γ are defined as shown in Fig. 1.19. The magnitudes a, b and c are called the lattice parameters. Fig. 1.20 shows the fourteen Bravais lattices.

The relation between a, b, c, α, β and γ for various lattices is given in Table 1.1. Out of these lattices, the cubic lattice with $a = b = c$; $\alpha = \beta = \gamma = 90°$ as all the angles is equal to 90°, and all sides are equal. The cubic lattices are further classified as (i) simple cubic in which all the lattice points are at the corner of the cube, and this is termed as cubic P; and (ii) body centred cubic in which the lattice point is at the centre, in addition to the points at the corners of the unit cell, and this is termed as cubic I; and (iii) face centred cubic in which the lattice points are at the centre of all faces in addition to the points at the corners of the unit cell, and this is termed as cubic F.

The next most symmetric lattice is tetragonal in which angles are of 90°. Two of its sides are equal that is $a = b$ but third side is different from the other two, that is, $a = b \neq c$ The tetragonal lattices are of two kinds (i) a simple tetragonal in which all the lattice points are at the corners of the unit cell, this is termed as tetragonal P, (ii) body centred

Table 1.1 Lattice parameter relationships for the seven crystal systems.

System	Restriction on axes and angles
Triclinic S	$a \neq b \neq c; \alpha \neq \beta \neq \gamma \neq 90^\circ$
Simple monoclinic Base centred monoclinic	$a \neq b \neq c; \alpha = \gamma = 90^\circ \neq \beta$
Simple orthorhombic Base centred orthorhombic Body centred orthorhombic Face centred orthorhombic	$a \neq b \neq c; \alpha = \beta = \gamma = 90^\circ$
Simple tetragonal Body centred tetragonal	$a = b \neq c; \alpha = \beta = \gamma = 90^\circ$
Trigonal	$a = b = c; \alpha = \beta = \gamma \neq 90^\circ$
Hexagonal	$a = b \neq c; \alpha = \beta = 90^\circ, \gamma = 120^\circ$
Simple cubic Body centred cubic Face centred cubic	$a = b = c; \alpha = \beta = \gamma = 90^\circ$

Fig. 1.19 Crystal axes a, b, c and interaxial angles (α, β and γ)

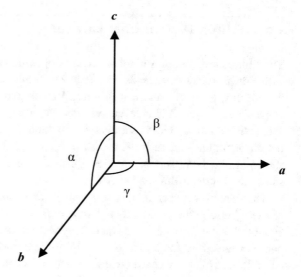

tetragonal lattice in which a lattice point is at the centre, in addition to the points at the corners. This is termed as tetragonal.

For an orthorhombic lattice, all sides are unequal but all angles are of 90°. The orthombic lattice can be further classified as (i) orthorhombic P in which all the lattice points are at the corners of the cell, (ii) orthorhombic I in which a lattice point is at the centre, in addition to the point at the corners, (iii) orthorhombic F in which all the lattice points are at the centre of the faces in addition to the points at the corners, and (iv) orthorhombic C

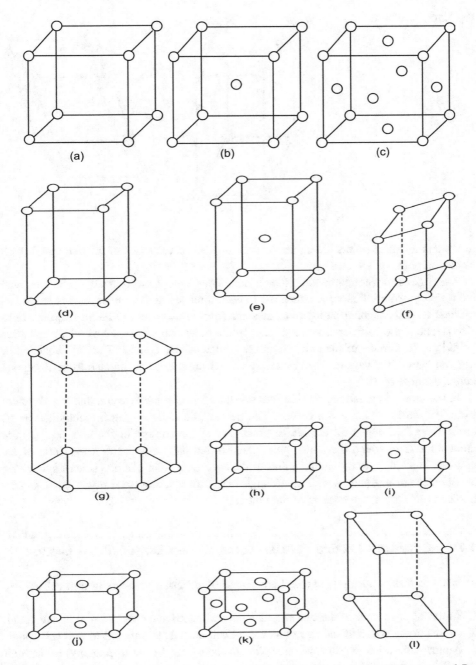

Fig. 1.20 Bravais lattices in three dimensions: **a** cubic P, **b** cubic I, **c** cubic F, **d** tetragonal P, **e** tetragonal I, **f** trigonal R, **g** hexagonal P, **h** orthorhombic P, **i** orthorhombic I, **j** orthorhombic C, **k** orthorhombic F, **l** monoclinic P, **m** monoclinic C, **n** triclinic

Fig. 1.20 (continued)

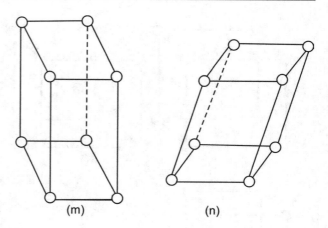

(m)   (n)

in which the lattice points are on the pairs of opposite faces (001) of the unit cell as well as having points at the corners.

The trigonal or rhombohedral lattice has all sides equal as well as all angles are equal but different from 90°. In this lattice the lattice points are at the corners and is termed as trigonal R. In hexagonal lattice two sides are equal ($a = b$) but different from the third side c. The angle between a and b is 120° while other two angles are of 90°. The lattice point is at the corners of the unit cell and is termed as hexagonal P. The rhombohedral or trigonal lattice is always denoted by the symbol R in order to distinguish from hexagonal lattice denoted by P.

In the monoclinic lattice, all sides are unequal, and angle between a and c is different from 90° while α and γ are of 90°. The monoclinic lattice is further classified as (i) monoclinic P in which all the lattice points are at the corners of the unit cell, (ii) the monoclinic C in which the lattice points are on one pair of opposite faces (001) of the unit cell in addition to the lattice points at the corners. Triclinic lattice is least symmetric as all sides are unequal as well as all angles are different and none of the angle is 90°. Lattice points are at the corners of the unit cell.

1.7   Counting Lattice Points/Atoms in Two Dimensional Lattice

Atoms in different positions in a cell are shared by differing numbers of unit cells:

1. Atoms at the corner of the two dimensional cell contribute only 1/4 to unit cell count.
2. Atoms at the edges of the two dimensional cell contribute only 1/2 to unit cell count
3. Atoms within the two dimensional cell contribute only 1 (that is uniquely) to that unit cell.

1.8   Counting Lattice Points/Atoms in Three Dimensional Lattice

Atoms in different positions in a cell are shared by differing numbers of unit cells:

1. Atoms at the corner points are shared by eight adjacent cell, that is, they contribute 1/8 atom to a cell.
2. Edge atom shared by four adjacent cell, that is, they contribute 1/4 atom to a cell.
3. Atoms at the face centres shared by two adjacent cell, that is, they contribute 1/2 atom to a cell.
4. Atoms within the cell contribute uniquely to that unit cell, that is, 1 atom per cell.

For *fcc* lattice, the number of atoms at the vertex of the cell is 8 and one atom each at the six faces. There is no atom at the edge or uniquely contributing to the unit cell. The number N of atoms in *fcc* lattices is thus

$$N = 8 \times \frac{1}{8} + 0 \times \frac{1}{4} + 6 \times \frac{1}{2} + 0 \times 1 = 4$$

For *bcc* lattice, the number of atoms at the vertex of the cell is 8. There is no atom at the edges or faces. One atom is uniquely contributing to the unit cell. The number N of atoms in *bcc* lattices is thus

$$N = 8 \times \frac{1}{8} + 0 \times \frac{1}{4} + 0 \times \frac{1}{2} + 1 \times 1 = 2$$

1.9   Point Coordinates

The structure of a crystal is defined with respect to a unit cell. The entire crystal consists of repeating unit cells. Within the unit cell the atomic arrangement is expressed using coordinates. The atomic coordinates are usually expressed in terms of fractional coordinates, (x, y, z). The coordinate system is coincident with the cell axes (a, b, c) and relate to the position of the atom in terms of the fraction along each axis. For example, consider a cubic cell with length of the edge 0.2 nm. The atom was 0.1 nm in the a direction away from the origin, 0.15 nm away from the origin in b direction and 0.16 nm in the c direction. As the a axis is 0.2 nm long the atom is (1/2) or 0.5 of the axis away from the origin. Similarly, it is (0.15/0.2) or 0.75 of the b axis and (0.16/0.2) or 0.8 of the c axis. The fractional coordinates of this atom are, therefore, (0.5, 0.75, 0.8). The coordinates of the equivalent atom in the next cell over in the a direction, however, are easily calculated as the atom is simply one unit cell away in a. Thus, all one has to do is add 1 to x coordinate (1.5, 0.75, 0.8).

1.10 Crystal Directions

The crystal direction is defined as a line between two points, or a vector. In order to determine the three directional indices, the following steps are utilized:

(a) Draw vector and find the coordinates of head $h_1\ k_1\ l_1$ and the tail $h_2\ k_2\ l_2$.
(b) Subtract coordinates of tail from coordinates of head.
(c) These three numbers so obtained are multiplied or divided by a common factor to reduce them to the smallest integer values.
(d) The three indices, not separated by commas, are enclosed in square brackets, thus [hkl]. The h, k and l integers correspond to reduced projection along a, b and c axes, respectively.

For example, the direction along the a axis would be [100] because this has a component only in the a direction and no component along either the b or c axial direction. A vector diagonally along the face defined by the b and c axis would be [011], while going from one corner of the unit cell to the opposite corner would be in the [111] direction. For each of the three axes, there will exist both positive and negative coordinates. Thus negative indices are also possible which are represented by a bar over the appropriate index. For example, the $\left[1\bar{1}1\right]$ direction would have a component in the $-b$ direction. Also changing the sign of all indices produces an antiparallel direction that is $\left[\bar{1}1\bar{1}\right]$ is directly opposite to $\left[1\bar{1}1\right]$.

For some crystal structures, several non-parallel directions with different indices are equivalent. For example in cubic crystals, all the directions represented by [100], [010], [001], $\left[0\bar{1}0\right]$, $\left[\bar{1}00\right]$ and $\left[00\bar{1}\right]$ are equivalent. These equivalent directions are grouped together into a family, which are enclosed in angle bracket, thus: $\langle 100\rangle$. Furthermore, directions in cubic crystals having the same indices irrespective to order or sign; for example, [123] and $\left[\bar{2}1\bar{3}\right]$ are equivalent.

1.11 Miller Indices

In a crystal, to express the inclination of lattice planes with respect to the three directions, Miller give a method according to which a crystal plane is represented by the three numbers h, k and l called the Miller indices. Miller indices are three smallest integer whose ratio is equal to the reciprocal of the intercept on the crystallographic axes by the plane. The mirror indices are determined by the following way

(a) First the plane should be displaced with a parallel displacement (if necessary) to a position where it does not pass through the origin of the coordinate system.
(b) Determine the intercepts of the plane along each of the three crystallographic axes.

Fig. 1.21 The plane A intercepts b axis at 2 and c axis at $\frac{1}{2}$, while plane B intercepts b and c axes at 1. Both the planes are parallel to a axis

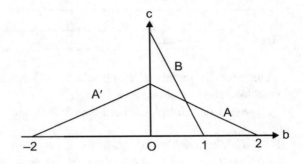

(c) Obtain the reciprocal.

(d) If fraction results, multiply each by the smallest common divisor.

Consider two planes A and B whose cross section is shown in Fig. 1.21. Both the planes are parallel to the a axis. Plane B cut b and c axes at 1 while plane A cut b axis at 2 and c axis at 1/2. For plane B

		a	b	c
Step 1	Intercept	∞	1	1
Step 2	Reciprocal	$1/\infty$	1/1	1/1
Step 3	Indices	0	1	1

The three integers are placed in ordinary parentheses to designate the plane (011) pronounced zero-one-four. Plane equivalent by symmetry may be denoted by curly bracket around indices; for example, the set of cube faces is {100}

For plane A

		a	b	c
Step 1	Intercept	∞	2	1/2
Step 2	Reciprocal	$1/\infty$	1/2	2/1
Step 3	Indices	0	1	4

Therefore, the indices of the plane A are (014). If the intercept is on the negative side of the axis, then a bar is put on the corresponding indices; for example, if the plane parallel to a axis cut b at -2 and c at 1/2, then

For plane B

		a	b	c
Step 1	Intercept	∞	-2	1/2
Step 2	Reciprocal	$1/\infty$	$-1/2$	2/1

		a	b	c
Step 3	Indices	0	-1	4

Therefore, the indices of the plane are $(0\bar{1}4)$. Miller indices of some important planes in a cubic lattice are shown in Fig. 1.22.

Some general principles about Miller indices are

(i)   If a Miller index is zero, the plane is parallel to that axis.
(ii) The smaller a Miller index, the more nearly parallel the plane to its axis.
(iii) The larger a Miller index, the more nearly perpendicular a plane to its axis.
(iv) Multiplying or dividing a Miller index by a constant has no effect on the orientation of the plane.
(v)   Miller indices are almost always small.

A point with coordinates x, y, z is in a plane (hkl) if and only if

$$hx + ky + lz = 1$$

For example, 1/4 1/4 1/4 is not on (111) while 1/3 1/3 1/3 is on the plane.

In order to represent a plane or a direction in a hexagonal crystal system, four axis coordinate system is used (Fig. 1.23), three axes (a_1, a_2, a_3) make angles of 120° with one another and lie in the base of hexagonal prism, the fourth axis, c, being perpendicular to the base plane. Every plane is denoted by four indices $hkil$. The additional label i occupies the third place. They are known as Miller–Bravais indices and are represented as $(hkil)$. The indices h, k and i correspond to three coplanar axes a_1, a_2, a_3 in the basal plane of the hexagon and the fourth index correspond to the fourth axis (c axis) perpendicular to the hexagonal prism. The relationship between the three equivalent axes is $a_1 + a_2 = - a_3$ and between the corresponding indices $h + k = - i$. The base plane parallel to the axes a_1, a_2, a_3 has the indices (0001). The plane parallel to the lateral face of the prism has indices of the $(10\bar{1}0)$ type.

A set of planes may have one direction in common. This common direction is the direction along which the planes intersect and is called a zone axis. The planes that share this zone axis are said to belong to the same zone. The indices of a zone axis $[uvw]$ and a plane (hkl) in this zone obey the relation

$$uh + vk + wl = 0 \tag{1.7}$$

The zone is $[uvw]$ of the two intersecting planes $(h_1k_1l_1)$ and $(h_2k_2l_2)$ can be determined from

$$u = \begin{vmatrix} k_1 & l_1 \\ k_2 & l_2 \end{vmatrix}, \quad v = \begin{vmatrix} l_1 & h_1 \\ l_2 & h_2 \end{vmatrix}, \quad w = \begin{vmatrix} h_1 & k_1 \\ h_2 & k_2 \end{vmatrix} \tag{1.8}$$

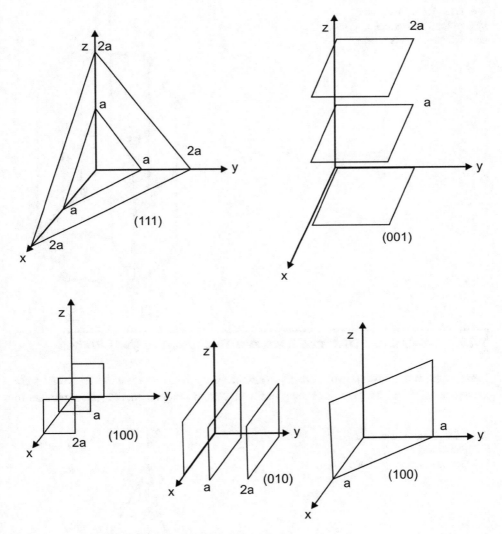

Fig. 1.22 Indices of important planes in a cubic crystal

The indices of a plane (hkl) belonging to two zones with zone axes $[u_1v_1w_1]$ and $[u_2v_2w_2]$ can be determined from

$$h = \begin{vmatrix} v_1 & w_1 \\ v_2 & w_2 \end{vmatrix}, \quad k = \begin{vmatrix} w_1 & u_1 \\ w_2 & u_2 \end{vmatrix}, \quad l = \begin{vmatrix} u_1 & v_1 \\ u_2 & v_2 \end{vmatrix}. \tag{1.9}$$

Fig. 1.23 Coordinate axis
system for a hexagonal unit
cell (Miller–Bravais scheme)

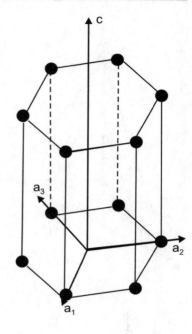

1.12   Interplaner Distance Between Two Nearest (*hkl*) Planes

Let the distances between two nearest (*hkl*) planes is d_{hkl}. Assume that the crystal axes are
orthogonal. Consider a plane as shown in Fig. 1.24 and imagine another plane parallel to

Fig. 1.24 Determination of
interplaner distance d_{hkl}
between two nearest planes

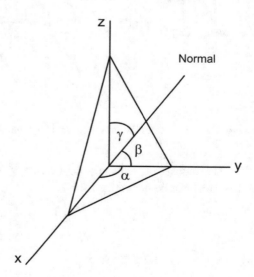

it and passing through the origin. Then the distance between these planes d_{hkl} is simply the length of the normal lines drawn from the origin to the plane shown in Fig. 1.24. Suppose the angle which the normal makes with the axes are α, β and γ and that the intercept of the plane (*hkl*) with the axes are x, y and z. From Fig. 1.24

$$d_{hkl} = x\cos\alpha = y\cos\beta = z\cos\gamma \qquad (1.10)$$

The direction cosines are related by

$$\cos\alpha = \frac{d_{hkl}}{x}, \quad \cos\beta = \frac{d_{hkl}}{y}, \quad \cos\gamma = \frac{d_{hkl}}{z} \qquad (1.11)$$

From Eqs. (1.10) to (1.11)

$$\frac{d_{hkl}^2}{x^2} + \frac{d_{hkl}^2}{y^2} + \frac{d_{hkl}^2}{z^2} = 1$$

$$d_{hkl}^2 \left(\frac{1}{x^2} + \frac{1}{y^2} + \frac{1}{z^2} \right) = 1 \qquad (1.12)$$

$$d_{hkl} = \frac{1}{\left(\frac{1}{x^2} + \frac{1}{y^2} + \frac{1}{z^2} \right)^{\frac{1}{2}}}$$

Now x, y and z are related to Miller indices h, k and l, that is, if x, y and z are intercepts of planes along a, b and c then the Miller indices are

$$h = n\frac{a}{x}, \quad k = n\frac{b}{y}, \quad l = n\frac{c}{z} \qquad (1.13)$$

where n is common factor used to reduce the indices to the smallest integer. Thus from Eq. (1.13)

$$\frac{1}{x} = \frac{h}{na}, \quad y = \frac{k}{nb}, \quad \frac{1}{z} = \frac{l}{nc} \qquad (1.14)$$

From Eqs. (1.12) and (1.14)

$$d_{hkl} = \frac{1}{\left(\frac{h^2}{n^2a^2} + \frac{k^2}{n^2b^2} + \frac{l^2}{n^2c^2} \right)^{1/2}} = \frac{n}{\left(\frac{h^2}{a^2} + \frac{k^2}{b^2} + \frac{l^2}{c^2} \right)^{1/2}} \qquad (1.15)$$

For simple cubic lattice $a = b = c$

$$d_{hkl} = \frac{na}{\left(h^2 + k^2 + l^2 \right)^{1/2}} \qquad (1.16)$$

The formula shows that the greater is the plane indices the shorter is the distance between them.

1.13   Density

A knowledge of the crystal structure permits computation of its density which is defined
as

$$\rho = \frac{nA}{VN_A} \qquad (1.17)$$

where

n	Number of atoms associated with each unit cell
A	Atomic weight
V	Volume of the unit cell
N_A	Avogadro number.

The theoretical density of a crystal can be found by calculating the mass of all the atoms
in the unit cell. The mass of an atom m_A is its molecular mass (gmol^{-1}) divided by the
Avogadro number N_A

$$m_A = \frac{\text{molecular mass (grams)}}{N_A} = \frac{\text{molecular mass}}{N_A} \text{ Kilograms}$$

The total mass of all the atoms in the unit cell is then

$$\frac{n_1 m_1 + n_2 m_2 + \cdots}{1000 \times N_A} \qquad (1.18)$$

where n_1 is the number of atoms of type 1 with a molecular mass m_1 and so on. Thus
Eq. (1.18) is written as

$$\sum_{i=1}^{q} \frac{m_i n_i}{1000 \times N_{AQ}}$$

where there are q different atom types in the unit cell. The density ρ is simply the total
mass divided by unit cell volume

$$\rho = \frac{\sum_{i=1}^{q} \frac{m_i n_i}{1000 \times N_{AQ}}}{V} \qquad (1.19)$$

1.14   Linear and Planer Densities

Sometimes it is important to know the atomic densities in a particular direction or in the
plane of a given lattice.

Linear density is defined as the number of atoms per unit length whose centres lie on the direction vector for a specific crystallographic direction, that is

$$\rho_{\text{linear}} = \frac{\text{Number of atoms centred on direction vector}}{\text{Length of direction vector}} \qquad (1.20)$$

The units of linear density are reciprocal length (e.g. nm^{-1}, m^{-1}). The planar density is defined as the ratio of number of atoms centred on a particular crystallographic plane to the area of the plane; that is

$$\rho_{\text{plane}} = \frac{\text{Number of atoms centred on a plane}}{\text{Area of the plane}} \qquad (1.21)$$

The units for planar density are reciprocal area (e.g. nm^{-2}, m^{-2}).

1.15 Some Simple Crystal Structures

Crystal structure may be described in a number of ways. The most common manner is to refer to the size and shape of the unit cell and the positions of the atoms (or ions) within the cell. Sometimes consideration of several unit cells, the arrangement of the atoms a with respect to each other, the number of other atoms they in contact with, and the distances to neighbouring atoms, often provide a better understanding. Many crystal structure can be described using the concept of close packing. The concept requires that the atoms (ions) are arranged so as to have a maximum density. When describing crystalline structures, atoms (or ions) are thought of being solid spheres having well defined diameters. This is termed as atomic hard sphere model in which sphere representing nearest neighbour atoms touch one another. The important characteristics of crystal structure are the coordination number and the atomic packing factor (APF). APF is a quantitative measure of the closeness of the packing in a crystal is provided by the packing fraction f and it is defined as 'the ratio of the volumes occupied by all the atoms, treated as hard sphere, to the volume of the unit cell of the structure'. The coordination number of an atom or ion within an extended structure is defined as the number of nearest neighbour atoms that are in contact with it. The common coordination number in solids is 3, 4, 6, 8 and 12.

1.15.1 Hexagonal and Cubic Closed Packed

Consider two dimensional space. The most efficient way for equal sized sphere to be packed is shown in Fig. 1.25. Each sphere is surrounded by, and is in contact with, six other spheres. Close packed layers A are formed by repetition to an infinite sheet. If a layer of sphere is placed on the layer of sphere A such that this new layer rest in the

Fig. 1.25 Close packed layer
of spheres is shown, with
centre at points marked A. A
second and identical layer of
spheres can be placed on top of
this, above and parallel to the
plane of the figure, over the
points marked B. There are two
choices for a third layer. It can
go over A or over C

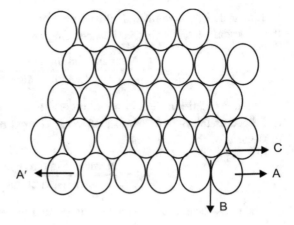

hollow (vacancy) between three of the spheres in layer A. The atoms in the second layer
may occupy one of two possible position. If a third layer is placed on top of the second
layer such that it exactly cover layer A, subsequent placement of layer will result in
the following sequence ABABAB.... This is known as close packing or hexagonal closed
packing. The hexagonal close packed cell is a derivative of the hexagonal Bravais lattice
with the addition of an atom inside the unit cell at the coordinates (1/3, 2/3, 1/2).

If the third layer C is added in such a way that it is over the holes in the first layer
A that are not occupied by layer B, then upon repetition the packing sequence will be
ABCABCABC.... This is known as cubic close packing or ccp. The unit cell of cubic
close packed structure is actually that of a face centred cubic Bravais lattice. In the *fcc*
lattice close packed layers constitute the {111} planes.

Octahedral and Tetrahedral Vacancies

In hexagonal and cubic closed-packed structure there is some unoccupied space between
the spheres (atoms). The unoccupied space is known as octahedral or tetrahedral site
depending upon the arrangement of atoms. The octahedral site differs from tetrahedral site
in the number of atoms surrounding the site. In tetrahedral site there is vacancy surrounded
by four atoms (three in one plane and single one in adjacent plane). This vacant space is
called tetrahedral site as straight line drawn from the centre of the surrounding spheres
(atoms) form a four sided tetrahedron. If the vacant space or site has six nearest atoms
surrounding and thus forming an octahedron, then the site is called octahedral site. The
size of octahedral site for a closed-packed lattice is larger than the tetrahedral site. In a
face centred cubic closed-packed structure there are eight tetrahedral sites at (n/4, n/4,
n/4) with n = 1, 3 for example (3/4, 3/4, 3/4), (1/4, 3/4, 1/4), etc. On the other hand
octahedral sites are face centred cubic closed structure which is at the centre of unit cell
as well as at each edge of the cell. However, if the structure is hexagonal closed packed,

then the tetrahedral sites are at (0.0, 3/8) and (1/3, 2/3, 7/8) and the octahedral sites are at (1/3, 1/3, 1/4) and all symmetry related equivalent positions.

1.15.2 Sodium Chloride Structure

The sodium chloride structure is shown in Fig. 1.26. The lattice is face centred cubic. The basis consists of one Na atom and one Cl atom. There are four units of NaCl in each unit of the atom in this structure is coordinated with six atoms of opposite kind. The separation between NaCl $= a/2 = 0.5a$ and

$$Na - Na = Cl - C = a/\sqrt{2} = 0.707\,a.$$

The Na and Cl atoms are located at

$$Na: \frac{1}{2}\frac{1}{2}\frac{1}{2}; 0\,0\frac{1}{2}; 0\frac{1}{2}0; \frac{1}{2}0\,0$$
$$Cl: 0\,0\,0; \frac{1}{2}\frac{1}{2}0; \frac{1}{2}0\frac{1}{2}; 0\frac{1}{2}\frac{1}{2}$$

The examples of NaCl structure are AgBr, KCl, LiH, PbS, etc.

Fig. 1.26 NaCl structure.
Blackened circles are Na and
blank circles are Cl

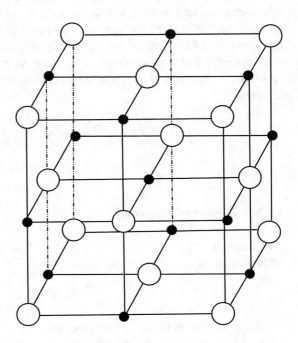

Fig. 1.27 CsCl structure.
Blackened circles are Cl and
blank circles are Cs

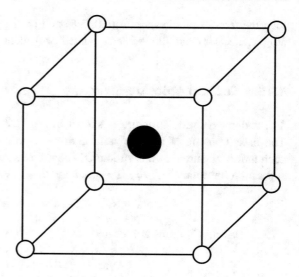

1.15.3 Cesium Chloride Structure

The cesium chloride structure is shown in Fig. 1.27. It has a simple (primitive) cubic cell with a chloride ions at the centre of the cube $\left(\frac{1}{2}\frac{1}{2}\frac{1}{2}\right)$ and the cesium ions at the corners of the cube (000). Each ion may be viewed as at the centre of a cube of ions of the opposite kind. The coordination number of both Cs and Cl is eight. Interchange of anions with cations and vice versa produces the same crystal structure. This is not a *bcc* crystal structure because ions of two different kinds are involved. The separations between the atoms are

$$\text{Cs} - \text{Cl} = \frac{a\sqrt{3}}{2} = 0.866a; \text{Cs} - \text{Cs} = \text{Cl} - \text{Cl} = a$$

The examples of CsCl like structure are NH_4Cl, TlI, CuPd, etc.

1.15.4 Diamond Structure

The diamond structure is shown in Fig. 1.28. The space lattice of diamond is *fcc*. The basis consists of two identical atoms at

$$0\,0\,0; \frac{1}{4}\frac{1}{4}\frac{1}{4}$$

The atoms are localized at the corner points and at the face centres. In the diamond lattice there are four atoms located inside the lattice. Their centre can be found on the body diagonal at a height of 1/4 and 3/4, respectively. The number of atoms belonging

Fig. 1.28 Diamond structure

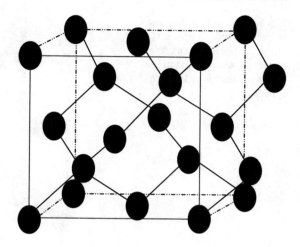

to a unit cell of the diamond crystal can be determined based on the *fcc* structure. The number of atoms is 4 in *fcc* crystal. In diamond lattice there are four further atoms inside the lattice; they belong fully to the unit cell. Hence the number of atoms belong to the unit cell in the diamond lattice is eight. The tetrahedral bonding is characteristic of diamond lattice. Each of the atom is coordinated with four atoms and the shortest interatomic distance is

$$C - C = a\frac{\sqrt{3}}{4} \approx 0.422a$$

Examples of diamond like structure are C, Si, Ge, Sn, etc.

1.15.5 Cubic Zinc Sulphide Structure

The cubic ZnS or zinc blende structure results when zinc atoms are placed on one *fcc* lattice and sulphur atoms on the other *fcc* lattice, which is shifted from the first by 1/4 the cube diagonal (Fig. 1.29). The zinc blende structure does not have an inversion centre at the midpoints of each line connecting nearest neighbour atom. The diamond structure on the other hand has an inversion centre. In the cell, the Zn atoms are located at

$$0\,0\,0;\ 0\frac{1}{2}\frac{1}{2};\ \frac{1}{2}0\frac{1}{2};\ \frac{1}{2}\frac{1}{2}0$$

and the S atoms are at

$$\frac{1}{4}\frac{1}{4}\frac{1}{4};\ \frac{1}{4}\frac{3}{4}\frac{3}{4};\ \frac{3}{4}\frac{1}{4}\frac{3}{4};\ \frac{3}{4}\frac{3}{4}\frac{1}{4}$$

Fig. 1.29 Unit cell structure of a zinc blende lattice. Zinc atoms are shown by small blank sphere, and sulphur atoms are shown by large sphere

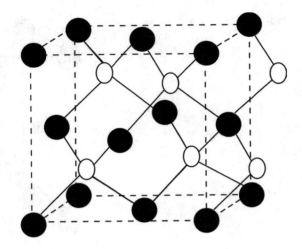

There are four molecules per unit cell. Each atom is coordinated at equal distance, with four atoms of opposite kind arranged at the corners of a regular tetrahedron. The separation between the atoms is

$$Zn - S = a\frac{\sqrt{3}}{4} \approx 0.422a; \ Zn - Zn = S - S = \frac{a}{\sqrt{2}} \approx 0.707a$$

The examples of ZnS like structure are CuF, AlP, GaAs, InSb, SiC, etc.

1.15.6 Wurtzite

This is hexagonal form of the zinc sulphide. It is identical in the number and types of atoms, but it is built from two independent *hcp* lattices as opposed to the *fcc* lattices in zinc blende. All the atoms in this structure are coordinated with the four atoms of other kind. The structure is shown in Fig. 1.30. The bond length in the structure is (when the ratio $c/a = 1.632$)

$$Zn - S = a\frac{\sqrt{3}}{8} \approx 0.612a = \frac{3c}{8} = 0.375c; \ Zn - Zn = S - S = 1.632c$$

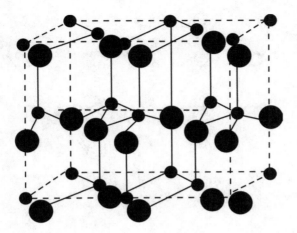

Fig. 1.30 Unit cell structure of a Wurtzite lattice. Zinc atoms are shown by small sphere while sulphur atoms shown by large sphere

Solved Examples

Example 1 How do we know whether a unit cell (or a set of translational vectors) is primitive?

Solution
The cell (and the translational vectors that generate it) is primitive if it contains lattice points at its corners. The translation vector defines the unit cell. The primitive cell is not unique, however all primitive cells have the same volume (or area in two dimension). The primitive cell is usually taken to be the one with the shortest translation vectors.

Example 2 Define the lattice translation vectors drawn in Fig. 1.31 in terms of the primitive vector pair a and b.

Solution
From Fig. 1.31

$$T_1 = a + 2b, T_2 = -2b, T_3 = -5a - 3b, T_4 = 2a - b$$

Example 3 Obtain the number of lattice points contained in each cell of Fig. 1.32.

Solution
Atoms in different positions in a cell are shared by differing numbers of unit cells. In a two dimensional cell. The atoms at the corner contribute only 1/4 to unit cell count, atoms at the

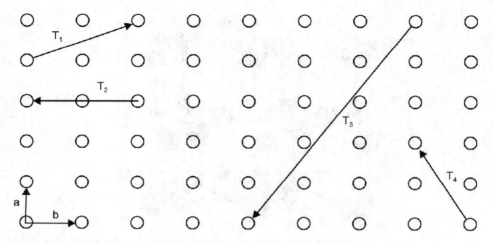

Fig. 1.31   A two dimensional lattice with primitive vectors a and b

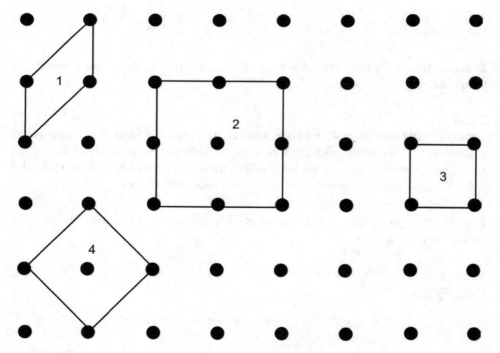

Fig. 1.32   Various unit cells of a two dimensional lattice

edges contribute only 1/2 to unit cell count and atoms within the cell contribute only 1 (that is uniquely) to that unit cell.

From these considerations, the number of lattice points in various cells is

Cell no.	Lattice points
1	1
2	4
3	1
4	2

Example 4 Explain the absence of face centred tetragonal lattice.

Solution

Consider a face centred tetragonal lattice as shown in Fig. 1.33. From this lattice a body centred lattice could be constructed as shown by dotted lines with the edges $1/\sqrt{2}$ times those of the original lattice. Thus face centred tetragonal on choosing a different set of axes is identical with the body centred tetragonal lattice.

Example 5 Explain the absence of C-centred tetragonal lattice.

Solution

Consider a C-centred tetragonal lattice as shown in Fig. 1.34. From this lattice a simple tetragonal lattice could be constructed as shown by dotted lines with edges $1/\sqrt{2}$ times those

Fig. 1.33 A face centred
tetragonal lattice

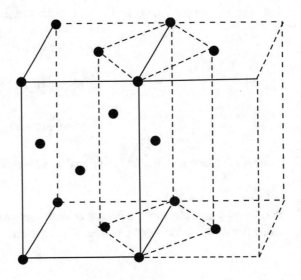

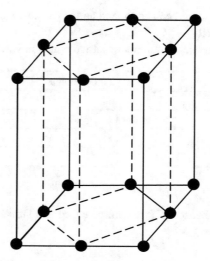

Fig. 1.34 A C-centred tetragonal lattice

of original lattice. Thus C-centred lattice on choosing a set of axes rotated through an angle of 45° with respect to c axis; primitive tetragonal lattice can be obtained.

Since the primitive cell is smaller than that of C-centred lattice, primitive cell is usually chosen.

Example 6 Find the primitive translation vectors of the face centred cubic lattice.

Solution

The primitive translation vectors a', b', c' from Fig. 1.35 are

$$a' = \frac{a}{2}[x+y] \tag{1.22}$$

$$b' = \frac{a}{2}[y+z] \tag{1.23}$$

$$c' = \frac{a}{2}[z+x] \tag{1.24}$$

Example 7 Show how fivefold rotational symmetry is not possible in a crystal lattice?

Solution

The fundamental requirement that all crystals must have periodic structure means that the symmetry elements present in crystal must conform to their translational periodicity. This

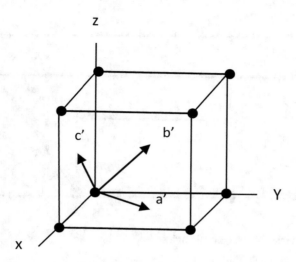

Fig. 1.35 Primitive translation vector of face centred cubic lattice. These connect the lattice points at the origin to the lattice points of the face of the cube

is reason why the number of symmetry elements found in crystals is limited. Consider a combination of n-fold (C_n) rotation axis with a translation a as shown in Fig. 1.36.

Figure 1.37 shows a rotation axis repeats the translation $360°/n$ away, and since n rotation causes superposition, it does not matter whether one rotates clockwise or counterclockwise. Beginning with the linear array, just two such rotations produce new lattice points labelled A and B. Because they are equidistant from the original lattice row by construction, the two lattice points A and B must be joined by the same translation a or some integral multiple of it, ma, depending on the magnitude of φ. This means that the allowed values that φ can have can be determined from Fig. 1.37.

$$ma = a + 2a\cos\varphi m = 0, \pm1, \pm2, \cdots$$

the value of m is positive or negative depends on whether the rotation is clockwise or anticlockwise. From the above equation

$$(m - 1)a = 2a\cos\varphi$$

Fig. 1.36 A lattice with lattice constant a having lattice points with n-fold rotation axis

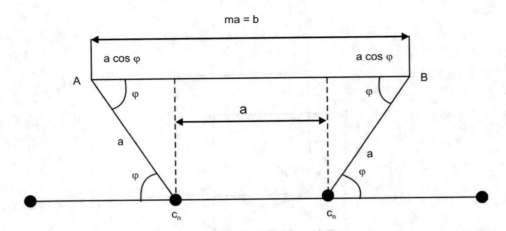

Fig. 1.37 Allowed values of rotational angles

$$m - 1 = 2\cos\varphi$$

Let $m - 1 = N$

$$N = 2\cos\varphi$$

$$\cos\varphi = \frac{N}{2}, N = 0, \pm1, \pm2, \ldots$$

Since the magnitude of cosine must be less than unity, the only possible values of N can have are

N	$\cos\varphi$	φ (°)	n
− 2	− 1	180	2
− 1	− 1/2	120	3
0	0	90	4
1	1/2	60	6
2	1	360 or 0	1

Thus only 1, 2, 3, 4 and 6-fold rotation symmetry is possible.

Example 8 Show for the body centred cubic crystal structure that the unit cell edge length a and atomic radius R are related through

$$a = \frac{4R}{\sqrt{3}}$$

Solution

In body centred cubic crystal structure, centre and corner atoms touch one another along the cube diagonal. Let R is the radius of the atom and therefore length of the diagonal is $4R$. In terms of unit cell edge, the length of the diagonal is $a\sqrt{3}$. On equating the two

$$a\sqrt{3} = 4R$$

$$a = \frac{4R}{\sqrt{3}} \tag{1.25}$$

Example 9 Show for the face centred cubic crystal structure that the unit cell edge length a and atomic radius R are related through

$$a = 2R\sqrt{2}$$

Solution

In face centred cubic crystal structure, the atom touch one another across a face diagonal. Let R is the radius of the atom, and therefore length of the face diagonal is $4R$. In terms of unit cell edge, the length of the face diagonal is $a\sqrt{2}$. On equating the two length

$$a\sqrt{2} = 4R$$

$$a = 2R\sqrt{2} \tag{1.26}$$

Example 10 What is the purpose of taking reciprocals in Miller indices?

Solution

By taking reciprocal, all the planes are brought inside a single unit cell, so that one can discuss all crystal planes in terms of the plane passing through a single unit cell. Since the entire crystal can be generated from it, the unit cell can be studied as representative of the whole sample. The unit cell is equivalent to the molecule in chemistry. One writes the chemical equation for a single molecule, but the equation applies to the entire sample.

Example 11 Show that the plane (123) contains the $\left[11\overline{1}\right]$ direction.

Solution

The indices of direction $[uvw]$ and crystal plane (hkl) must obey the relation (1.7)

$$uh + vk + wl = 0$$

we have $h = 1, k = 2, l = 3, u = 1, v = 1, w = -1$. Substituting these values in the above relation

$$1 \cdot 1 + 2 \cdot 1 - 3 \cdot 1 = 3 - 3 = 0$$

Thus plane (123) contains the $\left[11\bar{1}\right]$ direction.

Example 12 Which of the direction $\left[1\bar{1}0\right]$, [432], $\left[\bar{2}10\right]$, $\left[23\bar{1}\right]$ lie parallel to the plane (115)?

Solution
The indices of direction [*uvw*] and crystal plane (*hkl*) must obey the relation (1.7) we have
$h = 1, k = 1, l = 5$

(i) For direction $\left[1\bar{1}0\right]$; $u = 1, v = -1, w = 0$. Substituting the values of h, k, l, u, v, and w

$$1 \times 1 - 1 \times 1 + 5 \times 0 = 0$$

$\left[1\bar{1}0\right]$ direction lies in the plane (115).

(ii) For direction [432]; $u = 4, v = 3, w = 2$. Substituting the values of h, k, l, u, v, and w

$$1 \times 4 + 1 \times 3 + 5 \times 2 = 17 \neq 0$$

Thus [432] does not lie in (115) plane.

(iii) For direction $\left[\bar{2}10\right]$; $u = -2, v = 1, w = 0$. Substituting the values of h, k, l, u, v, and w

$$1 \times (-2) + 1 \times 1 + 5 \times 0 = -1 \neq 0$$

Thus $\left[\bar{2}10\right]$ does not lie in (115) plane.

(iv) For direction $\left[23\bar{1}\right]$; $u = 2, v = 3, w = -1$. Substituting the values of h, k, l, u, v, and w

$$1 \times 2 + 1 \times 3 - 5 \times 1 = 0$$

Thus $\left[23\bar{1}\right]$ direction lies in the plane (115).

Example 13 Find the direction which lies parallel to the planes $\left(2\,1\,\bar{3}\right)$, $(1\,1\,0)$.

Solution
We have $h_1 = 2, k_1 = 1, l_1 = -3; h_2 = 1, k_2 = 1, l_2 = 0$. Substituting these values in Eq. (1.8)

$$u = \begin{vmatrix} 1 & -3 \\ 1 & 0 \end{vmatrix} = 3, \quad v = \begin{vmatrix} -3 & 2 \\ 0 & 1 \end{vmatrix} = -3, \quad w = \begin{vmatrix} 2 & 1 \\ 1 & 1 \end{vmatrix} = 2 - 1 = 1$$

Thus the direction parallel to the plane is $[3\bar{3}1]$.

Example 14 Find the planes which lie parallel to the direction [102], $[\bar{1}11]$.

Solution

We have $u_1 = 1$, $v_1 = 0$, $w_1 = 2$, $u_2 = -1$, $v_2 = 1$, $w_2 = 1$. Substituting these values in Eq. (1.9)

$$h = \begin{vmatrix} 0 & 2 \\ 1 & 1 \end{vmatrix} = -2; \quad k = \begin{vmatrix} 2 & 1 \\ 1 & -1 \end{vmatrix} = -3; \quad l = \begin{vmatrix} 1 & 0 \\ -1 & 1 \end{vmatrix} = 1$$

Thus the indices of plane parallel to the direction [102], $[\bar{1}11]$ is $(\bar{2}31)$

Example 15 Copper has an atomic radius of 0.128 nm, atomic weight of 63.5 g/mol and density 8.94 g/cm³. Determine whether it has a face centred cubic or body centred cubic crystal structure.

Solution

From Eqs. (1.21), (1.25) and (1.26) for face centred cubic structure

$$\rho = \frac{nA}{16R^3 N_A \sqrt{2}}$$

and for body centred cubic structure

$$\rho = \frac{3\sqrt{3}}{16R^3 N_A}$$

for face centred cubic structure

$$n = \frac{\rho N_A 16R^3 \sqrt{2}}{A}$$

On substituting the values

$$n = \frac{8.94 \times 6.023 \times 10^{23} \times 16 \times (0.128 \times 10^{-7})^3 \times \sqrt{2}}{63.5} = 4.02$$

for body centred cubic structure

$$n = \frac{\rho N_A 16R^3}{3A\sqrt{3}}$$

On substituting the values

$$n = \frac{8.94 \times 6.023 \times 10^{23} \times 16 \times \left(0.128 \times 10^{-7}\right)^3}{3\sqrt{3} \times 63.5} = 5.47$$

From the values of n, it is seen that the structure is face centred cubic.

Example 16 Lead crystallizes in face centred cubic structure. If the atomic radius of lead is 0.175 *nm*, calculate the volume of the unit cell.

Solution
The side of the face centred cubic cell is given by the Eq. (1.26)

$$a = 2R\sqrt{2}$$

The volume V is

$$V = a^3 = 16R^3\sqrt{2}$$

On substituting the value of R

$$V = a^3 = 16(0.175 \text{ nm})^3 \sqrt{2} = 1.212 \times 10^{-28} \text{ m}^3$$

Example 17 Calculate the radius of a palladium atom given that palladium has a face centred cubic structure, a density of 12.0 g/cm^3 and an atomic weight of 106.4 g/mol.

Solution
The density is given by

$$\rho = \frac{n A_{Pd}}{V N_A}$$

$$\rho = \frac{n A_{Pd}}{V N_A} = \frac{4 \times 106.4}{12 \times 6.023 \times 10^{23}} = 5.888 \times 10^{-23} \text{cm}^3$$

For face centred cubic, volume and radius are related by

$$V = a^3 = 16R^3\sqrt{2}$$

$$R^3 = \frac{V}{16\sqrt{2}} = \frac{5.888 \times 10^{-23}\text{cm}^3}{16 \times 1.4142} = 2.6 \times 10^{-24}\text{cm}^3$$

$$R = 1.375 \times 10^{-8}\text{cm} = 0.1375\text{nm}$$

Example 18 Obtain spacing between (111) planes for a cubic lattice.

Solution
Substituting $h = k = l = 1$ in Eq. (1.16)

$$d_{111} = \frac{na}{(1+1+1)^{\frac{1}{2}}} = \frac{na}{\sqrt{3}}$$

Example 19 Obtain packing fraction for (1) simple cubic, (2) face centred cubic and (3) body centred cubic lattices.

Solution

(1) Simple cubic: In simple cubic lattice there is one lattice point in the unit cell of side a. In order to put one hard sphere closely packed inside the cube, the diameter of sphere should be equal to the side of the cube.

$$\text{The radius of the sphere} = \frac{a}{2}$$

$$\text{Volume of the sphere} = \frac{4\pi}{3}\left(\frac{a}{2}\right)^3 = \frac{\pi a^3}{6}$$

$$\text{Packing fraction } f = \frac{\text{Volume of the sphere}}{\text{Volume of the unit cell}}$$

$$= \frac{\left(\frac{\pi a^3}{6}\right)}{a^3} = \frac{\pi}{6} = 0.524$$

(2) Face centred cubic: In face centred cubic lattice, there are four lattice points. In order to closely pack the four spheres in the cubic cell, each sphere will have twice its diameter equal to the length of the face diagonal

$$\text{The length of the face diagonal} = \sqrt{a^2 + a^2} = a\sqrt{2}$$

$$\text{Length of the diagonal} = 2 \times \text{diameter of the sphere}$$

$$\text{Diameter of the sphere} = \frac{a\sqrt{2}}{2} = \frac{a}{\sqrt{2}}$$

$$\text{The radius of the sphere} = \frac{\text{Diameter of the sphere}}{2} = \frac{a}{2\sqrt{2}}$$

$$\text{Volume of the sphere} = \frac{4\pi}{3}\left(\frac{a}{2\sqrt{2}}\right)^3 = \frac{\pi a^3}{12\sqrt{2}}$$

Since there are four such spheres in face centred cubic lattice, volume occupied by the sphere is

$$4 \times \frac{\pi a^3}{12\sqrt{2}} = \frac{\pi a^3}{3\sqrt{2}}$$

$$\text{Packing fraction } f = \frac{\text{Volume of the sphere}}{\text{Volume of the unit cell}} = \frac{\pi a^3}{3\sqrt{2}} = \frac{\pi}{3\sqrt{2}} = 0.740$$

(3) Body centred cubic: In body centred cubic lattice, there are two lattice points. In order to closely pack the sphere in the cubic cell, each sphere will have twice its diameter equal to the length of the body diagonal.

$$\text{The length of the body diagonal} = \sqrt{2a^2 + a^2} = a\sqrt{3}$$
$$\text{Length of the diagonal} = 2 \times \text{diameter of the sphere}$$
$$\text{Diameter of the sphere} = \frac{a\sqrt{3}}{2}$$
$$\text{The radius of the sphere} = \frac{\text{Diameter of the sphere}}{2} = \frac{a\sqrt{3}}{4}$$
$$\text{Volume of the sphere} = \frac{4\pi}{3}\left(\frac{a\sqrt{3}}{4}\right)^3 = \frac{\pi a^3\sqrt{3}}{16}$$

Since there are two such spheres in body centred cubic, volume occupied by the spheres is

$$2 \times \frac{\pi a^3\sqrt{3}}{16} = \frac{\pi a^3\sqrt{3}}{8}$$

$$\text{Packing fraction } f = \frac{\text{Volume of the sphere}}{\text{Volume of the unit cell}} = \frac{\pi a^3\sqrt{3}}{8a^3} = \frac{\pi\sqrt{3}}{8} = 0.680$$

Example 20 Obtain packing fraction for diamond lattice.

Solution
The body diagonal of diamond lattice has a length $\sqrt{3}$ a. The radius of each sphere is R. Along the body diagonal there are four spheres. On equating the length of the body diagonal and diameter of four spheres

$$a\sqrt{3} = 8R \quad \text{or} \quad R = \frac{\sqrt{3}}{8}a$$

Each corner contains 1/8 of a sphere, each face centre contains 1/2 of a sphere and the cube contains four whole sphere. Therefore there are

$$8\left(\frac{1}{8}\right) + 6\left(\frac{1}{2}\right) + 4 = 8$$

spheres in the diamond lattice. The volume of the spheres is

$$V_{sphere} = 8\left(\frac{4\pi}{3}\right)R^3 = 8\left(\frac{4\pi}{3}\right)\left(\frac{a\sqrt{3}}{8}\right)^3 = \frac{\sqrt{3}\pi a^3}{16} \qquad (1.27)$$

The packing fraction f is

$$f = \frac{V_{sphere}}{V_{cube}} = \frac{\sqrt{3}\pi a^3}{16a^3} = \frac{\sqrt{3}\pi}{16} = 0.34$$

Example 21 Indium has a tetragonal unit cell with a = 0.459 nm and c = 0.495 nm. (a) If the atomic packing factor and atomic radius are 0.693 and 0.1625 nm, respectively, determine the number of atoms in each unit cell. (b) The atomic weight of indium is 114.82 g/mol; compute the density.

Solution

(a) The packing factor f is

$$f = \frac{\text{Volume of the sphere of radius } R}{\text{Total volume of the cell}}$$

$$f = \frac{n \times \left(\frac{4\pi}{3}\right)R^3}{a^2c}$$

$$\text{Number of atoms} = n = \frac{fV}{(4\pi/3)R^3} = \frac{0.693 \times (0.459 \text{ nm})^2 \times 0.495 \text{ nm}}{(4 \times 3.14/3)(0.1625 \text{ nm})^2} = 4.023 \approx 4$$

(b) The density is given by

$$\rho = \frac{n \times A}{V \times N_A} = \frac{4 \times 114.82 \text{g/mol}}{(0.459 \text{nm})^2 \times 0.495 \text{nm} \times 6.023 \times 10^{23}} = 7.31 \text{g/cm}^3$$

Example 22 Find the packing fraction for *hcp* lattice.

Solution

The hexagonal closed-packed structure is shown in Fig. 1.38. Each diagonal has a length

Fig. 1.38 Hexagonal closed
structure

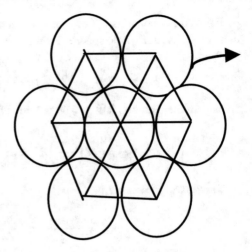

2a. The radius of the sphere R is given by

$$4R = 2a \quad \text{or} \quad R = a/2$$

Each corner contains 1/6 of a sphere and the top and bottom faces contains 1/2 of a sphere. The hexagonal structure also contains three whole sphere, thus there are $12(1/6) + 2(1/2) + 3 = 6$ sphere in the hexagonal close packed structure. The volume of the sphere is

$$V_{\text{sphere}} = 6\left(\frac{4\pi}{3}\right)R^3 = 8\pi\left(\frac{a}{2}\right)^3 = \pi a^3$$

The volume of the hexagonal structure is

$$V_{\text{hex}} = c \times \text{Area}$$

where c is the height of the hexagonal structure. The area of each face is given by

$$\text{Area} = \frac{6}{4}a^2\cot\left(\frac{\pi}{6}\right) = \frac{3\sqrt{3}}{2}a^2$$

The volume of the hexagonal structure is thus

$$V_{\text{hex}} = c \times \frac{3\sqrt{3}}{2}a^2 \tag{1.28}$$

Using

$$\frac{c}{a} = \sqrt{\frac{8}{3}} \quad \text{or} \quad c = \sqrt{\frac{8}{3}}a$$

$$V_{\text{hex}} = \sqrt{\frac{8}{3}}a \times \frac{3\sqrt{3}}{2}a^2 = 3\sqrt{2}a^3$$

The packing fraction f is

$$f = \frac{V_{\text{sphere}}}{V_{\text{hex}}} = \frac{\pi a^3}{3\sqrt{2}a^3} = \frac{\sqrt{2}\pi}{6} = 0.74$$

the *hcp* structure has the same packing fraction as the *fcc* structure. This is because both the *hcp* and *fcc* structure yield the maximum packing factor for the hard sphere.

Example 23 Explain that [100] is perpendicular to (100) in a cubic crystal.

Solution
The direction [100} represent a direction parallel to the a axis. On the other hand the (100) represent a plane parallel to b and c axis and cutting the a axis. Thus the direction [100] is perpendicular to (100) plane.

Example 24 Find the primitive translation vectors of the body centred cubic lattice.

Solution
The primitive cell is obtained on completing the rhombohedron (Fig. 1.39). In terms of the cube edge

$$a' = \frac{a}{2}[\hat{x} + \hat{y} - \hat{z}] \tag{1.29}$$

$$b' = \frac{a}{2}[-\hat{x} + \hat{y} + \hat{z}] \tag{1.30}$$

$$c' = \frac{a}{2}[\hat{x} - \hat{y} + \hat{z}] \tag{1.31}$$

Example 25 Show that if a two dimensional oblique lattice has a reflection symmetry then it results in two additional Bravais lattice: (a) rectangular primitive and (b) rectangular centred lattices.

Solution
Let us consider an oblique lattice ($a \neq b$, $\varphi =$ arbitrary) with fundamental translation vectors a and b. Consider a Cartesian coordinate system oriented in such a way that a is along the x-axis as shown in Fig. 1.37. We have

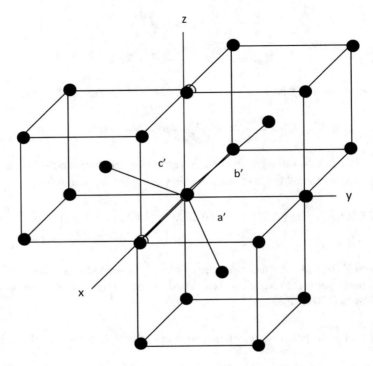

Fig. 1.39 Primitive translation vector of body centred cubic lattice. These connect the lattice points at the origin to the lattice points of the body centre of the cube

$$a = a_x \hat{i} \tag{1.32}$$

$$b = b_x \hat{i} + b_y \hat{j} \tag{1.33}$$

Let x-axis is a reflection line. On reflecting the lattice in this reflection line, the new vectors are

$$a' = a_x \hat{i} \tag{1.34}$$

$$b' = b_x \hat{i} + b_y \hat{j} \tag{1.35}$$

The vector a' is a translation vector as $a = a'$. However b' is not necessarily a translation vector. The vector b' will be a translation vector only if

$$b' = ua + vb \tag{1.36}$$

as required by the translation symmetry. If x-axis is true line of reflection symmetry, b' must be a translation vector. Therefore,

$$b' = ua + vb = ua_x\hat{i} + v\left(b_x\hat{i} + b_y\hat{j}\right) \tag{1.37}$$

Equating the coefficients of $\hat{i}$ and $\hat{j}$ in Eqs. (1.35) and (1.37)

$$b_x = ua_x + vb_x \tag{1.38}$$

$$-b_y = ub_y \tag{1.39}$$

or $u = -1$. Thus

$$b_x = ua_x - b_x \quad \text{or} \quad 2b_x = ua_x$$
$$b_x = \frac{u}{2}a_x \tag{1.40}$$

and b_y is arbitrary. If $u = 0$, $b_x = 0$ we have the primitive rectangular lattice ($a \neq b$, $\varphi = 90°$) and also for all even u. If $u = 1$, $b_x = a_x/2$ or $2b_x = a_x$. Thus we have the central rectangular lattice (also for all odd u).

Example 26 Show that *fcc* lattice is formed by all points (n_1, n_2, n_3) (all integers) provided the sum of all n_j to be even.

Solution

The basic lattice vectors of equal length a in the direction of the three cubic axes. Taking Cartesian coordinates along these axes, the basic vectors for the *fcc* lattice are

$$(a, 0, 0), (0, a, 0), (0, 0, a), \left(\frac{a}{2}, 0, \frac{a}{2}\right), \left(0, \frac{a}{2}, \frac{a}{2}\right), \left(\frac{a}{2}, \frac{a}{2}, 0\right)$$

The most general lattice point is

$$(n_1 a, n_2 a, n_3 a) \quad \text{or} \quad \left[\left(n_1 + \frac{1}{2}\right)a, \left(n_2 + \frac{1}{2}\right)a, n_3 a\right]$$

or

$$\left[\left(n_1 + \frac{1}{2}\right)a, n_2 a, \left(n_3 + \frac{1}{2}\right)a, n_3 a\right] \quad \text{or} \quad \left[n_1 a, \left(n_2 + \frac{1}{2}\right)a, \left(n_3 + \frac{1}{2}\right)a\right]$$

or, expressed differently

$$\left(\frac{n_1}{2}a, \frac{n_2}{2}a, \frac{n_3}{2}a\right)$$

with n_1, n_2, n_3 integers of which either one or all three are even, that is, $n_1 + n_2 + n_3$ must be even.

Example 27 Show that *bcc* lattice is formed by all points (n_1, n_2, n_3) (all integers) provided that n_j is all even or all odd.

Solution

The basic lattice vectors of equal length a in the direction of the three cubic axes. Taking Cartesian coordinates along these axes, the basic vectors for the *bcc* lattice are

$$(a, 0, 0), (0, a, 0), (0, 0, a), \left(\frac{a}{2}, \frac{a}{2}, \frac{a}{2}\right)$$

It is clear that twice the last vector equals the sum of the other three, so that the most general lattice point is either

$$(n_1 a, n_2 a, n_3 a) \quad \text{or} \quad \left[\left(n_1 + \frac{1}{2}\right)a, \left(n_2 + \frac{1}{2}\right)a, \left(n_3 + \frac{1}{2}\right)a\right]$$

The first set of points forms a simple cubic lattice, and the second consists of all the centres of the cubes formed by the adjacent points of the first set. It is evident that the set of points represent *bcc* lattice provided all n_j are even or odd.

Example 28 Show that for *hcp* $c/a = \sqrt{(8/3)}$.

Solution

Figure 1.41a show one third of an *hcp* unit cell. Consider the tetrahedron labelled as ABCD which is shown in Fig. 1.41b. The atom at point C is midway between the top and the bottom faces of the unit cell that is CE $= c/2$. Since atoms at points A, B and C all touché one another

$$AC = AB = 2R = a$$

where R is the radius of the atom. From the equilateral triangle ABD (Fig. 1.41c)

$$\frac{AO}{AE} = \cos 30 = \frac{\sqrt{3}}{2}$$

$$AO = \frac{a}{2} = AE \cos 30 = AE \frac{\sqrt{3}}{2}$$

$$AE = \frac{a}{\sqrt{3}}$$

From right angle triangle ACE

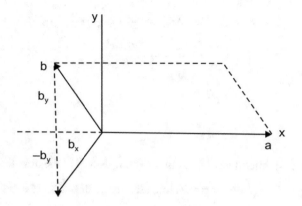

Fig. 1.40 An oblique lattice with reflection symmetry

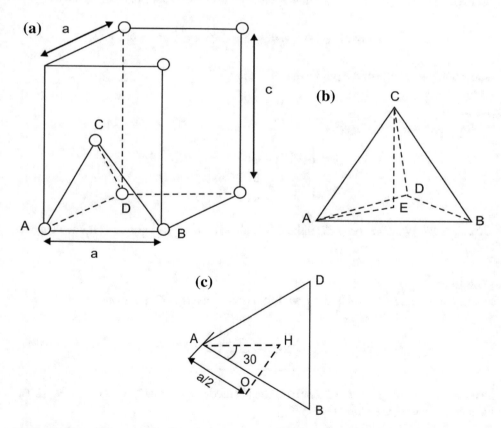

Fig. 1.41 For a hexagonal closed packed (*hcp*) unit cell: **a** one third of *hcp* unit cell, **b** tetrahedral ABCD of one third of *hcp* unit cell and **c** equilateral triangle ABD of tetrahedral ABCD

$$(AC)^2 = (AE)^2 + \frac{c^2}{4} = \frac{a^2}{3} + \frac{c^2}{4}$$

$$\frac{c^2}{4} = a^2 - \frac{a^2}{3} = \frac{2a^2}{3}$$

$$\frac{c^2}{a^2} = \frac{8}{3}$$

$$\frac{c}{a} = \sqrt{\frac{8}{3}}$$

If the c/a ratio is smaller than $\sqrt{\frac{8}{3}}$, atoms are compressed in the direction of c axis, while if the ratio is larger than $\sqrt{\frac{8}{3}}$, atoms are elongated in the direction of c axis.

Example 29 The primitive translation vectors of hexagonal space lattice is given by

$$a_1 = a\sqrt{\frac{3}{2}}\hat{x} + \frac{a}{2}\hat{y}, \, a_2 = -a\sqrt{\frac{3}{2}}\hat{x} + \frac{a}{2}\hat{y}, \, a_3 = c\hat{z}$$

show that the volume of the primitive cell is $a^2 c \frac{\sqrt{3}}{2}$.

Solution
Cell volume is $a_1 \cdot a_2 \times a_3$

$$\begin{vmatrix} +\frac{\sqrt{3}}{2}a & \frac{a}{2} & 0 \\ -\frac{\sqrt{3}}{2}a & \frac{a}{2} & 0 \\ 0 & 0 & c \end{vmatrix} = a^2 c \frac{\sqrt{3}}{2}.$$

Example 30 What are the $(hkil)$ indices of a plane in hexagonal crystal that intercept axes at $a_1 = 1$, $a_2 = 1$ and $c = 1/2$.

Solution
Plane cut a_1 and a_2 axes at 1 and 1 while plane cut c axis at 1/2. For plane

Intercept   1 1 $\frac{1}{2}$
Reciprocal 1 1 2
Indices   1 1 2

we have $h = 1$, $k = 1$ and $l = 2$. For hexagonal lattice $i = -(h+k) = -(1+1) = -2$ The Miller indices are therefore, $(11\bar{2}2)$.

Example 31 A plane includes points $0, 0, 0$; $1/2, 1/4, 0$; $1/2, 0, 1/2$. What are Miller indices.

Solution

Since the plane passes through the origin, shift the origin, for example, 1 unit in the x-direction. The intercepts are now $- a$, 0.5 a and a. Therefore, the reciprocals are $- 1$, 2 and 1. The Miller indices $i = - 1$, $k = 2$ and $l = 1$ or can be written as $(\overline{1}21)$..

Example 32 The <101> family of direction includes what individual directions in tetragonal crystal.

Solution

In tetragonal cell $a = b \neq c$, only the u and v indices of <uvw> are interchangeable, and w index is not. The directions are [101], [011], $[\overline{1}01]$, $[0\overline{1}1]$.

Example 33 Calculate the ionic packing factor of fcc NaCl.

Solution

We have

$$a_{fcc\,NaCl} = 2(r_{Na} + R_{Cl})$$
$$\text{Volume of the cell} = (2r_{Na} + 2R_{Cl})^3$$
$$= 8(0.097\,\text{nm} + 0.181\,\text{nm})^3$$
$$\text{Packing Fraction} = \frac{4(4\pi r^3/3) + 4(4\pi R^3/3)}{8(0.097\,\text{nm} + 0.181\,\text{nm})^3}$$
$$= \frac{16\pi\left[(0.097\,\text{nm})^3 + (0.181\,\text{nm})^3\right]}{3 \times 8(0.097\,\text{nm} + 0.181\,\text{nm})^3} = 0.67.$$

Example 34 Obtain the density of NaCl. The atomic mass of sodium is 22.99 g/mol and that of chlorine is 35.45 g/mol. The radius of Na and Cl ions are 0.102 nm and 0.181 nm, respectively.

Solution

NaCl crystallizes in fcc structure. The unit cell is cubic; therefore the volume V is a^3 (a is cell edge) For the face of cubic unit cell

$$a = 2(r_{Na} + r_{Cl}) = 2(0.102 + 0.181)\text{nm} = 0.566\text{nm}$$
$$\text{Volume } V = a^3 = (0.566\text{nm})^3$$

The density

$$\rho = \frac{n(A_{Na} + A_{Cl})}{V \times N_A}$$

Since *fcc* lattice contains four points, therefore $n = 4$ and

$$\rho = \frac{n(A_{Na} + A_{Cl})}{V \times N_A} = \frac{4 \times (22.99 + 35.45)\text{g/mol}}{(0.566\text{nm})^3 \times 6.023 \times 10^{23}} = 2.14\text{g/cm}^3.$$

Example 35 Show that a *bcc* lattice may be decomposed into two simple cubic lattices A and B with the property that none of the nearest-neighbour lattice points to a lattice point on A lie on A, and similarly for the B lattice.

Solution
The basic vectors are three vectors of equal length a in the direction of three cubic axes. Taking Cartesian coordinates along three axes, the basic lattice vectors are $(a, 0, 0)$; $(0, a, 0)$; $(0, 0, a)$. The general lattice vector and hence the position of any lattice point is $(n_1 a, n_2 a, n_3 a)$.

The basic vectors for the *bcc* lattice are

$$(a, 0, 0), (0, a, 0), (0, 0, a), \left(\frac{a}{2}, \frac{a}{2}, \frac{a}{2}\right)$$

It is clear that twice the last vector equals the sum of the other three, so that the most general lattice point is either

$$(n_1 a, n_2 a, n_3 a) \quad \text{or} \quad \left[\left(n_1 + \frac{1}{2}\right)a, \left(n_2 + \frac{1}{2}\right)a, \left(n_3 + \frac{1}{2}\right)a\right]$$

The first set of points forms a simple cubic lattice, the second consists of all the centres of the cubes formed by the adjacent points of the first set. Hence one can look at this lattice having the translation group of the simple cubic lattice and two atoms in the unit cell.

Objective Type Questions

1. The number of crystal system is
    (a) 5 (b) 6 (c) 7
2. The angle between the *a* and *c* axes in a unit cell is labelled as
    (a) γ (b) α (c) β (d) θ
3. A tetragonal unit cell is defined by
    (a) $a = b = c; \alpha = \beta = \gamma = 90^\circ$
    (b) $a = b \neq c; \alpha = \beta = \gamma = 90^\circ$
    (c) $a = b \neq c; \alpha = \beta = \gamma = 60^\circ$
    (d) $a = b = c; \alpha = \beta = \gamma = 120^0$
4. A crystal is built by the stacking of unit cell with
    (a) Orientational and translational long range order
    (b) Orientational long range order
    (c) Translational long range order

5. Miller indices are used to label
   (a) Crystal shape (b) Crystal faces (c) Crystal size
6. In crystallography the letter Z specifies
   (a) The number of atoms in a unit cell
   (b) The number of formula units in a unit cell
   (c) The number of molecules in a unit cell
7. The number of atoms in a unit cell of NaCl structure is
   (a) 2 (b) 4 (c) 8
8. When determining the number of atoms in a unit cell an atom in a face counts as
   (a) 1/2 (b) 1/4 (c) 1/8
9. A lattice is
   (a) a crystal structure (b) an ordered array of points (c) a unit cell
10. The basis vectors in a lattice define
    (a) The crystal structure
    (b) The atom positions
    (c) The unit cell
11. The number of different two dimensional plane lattices is
    (a) 5 (b) 6 (c) 7
12. A rectangular primitive planer lattice has lattice parameters
    (a) $a \neq b, \gamma = 90°$ (b) $a = b, \gamma = 90°$ (c) $a \neq b, \gamma \neq 90°$
13. The number of Bravais lattices is
    (a) 12 (b) 13 (c) 14
14. An orthorhombic body centred Bravais lattice has lattice parameters
    (a) $a \neq b \neq c; \alpha = \beta = \gamma = 90°$
    (b) $a \neq b \neq c; \alpha = \beta = \gamma \neq 90°$
    (c) $a = b \neq c; \alpha = \beta = 90°, \gamma = 120°$
15. A face centred lattice unit cell contains
    (a) One lattice point
    (b) Two lattice points
    (c) Four lattice points
16. A unit cell with a lattice point at each corner and one at the centre of the cell is labelled
    (a) B (b) C (c) I
17. The notation $[uvw]$ means
    (a) A single direction in a crystal
    (b) A set of parallel directions in a crystal
    (c) A direction perpendicular to a plane $[uvw]$
18. The notation $[hkl]$ represents
    (a) A set of directions that are identical by virtue of the symmetry of the crystal
    (b) A set of planes that are identical by virtue of the symmetry of the crystal
    (c) Both a set of planes or directions that are identical by virtue of the symmetry of the crystal
19. A plane intercepts the crystal axes at $a = 1, b = 2$ and $c = 1$. What are the Miller indices.

(a) (212) (b) (211) (c) (111) (d) (222)

20. We can identify the structure of a *bcc* metal by placing atoms at the location

(a) (000); $\left(\frac{1}{2}\frac{1}{2}\frac{1}{2}\right)$ (b) (000); $\left(\frac{1}{2}\frac{1}{2}0\right)$; $\left(\frac{1}{2}0\frac{1}{2}\right)$; $\left(0\frac{1}{2}\frac{1}{2}\right)$ (c) (000); $\left(\frac{1}{2}0\frac{1}{2}\right)$; $\left(\frac{1}{2}\frac{1}{2}0\right)$ (d) (000); $\left(\frac{1}{2}\frac{1}{2}0\right)$

21. We can identify the structure of a *fcc* metal by placing atoms at the location

(a) (000); $\left(\frac{1}{2}\frac{1}{2}\frac{1}{2}\right)$ (b) (000); $\left(\frac{1}{2}\frac{1}{2}0\right)\left(\frac{1}{2}0\frac{1}{2}\right)$; $\left(0\frac{1}{2}\frac{1}{2}\right)$ (c) (000); $\left(\frac{1}{2}\frac{1}{2}0\right)\left(\frac{1}{2}0\frac{1}{2}\right)$ (d) (000); $\left(\frac{1}{2}\frac{1}{2}0\right)$

22. What are the indices for a plane with intercepts at $a = 1$, $b =- 3/2$, and $c = 2/3$.

(a) (649) (b) (669) (c) $(6\bar{4}9)$ (d) (469)

23. For tetragonal crystal the indices of direction that is not equivalent to [100] is

(a) $\left[\bar{1}00\right]$ (b) [010] (c) $\left[0\bar{1}0\right]$ (d) [001]

24. The indices of direction that results from the intersection of (001)and$(11\bar{1})$ planes within a cubic crystal is

(a) [110] (b) $\left[\bar{1}10\right]$ (c) [011] (d) $\left[\bar{1}01\right]$

25. Atomic packing factor of simple cubic lattice is

(a) 0.74 (b) 0.68 (c) 0.34 (d) 0.52

26. Atomic packing factor of diamond lattice is

(a) 0.74 (b) 0.68 (c) 0.34 (d) 0.52

27. Atomic packing factor of *fcc* lattice is

(a) 0.74 (b) 0.68 (c) 0.34 (d) 0.52

28. Atomic packing factor of *bcc* lattice is

(a) 0.74 (b) 0.68 (c) 0.34 (d) 0.52

29. Atomic packing factor of *hcp* lattice is

(a) 0.74 (b) 0.68 (c) 0.34 (d) 0.52

Problems

1. Determine the area of a primitive rectangle whose sides are 0.3 nm and 0.4 nm, respectively. Consider another primitive cell of equal area whose sides are 0.5 and 0.4 nm. Determine the angle between them.

2. Construct Wigner–Seitz cells of plane lattices.

3. Show that distance AB between two lattice points $A(x_1, y_1)$ and $B(x_2, y_2)$ of an oblique plane lattice is given by

$$AB = \left[(x_2 - x_1)^2 a^2 + (y_2 - y_1)^2 b^2 + 2(x_2 - x_1)(y_2 - y_1)ab \, \cos\gamma\right]^{\frac{1}{2}}$$

4. Show that volume of a unit cell is given by

$$V = abc\left(1 - \cos^2\alpha - \cos^2\beta - \cos^2\gamma + 2\cos\alpha \, \cos\beta \, \cos\gamma\right)^{\frac{1}{2}}$$

5. What is the difference between atomic structure and crystal structure?

6. Differentiate between primitive and non-primitive cell.
7. For tetragonal crystal cite the indices of directions that are equivalent to the direction [100].
8. Determine the Miller indices for the plane shown in the following unit cell.

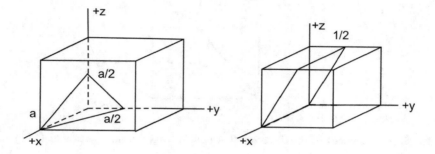

9. Determine the Miller indices for the plane shown in the following unit cell.

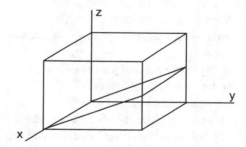

10. For a cubic cell of edge length a, determine the indices for the direction shown in the unit cell below

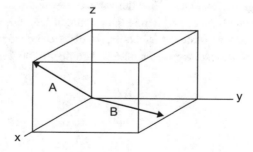

11. For a cubic cell of edge length a, determine the indices for the direction shown in the unit cell below

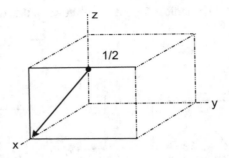

12. Cite the indices of the direction that results from the intersection of $(11\bar{1})$ and (001) planes.
13. Titanium has an *hcp* crystal structure and a density of 4.51 g/cm^3 (a) what is the volume of its unit cell (b) if $c/a = 1.58$, compute the values of c and a.
14. Cobalt has an *hcp* crystal structure, an atomic radius of 0.1253 nm and a c/a ratio of 1.623. Compute the volume of the unit cell for Co.
15. Specify point coordinates for all atom positions for a cubic unit cell with atoms at cube corners and cube centre.
16. Show that volume of primitive cell of *fcc* lattice is $a^3/4$
17. Draw a $(1\bar{1}0)$ direction within a cubic unit cell.
18. Construct a $(0\bar{1}1)$ plane within a cubic unit cell.
19. List the point coordinates for all atoms that are associated with *fcc*.
20. Sketch [111], [110] and [001] directions in a cubic unit cell.
21. Show that for *fcc* lattice, the second neighbours are at a distance a
22. Show that volume of primitive cell of *bcc* lattice is $a^3/2$
23. Show that for *fcc*, the number of second neighbours is six.
24. The angle between the tetrahedral bonds in diamond is the same as the angles between the body diagonals of a cube. Find the value of the angle.
25. Potassium crystallizes in *bcc* lattice with unit cell length $a = 520$ pm. Determine (i) the distance between nearest neighbours, (ii) distance between next nearest neighbours, (iii) number of nearest neighbours, and (iv) number of next nearest neighbours.

Answers
Objective Type Questions

1. (c)	2. (c)	3. (b)	4. (c)	5. (b)	6. (b)	7. (b)	8. (a)
9. (b)	10. (c)	11. (a)	12. (a)	13. (c)	14. (a)	15. (c)	16. (c)
17. (a)	18. (a)	19. (a)	20. (a)	21. (b)	22. (c)	23. (d)	24. (b)
25. (d)	26. (c)	27. (a)	28. (b)	29. (a)			

Chemical Bonding in Solids

2

2.1 Attractive and Repulsive Forces

Solids are stable structure; for example, a crystal of KCl is more stable than a collection of free K and Cl atoms. Similarly Cu metal is more stable than a collection of free Cu atoms. This implies that K and Cl atoms (or Cu atoms) attract each other when they approach each other; that is, an attractive interatomic force exists which holds the atoms together. It means that energy of the solid is lower than that of free atoms. The difference in energy is needed to pull apart solid into neutral free atoms at rest at infinite separation. This is called binding energy (or cohesive energy) of the crystal.

To form a solid, two different forces must exists. An attractive force which is necessary for any bonding, a repulsive force on the other hand, is required in order to keep the atoms from getting collapsed. Magnetic forces have only a weak effect in bonding and gravitational forces are negligible. An expression for an interatomic potential can be written as

$$V(r) = \frac{A}{r^n} - \frac{B}{r^m}$$

where A and B are constants and $n > m$, that is, the repulsive part has to prevail for short distances. Such a potential and resulting force is shown in Fig. 2.1a, b, respectively. Figure 2.1 has a minimum at some distance r_0. For $r > r_0$ the potential increases gradually, approach 0 as $r \to \infty$, while $r < r_0$, the potential increases very rapidly, approaches ∞ at small value of r. The system is most stable at the minimum point which represents the equilibrium position. At equilibrium position the attractive and repulsive forces cancel each other. The binding energy E_0 corresponds to the energy at this minimum point. A number of solid properties depend on E_0, the curve shape bonding type.

© The Author(s) 2022
V. K. Jain, *Solid State Physics*, https://doi.org/10.1007/978-3-030-96017-9_2

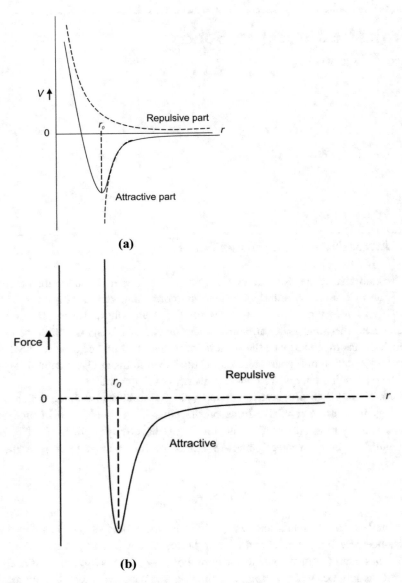

Fig. 2.1   **a** Typical interatomic potential for bonding in solids, **b** typical interatomic force for bonding in solids

For example solids having large bonding energies typically also have high melting temperature, at room temperature, solid substances are formed for large bonding energies whereas for small energies gaseous state is favoured, liquids prevail when the energies are of intermediate magnitude. The mechanical stiffness (modulus of elasticity) of a material is dependent on the shape of its force versus interatomic separation curve. For relatively stiff material the slope for such a curve at $r = r_0$ position will be quite steep; slopes are shallower for more flexible material. For V versus r curve, a deep and narrow trough which typically occurs for materials having large bonding energies normally correlate with a low coefficient of thermal expansion.

In solids, the atoms and molecules are closely packed and held together by force of attraction which is electrostatic in origin. The rearrangement of electrons gives rise to different types of bonds which holds the atom together in solids. There are several types of bonding depending on the physical origin and nature of bonding forces involved. The three main types of bonding are

(i)   Ionic bonding
(ii) Covalent bonding
(iii) Metallic bonding

In addition to the three main types of bonding there are other weaker bonds which often play important roles in explaining some of the bonding properties. These are

(i) Hydrogen bonding
(ii) van der Waals bonding

Often mixed bonding types are found in solids.

Thus on the basis of binding type we have following five categories of solids.

(i)   Ionic solids
(ii) Covalent solids
(iii) Metallic solids
(iv) Hydrogen bonded solids
(v)   van der Waals bonded solids.

2.2   Ionic Bonding

Ionic bonding is always found in compounds that are composed of both metallic and non-metallic elements, elements that are situated at the horizontal extremities of the periodic table. Atoms of metallic element have low ionization energies and hence can loose electrons readily interact with atoms of non-metallic element, that have high electron affinity.

The former atom which loses electron(s) becomes positive ion and the latter atom which gains electron(s) becomes negative ion. In the process all the atoms acquire stable or inert gas configuration and in addition an electric charge; that is they becomes ions.

Consider the formation of ionic bond in NaCl. The electron configuration of Na atom is (At. No. 11) 2, 8, 1 with having one valence electron. The electronic configuration of Cl atom (At. No. 17) is 2, 8, and 7 having 7 electrons in the outermost orbit. Cl requires one more electron to attain a stable configuration. In NaCl, Na atom loses electron and becomes positive ion while Cl atom gain electron and becomes negative ion. The Coulomb attraction of the oppositely charged ions produces a stable union. In the case of NaCl, sodium and chlorine ions have acquired the configuration of nearest inert gas. The force F_a of attraction is

$$F_a = -\frac{e^2}{r^2} \tag{2.1}$$

where e is the electronic charge and r is the distance between the ions. Besides the force of attraction, a force of repulsion will also come into play when the ions are brought nearer, due to the Pauli exclusion principle. The force of repulsion F_r is

$$F_r = \frac{mb}{r^{m+1}} \tag{2.2}$$

where b is a constant and m is a number. Figure 2.2 shows the NaCl structure. The bond energy U is

$$U = -\int_{-\infty}^{r} \left[-\frac{e^2}{r^2} - \frac{mb}{r^{m+1}} \right] dr \tag{2.3}$$

$$U_{\text{attraction}} = -\frac{6e^2}{r} + \frac{12e^2}{\sqrt{2}r} - \frac{8e^2}{\sqrt{3}r} + \frac{6e^2}{2r} - \frac{24e^2}{\sqrt{5}r} + \cdots$$

$$U_{\text{attraction}} = -\frac{e^2}{r}\left(6 - \frac{12}{\sqrt{2}} + \frac{8}{\sqrt{3}} - \frac{6}{2} + \frac{24}{\sqrt{5}} - \cdots \right) \tag{2.4}$$

The term within brackets is called the Madelung constant α of the crystal and represents the geometrical arrangement of ions.

The potential energy contribution of the repulsive forces has the form

$$U_{\text{repulsion}} = -\frac{B}{r^m} \tag{2.5}$$

The r^m dependence implies a short range force; it increases as the separation distance r increases. The total potential energy U is

$$U = U_{\text{attraction}} + U_{\text{repulsion}} = -\frac{\alpha e^2}{r} + \frac{B}{r^m} \tag{2.6}$$

Fig. 2.2 NaCl structure. From the structure it is seen that each Na$^+$ ion is surrounded by (i) 6 nearest Cl$^-$ ions at a distance r (ii) 12 next nearest Na$^+$ ions at $\sqrt{2}r$ (iii) 8 Cl$^-$ ions at a distance $\sqrt{3}r$ (iv) 6 Na$^+$ ions at a distance r (v) 24 Cl$^-$ ions at a distance $\sqrt{5}r$.

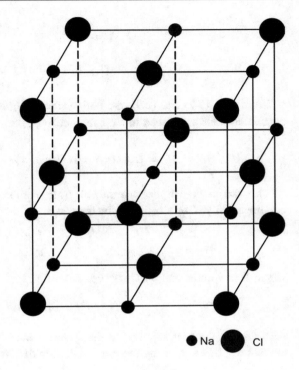

● Na ● Cl

Since there are N positive and N negative ions, the total lattice energy U_{total} of a crystal composed of N molecule may be written as

$$U_{\text{total}} = NU(r) \tag{2.7}$$

At the equilibrium separation r_0, the potential energy must be minimum and the first derivative of Eq. (2.6) must vanish or

$$\left(\frac{dU}{dr}\right)_{r=r_0} = 0$$

or

$$\left(\frac{\alpha e^2}{r^2} - \frac{mB}{r^{m+1}}\right)\Bigg|_{r=r_0} = 0$$

$$\frac{\alpha e^2}{r_0^2} = \frac{mB}{r_0^{m+1}} \tag{2.8}$$

$$B = \frac{\alpha e^2}{m} r_0^{m+1}$$

From Eqs. (2.6) and (2.8)

$$U(r)|_{r=r_0} = -\frac{\alpha e^2}{r_0} + \frac{\alpha e^2 r_0^{m-1}}{m r_0^m}$$

$$U(r_0) = -\frac{\alpha e^2}{r_0}\left(1 - \frac{1}{m}\right) \tag{2.9}$$

From Eqs. (2.7) and (2.9), the lattice energy U_L defined as energy released when the constituent ions are placed in their respective position in the crystal lattice

$$U_L = U_{\text{total}}(r_0) = NU(r_0) = -\frac{N\alpha e^2}{r_0}\left(1 - \frac{1}{m}\right) \tag{2.10}$$

The factor m can be determined from compressibility data of the crystal. The compressibility X of a solid can be expressed as

$$X = -\frac{1}{V}\frac{dV}{dp} \tag{2.11}$$

where V is volume and p is pressure. From first law of thermodynamics

$$dQ = dU + dW \tag{2.12}$$

where dQ is the heat supplied to the system which is consumed in performing work of amount dW and increasing the internal energy dU. We have

$$dQ = C_V dT \tag{2.13}$$

$$dW = p dV \tag{2.14}$$

At 0 K, C_V is very small and hence dQ can be neglected. Thus from Eq. (2.12)

$$dU = -dW = -p dV$$

$$\frac{dU}{dV} = -p \tag{2.15}$$

$$\frac{d^2U}{dV^2} = -\frac{dp}{dV}$$

The compressibility X_0 at 0 K using Eqs. (2.11) and (2.15)

$$\frac{1}{X_0} = \left(-\frac{1}{V}\frac{dV}{dp}\right)^{-1}_{t=0} = V_0\frac{d^2U}{dV^2} \tag{2.16}$$

The relation between volume V and interatomic distance r is given by

$$V = CNr^3 \tag{2.17}$$

where C is a constant depending on the type of lattice. For NaCl, $C = 2$ hence

$$V = 2Nr^3 \tag{2.18}$$

Now

$$\frac{dU}{dV} = \frac{dU}{dr}\frac{dr}{dV}$$

$$\frac{d^2U}{dV^2} = \frac{dU}{dr}\frac{d^2r}{dV^2} + \frac{d^2U}{dr^2}\left(\frac{dr}{dV}\right)^2$$

For $r = r_0$

$$\frac{dU}{dr}\bigg|_{r=r_0} = 0$$

$$\frac{d^2U}{dV^2} = \frac{d^2U}{dr^2}\bigg|_{r=r_0}\left(\frac{dr}{dV}\right)^2\bigg|_{r=r_0}$$

From Eq. (2.18)

$$\frac{dV}{dr} = 6Nr^2; \quad \left(\frac{dr}{dV}\right)^2\bigg|_{r=r_0} = \frac{1}{36N^2r_0^4} \tag{2.19}$$

From Eqs. (2.15), (2.16) and (2.19) we have

$$\frac{1}{X_0} = V_0\left(\frac{d^2U_{\text{total}}}{dr^2}\right)\bigg|_{r=r_0} \times \frac{1}{36N^2r_0^4} \tag{2.20}$$

From Eqs. (2.5) and (2.8) we have

$$U_{\text{total}} = -\frac{N\alpha e^2}{r}\left[1 - \frac{1}{m}\left(\frac{r_0}{r}\right)^{m-1}\right] \tag{2.21}$$

$$\frac{dU_{\text{total}}}{dr} = \frac{N\alpha e^2}{r^2} - \frac{N\alpha e^2 r_0^{m-1}}{r^{m+1}} \tag{2.22}$$

$$\frac{d^2U_{\text{total}}}{dr^2} = -\frac{2N\alpha e^2}{r^3} + \frac{N\alpha e^2(m+1)r_0^{m-1}}{r^{m+2}} \tag{2.23}$$

$$\frac{d^2U_{\text{total}}}{dr^2}\bigg|_{r=r_0} = -\frac{2N\alpha e^2}{r_0^3} + \frac{N\alpha e^2(m+1)}{r_0^3}$$

$$= \frac{N\alpha e^2}{r_0^3}(-2 + m + 1) = \frac{N\alpha e^2}{r_0^3}(m-1) \tag{2.24}$$

From Eqs. (2.20) and (2.24)

$$\frac{1}{X_0} = 2Nr_0^3 \frac{N\alpha e^2}{r_0^3}(m-1)\frac{1}{36N^2 r_0^4} = \frac{\alpha e^2(m-1)}{18r_0^4}$$

$$m = 1 + \frac{18r_0^4}{X_0\alpha e^2} \tag{2.25}$$

Thus the value of m may be determined. For NaCl crystal, $\alpha = 1.747558$, $r_0 = 0.281$ nm and $m = 9$. Substituting the value of m in Eq. (2.9)

$$U_L = NU(r_0) = -\frac{N\alpha e^2}{r_0}\left[1 - \left(1 + \frac{18r_0^4}{X_0 e^2\alpha}\right)^{-1}\right] \tag{2.26}$$

In ionic crystal, it is not possible for the electron to move about freely between ions unless a large amount of energy (~ 10 eV) is supplied. Solids with ionic bonding are therefore non-conducting. However, the presence of defects means that at high temperatures the ions themselves can move around giving rise to ionic conduction. The ionic crystals are transparent to visible light but strongly absorb infrared radiation at the frequencies at which the ions vibrate about their equilibrium positions. At frequencies higher than this they are opaque. The ionic crystals have high melting points and boiling points. They are highly soluble in solvents like water and liquid ammonia and insoluble in non-polar solvents. The ionic bonding is termed non-directional; that is the magnitude of the bond is equal in all direction around an ion. The examples of ionic crystals are LiF, LiCl, NaCl, LiBr, NaBr, KOH, etc.

2.3   Covalent Bonding

In covalent bonding, stable electron configuration is assumed by the sharing of electrons between adjacent atoms. Two atoms that are covalently bonded will each contribute at least one electron to the bond, and the shared electron may be considered to belong to both atoms. The number of covalent bonds that is possible for a particular atom is determined by the number of valence electrons. In a covalent bond, the spins of the two electrons are antiparallel.

The elements C, Si and Ge lack four electrons with respect to filled shell and thus these elements can have an attractive interaction associated with the charge overlap. Consider the case of diamond. The electron configuration of C is $1s^2 2s^2 2p^2$. Thus carbon can form two bonds. However, carbon forms four bonds. The reason why carbon does form four bonds can be understood when the energies of the $2s$ and $2p$ are considered. It turns out that the energy difference between $2s$ and $2p$ is very small so that one of the $2s$ electron of the atom can be promoted to $2p$ to make the configuration $1s^2 2s^1 2p^3$ resulting in four unpaired electrons. This promotion of electron from the ground state require 4 eV an amount more than regained when the bonds are formed. The orbital of these four electrons is disposed towards the four corners of the tetrahedral. The tetrahedral bond allows only four nearest neighbours, whereas a close packed structure has 12. The covalent bonded solids are therefore less dense than the ionic solids. Characteristic properties of the covalent bond, which distinguished it from the bonds of other types are its saturability and directionality. Saturability means that each atom can form covalent bonds only with a limited number of neighbours. For example, each hydrogen atom can form covalent bonds only with one of its neighbours. The electron pair contributing such a bond has antiparallel spins. A third atom in this case instead of being attracted will be repelled.

Some common features of materials with covalent bonds are:

(i) covalent bond crystals are usually hard and brittle; (ii) binding energy is high so that their melting and boiling points are high, but low compared to ionic crystals; (iii) covalent bonds are highly directional in character; (iv) these bonds have saturation properties; (v) covalent substances are insoluble in water; (vi) these materials are soluble in non-polar solvents like benzene; (vii) the conductivity of covalent crystals varies over a wide range. Some are excellent insulators, others are medium conductors like Si and Ge, and some behave as poor metals like grey tin; (viii) the conductivity increases with increase in temperature; (ix) these are transparent for wavelength but opaque to shorter wavelength. Ge and Si are transparent for wavelength longer than the infrared radiation; (x) carbon and diamond structure is the hardest substance and has high melting point of 3820 K. Compounds with covalent bonds may be solid, liquid or gas at room temperature depending on the number of atoms in the compound. The more atoms in each molecule, the higher a compound's melting and boiling temperature will be. Since most covalent compounds contain only a few atoms and the forces between molecules are weak, most covalent compounds have low melting and boiling points. Examples of covalent crystals are: CO, N_2, H_2, diamond, methane, silicon, germanium, rubber, etc.

2.4 Metallic Bonding

The valence electrons in a metal are rather loosely bound and frequently the electronic shells are only partially filled, so that metals bond not to form covalent bonds. The basic structure of metals is a gas of approximately free electrons surrounding a lattice of positive ions. The metal is held together by the interaction of positive ions with the electron gas. The electrons moving between the ions compensate the repulsive forces existing between the positively charged ions and bring them closer together. As the distance between the ions becomes smaller the density of electron gas increases and this leads to an increase in force drawing the ions together. On the other hand, in this case the repulsive force acting between the ions tends to move them away from each other. When the distance between the ions become such that the force of attraction are compensated by the force of repulsion, a stable lattice is formed.

In metals, the outer valence electrons are removed from the ion cores. They are free to move between the remaining ion cores. These delocalized valence electrons are involved in the conduction of electricity and are therefore called conduction electrons. Thus the metals are expected to be formed from those elements for which the expenditure of energy for removing the electron is small. This expenditure of energy can be more than compensated by the bonding. The energy of the electron is kinetic plus potential energy. The kinetic energy T is given by $T = -\frac{\hbar^2}{2m}\nabla^2$. The quantum mechanical average kinetic energy is $\int \psi^* T \psi \, d\tau$.

Where $d\tau$ is volume element, and ψ is wave function. $T\psi$ or $T = -\frac{\hbar^2}{2m}\nabla^2\psi$ is proportional to the second derivative of the wave function, that is the curvature. For an electron, localized to an atom the curvature is much higher than that for a nearly free electron in a metal. This result in a gain of energy. The potential energy gain comes mostly from Pauli exclusion principle which does not allow two electrons with the same spin direction to be at the same place and therefore the electron go out of each other way. Further, there is also Coulomb repulsion between the electrons themselves.

Consider sodium metal. It crystallizes in *bcc* structure. In the structure each Na is surrounded by eight Na atoms. The electronic configuration of Na ($Z = 11$) is $1s^2 2s^2 2p^6 3s^1$. Each atom has complete K and L shell and one unpaired $3s$ electron in its outer shell. When Na atoms come together, the electrons in the $3s$ atomic orbital of one sodium atom shares space with the corresponding electron on a neighbouring atom to form a molecular orbital. Each sodium atom is being touched by eight other sodium atoms and the sharing occurs between the central atom and the $3s$ orbital on all of the eight other atoms. And each of these eight is in turn being touched by eight sodium atoms and so on. All the $3s$ orbital on all of the atoms overlap to give a large number of molecular orbital which extend over the whole of the metal piece. An orbital can accommodate two electrons in accordance with Pauli exclusion principle. As a result there are a large number of orbital. The electron can move freely within these molecular orbital. Hence each electron becomes detached from its parent atom. The electrons are said to be delocalized. The metal is held together by the strong force of attraction between the positive nuclei and the delocalized electrons. Thus there is an array of positive ions in a sea of electrons.

Consider magnesium (Mg) metal. It crystallizes in *fcc* structure. The electronic configuration of Mg($Z = 12$) is $1s^2 2s^2 2p^6 3s^2$. Mg has two outer electrons. Both of these electrons become delocalized, so these has twice the electron density as it does in Na. Since Mg ion is a divalent ion and therefore it has twice the charge than the Na ion. Thus there will be greater attraction between the electrons and Mg ions. Further, magnesium atom has slightly smaller radius than sodium, so the delocalized electrons are closer to the nucleus. The magnesium atom also has twelve near neighbour rather than eight as in the case of sodium. Both of these factors increase the strength of the bond.

Consider the case of transition metals. Stronger bonding is found in transition metals, that is, metals with both s and p conduction electrons and a partially filled d shell. In such a case there is a mixed bonding. The s and p electrons turn into delocalized metallic conduction electrons whereas the d electrons create much more localized covalent type bonds.

The most common crystal structures of metallic solids are *fcc*, *bcc* or hexagonal close packed. The coordination number of atoms in typical metal can be as high as 12, because a pure metal has atoms of only one size. Approximately 40% of the pure metals solidify with coordination number 8. The cohesive energy of metal bonds tends to fall in the range (1–4 eV/atom) making the metal less strongly bound than ionic or covalent solids. Therefore, metals have relatively low melting points. The relatively free electrons in the metal interact readily with photons of visible light, so metals are not transparent. The free electrons are responsible for the high electrical and thermal conductivity of metals. Metallic bonding is found in the periodic table for group IA and IIA elements and in fact for all elemental metals.

2.5   The Hydrogen Bonding

The hydrogen bond occurs when a hydrogen atom is bound to two other atoms. When hydrogen take part in a covalent bond, with a strongly electronegative atom, for example, oxygen, its single electron is almost completely transferred to the partner atom. The proton which remains can then exert an attractive force on the second negatively charged atom. Because of the extended electron cloud of the electronegative atom and the extremely small size of the proton with its strongly reduced electron screening it is not possible for a third atom to be bond. The hydrogen atom is thus always doubly coordinated in a hydrogen bond. Such bonds are most common between strongly electronegative atoms. The binding energies of hydrogen bonds are of the order of 0.1 eV per bond. The hydrogen bond is directional and its range of action is ~ 300 pm. The hydrogen bonded solids are non-conducting, transparent optically. The example of hydrogen bonding is water, in particular when it is in the form of ice, HF_2, KH_2PO_4 etc.

2.6   The Van Der Waals Bonding

The van der Waals bond is only significant where other types of bonding are not present, for example, between atoms with closed electron shells, or between saturated molecules. The source of this bonding is the charge fluctuations in the atom due to zero point motion. The dipole moment which thereby arise causes an additional attractive force. Van der Waals forces are responsible for the bonding in molecular crystals.

The van der Waals bonding arises from atomic or molecular dipoles. An electric dipole exists whenever there is some separation of positive and negative portion of an atom or molecule. The bonding results from the coulombic attraction between the positive end of one dipole and the negative region of an adjacent as shown in Fig. 2.3. Dipole interaction occurs between induced dipoles, between induced dipoles and polar molecules (which have permanent dipoles) and between polar molecules. A dipole may be induced or created in an atom or molecule that is normally electric symmetric (Fig. 2.4).

All atoms are experiences constant vibrational motion. As a result the centre of negative charges for the electrons is displaced momentarily ($< 10^{-15}$ s), but repeatedly, from the centre of positive charges that accompanies the proton in the nuclei. These oscillations produce small electric dipole, when the positive side of one atom or molecule is attracted to the negative side of an adjacent atom or molecule (Fig. 2.5).

The resulting dipoles provide the force of attraction. At very low temperatures (where the thermal energy is small) these attractive force are able to condense and solidify the materials.

Fig. 2.3 Schematic representation of van der Waals bonding between two dipoles

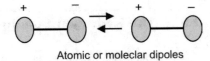

Atomic or moleclar dipoles

Fig. 2.4 Schematic representation of an electrically symmetric atom

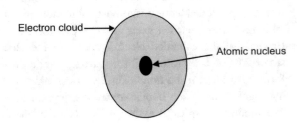

Fig. 2.5 Schematic representation of an induced atomic dipole

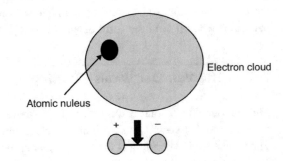

Fig. 2.6 Schematic
representation of a polar HCl
molecule

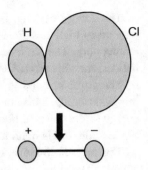

Permanent dipole moments exist in some molecules because of asymmetrical arrangements of positively and negatively charged regions, such molecules are termed polar molecules, for example, HCl (Fig. 2.6).

The polar molecules also induce dipoles in adjacent non-polar molecules and a bond will form as a result of attractive forces between the two molecules. van der Waal forces will also exists between adjacent polar molecules. The associated binding energies are significantly greater than for bonds involving induced dipoles.

The van der Waals bonds are weak and the bonding energy is ~ 0.1 eV. The range of van der Waals force ~ 300–500 pm. The van der Waals potential is found to decrease with distance as $1/r^6$, far more rapidly than the ionic potential, which decreases as $1/r$. Inert gas (Ar, Ne, Xe, Kr, He) solids, solid methane, etc. are example of van der Waals bonding. These solids have low melting points, little mechanical strength, are insulator and transparent to far-ultraviolet region.

2.7 Comparison Between Bonds of Various Kinds

The ionic bond involves electron transfer and is non-directional. The covalent bond involves electron sharing and is directional. The third type of bond, that is, metallic bond involves electron sharing and is non-directional. In this case the valence electrons are said to be delocalized electrons; that is they have an equal probability of being associated with any of a large number of adjacent atoms.

The van der Waals bond is the most universal one. It exists in all the cases. This is the weakest having an energy of the order of 10^4 J/mol. Ideally, it operates between neutral atoms, or molecules with closed inner electron shell. These forces are responsible for the existence of liquid and solid state of inert gases, hydrogen, oxygen, nitrogen and many other organic and inorganic compounds. Because of low energy values of the van der Waals bond all structure based on it are unstable, volatile and have low melting temperatures.

The energy of ionic bond is much higher than that of the van der Waals bond and may be as high as 10^6 J/mol. Therefore, solid based on the ionic bond has high melting temperature. The ionic bond is very frequent among inorganic compounds such as metal halides, metallic oxides, sulphides and other polar compounds.

The covalent bond is very frequent among organic compounds but it also present in inorganic compounds, in some metals and in numerous intermetallic compounds. The energy of the covalent bond is also high (10^6 J/mol) resulting in high melting temperature of the solids. This bond is responsible for the existence of diamond, Ge, Si etc.

The metallic bond is characteristic of typical metals and various intermetallic compounds. The order of the magnitude of the energy of this type of bond is comparable to that of energy of covalent bond. The hydrogen bond is relatively weak and plays an important part.

Solved Examples

Example 1 Calculate the Madelung constant in a linear chain of ions of alternate sign.

Solution
Consider the case of a one-dimensional linear chain consisting of alternative positive and negative ions as shown in Fig. 2.7. Consider a negative ion as a reference and R is the shortest distance between the two ions. The reference ion has two positive ions as its neighbours at a distance R, then two next neighbouring negative ions at a distance $2R$, then the two positive ions at distance $3R$ and so on.

Thus, total energy due to all ions in the linear chain is

$$U = -2\frac{e^2}{(4\pi\varepsilon_0)R} + 2\frac{e^2}{(4\pi\varepsilon_0)2R} - 2\frac{e^2}{(4\pi\varepsilon_0)3R} + \cdots$$

$$U = -2\frac{e^2}{(4\pi\varepsilon_0)R}\left[1 - \frac{1}{2} + \frac{1}{3} - \cdots\right]$$

We have the series expansion

$$\ln(1+x) = x - \frac{x^2}{2} + \frac{x^3}{3} - \frac{x^4}{4} + \cdots$$

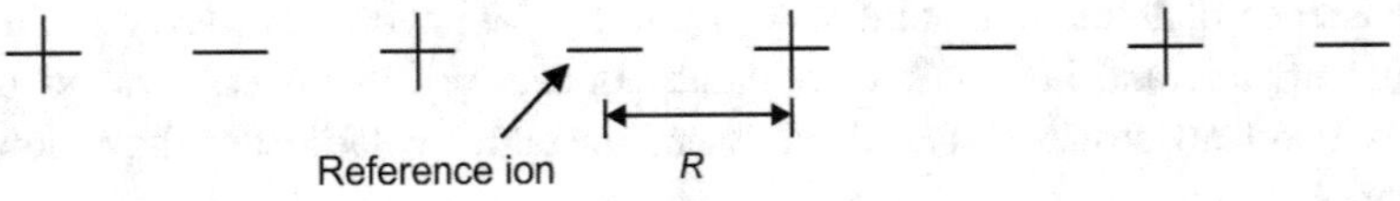

Reference ion R

Fig. 2.7 Line of ions of alternating signs, with distance R between the ions

Putting $x = 1$

$$\ln 2 = 1 - \frac{1}{2} + \frac{1}{3} - \cdots$$

Then we have

$$U = -2\frac{e^2}{(4\pi\varepsilon_0)R}\ln 2$$

$$\alpha = 2\ln 2 = 1.38$$

Example 2 In NaCl crystal, the equilibrium distance between ions is $r_0 = 0.281$ nm and $\alpha = 1.748$. Taking $m = 9$, calculate the potential energy per ion pair.

Solution
We have

$$U = -\frac{\alpha e^2}{(4\pi\varepsilon_0)r_0}\left(1 - \frac{1}{m}\right)$$

$$= -\frac{(9 \times 10^9 \text{ Nm}^2/\text{C}^2) \times 1.748 \times (1.6 \times 10^{-19} \text{ C})^2}{2.81 \times 10^{-10} \text{ m}}\left(1 - \frac{1}{9}\right)$$

$$= -1.27 \times 10^{-18} \text{ J} = -\frac{12.7 \times 10^{-19}}{1.6 \times 10^{-19}}\text{eV} = -7.94 \text{ eV}$$

The contribution per ion is half of this, that is, -3.96 eV. This is called as the lattice energy per ion in the NaCl crystal.

Example 3 How much energy per neutral atom would be needed to take apart a crystal of NaCl. The binding energy of an ion pair in NaCl is 7.98 eV.

Solution
If we supply an energy E to a mole of NaCl, we obtain $N_A\text{Na}^+$ ions and $N_A\text{Cl}^-$ ions (N_A is Avogadro number). To convert these to neutral atoms, an electron must be removed from each Cl^-. This required an amount of energy which is equal to electron affinity of Cl that is 3.61 eV. This removed electron must be attached to Na^+ which returns the ionization energy of Na (5.14 eV). The net cost per pair of Na and Cl atom is

$$7.98 \text{ eV} + 3.61 \text{ eV} - 5.14 \text{ eV} = 6.45 \text{ eV}$$

Since expending of 6.45 eV gives two neutral atoms (Na and Cl), the net cost per atom is half that amount or 3.23 eV.

Example 4 Calculate the bond energy and bond length for Argon if the potential is given by

$$V = -\frac{K_A}{r^6} + \frac{K_R}{r^{12}}$$

where K_A and K_R are constants for attraction and repulsion, respectively. The values of $K_A = 10.37 \times 10^{-78}$ Jm^6 and $K_R = 16.16 \times 10^{-135}$ Jm^{12}.

Solution

The equilibrium bond length occurs at $dV/dr = 0$

$$\left(\frac{dV}{dr}\right)_{r=r_0} = 0 = \frac{6K_A}{r_0^7} - \frac{12K_R}{r_0^{13}}$$

Rearranging gives us

$$r_0 = \left(\frac{2K_R}{K_A}\right)^{\frac{1}{6}} = \left(\frac{2 \times 16.16 \times 10^{-135}}{10.37 \times 10^{-78}}\right)^{\frac{1}{6}}$$

$$m = 0.382 \times 10^{-9}\,\text{m} = 0.382\,\text{nm}$$

The bond energy $= E(r_0)$

$$E(r_0) = -\frac{K_A}{(0.382\,\text{nm})^6} + \frac{K_R}{(0.382\,\text{nm})^{12}}$$

$$E(r_0) = -\frac{10.37 \times 10^{-78}\,\text{Jm}^6}{(0.382\,\text{nm})^6} + \frac{16.16 \times 10^{-135}\,\text{Jm}^{12}}{(0.382\,\text{nm})^{12}}$$

$$= -1.66 \times 10^{-21}\,\text{J}$$

For one mol of argon

$$E_{\text{bonding}} = -1.66 \times 10^{-21}\,(\text{J/bond}) \times 6.02 \times 10^{24}\,(\text{bonds/mol})$$

$$= -0.999 \times 10^3\,\text{J/mol}$$

$$E_{\text{bonding}} = -0.999\,\text{kJ/mol}$$

Example 5 The potential energy between the two adjacent atoms may be represented by

$$V = -\frac{A}{r} + \frac{B}{r^n}$$

calculate the binding energy E_0.

Solution

The equilibrium bond length occurs at $dV/dr = 0$

$$\left(\frac{dV}{dr}\right)_{r=r_0} = 0 = \frac{A}{r_0^2} - \frac{nB}{r_0^{n+1}}$$

$$r_0^{-1} = \frac{nB}{A}$$

$$r_0 = \left(\frac{nB}{A}\right)^{\frac{1}{n-1}} = \left(\frac{A}{nB}\right)^{\frac{1}{1-n}}$$

The binding energy

$$E_0 = \frac{A}{r_0} + \frac{B}{r_0^n} = -\frac{A}{\left(\frac{A}{nB}\right)^{\frac{1}{1-n}}} + \frac{B}{\left(\frac{A}{nB}\right)^{\frac{n}{1-n}}}$$

Objective Type Questions

1. The lattice energy in an ionic crystal depend on the interatomic distance r as
   (a) r^{-1} (b) r^{-2} (c) r^{-6} (d) r^2

2. The force important for ionic bonding is
   (a) electric force (b) nuclear forces (c) magnetic force (d) gravitational force

3. The elements which are likely to form crystal through the ionic bonding are
   (a) metal–metal (b) metal–non-metal (c) non-metal-non-metal (d) closed shell elements

4. The covalent bond
   (a) involves electron sharing and is directional
   (b) involves electron transfer and is non-directional
   (c) involves electron sharing and is non-directional
   (d) none of these

5. The ionic bond
   (a) involves electron sharing and is directional
   (b) involves electron transfer and is non-directional
   (c) involves electron sharing and is non-directional
   (d) none of these

6. The metallic bond
   (a) involves electron sharing and is directional
   (b) involves electron transfer and is non-directional
   (c) involves electron sharing and is non-directional
   (d) none of these

7. In polymers the bonding is mainly
   (a) ionic (b) covalent (c) metallic (d) van der Waals

8. In Cu the bonding is mainly
   (a) involves electron sharing and is directional
   (b) involves electron transfer and is non-directional
   (c) involves electron sharing and is non-directional
   (d) none of these
9. In KOH the bonding is mainly
   (a) involves electron sharing and is directional
   (b) involves electron transfer and is non-directional
   (c) involves electron sharing and is non-directional
   (d) none of these
10. The force necessary for binding of atoms into solid is
    (a) magnetic (b) nuclear (c) electrical (d) gravitational
11. What type of bonding would be expected for solid xenon?
    (a) covalent (b) ionic (c) metallic (d) van der Waals
12. What type of bonding would be expected for tungsten?
    (a) covalent (b) ionic (c) metallic (d) van der Waals
13. What type of bonding would be expected for rubber?
    (a) covalent (b) ionic (c) metallic (d) van der Waals
14. What type of bonding would be expected for CdTe?
    (a) covalent (b) ionic (c) metallic (d) van der Waals
15. What type of bonding would be expected for CaF_2?
    (a) covalent (b) ionic (c) metallic (d) van der Waals
16. What type of bonding would be expected for ice?
    (a) covalent (b) hydrogen (c) metallic (d) van der Waals
17. The coordination number of atoms in typical metal can be as high as
    (a) 4 (b) 6 (c) 8 (d) 12
18. Other factors being equal, materials with strong bonds and high melting temperature have
    (a) low thermal expansion (b) high thermal expansion
19. Electrical and thermal conductivity are greater in those material having
    (a) localized electrons (b) delocalized electrons (c) completely filled shell (d) ionic bonding

Problems

1 Which elements are likely to form metals?
2 Where does the energy gain in metallic bonding comes from?
3 What is the difference between a simple metal like Na and transition metal?
4 Why is the van der Waals bonding much weaker than most other bonding types?
5 Why covalently bonded materials are generally less dense than ionically or metallically bonded ones?

6 Briefly cite the main differences between ionic, covalent and metallic binding.

Answers
Objective Type Questions

1. (a)	2. (a)	3. (b)	4. (a)	5. (b)	6. (c)	7. (b)	8. (c)
9. (b)	10. (c)	11. (d)	12. (c)	13. (a)	14. (a)	15. (a)	16. (b)
17. (d)	18. (a)	19. (b)					

Defects in Solids

3

3.1 Classification of Defects

In an ideal crystal structure every unit cell is identical and each has a specified shape, size and cell contents. The concept of an ideal crystal with a perfect arrangement of atoms is valid only at 0 K, and then there is no entropy contribution. However, at a finite temperature, a certain disorder is introduced into the structure of solids and solid becomes structurally imperfect. Some cells may have one or more atoms less whereas others may have more atoms than the ideal unit cell. The imperfections of the crystal are called crystal defects.

Imperfection in crystals may be classified on the basis of their geometry such as point, line and surface imperfection. Point defects are zero dimensional dimension defects confined over a few interatomic distances other defects such as line defects and surface defect extend through microscopic region in crystal.

1. Point defects (zero dimensional defects): The important point defects are:
   - (i) Vacancies: Whenever one or more atoms are missing from a normally occupied position, as shown in Fig. 3.1, the defect caused is known as vacancy.
   - (ii) Interstitial defects: Whenever an extra atom (generally smaller than the parent atoms) occupies interstitial position in the crystal system without dislodging the parent atom as shown in Fig. 3.2, the defect caused is known as interstitial defect
   - (iii) Substitutional defect: Whenever an impurity atom (other than the parent atoms) occupies a position which was initially meant for a parent atom, as shown in Fig. 3.3, the defect caused is known as substitutional defect.
   - (iv) Frankel defect: It consists of an atom displaced from its regular site to an interstitial site. The atom in the interstitial site and the vacancy caused by the

© The Author(s) 2022
V. K. Jain, *Solid State Physics*, https://doi.org/10.1007/978-3-030-96017-9_3

Fig. 3.1 An empty lattice site
is a vacancy

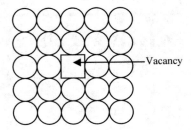

Fig. 3.2 Interstitial is an atom
on a non-lattice site

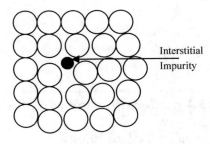

Fig. 3.3 Substitutional
impurity

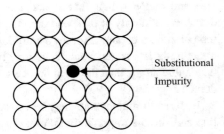

displacement of atom from the regular site (Fig. 3.4) together are named Frankel
pair.

Fig. 3.4 Frenkel defect

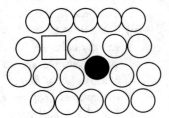

Fig. 3.5 Schottky defect

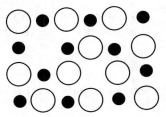

(v) Schottky defect: In ionic crystals if there is a vacancy in positive ion site, then charge neutrality of crystal can be maintained by creating a vacancy in the neighbouring negative ion site (Fig. 3.5), such a pair of vacant site is called Schottky defect.

2. Line defects (One dimensional defects): These defects are mostly due to misalignment of ions or presence of vacancies along a line. Line defects in crystalline solids are termed as dislocations. The dislocations are classified into two categories
   (i) Edge dislocation:
   (ii) Screw dislocation:
3. Surface defects (two dimensional defects)
   (i) Grain boundary
   (ii) Tilt boundaries
   (iii) Twin boundaries
   (iv) Stacking faults
4. Volume defects (three dimensional defects)
   (a) Foreign particle inclusion
   (b) Large voids or pores: Voids are small regions where there are no atoms and can be thought of as clusters of vacancies.

3.2 Point Defects

3.2.1 Vacancy Defects

The vacancies are created and destroyed constantly in the crystal due to thermal fluctuations. The number of vacancies (n) is determined by the temperature (T), total number of atoms (N) and average energy (E_V) required to create a vacancy.

The number of possibilities to distribute n vacancies on N regular atom sites is given by

$$W = \frac{N(N-1)(N-2)\cdots(N-n+1)}{n!} = \frac{N!}{n!(N-n)!} \tag{3.1}$$

The creation of vacancy inside the crystal results in an increase in disorder, resulting an increase in entropy. By entropy probability relation

$$S = k_B \ln W \tag{3.2}$$

where k_B is Boltzmann constant. From Eqs. (3.1) and (3.2)

$$S = k_B \ln \frac{N!}{n!(N-n)!} \tag{3.3}$$

$$S = k_B \ln N! - k_B \ln n! - k_B \ln(N-n)! \tag{3.4}$$

The Helmholtz free energy F is

$$F = U - TS$$

where U is internal energy and S is entropy.

The change in Helmholtz free energy F is

$$\Delta F = \Delta U - T\Delta S \tag{3.5}$$

From Eqs. (3.4) and (3.5)

$$F = nE_V - Tk_B \ln N! + Tk_B \ln n! + Tk_B \ln(N-n)! \tag{3.6}$$

Using Stirling's formula valid for large x

$$\ln x! = x(\ln x - 1) = x \ln x - x \tag{3.7}$$

$$
\begin{aligned}
F &= nE_V - Tk_B \ln N + Tk_BN + nTk_B \ln n - nTk_B \\
&\quad + (N-n)Tk_B \ln(N-n) - Tk_B(N-n) \\
F &= nE_V - Tk_B[N \ln N - nT \ln n - (N-n)\ln(N-n)]
\end{aligned} \tag{3.8}
$$

In equilibrium the system is in the state of lowest free energy. The corresponding equilibrium concentration is obtained by differentiating the free energy with respect to n, that is,

$$\frac{\partial F}{\partial n} = 0$$

$$\frac{\partial F}{\partial n} = E_v - k_B T\left[0 + \ln(N-n) + \frac{N-n}{N-n} - \frac{n}{n} - \ln n\right] = 0$$

$$E_v = k_B T[\ln(N-n) - \ln n] \tag{3.9}$$

$$E_V = k_B T \ln \frac{N-n}{n}$$

$$\frac{E_V}{k_B T} = \ln \frac{N-n}{n}$$

since $n \ll N$, the Eq. (3.9) is

$$\frac{E_V}{k_B T} = \ln \frac{N}{n}$$

$$n = N \exp\left(-\frac{E_V}{k_B T}\right) \tag{3.10}$$

The number of vacancies increases with increase in temperature. The higher the temperature more often atoms are jumping from one equilibrium position to another and large number of vacancies can be found in a crystal. At ~ 300 K in copper there is one vacancy per 10^{15} lattice atoms whereas at higher temperature there is one vacancy for every 10^4 atoms. In Eq. (3.5) the enthalpy and entropy terms are not in themselves very temperature dependent, so at higher temperature $T\Delta S$ term becomes greater and the free energy minimum occurs at a higher defect concentration. This is in accord with the general expectation that crystal structure will have more defects and to more disorder states at higher temperature. Thus at higher temperature there are more vacancies, a greater accommodation of impurity atoms and a degree of cation disorder. The actual concentration of vacancies will be higher than the equilibrium value if the crystal is grown at an elevated temperature and then cooled suddenly, thereby freezing in the vacancies. Lattice vacancies are present in alkali halides when divalent impurities are incorporated. If a crystal of KCl is grown with a small amount of $CaCl_2$, the density of crystal varies as if a K^+ lattice vacancy were formed for each Ca^{2+} ion in the crystal. The Ca^{2+} substitutes for K^+ and two Cl^- ion enter replace two Cl^- sites in KCl. For charge compensation, a vacancy is created at K^+ site. Experimental results indicates that the addition of $CaCl_2$ in KCl lower the density of the crystal. If no vacancy were present the density would increase because Ca^{2+} is heavier than K^+.

3.2.2 Schottky Defects

Consider an ionic crystal having equal number of positively and negatively charged ions. Let the crystal contain N ions and n Schottky defects (n cation vacancy and n anion vacancies). The number of ways in which each kind of vacancy is produced is

$$\frac{N!}{n!(N-n)!} \tag{3.11}$$

The different ways W in which n vacancies pair can be produced is obtain by squaring Eq. (3.11) since the number of cation and anion vacancies is equal

$$W = \left(\frac{N!}{n!(N-n)!}\right)^2 \tag{3.12}$$

On substituting the value of W from Eq. (3.12) in Eq. (3.2)

$$S = k_B \ln\left(\frac{N!}{n!(N-n)!}\right)^2$$

$$S = 2k_B \ln \frac{N!}{n!(N-n)!} \tag{3.13}$$

Let E_V is the energy required to produce a pair of vacancies, and thus nE_V represent the total change in its internal energy U of the crystal. Thus the Helmholtz free energy F of the crystal is

$$F = U - TS = nE_V - 2k_B T \ln \frac{N!}{n!(N-n)!} \tag{3.14}$$

$$F = nE_V - 2k_B T[\ln N! - \ln n! - \ln(N-n)!]$$

Using Sterling's approximation

$$F = nE_V - 2k_B T[N \ln N - N - n \ln n + n - (N-n)\ln(N-n) + (N-n)]$$
$$F = nE_V - 2k_B T[N \ln N - n \ln n - (N-n)\ln(N-n)] \tag{3.15}$$

In equilibrium the system is in the state of lowest free energy. When equilibrium is attained at a given temperature, the Helmholtz free energy is constant and its first derivative relative to n (since the total number of atomic positions in the crystal N is not changed) is

$$\frac{\partial F}{\partial n} = 0$$

$$\frac{\partial F}{\partial n} = nE_V - 2k_B T\left[0 - \ln n + \frac{N-n}{N-n} - \frac{n}{n} + \ln(N-n)\right] = 0$$

$$E_V = 2k_B T \ln \frac{N-n}{n}$$

Since $n \ll N$

$$E_V = 2k_B T \ln \frac{N}{n}$$

$$n = N \exp\left[-\frac{E_V}{2k_B T}\right] \tag{3.16}$$

The number of Schottky defects depends on (a) total number of ion pairs N (that is on mass of the ionic crystal), (b) the average energy required to produce a Schottky defect, and (c) temperature: the fraction of Schottky defect increases with increasing temperature.

3.2.3 Frenkel Defects

Consider a crystal composed of equal numbers of positively and negatively charged ions. Let N be the total number of atoms, N_i the number of interstitial sites and E_i is the energy required to displace a certain atom from its regular position to an interstitial position. The total number of ways in which an atom can be removed is given by

$$\frac{N!}{n!(N-n)!} \tag{3.17}$$

where n is the number of interstitial defects. Since N_i is the total number of interstitial sites, out of which n are occupied; therefore, the number of different ways in which n interstitial defects can be formed is

$$W_i = \frac{N!}{n!(N-n)!} \times \frac{N_i!}{n!(N_i-n)!} \tag{3.18}$$

The corresponding increase in entropy from Boltzmann relation (3.2) is

$$S = k_B \ln W_i = k_B \ln \left[\frac{N!}{n!(N-n)!} \times \frac{N_i!}{n!(N_i-n)!} \right]$$

$$S = k_B \ln W_i = k_B \left[\ln \frac{N!}{n!(N-n)!} + \ln \frac{N_i!}{n!(N_i-n)!} \right] \tag{3.19}$$

$$S = k_B [\ln N! - \ln(N-n)! - \ln n! + \ln N_i! - \ln n! - \ln(N_i-n)]$$

Using Stirling approximation

$$S = k_B \left[\begin{array}{l} N \ln N - N - (N-n)\ln(N-n) + (N-n) - n\ln n + n + N_i \ln N_i - N_i \\ -n \ln n + n - (N_i-n)\ln(N_i-n) + (N_i-n) \end{array} \right]$$

$$S = k_B [N \ln N - (N-n)\ln(N-n) - 2n\ln n + N_i \ln N_i - (N_i-n)\ln(N_i-n)]$$

The change in Helmholtz free energy is

$$F = U - TS$$

$$F = nE_i - k_B T [N \ln N - (N-n)\ln(N-n) - 2n\ln n \tag{3.20}$$

$$+ N_i \ln N_i - (N_i-n)\ln(N_i-n)]$$

In equilibrium the system is in the state of lowest free energy. The corresponding equilibrium concentration is obtained by differentiation the free energy with respect to n, that is,

$$\frac{\partial F}{\partial n} = E_i - k_B T$$

$$\left[0 + \frac{(N - n)}{(N - n)} + \ln(N - n) - 2\ln n - 2 \right.$$

$$\left. + 0 + \frac{(N_i - n)}{(N_i - n)} + \ln(N_i - n) \right] \tag{3.21}$$

$$E_i = k_B T \left[\ln(N - n) - \ln n^2 + \ln(N_i - n) \right]$$

$$E_i = k_B T \ln \frac{(N - n)(N_i - n)}{n^2}$$

For $n \ll N, N_i$

$$E_i = k_B T \ln \frac{N N_i}{n^2}$$

$$\frac{n^2}{N N_i} = \exp\left(-\frac{E_i}{k_B T} \right) \tag{3.22}$$

$$n = \sqrt{N N_i} \exp\left(-\frac{E_i}{2 k_B T} \right)$$

The concentration rises exponentially with the temperature. The factor of two in the denominator of the exponent arises because vacancies as well as interstitial atoms are distributed independently in the crystal.

Point defects are common in crystals with large anions such as AgBr, AgI, RbAgI$_4$ etc. Due to their defects the ions have some freedom to move about in crystal making them relatively good conductor. Frenkel and Schottky defects are common in ionic crystals. In pure alkali halides the most common defects are Schottky defects, and in pure silver halides the most common vacancies are Frenkel defects. The Frenkel defects occur more often in lattices with low coordination numbers. The ratio of cation to anion is not altered by the formation of either a Frenkel or Schottky defect. In both Frenkel and Schottky defects the pair of point defects stay near each other because of strong coulombic attraction of opposite charges.

The Frenkel and Schottky defects play an important part in many processes in crystals. They act as carrier scattering centres reducing their mobility. They can also act as sources of carrier productions, that is, play the role of donor and acceptors. They can also appreciably affect the optical, magnetic, mechanical and thermodynamic property of the crystal.

3.2.4 Colour Centres

A colour centre is a lattice defect that absorbs visible light. Pure alkali halides are transparent throughout the visible part of the electromagnetic spectrum. The crystal may be coloured by introduction of chemical impurity, by irradiation with x-rays, γ-rays, neutron, electron bombardment, by introducing an excess of metal ion for example heating NaCl crystal in the vapour of Na metal then cooling it quickly. NaCl crystal heated in Na vapour becomes yellow while KCl heated in vapours of K becomes magenta, LiF heated in vapours of Li turn pink and ZnO in the presence of excess Zn turn yellow. These centres known as F centres are due to electron bound at a negative ion vacancy as suggested by de Boer. When excess alkali atoms are added to alkali halide crystal a corresponding number of negative ion vacancies are created. The valence electron of the alkali atom is not bound to the atom; the electron migrates in the crystal and becomes bound to a negative ion site. The F centre is the simplest trapped electron centre in the alkali halide crystals. Another kind of centre known as V centre is also produced in alkali halides crystals containing excess halogen ions. Such excess ions are accompanied by positive ion vacancies which serve to bound holes.

3.3   Line Defects

3.3.1 Edge Dislocation

In this kind of dislocation an extra portion of a plane of atoms, or half plane, the edge of which terminates within the crystal. The edge dislocation centres around the line that is defined along the end of extra half plane of atom. This is sometimes termed the dislocation line and is perpendicular to the plane of the page. The horizontal plane in which the extra plane of atom terminates is called slip plane. The atoms above the dislocation line are squeezed together and those below are pulled apart. The magnitude of this distortion decreases with distance away from the dislocation line. The edge dislocation is represented by the symbol ⊥, which also indicates the position of dislocation line. An edge dislocation may also be formed by an extra half plane of atoms that is included in the bottom portion of the crystal, its designation is T. If the extra half plane is in the upper part of the lattice, the edge dislocation is assumed to be positive. But if the extra plane is in the lower part of the lattice, the dislocation is assumed to be negative (Fig. 3.6).

Fig. 3.6 Edge dislocation

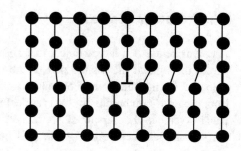

Fig. 3.7 Screw dislocation

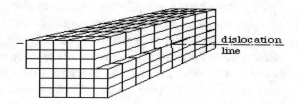

3.3.2 Screw Dislocation

It forms when one part of crystal lattice is shifted (through shear) relative to the other crystal part. It is called screw as atomic planes form a spiral surface around the dislocation line (Fig. 3.7).

Both screw and edge dislocation may occur together in the crystal forming a curved dislocation.

3.3.3 Burger Vector

To describe the size and direction of the main lattice distortion caused by a dislocation, the Burger vector **b** is introduced. To find the Burger vector we make a circuit from atom to atom counting the same number of atomic distances in all directions. If the circuit encloses a dislocation it will not close. The vector that closes the loop is the Burger vector. Consider a path starting from any point S, tracing out a circuit with equal number of atomic distances in opposite direction, if the crystal is perfect we return to the original position S. If a dislocation is present as shown in Fig. 3.8, the same circuit fails to close and will finish at point F. The closure failure SF defines the magnitude and direction of the Burgers vector and returning to the original position.

Fig. 3.8 Burger vector **b**

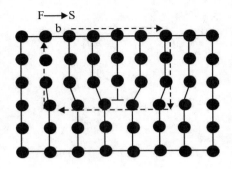

3.4 Surface Defects

3.4.1 Grain Boundaries

Although the material such as copper in an electric wire may contain only one phase—that is only one structure (*fcc*)—it contains many crystals of various orientations. These individual crystals are called grains. The shape of a grain in a solid usually is controlled by the presence of surrounding grains. Within any particular grain all of the unit cells are arranged with one orientation and are pattern. The crystalline orientations vary from grain to grain. There exist some atomic mismatch within the region where the two grains meet; this area is called a grain boundary (Fig. 3.9).However at the grain boundary between two adjacent grains, there is a transition zone that is not aligned with either grain. Atoms at grain boundaries tend to be loosely bound. Typical grain size ~ 0.01 to ~ 10 μm. Grain boundaries are called large angle boundaries if misorientation of two neighbouring grains exceeds 10°–15°, and if misorientation of two neighbouring grains is 5° or less then it is called small angle boundaries. Grain boundaries are more chemically reactive than the grain themselves. Grain boundaries usually occur when two crystals begin growing separately and then meet. Grain size varies inversely with the grain boundary area. The growth of grains depends on temperature. An increase in temperature increases the thermal vibrational energy, which in turn accelerate the net diffusion of atoms across the boundary from small to large grain. A subsequent decrease in temperature slows down

Fig. 3.9 Grain boundary

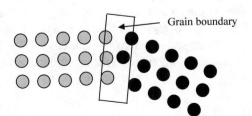

Grain boundary

Fig. 3.10 Tilt boundaries

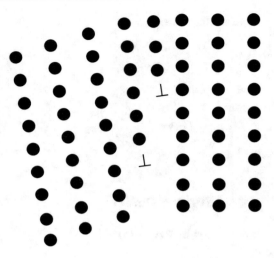

the boundary movement, but does not reverse it. The grain size in alloys having a single phase can be reduced by deforming the grains plastically and to start new grains by recrystallization.

3.4.2 Tilt Boundaries

A tilt boundary is defined as the boundary between two adjacent perfect regions in the same crystal that are slightly tilted relative to one another (Fig. 3.10). In other words, when the angle between the crystalline orientations of the two grains is small (less than 10°), the boundary is said to be tilt boundary. Their structure of such boundaries appears as an array of edge dislocations of the same sign located one above the other.

3.4.3 Twin Boundaries

A twin boundary occurs when the crystal on either side of a plane are mirror images of each other. The boundary between the twinned crystals will be a single plane of atoms. There is no region of disorder and the boundary atoms can be viewed as belonging to the crystal structure of both twins (Fig. 3.11). Twins are either grown in during the crystallization, or the result of mechanical or thermal work.

Fig. 3.11 Twin boundaries

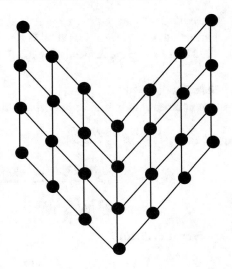

3.4.4   Stacking Fault

Stacking fault occurs in a number of crystal structures, but the common example is closed packed structure. *fcc* structure differs from *hcp* structure only in stacking order. A stacking fault is a surface defect that results from the stacking of one atomic plane out of the sequence on another while lattice on either side of the fault is perfect. For example, in *fcc* crystal, stacking is *ABC ABC ABC*......... A stacking fault can change the stacking to *ABC AB ABC*....., that is, one *hcp* layer in a *fcc* crystal. Stacking faults occur during the growth of the crystal, or during a structural transformation from one layer structure to a different structure made of similar layer, or during deformation by the glide of partial dislocation.

Solved Examples

Example 1 Calculate the fraction of atom sites that are vacant for copper at 1357 K. The energy for vacancy is 0.90 eV/atom.

Solution
We have

$$\frac{n}{N} = \exp\left(-\frac{E_V}{k_B T}\right) = \exp\left(-\frac{0.90\,\text{eV/atom}}{8.62 \times 10^{-5}\,\frac{\text{eV}}{\text{K}} \times 1357\,\text{K}}\right)$$

$$\frac{n}{N} = \exp\left(-\frac{0.90\,\text{eV}}{0.11697}\right) \approx \exp(-7.69428) \approx 4.55 \times 10^{-4}$$

Example 2 Calculate the energy for vacancy formation in silver (Ag), given that the equilibrium number of vacancies at 1073 K is 3.6×10^{23} m^{-3}. The atomic weight and density at 1073 K for silver are respectively, 107.9 g/mol and 9.5 g/cm^3.

Solution

The number of atomic sites per cubic metre for Ag is

$$N = \frac{N_A \, \rho}{A_{Ag}}$$

$$= \frac{\left(6.023 \times 10^{23} \text{ atoms/mol}\right)\left(9.5 \text{ g/cm}^3\right)\left(10^6 \text{ cm}^3/\text{m}^3\right)}{107.9 \text{ g/mol}}$$

$$= 5.299 \times 10^{28}$$

From Eq. (3.10)

$$n = N \exp\left(-\frac{E_V}{k_B T}\right)$$

$$\ln \frac{N}{n} = \frac{E_V}{k_B T}$$

$$E_V = k_B T \, \ln \frac{N}{n}$$

$$E_V = \left(8.62 \times 10^{-5} \text{ eV/K}\right)(1073 \text{ K}) \ln \frac{5.299 \times 10^{28} \text{ m}^{-3}}{3.6 \times 10^{23} \text{ m}^{-3}}$$

$$E_V = \left(8.62 \times 10^{-5} \text{ eV/K}\right)(1073 \text{ K}) \ln\left(1.47 \times 10^5\right)$$

$$E_V = \left(8.62 \times 10^{-5} \text{ eV/K}\right)(1073 \text{ K}) \times 11.899 = 1.1 \text{ eV}$$

Example 3 Obtain the number of Schottky defects in KCl at 773 K. The energy required to form each Schottky defect is 2.6 eV. The density of KCl at 773 K is 1.955 g/cm^3. The atomic weight of K is 39.10 g/mol and of Cl is 35.45 g/mol.

Solution

The number of lattice sites per cubic metre for KCl is

$$N = \frac{N_A \, \rho}{(A_K + A_{Cl})}$$

$$N = \frac{\left(6.023 \times 10^{23} \text{ atoms/mol}\right)\left(1.955 \text{ g/cm}^3\right) \times 10^6 \text{ cm}^3/\text{m}^3}{(39.10 + 35.45) \text{ g/mol}}$$

$$= 1.58 \times 10^{28} \text{ m}^{-3}$$

From Eq. (3.16)

$$n = N \exp\left(-\frac{E_V}{2k_B T}\right)$$

$$n = 1.58 \times 10^{28}\,\text{m}^{-3} \times \exp\left(-\frac{2.6\ \text{eV}}{2 \times 8.62 \times 10^{-5}\,\text{eV/K} \times 773\,\text{K}}\right)$$

$$= 5.31 \times 10^{10}\ \text{defects/m}^3$$

Example 4 Suppose that CaO is added as an impurity to Li_2O. If the Ca^{2+} substitutes for Li^+ what kind of vacancies would be created? How many of these vacancies are created for every Ca^{2+} added.

Solution
In the lattice of Li_2O, Ca^{2+} substitutes for Li^+. Since divalent cation is substituting for a monovalent cation, therefore to maintain neutrality of the crystal a positive ion vacancy should be created. Thus for every Ca^{2+} substituting Li^+ site, a Li^+ ion vacancy is created.

Objective Type Questions

1. The grain size increases as the grain boundary area
      (a) Increases (b) decreases (c) independent of grain boundary area
2. In screw dislocation, the burger vector
      (a) is parallel to dislocation (b) is at right angle to dislocation (c) is $45°$ to dislocation
   (d) is $30°$ to dislocation
3. In edge dislocation, the burger vector
      (a) is parallel to dislocation (b) is at right angle to dislocation (c) is $45°$ to dislocation
   (d) is $30°$ to dislocation
4. Colour centre is an example of
      (a) Zero dimensional defect (b) one dimensional defect (c) two dimensional defect
   (d) three dimensional defect
5. In both Frenkel and Schottky defects, the pair of point defects stay near each other because of
      (a) Nuclear forces (b) magnetic forces (c) gravitational forces (d) electric forces
6. A Frenkel defect is
      (a) A pair of cation vacancy and a cation interstitial (b) A pair of cation vacancy and a anion interstitial (c) A pair of anion vacancy and a cation interstitial (d) a pair of anion vacancy and cation vacancy
7. A Schottky defect is
      (a) A pair of cation vacancy and a cation interstitial (b) a pair of cation vacancy and a anion interstitial (c) a pair of anion vacancy and a cation interstitial (d) a pair of anion vacancy and cation vacancy

Problems

1. State and explain various point defects that are possible in solids.
2. Classify the imperfections according to their dimensions.
3. Obtain an expression for the concentration of Frenkel defects in an ionic solid.
4. Obtain an expression for the concentration of vacancies in a monoatomic solid.
5. Calculate the number of Schottky defects per cubic metre in KCl at 500 °C. The energy required to form each Schottky defect is 2.6 eV, while the density for KCl at 500 °C is 1.955 g/cm^3.

Answers
Objective Type Questions

1. (b)   2. (a)   3. (b)   4. (a)   5. (d)   6. (a)   7. (d)

Elements of Quantum Mechanics

4

4.1 De-Broglie Hypothesis

In 1923–24 L. de-Broglie proposed that just as radiation have particle like properties, electrons and other material particle posse's wave like properties. He assumed that wavelength associated with any particle is related to the magnitude of its momentum by the relation

$$\lambda = \frac{h}{p_{particle}} \tag{4.1}$$

For a particle of mass m moving with non-relativistic speed

$$p_{particle} = mv \tag{4.2}$$

The wavelength associated with a particle is called de Broglie wavelength (it is not a wavelength of light nor it is associated with Compton wavelength of a particle).

He also assumed that the energy E of the particle is proportional to the frequency v of the associated wave given by

$$E = hv \tag{4.3}$$

just as for photon. For a photon

$$p_{photon} = \frac{h}{\lambda} \text{ or } \lambda = \frac{h}{p_{photon}} \tag{4.4}$$

The particle speed v (which determines its momentum) is not the same as the speed v' of the de Broglie wave associated with the particle. The de Broglie hypothesis implies that wave particle duality has a universal and symmetrical character. Waves have particle

© The Author(s) 2022
V. K. Jain, *Solid State Physics*, https://doi.org/10.1007/978-3-030-96017-9_4

property, particles have wave property. The very small value of h explains why the wave like nature of the matter is very difficult to demonstrate on a macroscopic scale. Thus the particle mass must be sufficiently small to obtain a measurable wavelength.

4.2   Uncertainity Relation

In classical mechanics the equation of motion of a system with a given force can be solved to give the position and momentum of a particle at all values of time t. The precise position and momentum of the particle at any time t, for example, at $t = 0$, is known then future motion is determined exactly. During the process of making observations, the observer interact with the system and thus the motion of the particle is disturbed, however, for macroscopic objects the disturbance can be ignored and one can take the measurement very precisely. However, same is not true for microscopic system. The wave particle duality limits the simultaneous measurement of position and momentum of particle exactly.

The Heisenberg uncertainty principle states that experiment cannot simultaneous determine the exact value of a component of momentum say p_x, of a particle and also the exact value of the corresponding coordinate x. Instead, our precision of measurement is inherently limited by the measurement process itself such that

$$\Delta x \Delta p_x \geq \frac{\hbar}{2} \tag{4.5}$$

where the momentum p_x is known with an uncertainty Δp_x and position x at the same time with an uncertainty Δx. Similarly

$$\Delta y \Delta p_y \geq \frac{\hbar}{2} \tag{4.6}$$

$$\Delta z \Delta p_z \geq \frac{\hbar}{2} \tag{4.7}$$

there is likewise a minimum for the product of uncertainties of the energy and time

$$\Delta E \Delta t \geq \frac{\hbar}{2} \tag{4.8}$$

This is neither a statement about the inaccuracy of measurement instruments, nor a reflection on the quality of experimental methods, it arises from the wave properties inherent in quantum mechanical description of nature. Even with perfect instruments and techniques, the uncertainties are inherent in the nature of things.

4.2.1 The Time Dependent Schrödinger Equation in One Dimension

The various experimental findings reveal that electron and other subatomic particles exhibit properties similar to those commonly associated with waves. Let this wave be $\Psi\ (r, t)$ of free electrons of momentum p and energy $E = (p^2/2m)$, one can consider these to be free plane waves, that is, $\Psi\ (r, t)$ takes the form

$$\Psi(r, t) = A \exp[i(k \cdot r - \omega t)] \tag{4.9}$$

$$E = \hbar\omega \tag{4.10}$$

$$p = \hbar k \tag{4.11}$$

where $k = (2\pi/\lambda)$. In one dimension Eqs. (4.9) and (4.11) takes the form

$$\Psi(x, t) = A \exp[i(kx - \omega t)] \tag{4.12}$$

$$p = \hbar k \tag{4.13}$$

Equation (4.12) is a solution to the wave equation and form of the wave equation applicable to many classical waves, for example, transverse waves on a string or plane sound waves as in a gas, is

$$\frac{\partial^2 \Psi}{\partial t^2} = c^2 \frac{\partial^2 \Psi}{\partial x^2} \tag{4.14}$$

where c is real constant equal to the wave velocity. Differentiating Eq. (4.12) with respect to x and t,

$$\frac{\partial \Psi}{\partial x} = ik\Psi$$

$$\frac{\partial^2 \Psi}{\partial x^2} = -k^2\Psi \tag{4.15}$$

$$\frac{\partial \Psi}{\partial t} = -i\omega\Psi \tag{4.16}$$

$$\frac{\partial^2 \Psi}{\partial t^2} = -\omega^2\Psi \tag{4.17}$$

Substituting Eqs. (4.15) and (4.17) in Eq. (4.14)

$$-\omega^2 \Psi = -c^2 k^2 \Psi$$

$$c^2 = \frac{\omega^2}{k^2}$$

$$\omega = c|k| \tag{4.18}$$

From Eqs. (4.10) and (4.18)

$$E = \hbar c |k| \tag{4.19}$$

for non-relativistic free particles of mass m we have

$$E = \frac{p^2}{2m} \tag{4.20}$$

Therefore, Eq. (4.14) cannot be wave equation governing the matter waves. We therefore discard the differential Eq. (4.14).

We note that differentiation with respect to x of wave function like those of Eq. (4.12) has the general effect of multiplication of the function by k whereas the differentiation with respect to t has the general effect of multiplication by ω. Then from Eqs. (4.10) and (4.20), it is observed that the differential equation should have first derivative with respect to t and second derivative with respect to x, that is

$$\frac{\partial \Psi}{\partial t} = \alpha \frac{\partial^2 \Psi}{\partial x^2} \tag{4.21}$$

where α is a constant. Substituting Eqs. (4.15) and (4.16) in Eq. (4.21)

$$-i\omega \Psi = -\alpha k^2 \Psi$$

$$\alpha = \frac{i\omega}{k^2} \tag{4.22}$$

From Eqs. (4.10), (4.13) and (4.22) we have

$$\alpha = \frac{i\hbar E}{p^2} = \frac{i\hbar p^2}{2mp^2} = \frac{i\hbar}{2m} \tag{4.23}$$

Substituting Eq. (4.23) in (4.21)

$$\frac{\partial \Psi}{\partial t} = \frac{i\hbar}{2m} \frac{\partial^2 \Psi}{\partial x^2}$$

or

$$i\hbar \frac{\partial \Psi}{\partial t} = -\frac{\hbar^2}{2m} \frac{\partial^2 \Psi}{\partial x^2} \tag{4.24}$$

This is one dimensional form of time dependent Schrödinger wave equation. The Schrödinger equation is linear in Ψ and the coefficients of the equation involves constants such as $\hbar$ and mass m and is independent of parameters of a particular kind of motion of the particle, for example, momentum, energy, k and frequency. The Schrödinger equation is fundamental equation of quantum mechanics in the sense that the second law of motion is the fundamental equation of Newtonian mechanics.

Suppose that particle is not free but moves in a potential so that instead of $E = (p^2/2m)$ we have

$$E = \frac{p^2}{2m} + V(x, t))$$

The potential energy $V(x, t)$ does not depend on p and E. The equation

$$E\Psi = \frac{p^2}{2m}\Psi$$

suggest that Eq. (4.24) is generalized to

$$i\hbar\frac{\partial\Psi}{\partial t} = -\frac{\hbar^2}{2m}\nabla^2\Psi + V(x, t)\Psi \tag{4.25}$$

which is one dimensional time dependent Schrödinger wave equation for a particle in a field characterized by the potential energy $V(x, t)$. If the particle cannot move to infinity then bound state occurs. That is, the particle is bound at all energies to move within a finite and limited region of space. The Schrödinger equation in this region admits only solutions that are discrete. Unbound state occurs in those cases where the motion of the system is not confined; for example, a free particle. Potentials that confine the particle for only some energies give rise to mixed spectra. For example, the mixed spectra are observed for finite square well potential.

In obtaining the Schrödinger wave equation we started with the experimental knowledge concerning the properties of free particle and their associated plane waves and ended with equation for wave function associated with a particle moving under the influence of a potential. Equation (4.25) is obtained by method of induction, where we start from a particular example to a more general law in contrast with deduction where a particular result is obtained from a general law.

Equation (4.25) is linear in Ψ. As a result, it satisfies the superposition principle. According to superposition principle, linear combination of possible wave functions is also a possible wave function. If $\Psi_1(x, t)$ and $\Psi_2(x, t)$ separately satisfy Eq. (4.25), then the wave function

$$\Psi(x, t) = c_1\Psi_1(x, t) + c_2\Psi_2(x, t) \tag{4.26}$$

also satisfy the Schrödinger wave Eq. (4.25), where c_1 and c_2 are complex numbers. In general

$$\Psi(x, t) = \sum_i c_i \Psi_i \tag{4.27}$$

also satisfy Eq. (4.25). The fact that the superposition principle applies is directly related to the wave nature of the matter, and in particular to the existence of interference effects for de Broglie waves. Since Eq. (4.25) is of first order in time derivative therefore, once the initial value of the wave function Ψ is given at some time, its value at all times can be found.

4.2.2   The Time Dependent Schrödinger Equation in Three Dimension

The one dimensional treatment can be extended to three dimensions. The wave function is then given by Eq. (4.9). Using Eqs. (4.10) and (4.11) the Eq. (4.9) is written as

$$\Psi(r, t) = A \exp\left[\frac{i}{\hbar}(p \cdot r - Et)\right] \tag{4.28}$$

$$\Psi(r, t) = A\exp\left[\frac{i}{\hbar}(p_x x + p_y y + p_z z - Et)\right] \tag{4.29}$$

Differentiating Eq. (4.29) with respect to t

$$\frac{\partial \Psi}{\partial t} = -\frac{i}{\hbar} E \Psi \tag{4.30}$$

Differentiating Eq. (4.29) with respect to x, y and z twice, we have

$$\frac{\partial \Psi}{\partial x} = \frac{i}{\hbar} p_x \Psi$$

$$\frac{\partial^2 \Psi}{\partial x^2} = -\frac{p_x^2}{\hbar^2} \Psi \tag{4.31}$$

Similarly

$$\frac{\partial^2 \Psi}{\partial y^2} = -\frac{p_y^2}{\hbar^2} \Psi \tag{4.32}$$

$$\frac{\partial^2 \Psi}{\partial z^2} = -\frac{p_z^2}{\hbar^2} \Psi \tag{4.33}$$

On adding Eqs. (4.31)–(4.33)

$$\frac{\partial^2 \Psi}{\partial x^2} + \frac{\partial^2 \Psi}{\partial y^2} + \frac{\partial^2 \Psi}{\partial z^2} = \frac{p_x^2 + p_y^2 + p_z^2}{\hbar^2} \tag{4.34}$$

where

$$\frac{\partial^2}{\partial x^2} + \frac{\partial^2}{\partial y^2} + \frac{\partial^2}{\partial z^2} = \nabla^2$$

From Eqs. (4.30) and (4.34)

$$\nabla^2 \Psi = -\frac{2m}{\hbar^2} i\hbar \frac{\partial \Psi}{\partial t}$$

or

$$i\hbar \frac{\partial \Psi}{\partial t} = -\frac{\hbar^2}{2m} \nabla^2 \Psi \tag{4.35}$$

which is three dimensional time dependent Schrödinger equation for a free particle? A comparison of Eqs. (4.30) and (4.35) and classical energy equation

$$E = \frac{p^2}{2m} \tag{4.36}$$

suggests that for a free particle the energy and momentum can be represented by differential operators that act on the wave function Ψ. The operators for E and p are

$$E \rightarrow i\hbar \frac{\partial}{\partial t} \tag{4.37}$$

$$p \rightarrow \frac{\hbar}{i} \nabla = -i\hbar \nabla \tag{4.38}$$

It is a postulate of wave mechanics that when the particle is not free, the dynamical variables E and p are still represented by Eqs. (4.37) and (4.38), respectively. Suppose that the particle is not free, so that instead of Eq. (4.36) we have

$$E = \frac{p^2}{2m} + V(r, t) \tag{4.39}$$

Since potential energy $V(r, t)$ does not depend on p and E, then Eq. (4.39) suggests

$$E\Psi = \left[\frac{p^2}{2m} + V(r, t) \right] \Psi \tag{4.40}$$

this in turn suggests that the wave Eq. (4.35) is generalized to

$$i\hbar \frac{\partial \Psi}{\partial t} = \left[\frac{p^2}{2m} + V(r, t) \right] \Psi = -\frac{\hbar^2}{2m} \nabla^2 \Psi + V(r, t)\Psi \tag{4.41}$$

this is time dependent Schrödinger equation for a particle in a field characterized by potential energy $V(\mathbf{r}, t)$. Equation (4.41) is written as

$$i\hbar\frac{\partial \Psi}{\partial t} = H\Psi \tag{4.42}$$

$$H = -\frac{\hbar^2}{2m}\nabla^2 + V(\mathbf{r}, t) \tag{4.43}$$

is known as Hamiltonian operator.

4.3   The Wave Function

The wave function is complex consisting of real and imaginary part. Neither real nor the imaginary part of the wave function, but only the full complex expression, is a solution to the Schrödinger equation. Since the wave function is complex, it cannot be measured by any actual physical instrument. A physical system is completely described by the wave function. The wave function contains all possible information that can be obtained about the system.

4.3.1   Statistical Interpretation

The wave function $\Psi(\mathbf{r}, t)$ is assumed to provide a quantum mechanically complete description of the behaviour of a particle of mass m with the potential energy $V(\mathbf{r}, t)$. Since the motion of the particle is connected with the propagation of an associated wave function, (the de Broglie connection), these two entities must be associated in space. That is particle must be at some location where the wave function has appreciable amplitude. Thus $\Psi(\mathbf{r}, t)$ is to be large where the particle is likely to be and small elsewhere. This indicates that $\Psi(\mathbf{r}, t)$ can be interpreted in statistical terms. M. Born in 1926 made the fundamental postulate that if the particle is described by a wave function $\Psi(\mathbf{r}, t)$, the probability of finding the particle within the volume element $d\tau = dx\,dy\,dz$ about the point $\mathbf{r}$ at the time t is

$$P(\mathbf{r}, t)d\tau = |\Psi(\mathbf{r}, t)|^2 d\tau \tag{4.44}$$

$$P(\mathbf{r}, t) = |\Psi(\mathbf{r}, t)|^2 = \Psi^*(\mathbf{r}, t)\Psi(\mathbf{r}, t) \tag{4.45}$$

where $\Psi^*(\mathbf{r}, t)$ is complex conjugate of $\Psi(\mathbf{r}, t)$ and $P(\mathbf{r}, t)$ is the probability density. Thus if we know the wave function associated with a physical system, we can calculate the probability of finding the particle in the vicinity of particular point.

4.3.2 Normalization of the Wave Function

The probability of finding the particle somewhere in the region must be unity, therefore we can write

$$\int\limits_{-\infty}^{+\infty} P(\boldsymbol{r}, t)\mathrm{d}\tau = 1 \tag{4.46}$$

or

$$\int\limits_{-\infty}^{+\infty} \Psi^*(\boldsymbol{r}, t)\Psi(\boldsymbol{r}, t)\mathrm{d}\tau = 1 \tag{4.47}$$

If Ψ is a solution of the Schrödinger wave equation, then $A\Psi$ is also a solution where A is any constant. The scale of wave function can therefore always be chosen to ensure that Eq. (4.46) as well as Eq. (4.44) holds at the same time. This process is known as normalization and the wave function which obeys these are said to be normalized. Thus a wave function can always be multiplied by a phase factor of the form $\exp(i\,\alpha)$ where α is arbitrary real constant, without affecting the value of any physically significant quantity. In contrast to bound states, unbound states cannot be normalized.

4.3.3 Probability Current Density

We have from Eq. (4.45)

$$P(\boldsymbol{r}, t) = \Psi^*(\boldsymbol{r}, t)\Psi(\boldsymbol{r}, t)$$

$$\frac{\partial}{\partial t}\int P(\boldsymbol{r}, t)\mathrm{d}\tau = \frac{\partial}{\partial t}\int \Psi^*(\boldsymbol{r}, t)\Psi(\boldsymbol{r}, t)\mathrm{d}\tau = \int \left(\frac{\partial \Psi^*}{\partial t}\Psi + \Psi^*\frac{\partial \Psi}{\partial t}\right)\mathrm{d}\tau \tag{4.48}$$

From Schrödinger wave Eq. (4.41) we have

$$\frac{\partial \Psi}{\partial t} = \frac{i\hbar}{2m}\nabla^2\Psi - \frac{i}{\hbar}V\Psi$$

The complex conjugate of above relation is

$$\frac{\partial \Psi^*}{\partial t} = -\frac{i\hbar}{2m}\nabla^2\Psi^* + \frac{i}{\hbar}V\Psi^*$$

Substituting $\frac{\partial \Psi}{\partial t}$ and $\frac{\partial \Psi^*}{\partial t}$ in Eq. (4.48)

$$\frac{\partial}{\partial t}\int P(\boldsymbol{r}, t)\mathrm{d}\tau = \int \left[\left(-\frac{i\hbar}{2m}\nabla^2\Psi^* + \frac{i}{\hbar}V\Psi^*\right)\Psi + \Psi^*\left(\frac{i\hbar}{2m}\nabla^2\Psi - \frac{i}{\hbar}V\Psi\right)\right]\mathrm{d}\tau$$

$$\frac{\partial}{\partial t}\int P(\boldsymbol{r}, t)\mathrm{d}\tau = -\frac{i\hbar}{2m}\int \left[\Psi(\nabla^2\Psi^*) - \Psi^*(\nabla^2\Psi)\right]\mathrm{d}\tau \tag{4.49}$$

using the vector identity

$$\mathrm{div}\left[\Psi\,\mathrm{grad}\Psi^* - \Psi^*\mathrm{grad}\Psi\right] = \left[\Psi\left(\nabla^2\Psi^*\right) - \Psi^*\left(\nabla^2\Psi\right)\right]$$

in Eq. (4.49) we have

$$\frac{\partial}{\partial t}\int P(r,t)\mathrm{d}\tau = -\frac{i\hbar}{2m}\int \mathrm{div}\left[\Psi\,\mathrm{grad}\Psi^* - \Psi^*\mathrm{grad}\Psi\right]\mathrm{d}\tau$$

$$\frac{\partial}{\partial t}\int P(r,t)\mathrm{d}\tau = -\int \mathrm{div}\left[\frac{i\hbar}{2m}\left(\Psi\,\mathrm{grad}\Psi^* - \Psi^*\mathrm{grad}\Psi\right)\right]\mathrm{d}\tau \qquad (4.50)$$

probability current density $J(r, t)$ is

$$\mathbf{J}(\mathbf{r},t) = \frac{i\hbar}{2m}(\Psi\,\mathrm{grad}\Psi^* - \Psi^*\mathrm{grad}\Psi)$$

Substituting $J(r, t)$ from the above relation in Eq. (4.50)

$$\frac{\partial}{\partial t}\int P(\mathbf{r},t)d\tau + \int \nabla \cdot \mathbf{J}d\tau = 0 \qquad (4.51)$$

$J(r, t)$ is also called current density or particle density flux.

According to Gauss theorem, the surface integral of the component of any vector A along the outward normal taken over a closed surface S is equal to the integral of the divergence of A taken over the volume V enclosed by the surface S, that is,

$$\int_V \nabla \cdot \mathbf{A}\, d\tau = \int_S \mathbf{A} \cdot ds \qquad (4.52)$$

where ds is a vector whose magnitude is equal to an element ds of the surface S. Using Eq. (4.52) in (4.50) we have

$$\frac{\partial}{\partial t}\int P(\mathbf{r},t)d\tau + \int_S \mathbf{J} \cdot ds = 0 \qquad (4.53)$$

Let the volume is extended to the entire space. The surface S in Eq. (4.53) then recedes to infinity. Since Ψ is a square integrable it vanishes at larger distances so that surface integral in Eq. (4.53) is equal to zero. Equation (4.53) is then

$$\frac{\partial}{\partial t} \int P(\mathbf{r}, t) d\tau = 0 \tag{4.54}$$

that is, normalization integral $\int P(\mathbf{r}, t) d\tau$ is independent of time. We have from Eq. (4.51)

$$\frac{\partial}{\partial t} P(\mathbf{r}, t) + \nabla \cdot \mathbf{J} = 0 \tag{4.55}$$

By analogy with charge conservation in electrodynamics, Eq. (4.55) is interpreted as the conservation of probability. The equation is known as equation of continuity.

4.4   Time Independent Schrödinger Equation

Consider a closed system in which energy is conserved and the potential energy is time independent, that is, $V = V(r)$. The Schrödinger Eq. (4.41) is then

$$i\hbar \frac{\partial \Psi}{\partial t} = -\frac{\hbar^2}{2m} \nabla^2 \Psi + V(\mathbf{r}) \Psi \tag{4.56}$$

For convenience, we consider one dimensional case for which $V(r) = V(x)$ and Schrödinger Eq. (4.56)

$$i\hbar \frac{\partial \Psi}{\partial t} = -\frac{\hbar^2}{2m} \frac{\partial^2 \Psi}{\partial x^2} + V(x) \Psi \tag{4.57}$$

Assuming that the wave function can be expressed as a product

$$\Psi(x, .t) = \psi(x) f(t) \tag{4.58}$$

On differentiating $\Psi(x, t)$ with respect to x and t

$$\frac{\partial \Psi}{\partial t} = \psi \frac{df}{dt} \tag{4.59}$$

$$\frac{\partial \Psi}{\partial x} = f(t) \frac{d\psi}{dx}$$

$$\frac{\partial^2 \Psi}{\partial x^2} = f(t) \frac{d^2 \psi}{dx^2} \tag{4.60}$$

Substituting Eqs. (4.59) and (4.60) in Eq. (4.57)

$$i\hbar \psi \frac{df}{dt} = -\frac{\hbar^2}{2m} \frac{d^2 \psi}{dx^2} f + V(x) f \psi \tag{4.61}$$

Dividing both side of Eq. (4.61) by $\psi(x) f(t)$

$$i\hbar\frac{1}{f}\frac{df}{dt} = -\frac{\hbar^2}{2m}\frac{1}{\psi}\frac{d^2\psi}{dx^2} + V(x) \tag{4.62}$$

Each side of Eq. (4.62) is equal to certain function of x and t. The left hand side of Eq. (4.62) does not depend on x, so this function is independent of x. The right hand side does not depend on t, this function must be independent of t. Since the function is independent of both variables x and t, it must be a constant. We denote this constant by E. Equating left hand side of Eq. (4.62) to E, we have

$$i\hbar\frac{1}{f}\frac{df}{dt} = E \tag{4.63}$$

$$\frac{df}{dt} = -\frac{i}{\hbar}Ef \tag{4.64}$$

On integration

$$\ln f(t) = -\frac{i}{\hbar}Et \tag{4.65}$$

$$f(t) = \exp\left(-\frac{i}{\hbar}Et\right) \tag{4.66}$$

Equating right hand side of Eq. (4.62) to E

$$-\frac{\hbar^2}{2m}\frac{d^2\psi}{dx^2} + V(x)\psi = E\psi \tag{4.67}$$

Equation (4.67) is known as time independent Schrödinger equation for a particle of mass m moving in time independent potential $V(x)$. $\psi(x)$ are not necessarily complex and time independent Schrödinger equation does not contain the imaginary number. From Eq. (4.67) it is seen that E has dimension of energy. We assume that E is the energy of the system.

From Eqs. (4.58) and (4.66)

$$\Psi(x, t) = \exp\left(-\frac{i}{\hbar}Et\right)\psi(x) \tag{4.68}$$

The wave function $\Psi(x, t)$ correspond to states of constant energy.

4.5   Stationary States

The wave function in Eq. (4.68) is complex, but the quantity that is experimentally determined is the probability density $|\Psi(x, t)|^2$. The probability density is

$$|\Psi(x,t)|^2 = \Psi^*(x,t)\Psi(x,t) = \left[\exp\left(-\frac{i}{\hbar}Et\right)\psi(x)\right]^*\left[\exp\left(-\frac{i}{\hbar}Et\right)\psi(x)\right]$$

$$|\Psi(x,t)|^2 = \psi^*(x)\psi(x) = |\psi(x)|^2 \tag{4.69}$$

The probability density is then independent of time. A particle in such a state will remain in that state until acted upon by some external entity that forces it out of that state. The solution (4.68) of the Schrödinger Eq. (4.62) for time independent potential is called stationary state. The stationary state is a state of well-defined energy, E being the definite value of its energy and not only it's the expectation value. For a stationary state $\Psi(x,.t)$ equal a function of time multiplied by a function of particle coordinate. For a stationary state $\Psi(x,.t)$ is an eigenfunction of Hamiltonian. Any determination of the energy of a particle which is in stationary state always yield a particular value E and only that value. Such an interpretation is in line with the Heisenberg uncertainty relation

$$\Delta E \Delta t > \hbar$$

which implies that a quantum state with a precise energy ($\Delta E = 0$) is possible only if there is an infinite time available to determine that energy. Stationary states are of just such nature in view of constancy of $|\Psi(x,t)|^2$ in time.

4.6 Boundary Conditions

The wave function itself has no physical interpretation, however, the square of its absolute magnitude $|\Psi(x,t)|^2$ evaluated at a particular place and at a particular time is proportional to the possibility of finding the particle at that time. The probability density $|\Psi(x,t)|^2$ is positive and real and is taken equal to $\Psi^*(r,t)\Psi(r,t)$. The wave function Ψ can take on negative values but probability density is always be positive. Besides fulfilling the normalization condition a solution of the time independent Schrödinger equation must obey the following boundary conditions.

1. The wave function must be continuous and single valued.
2. $\frac{\partial \psi}{\partial x}$, $\frac{\partial \psi}{\partial y}$ and $\frac{\partial \psi}{\partial z}$ must be continuous and single valued everywhere.
3. The integral of the square modulus of the wave function over all values x must be finite

$$\int \psi^*\psi \, d\tau = \text{finite} \tag{4.70}$$

that is the wave function must be square integrable. This condition means that wave function must be normalizable that is wave function must go to zero as $x(y,z) \to \pm\infty$ in order that $\int |\Psi|^2 d\tau$ over all space is finite constant.

The boundary conditions ensure that the probability of finding the particle in the vicinity of any point is unambiguously defined rather than having two or more possible values. Thus the wave function is single valued and continuous. If $\psi(x)$ and $(d\psi/dx)$ are not single valued, finite then the same is true for $\Psi(x, t)$. Since the given formula for calculating the expectation values of position and momentum contains $\Psi(x, t)$ and $\frac{\partial \Psi}{\partial t}$. We observe that in any of these cases we might not obtain finite and definite values when we evaluate measured quantities.

The first derivative of the wave function with respect to position coordinates must be continuous every where except where there is an infinite discontinuity in the potential. We know any function always has an infinite derivative whenever it has a discontinuity. Let us consider the time independent Schrödinger Eq. (4.67) in one dimension

$$\frac{d^2\psi}{dx^2} = \frac{2m}{\hbar^2}(V - E)\psi$$

for finite V, E and ψ, $(d^2\psi/dx^2)$ is finite. This in turn requires $(d\psi/dx)$ to be continuous. A finite discontinuity in $(d\psi/dx)$ implies an infinite discontinuity in $(d^2\psi/dx^2)$ and from the Schrödinger equation in $V(x)$.

4.7   Hydrogen Atom

Consider the hydrogen atom as a system of two interacting particles, the interaction being due to Coulomb attraction of their electrical charges. Let the charge on the nucleus is Zq and the charge on the electron is $- q$. The potential energy of the system in the absence of the external field is

$$V(r) = -\frac{Zq^2}{(4\pi \varepsilon_0)r} \tag{4.71}$$

in which r is the distance between the electron and the nucleus.

Let m_1 and m_2 are the masses of nucleus and the electron, respectively. If we write for the Cartesian coordinates of the nucleus and the electrons x_1, y_1, z_1 and x_2, y_2, z_2, respectively, the Hamiltonian of the hydrogenic atoms has the form

$$H = \frac{p_1^2}{2m_1} + \frac{p_2^2}{2m_2} + V(r) = E \tag{4.72}$$

The Schrödinger wave equation is

$$\frac{1}{m_1}\left(\frac{\partial^2\Psi}{\partial x_1^2}+\frac{\partial^2\Psi}{\partial y_1^2}+\frac{\partial^2\Psi}{\partial z_1^2}\right)+\frac{1}{m_2}\left(\frac{\partial^2\Psi}{\partial x_2^2}+\frac{\partial^2\Psi}{\partial y_2^2}+\frac{\partial^2\Psi}{\partial z_2^2}\right)+\frac{2}{\hbar^2}[E-V]\Psi=0$$

$$(4.73)$$

Here wave function Ψ refers to the complete system with six coordinates. Equation (4.73) can be separated into two, one of which represents the translational motion of a molecule as a whole and the other, the relative motion of the two particles. For this, consider new variables X, Y, Z which are Cartesian coordinates of the centre of mass of the system and r, θ and φ of the polar coordinates of the second particle relative to the first. These coordinates are related to the Cartesian coordinates of the two particles by the equations

$$X=\frac{m_1x_1+m_2x_2}{m_1+m_2} \tag{4.74}$$

$$Y=\frac{m_1y_1+m_2y_2}{m_1+m_2} \tag{4.75}$$

$$Z=\frac{m_1z+m_2z_2}{m_1+m_2} \tag{4.76}$$

$$x=x_2-x_1=r\sin\theta\cos\varphi \tag{4.77}$$

$$y=y_2-y_1=r\sin\theta\sin\varphi \tag{4.78}$$

$$z=z_2-z_1=r\cos\theta \tag{4.79}$$

Differentiating Ψ with respect to x_1, y_1, z_1, x_2, y_2 and z_2 and putting in Eq. (4.73) we have

$$-\frac{\hbar^2}{2(m_1+m_2)}\left(\frac{\partial^2}{\partial X^2}+\frac{\partial^2}{\partial Y^2}+\frac{\partial^2}{\partial Z^2}\right)\Psi-\frac{\hbar^2}{2\mu}\left(\frac{\partial^2}{\partial x^2}+\frac{\partial^2}{\partial y^2}+\frac{\partial^2}{\partial z^2}\right)\Psi+V\Psi=E\Psi$$

$$(4.80)$$

where the reduced mass μ is

$$\mu=\frac{m_1m_2}{m_1+m_2}$$

Since the potential does not depend on $R(=X,Y,Z)$, the wave function is written as

$$\Psi=\psi(R)\psi(r) \tag{4.81}$$

and energy E is

$$E = E_R + E_r \tag{4.82}$$

Equation (4.80) is then

$$-\frac{\hbar^2 \psi(r)}{2M}\left(\frac{\partial^2}{\partial X^2} + \frac{\partial^2}{\partial Y^2} + \frac{\partial^2}{\partial Z^2}\right)\psi(R) - \frac{\hbar^2 \psi(R)}{2\mu}\left(\frac{\partial^2}{\partial x^2} + \frac{\partial^2}{\partial y^2} + \frac{\partial^2}{\partial z^2}\right)\psi(r)$$

$$+ V(r)\psi(R)\psi(r) = (E_R + E_r)\psi(R)\psi(r) \tag{4.83}$$

where $M = m_1 + m_2$.

Dividing Eq. (4.83) by $\Psi = \psi(R)\psi(r)$ and simplifying

$$-\frac{\hbar^2}{2M}\left(\frac{\partial^2}{\partial X^2} + \frac{\partial^2}{\partial Y^2} + \frac{\partial^2}{\partial Z^2}\right)\psi(R) = E_R\psi(R) \tag{4.84}$$

$$-\frac{\hbar^2}{2\mu}\left(\frac{\partial^2}{\partial x^2} + \frac{\partial^2}{\partial y^2} + \frac{\partial^2}{\partial z^2}\right)\psi(r) + V(r)\psi(r) = E_r\psi(r) \tag{4.85}$$

The solution of the Eq. (4.84) is

$$\psi(R) \sim \exp\left(\frac{i\,\vec{P}\cdot\vec{R}}{\hbar}\right) \tag{4.86}$$

with

$$E_R = \frac{\hbar^2 P^2}{2M}$$

Equation (4.84) represents the motion of a free particle; hence the translational motion of the system is same as that of a particle with masses $m_1 + m_2 = M$ equal to the sum of the masses of the two particles.

Equation (4.85) is

$$\frac{\partial^2 \psi}{\partial x^2} + \frac{\partial^2 \psi}{\partial y^2} + \frac{\partial^2 \psi}{\partial z^2} + \frac{2\mu}{\hbar^2}[E_r - V(r)]\psi = 0 \tag{4.87}$$

$$\nabla^2 \psi + \frac{2\mu}{\hbar^2}[E - V(r)]\psi = 0 \tag{4.88}$$

where $\nabla^2 = \frac{\partial^2}{\partial x^2} + \frac{\partial^2}{\partial y^2} + \frac{\partial^2}{\partial z^2}$ is called Laplacian operator or 'del squared' in rectangular coordinates.

Equation (4.88) is a partial differential equation because it contains three independent variables, the space coordinates x, y, z. The time independent Schrödinger equation for the Coulomb potential can be solved by making application of technique of separation of variable to the partial differential equation into set of three ordinary differential equation, each involving one coordinate. However, separation of variable cannot be carried out when rectangular coordinates are employed because the Coulomb potential energy is function of all three x, y and z coordinates. Hence separation of variable will not work in rectangular coordinates because the potential itself cannot be split into terms each of which involves only one such coordinate. Therefore, it is convenient to work with spherical polar coordinates r, θ and φ.

In spherical polar coordinates the Coulomb potential can be expressed as a function of a single coordinate r. It is then possible to carry out the separation of variables on the time independent Schrödinger equation. The Schrödinger Eq. (4.88) in spherical polar coordinates can be written as

$$\frac{1}{r^2}\frac{d}{dr}\left(r^2\frac{\partial\Psi}{\partial r}\right) + \frac{1}{r^2}\frac{1}{\sin\theta}\frac{d}{d\theta}\left(\sin\theta\frac{d\Psi}{d\theta}\right) + \frac{1}{r^2\sin^2\theta}\frac{\partial^2\Psi}{\partial\varphi^2} + \frac{2\mu}{\hbar^2}[E-V(r)]\Psi = 0$$

(4.89)

where $\Psi = \Psi(r,\theta,\varphi)$ and it can be written as a product of three functions, each of which depends only on one coordinate

$$\Psi(r,\theta,\varphi) = R(r)\Theta(\theta)\Phi(\varphi) = R(r)Y(\theta,\varphi)$$

(4.90)

$$Y(\theta,\varphi) = \Theta(\theta)\Phi(\varphi)$$

(4.91)

Substituting the values of Ψ from Eq. (4.90) into Eq. (4.89) we obtain

$$\frac{Y}{r^2}\frac{\partial}{\partial r}\left(r^2\frac{\partial R}{\partial r}\right) + \frac{R}{r^2\sin\theta}\frac{\partial}{\partial\theta}\left(\sin\theta\frac{\partial Y}{\partial\theta}\right) + \frac{R}{r^2\sin^2\theta}\frac{\partial^2 Y}{\partial\varphi^2} + \frac{2\mu}{\hbar^2}[E-V(r)]RY = 0$$

Dividing the above equation by Ψ

$$\frac{1}{R}\frac{d}{dr}\left(r^2\frac{dR}{dr}\right) + \frac{2\mu}{\hbar^2}[E-V(r)] = -\frac{1}{Y}\left[\frac{1}{\sin\theta}\frac{d}{d\theta}\left(\sin\theta\frac{\partial Y}{\partial\theta}\right) + \frac{1}{\sin^2\theta}\frac{\partial^2 Y}{\partial\varphi^2}\right]$$

(4.92)

Using

$$L^2 = -\hbar^2\left[\frac{1}{\sin\theta}\frac{\partial}{\partial\theta}\left(\sin\theta\frac{\partial Y}{\partial\theta}\right) + \frac{1}{\sin^2\theta}\frac{\partial^2 Y}{\partial\varphi^2}\right]$$

In Eq. (4.92)

$$\left[-\frac{\hbar^2}{2\mu r^2}\frac{\partial}{\partial r}\left(r^2\frac{\partial}{\partial r}\right) + \frac{L^2}{2\mu r^2} + V\right]\Psi(r,\theta,\phi) = E\Psi(r,\theta,\phi)$$

(4.93)

The eigenvalues of L^2 are given by

$$L^2 Y_{lm}(\theta,\phi) = \hbar^2 l(l+1)Y_{lm}(\theta,\phi)$$

where $Y_{lm}(\theta, \varphi)$ are eigenfunctions of L^2. Thus

$$\Psi(r, \theta, \phi) = R(r)Y_{lm}(\theta, \phi)$$

Using Eq. (4.90) and eigenvalues of L^2 in Eq. (4.93) we have

$$\frac{d^2R}{dr^2} + \frac{2}{r}\frac{dR}{dr} + \left[-\frac{l(l+1)}{r^2} + \frac{2\mu}{\hbar^2}[E - V(r)]\right]R = 0 \qquad (4.94)$$

Let $u(r) = rR(r)$, we have

$$\frac{du}{dr} = R + r\frac{dR}{dr}$$

$$\frac{d^2u}{dr^2} = r\frac{d^2R}{dr^2} + 2\frac{dR}{dr} = r\left(\frac{d^2R}{dr^2} + \frac{2}{r}\frac{dR}{dr}\right)$$

Substituting u for R in Eq. (4.94)

$$\frac{d^2u}{dr^2} + \left[-\frac{l(l+1)}{r^2} + \frac{2\mu}{\hbar^2}[E - V(r)]\right]u = 0$$

Writing $u(\rho) = u(r)$ with

$$\rho = \left(-\frac{8\mu E}{\hbar^2}\right)^{1/2}r$$

we have

$$\frac{du}{dr} = \frac{du}{d\rho}\frac{d\rho}{dr} = \left(-\frac{8\mu E}{\hbar^2}\right)^{1/2}\frac{du}{d\rho}$$

$$\frac{d^2u}{dr^2} = \frac{d^2u}{d\rho^2}\frac{d\rho}{dr} = \left(-\frac{8\mu E}{\hbar^2}\right)\frac{d^2u}{d\rho^2}$$

Substituting the value of $\frac{d^2u}{dr^2} = \left(-\frac{8\mu E}{\hbar^2}\right)\frac{d^2u}{d\rho^2}$ we have

$$\left(-\frac{8\mu E}{\hbar^2}\right)\frac{d^2u}{d\rho^2} + \left[-\frac{l(l+1)}{r^2} + \frac{2\mu}{\hbar^2}[E - V(r)]\right]u(\rho) = 0$$

Substituting the value of V from Eq. (4.71) and writing r in terms of ρ in the above equation

$$\left(-\frac{8\mu E}{\hbar^2}\right)\frac{d^2u}{d\rho^2} + \left[-\frac{l(l+1)}{\rho^2}\left(-\frac{8\mu E}{\hbar^2}\right) + \frac{2\mu}{\hbar^2}E\right.$$

$$\left. + \frac{2\mu}{\hbar^2}\frac{Ze^2}{(4\pi\varepsilon_0)\rho}\left(-\frac{8\mu E}{\hbar^2}\right)^{1/2}\right]u(\rho) = 0$$

Let

$$\beta = \frac{Ze^2}{\hbar}\left(-\frac{\mu}{2E}\right)^{1/2} \tag{4.95}$$

Substituting the value of β and simplifying, we have

$$\frac{d^2u(\rho)}{d\rho^2} + \left[-\frac{l(l+1)}{\rho^2} + \frac{\beta}{\rho} - \frac{1}{4}\right]u(\rho) = 0 \tag{4.96}$$

For $\rho \to \infty$, Eq. (4.96) is

$$\frac{d^2u(\rho)}{d\rho^2} - \frac{u(\rho)}{4} = 0 \tag{4.97}$$

The solution of the above equation is

$$u(\rho) = \exp\left(\pm\frac{\rho}{2}\right) \tag{4.98}$$

the boundary condition at $\rho \to \infty$ allows only negative exponential

$$u(\rho) = \exp\left(-\frac{\rho}{2}\right) \tag{4.99}$$

when $\rho \to 0$, the radial Eq. (4.96) becomes

$$\frac{d^2u(\rho)}{d\rho^2} - \frac{l(l+1)}{\rho^2}u(\rho) = 0 \tag{4.100}$$

The solution of Eq. (4.100) has the form

$$u(\rho) \to \rho^{l+1} \tag{4.101}$$

as function must vanish at the origin; otherwise the integrated probabilities over a sphere of vanishingly small radius would remain finite. The general solution of radial Eq. (4.96) can be written as

$$u(\rho) = \rho^{l+1} \exp\left(-\frac{\rho}{2}\right)L(\rho) \tag{4.102}$$

$$\frac{du}{d\rho} = u' = (l+1)\rho^l \exp\left(-\frac{\rho}{2}\right)L(\rho)$$

$$-\frac{1}{2}\rho^{l+1}\exp\left(-\frac{\rho}{2}\right)L(\rho) + \rho^{l+1}\exp\left(-\frac{\rho}{2}\right)L'(\rho)$$

$$u'' = l(l+1)\rho^{l-1}\exp\left(-\frac{\rho}{2}\right)L(\rho)$$

$$-\frac{1}{2}(l+1)\rho^l \exp\left(-\frac{\rho}{2}\right)L(\rho) + (l+1)\rho^l \exp\left(-\frac{\rho}{2}\right)L'$$

$$-\frac{1}{2}(l+1)\rho^l \exp\left(-\frac{\rho}{2}\right)L(\rho)$$

$$+ \frac{1}{4}\rho^{l+1} \exp\left(-\frac{\rho}{2}\right) L(\rho) - \frac{1}{2}\rho^{l+1} \exp\left(-\frac{\rho}{2}\right) L'(\rho)$$

$$+ (l+1)\rho^{l} \exp\left(-\frac{\rho}{2}\right) L'(\rho)$$

$$- \frac{1}{2}\rho^{l+1} \exp\left(-\frac{\rho}{2}\right) L'(\rho) + \rho^{l+1} \exp\left(-\frac{\rho}{2}\right) L''(\rho)$$

$$u'' = \exp\left(-\frac{\rho}{2}\right)\left[\rho^{l+1} L'' + 2(l+1)\rho^{l} L' - \rho^{l+1} L'\right.$$

$$\left. + l(l+1)\rho^{l-1} L - (l+1)\rho^{l} L + \frac{1}{4}\rho^{l+1} L\right]$$

Substituting the value of u and u'' in Eq. (4.96)

$$\rho L''(\rho) + [2(l+1) - \rho]L'(\rho) + [\beta - l - 1]L(\rho) = 0 \qquad (4.103)$$

Let us try a power series solution for Eq. (4.103) by putting

$$L(\rho) = \sum_{s=0}^{\infty} a_s \rho^s \qquad (4.104)$$

in Eq. (4.103) we obtain

$$\sum \{s[(s-1) + 2l + 2]a_s \rho^{s-1} + (\beta - l - 1 - s)a_s \rho^s\} = 0 \qquad (4.105)$$

For this expression to be true the coefficient of each term must vanish which leads to

$$(s+1)(s+2l+2)a_{s+1} + (\beta - l - 1 - s)a_s = 0 \qquad (4.106)$$

$$\frac{a_{s+1}}{a_s} = \frac{l+s+1-\beta}{(s+1)(s+2l+2)} \qquad (4.107)$$

For large value of s, and fixed l, Eq. (4.107) is

$$\frac{a_{s+1}}{a_s} \to \frac{1}{s} \qquad (4.108)$$

but this is the behaviour of an exponential series

$$\exp(\rho) = 1 + \rho + \frac{\rho^2}{2!} + \cdots + \frac{\rho^{s-1}}{(s-1)!} + \frac{\rho^s}{s!} + \frac{\rho^{s+1}}{(s+1)!} + \cdots \qquad (4.109)$$

the coefficient

$$\frac{a_{s+1}}{a_s} \mapsto \frac{1}{s}$$

Thus the solution of Eq. (4.102) is of the form

$$u(\rho) = \rho^{l+1} \exp\left(-\frac{\rho}{2}\right) \exp(\rho) = \rho^{l+1} \exp\left(\frac{\rho}{2}\right) \tag{4.110}$$

the boundary condition $u(\rho) \to 0$ as $\rho \to \infty$ is violated. Therefore, the series has to be terminated and this is possible only if for some value of $s = s_{max}$, we have

$$s_{max} + l + 1 = \beta \tag{4.111}$$

$s_{max} = 0, 1, 2, 3, \ldots$ is radial quantum number. Since both s_{max} and l takes only integer value, we can write Eq. (4.111) as

$$s_{max} + l + 1 = \beta = n \tag{4.112}$$

Replacing in Eq. (4.95) the quantity β by its value given by Eq. (4.112), we obtain the bound state energy eigenvalue

$$\beta^2 = -\frac{\mu Z^2 q^4}{2E(4\pi\varepsilon_0)^2 n^2 \hbar^2} = n^2$$

$$E = -\frac{\mu Z^2 q^4}{2(4\pi\varepsilon_0)^2 n^2 \hbar^2} = -\frac{1}{2}\frac{Z^2 q^2}{a_0 n^2}$$

$$= -\frac{R_\infty ch Z^2}{n^2} = -\frac{13.6(\text{eV})Z^2}{n^2} \tag{4.113}$$

where R_∞ is Rydberg constant given by $R_\infty = 109{,}737\,\text{cm}^{-1}$ and $a_0 = [(4\pi\varepsilon_0)\hbar^2/\mu q^2]$.

Following Wolfgang Grotrian, Bohr's permitted values of energy (Eq. 4.113) for hydrogen atom ($Z = 1$) are represented graphically on an energy level diagram (Fig. 4.1). The vertical axis of the graph is an energy (eV or wave number) scale, there is no horizontal scale, but horizontal lines are drawn to show the position of energy levels.

The ground level ($n = 1$ for hydrogen) lies near the bottom of the diagram, and the remaining (excited) levels are ranked by energy above this level. The distance between energy level rapidly decreases. Each spectral line is represented by a vertical line joining two energy levels The length of any arrow is proportional to the frequency (or the reciprocal wavelength) for the corresponding spectral lines. The ionisation limit is defined as the zero point of the energy scale. With this energies of the discrete excited states are negative numbers, whereas the positive energies occur only when the atom is ionised.

An electron in an orbit with $n_i > 1$ and energy E_i, in making a transition to an orbit of lower energy E_f produces a photon with specific energy, equivalently with a definite frequency or wavelength. The discrete emission spectrum of hydrogen therefore corresponds to electrons cascading down to lower energy levels in hydrogen atoms. All their spectral lines can be grouped in to five series depending on the values of n_f and n_i. These series are named after their discoverer. In a series spacing and intensity of lines decrease regularly. The series limit is derived from Eq. (4.113) for every series by putting $n_i = \infty$. The spectral lines lie in various spectral regions as shown in Table 4.1.

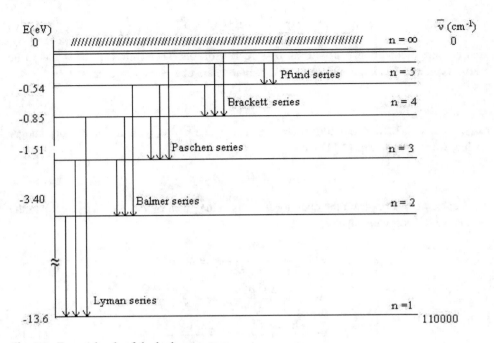

Fig. 4.1 Energy levels of the hydrogen atom

Table 4.1 Spectral lines in various spectral regions

Series	$\bar{v} = \frac{1}{\lambda} = R_\infty \left(\frac{1}{n_f^2} - \frac{1}{n_i^2} \right)$		Spectral region
	n_f	n_i	
Lyman	1	2, 3, 4 ..., ∞	Far UV
Balmer	2	3, 4, 5, ..., ∞	Near UV and visible
Paschen	3	4, 5, 6, ..., ∞	Infrared
Brackett	4	5, 6, 7, ..., ∞	Infrared
Pfund	5	6, 7, 8, ..., ∞	Infrared

Normally, only lines corresponding to Lyman Series appear in the absorption spectrum as atom is always initially in the ground state $n = 1$, so that only absorption process from $n = 1$ to $n > 1$ can occur. However at high temperature ($T \sim 10^{15}$ K) owing to collisions some of the atoms will initially be in the first excited state ($n = 2$) and absorption lines corresponding to the Balmer Series will be observed. It is observed that for every line in the absorption spectrum there is a corresponding (same wavelength) line in its emission spectrum, however the reverse is not true. If hydrogen atom is initially in excited state n

> 1 then in going to its ground state it can follow different paths, and as a result of this it will emit $[n\,(\,n-1)]/2$ number of different wavelengths.

Eigenfunctions

The polynomial solution of Eq. (4.103) can be expressed in terms of Laguerre polynomial $L_q^p\,(\rho)$ that are known to satisfy the equation

$$\rho \frac{d^2 L_q^p}{d\rho^2} + (p+1-\rho)\frac{dL_q^p}{d\rho} + (q-p)L_q^p = 0 \tag{4.114}$$

The equation has precisely this form with $2l+1 = p$ and $q = n+l$. Therefore, its polynomial solution is the associated Laguerre polynomial apart from an arbitrary constant factor N_{nl}. The radial function R of Eq. (4.90) is labelled by two indices n and l. The radial wave function after normalization is

$$R_{nl}(r) = -\left[\left(\frac{2Z}{na_0}\right)^3 \frac{(n-l-1)!}{2n(n+l)!}\right]^{1/2} \exp\left(-\frac{\rho}{2}\right)\rho^l L_{n+l}^{2l+1}(\rho) \tag{4.115}$$

where $\rho = \frac{2Zr}{na_0}$ and

$$L_{n+l}^{2l+1}(\rho) = \sum_{k=0}^{n-l-1} (-1)^{k+2l+1}\frac{(n+l)^2 \rho^k}{(n-l-1-k)!(2l+1+k)!k!} \tag{4.116}$$

The wave function $\Psi(r, \theta, \varphi)$ given by Eq. (4.90) to which we attach subscript nlm is given by

$$\Psi_{nlm}(r, \theta, \varphi) = R_{nl}(r)Y_{lm}(\theta, \varphi) \tag{4.117}$$

with $R_{nl}(r)$ and $Y_{lm}(\theta, \varphi)$ given from Eqs. (4.115) and (6.82), respectively. Explicit form of $R_{nl}(r)$ is given in Table 4.2.

Plots of $R_{nl}(r)$ are shown in Fig. 4.2a. It is observed that due to properties of the associated Laguerre polynomial, $R_{nl}(r)$ will change its sign $n-l-1$ times. Further,

Table 4.2 Radial wave functions R_{nl} in hydrogenic atom

$R_{10}(r)$	=	$2(Z/a_0)^{3/2}\exp(-Zr/a_0)$
$R_{20}(r)$	=	$2(Z/a_0)^{3/2}(1 - Zr/2a_0)\exp(-Zr/2a_0)$
$R_{21}(r)$	=	$(1/\sqrt{3})^{1/2}(Z/2a_0)^{3/2}(Zr/a_0)\exp(-Zr/2a_0)$
$R_{30}(r)$	=	$2(Z/3a_0)^{3/2}(1 - Zr/3a_0 + 2Z^2r^2/27a_0^2)\exp(-Zr/3a_0)$
$R_{31}(r)$	=	$(4\sqrt{2}/9)(Z/3a_0)^{3/2}(1 - Zr/6a_0)\,(Zr/a_0)\exp(-Zr/3a_0)$
$R_{32}(r)$	=	$(4/27\sqrt{10})(Z/3a_0)^{3/2}(Zr/a_0)^2\exp(-Zr/3a_0)$

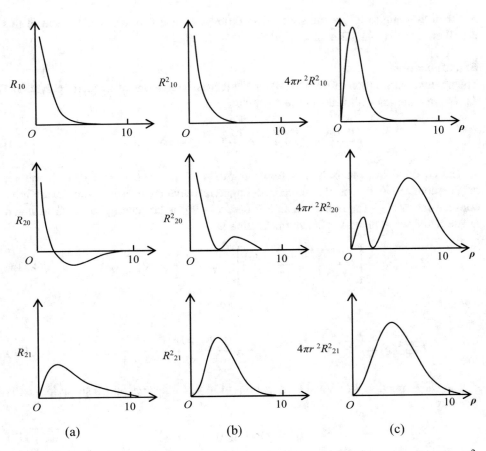

(a)   (b)   (c)

Fig. 4.2 Plots of **a** the radial wave function R_{nl}, **b** the radial probability distribution function R_{nl}^2, and **c** the radial charge density function $4\pi r^2 R_{nl}^2$ against ρ

the states with $l = 0$ are finite at the origin, whereas the states with $l \geq 1$ vanish there. Figure 4.2b shows the plot of $R_{nl}^2(r)$ against ρ (or r). Since $R_{nl}(r)$ is a real function $R_{nl}(r) = R_{nl}^*(r)$ and $R_{nl}^2(r)dr$ is the probability of finding the electron between r and $r + dr$. A plot of $R_{nl}^2(r)$ against ρ (or r) represents the radial probability distribution of the electron. Figure 4.2c shows the plot of radial charge density $4\pi r^2 R_{nl}^2(r)$ against ρ (or r).This plot has $n - l$ bumps. The quantity $4\pi r^2 R_{nl}^2(r)$ is the probability of finding an electron within an element of volume that consists of spherical shell of thickness dr and a volume element $4\pi r^2 dr$. The dependence of the function $4\pi r^2 R_{nl}^2(r)$ on r differs from that of $R_{nl}^2(r)$ function. The $R_{nl}^2(r)$ function is maximum in the neighbourhood of the nucleus. The $4\pi r^2 R_{nl}^2(r)$ changes, however, from zero at $r = 0$ passes through the maximum and gradually decreases. This behaviour is due to dependence on the factors

$4\pi r^2$ and $R_{nl}^2(r)$. It starts from zero (at $r = 0$) and increases with r. For small values of r, $4\pi r^2$ value increases more rapidly than $R_{nl}^2(r)$ decreases, thus leading to an increase of their product; then $R_{nl}^2(r)$ starts decreasing more rapidly than the $4\pi r^2$ increases. As a result their product passes the maximum and shows a further gradual decrease. The maximum of the radial distribution corresponds to the atomic radius. The average value of various power of r is defined as

$$\left\langle r^k \right\rangle_{nl} = \int r^k [R_{nl}(r)]^2 \mathrm{d}r \tag{4.118}$$

The forms of a few $Y_{lm}(\theta, \varphi)$ are given in Table 4.3. The value $|Y_{lm}(\theta, \varphi)|^2$ is plotted as a function of angle θ (Fig. 4.3).

The wave function of an atom can also be characterized by its parity. The parity operator P transforms the position vector $\mathbf{r}$ into $-\mathbf{r}$, or acting on a state Ψ, one has $P\Psi(\mathbf{r}) =$

Table 4.3 Angular wave function $Y_{lm}(\theta, \varphi)$

$Y_{0,0}$	$=$	$(1/4\pi)^{1/2}$
$Y_{1,0}$	$=$	$(3/4\pi)^{1/2} \cos\theta$
$Y_{1,\pm1}$	$=$	$+ (3/8\pi)^{1/2} \sin\theta \exp(\pm i\varphi)$
$Y_{2,0}$	$=$	$(5/16\pi)^{1/2} (3\cos^2\theta - 1)$
$Y_{2,\pm1}$	$=$	$+ (15/8\pi)^{1/2} \sin\theta \cos\theta \exp(\pm i\varphi)$
$Y_{2,\pm2}$	$=$	$(15/32\pi)^{1/2} \sin^2\theta \exp(\pm 2i\varphi)$
$Y_{3,0}$	$=$	$(7/16\pi)^{1/2} (5\cos^3\theta - 3\cos\theta)$
$Y_{3,\pm1}$	$=$	$- 1x(\pm) (21/64\pi)^{1/2} \sin\theta(5\cos^2\theta - 1) \exp(\pm i\varphi)$
$Y_{3,\pm2}$	$=$	$(105/32\pi)^{1/2} \sin^2\theta \cos\theta \exp(\pm 2i\varphi)$
$Y_{3,\pm3}$	$=$	$- 1x(\pm) (35/64\pi)^{1/2} \sin^3\theta \exp(\pm 3i\varphi)$

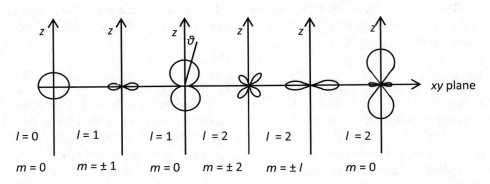

Fig 4.3 Polar diagrams of the $|Y_{lm}|^2$ with $l = 0, 1, 2$

$\Psi(-r)$. The vector $-r$ has polar coordinates $\pi - \theta$, $\pi + \varphi$. Because of $\cos(\pi - \theta) = -\cos\theta$ we have

$$PY_{lm}(\theta, \varphi) = Y_{lm}(\pi - \theta, \pi + \varphi) = \exp(im\pi)(-1)^{l+m}Y_{lm}(\theta, \varphi) \qquad (4.119)$$

$$PY_{lm}(\theta, \varphi) = (-1)^l Y_{lm}(\theta, \varphi) \qquad (4.120)$$

If the multiplying factor is $+1$ or -1, the parity of the wave function is said to be even or odd, respectively.

Quantum Numbers for the Hydrogen Atom Wave Function

In the process of solving the Schrödinger wave equation for hydrogen like atoms, the integer n, l and m are introduced in a logical way. These integers are called quantum numbers. In Bohr's theory the quantum number n is introduced arbitrarily in the form of the quantized condition. The wave function $\Psi_{nlm}(\theta, \varphi)$ is description of states of the system. They are related to n, l and m. Thus the quantum number themselves may be said to describe the state of the system. The values of these quantum numbers are

$$n = 1, 2, 3, \ldots$$
$$l = 0, 1, 2, \ldots, n-1$$
$$m = 0, \pm 1, \pm 2, \ldots, \pm l \qquad (4.121)$$

It is now shown that the quantum numbers l and m are related to the magnitude L of the orbital angular momentum and m is the z component L_z. $Y_{lm}(\theta, \varphi)$ is simultaneously the eigenfunction of L^2 and L_z operator. Therefore, L^2 and L_z commute, i.e. corresponding observable can be precisely measured simultaneously. The magnitude of the orbital angular momentum is $\hbar\sqrt{l(l+1)}$ while that of L_z is $m \hbar$. Since L^2 and L_z do not operate on the radial part of the wave function, $\Psi_{nlm}(\theta, \varphi)$ itself is a simultaneous eigenfunction of L^2 and L_z. The quantum number l gives the magnitude of the angular momentum, m the orientation of the angular momentum and n gives the quantization of energy. The quantum numbers l and m can be equal only for $l = 0$.

Degeneracy

Energy depends only on the index n and not on l, that is, for $n > 1$ we have states with different values of l corresponding to the same value of n and hence these states are degenerate. From Eq. (4.112) $s_{max} = n - l - 1$ and $s_{max} \geq 0$, we obtain $l \leq n - 1$, so for a given value of n, l cannot exceed the value $l_{max} = n - 1$. This corresponds to the value $l = 0, 1, \ldots (n-1)$. Each of these values correspond to different values of s_{max}, therefore different wave functions. The energy E depends only on n whereas the wave function depends on n, l and m. For each value of n, l can take value 0, 1, 2, ... $(n-1)$ and for each value of l, there are $(2l + 1)$ possible values of m, i.e. from $-l, -l + 1, \ldots, +l$. The total degeneracy of the energy level is therefore given by

$$\text{Degeneracy} = \sum_{l=0}^{n-1} (2l + 1) = \sum_{l=0}^{n-1} 2l + \sum_{l=0}^{n-1} 1$$

$$= 2\left[\frac{n}{2}(n - 1)\right] + n = n^2 \qquad (4.122)$$

The energy levels are said to be n^2-fold degenerate. The degeneracy with respect to m is obvious, for m describes the projection of vector l on a coordinate axis (the z-axis) in space, if this axis is not defined by some physical criterion but is arbitrarily chosen then the physical observable energy cannot depend on this arbitrary choice. We are assuming x, y, z space to be isotropic. The degeneracy with respect to l arising from the fact that for hydrogen like atom $V(r)$ is purely Coulomb field, i.e. $V(r)$ is proportional to r^{-1} and further, that Coulomb's law is valid over the entire range of r of interest, in atoms. This is called accidental degeneracy. The degeneracy is $2n^2$ if spin is taken into account.

4.8 Harmonic Oscillator

The harmonic oscillator is a system in which a particle of mass m subject to a linear restoring force **F** proportional to the displacement x from the equilibrium position

$$F = -kx \qquad (4.123)$$

The proportionality constant k is known as force constant. The minus sign indicates that force is in the direction opposite to the direction of the displacement.

The potential energy is given by

$$V = \frac{1}{2}kx^2 = \frac{1}{2}m\omega^2 x^2 \qquad (4.124)$$

$$\omega = \sqrt{\frac{k}{m}} \qquad (4.125)$$

One Dimensional Harmonic Oscillator

The Schrödinger wave equation in one dimension for a particle of mass m is

$$\frac{d^2\psi}{dx^2} + \frac{2m}{\hbar^2}[E - V]\psi = 0 \qquad (4.126)$$

where E is the energy and V is given by Eq. (4.124). Substituting the value of V in Eq. (4.126)

$$\frac{d^2\psi}{dx^2} + \frac{2m}{\hbar^2}\left[E - \frac{1}{2}kx^2\right]\psi = 0 \tag{4.127}$$

Let

$$\lambda = \frac{2m}{\hbar^2}E \tag{4.128}$$

$$\alpha^2 = \frac{mk}{\hbar^2} = \frac{m^2\omega^2}{\hbar^2} \tag{4.129}$$

Equation (4.127) is then written as

$$\frac{d^2\psi}{dx^2} + (\lambda^2 - \alpha^2 x^2)\psi = 0 \tag{4.130}$$

with the boundary condition $\psi \to 0$ as $|x| \to \infty$. Let us suppose αx to be very large in particular $\alpha x > > 1$ and $\alpha x \gg \lambda$. Equation (4.130) becomes

$$\frac{d^2\psi}{dx^2} - \alpha^2 x^2\psi = 0 \tag{4.131}$$

The equation is satisfied asymptotically by the exponential function

$$\psi = \exp(\pm \alpha x^2/2) \tag{4.132}$$

Of the two asymptotic solutions $\exp(-\alpha x^2/2)$ and $\exp(\alpha x^2/2)$, the latter is unsatisfactory as a wave function, since it tends rapidly to infinity with increasing value of x. Let

$$\psi = \exp(-\alpha x^2/2)f(x) \tag{4.133}$$

Equation (4.130) becomes

$$f'' - 2\alpha f' + (\lambda - \alpha)f = 0 \tag{4.134}$$

Let

$$\xi = \sqrt{\alpha}x \tag{4.135}$$

$$f(x) = H(\xi) \tag{4.136}$$

Using Eqs. (4.135) and (4.136) in Eq. (4.134)

$$H''(\xi) - 2\xi H'(\xi) + \left(\frac{\lambda}{\alpha} - 1\right)H(\xi) = 0 \tag{4.137}$$

Equation (4.137) can be solved by assuming a power series of the form

$$H(\xi) = \sum_{s=0}^{\infty} a_s \xi^s \tag{4.138}$$

Substituting Eq. (4.138) in Eq. (4.137)

$$\sum_{s=0}^{\infty} s(s-1)a_s \xi^{s-2} + \sum_{s=0}^{\infty} \left(\frac{\lambda}{\alpha} - 1 - 2s\right)a_s \xi^s = 0 \tag{4.139}$$

In order for Eq. (4.139) to vanish for all values of ξ, that is, for $H(\xi)$ to be a solution of Eq. (4.137) the coefficients of individual power of ξ must vanish separately, that is,

$$1.2a_2 + \left(\frac{\lambda}{\alpha} - 1\right)a_0 = 0$$

$$2.3a_3 + \left(\frac{\lambda}{\alpha} - 1 - 2\right)a_1 = 0 \text{ etc.}$$

In general

$$(s+1)(s+2)a_{s+2} + \left(\frac{\lambda}{\alpha} - 1 - 2s\right)a_s = 0$$

or

$$\frac{a_{s+2}}{a_s} = \frac{\left(2s + 1 - \frac{\lambda}{\alpha}\right)}{(s+1)(s+2)} \tag{4.140}$$

This expression is called recursion formula. Using this, a_2, a_3, ... can be calculated in terms of a_0 and a_1, which are arbitrary. If a_0 is set equal to zero only odd powers appear; with $a_1 = 0$, the series contains only even powers. For large values of s in Eq. (4.140)

$$\frac{a_{s+2}}{a_s} = \lim_{s\to\infty} \frac{\left(2s + 1 - \frac{\lambda}{\alpha}\right)}{(s+1)(s+2)} = \frac{2s}{s^2} = \frac{2}{s} \tag{4.141}$$

It is seen that this behaviour is the same as that of the series for $\exp(\xi^2)$

$$\exp(\xi^2) = 1 + \xi^2 + \frac{\xi^4}{2!} + \cdots + \frac{\xi^s}{(s/2)!} + \frac{\xi^{s+2}}{[(s/2) + 1]!} + \cdots$$

The ratio of the coefficients of ξ^{s+2} and ξ^s is $2/s$ for large values of s. The $H(\xi)$ behaves like $\exp(\xi^2)$. The function

$$\psi = \exp(-\alpha x^2/2) f(x) = \exp(-\xi^2/2) H(\xi)$$

behaves like an $\exp(\xi^2/2)$, which increases with increasing value of ξ^2 or increasing value of x thus making it unacceptable as a wave function. Therefore, we choose the values of the energy parameter, which will cause the series for $H(\xi)$ to break off after a finite number of terms. This yields a satisfactory wave function, because the negative exponential factor $exp(-\xi^2/2)$ will cause the function to approach zero for large value of ξ. The value of λ which causes the series to break off after the n-th term as is seen from Eq. (4.141) is

$$\lambda = (2n + 1)\alpha \tag{4.142}$$

$$\frac{2mE}{\hbar^2} = (2n + 1)\frac{m\omega}{\hbar}$$

$$E = \left(n + \frac{1}{2}\right)\hbar\omega \tag{4.143}$$

$$E = \left(n + \frac{1}{2}\right)h\nu \tag{4.144}$$

where $\omega = 2\pi\nu$. The possible energy levels of the harmonic oscillator are restricted to the infinite discrete set of values given by the Eq. (4.143) with $n = 0, 1, 2, \ldots$

Thus the wave function (eigenfunction) is

$$\psi_n(x) = N_n \exp(-\xi^2/2) H_n(\xi) \tag{4.145}$$

where $H_n(\xi)$ is Hermite polynomial of n-th degree in ξ and N_n is a constant which is adjusted so that ψ_n is normalized, and ψ_n satisfies the equation

$$\int_{-\infty}^{+\infty} \psi_n^* \psi_n dx = 1 \tag{4.146}$$

The value of N_n which makes the above condition true is

$$N_n = \left[\left(\frac{\alpha}{\pi}\right)^{\frac{1}{2}} \frac{1}{2^n n!}\right]^{\frac{1}{2}} \tag{4.147}$$

The first few Hermite polynomials are given in Table 4.4. From Eqs. (4.145), (4.147) and Table 4.4 ψ_n is obtained for various values of n as given in Table 4.5. The wave function is either even or odd depending on n. The functions ψ_{2n} are even [that is,

Table 4.4 Hermite polynomials

$H_0(\xi)$	$=$	1
$H_1(\xi)$	$=$	2ξ
$H_2(\xi)$	$=$	$4\xi^2 - 2$
$H_3(\xi)$	$=$	$8\xi^3 - 12\xi$
$H_4(\xi)$	$=$	$16\xi^4 - 48\xi^2 + 12$

Table 4.5 Wave functions of simple harmonic oscillator

ψ_0	$=$	$(\alpha/\pi)^{1/4} \exp(-\xi^2/2)$
ψ_1	$=$	$(\alpha/4\pi)^{1/4} 2\xi \exp(-\xi^2/2)$
ψ_2	$=$	$(\alpha/64\pi)^{1/4} (4\xi^2 - 2) \exp(-\xi^2/2)$
ψ_3	$=$	$(\alpha/2304\pi)^{1/4} (8\xi^3 - 12\xi) \exp(-\xi^2/2)$

$\psi_{2n}(-x) = \psi_{2n}(x)]$ and $\psi_{2n+1}(x)$ are odd that is $\psi_{2n}(-x) = -\psi_{2n}(x)$ as Hermite polynomial H_{2n} are even and H_{2n+1} are odd.

According to Eq. (4.144) the energy level diagram consists of a series of equidistant levels. Figure 4.4 shows the simple harmonic oscillator potential energy curve with the energy levels indicated by the horizontal lines for each value of n. The points where this line intersects the curve represent $E = V$, and hence, kinetic energy is zero. These points are called turning points of vibration. At the mid-point of vibrational energy level (oscillator energy levels) all the energy is kinetic. For $n = 0$ level, which has an energy $E = (1/2)h\nu$, is called zero-point level and $(1/2)h\nu$ the zero point energy. This is the minimum vibrational energy, that the oscillator may have at 0 K, and is the consequence of uncertainty principle.

The wave function (4.145) is

Fig. 4.4 Energy levels of harmonic oscillator

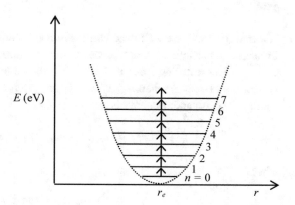

Fig. 4.5 Harmonic oscillator
wave functions $\psi_n(x)$ and
$|\psi_n(x)|^2$ for $n = 1, 2, 3$

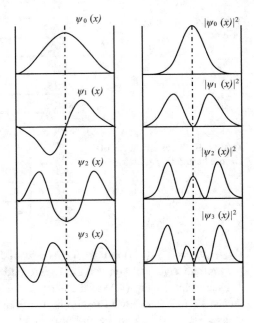

$$\psi_n(x) = \left[\left(\frac{\alpha}{\pi}\right)^{\frac{1}{2}} \frac{1}{2^n n!}\right]^{\frac{1}{2}} \exp\left(-\frac{\xi^2}{2}\right) H_n(\xi) \tag{4.148}$$

The four lowest harmonic oscillator eigenfunctions are plotted in Fig. 4.5.

The wave function (4.148) is either even or odd. The function $\psi_{2n}(x)$ are even, that is, $\psi_{2n}(-x) = \psi_{2n}(x)$ and $\psi_{2n+1}(x)$ are odd, that is, $\psi_{2n+1}(-x) = -\psi_{2n+1}(x)$ as Hermite polynomial $H_{2n}(x)$ are even and $H_{2n+1}(x)$ are odd. The wave function thus has definite parity.

Harmonic Oscillator in Three Dimensions (Cartesian Coordinates)

Consider a particle of mass m oscillating in three dimensions about the equilibrium position. The restoring force is $-k\mathbf{r}$ with force component along x, y and z axes are $-k_x x$, $-k_y y$ and $-k_z z$, where k_x, k_y and k_z are the force constants in the three directions x, y and z. The potential energy is

$$V = \frac{1}{2}k_x x^2 + \frac{1}{2}k_y y^2 + \frac{1}{2}k_z z^2 \tag{4.149}$$

where

$$k_x = m\omega_x^2, k_y = m\omega_y^2, k_z = m\omega_z^2 \tag{4.150}$$

The potential energy is therefore,

$$V = \frac{m}{2}\left(\omega_x^2 x^2 + \omega_y^2 y^2 + \omega_z^2 z^2\right) \tag{4.151}$$

The Schrödinger equation is now

$$\frac{\partial^2 \Psi}{\partial x^2} + \frac{\partial^2 \Psi}{\partial y^2} + \frac{\partial^2 \Psi}{\partial z^2} + \frac{2m}{\hbar^2}\left[E - \frac{m}{2}\left(\omega_x^2 x^2 + \omega_y^2 y^2 + \omega_z^2 z^2\right)\right]\Psi = 0 \tag{4.152}$$

Let

$$\lambda = \frac{2m}{\hbar^2} E \tag{4.153}$$

$$\alpha_x = \frac{m}{\hbar}\omega_x \tag{4.154}$$

$$\alpha_y = \frac{m}{\hbar}\omega_y \tag{4.155}$$

$$\alpha_z = \frac{m}{\hbar}\omega_z \tag{4.156}$$

Substituting Eqs. (4.153)–(4.156) in Eq. (4.151)

$$\frac{\partial^2 \Psi}{\partial x^2} + \frac{\partial^2 \Psi}{\partial y^2} + \frac{\partial^2 \Psi}{\partial z^2} + (\lambda - \alpha_x^2 x^2 - \alpha_y^2 y^2 - \alpha_z^2 z^2)\Psi = 0 \tag{4.157}$$

Using the method of separation of variables

$$\Psi(x, y, z) = X(x)Y(y)Z(z) \tag{4.158}$$

Equation (4.157) is

$$YZ\frac{\partial^2 X}{\partial x^2} + XZ\frac{\partial^2 Y}{\partial y^2} + XY\frac{\partial^2 Z}{\partial z^2}$$
$$+ (\lambda - \alpha_x^2 x^2 - \alpha_y^2 y^2 - \alpha_z^2 z^2)XYZ = 0$$

Dividing by XYZ

$$\left(\frac{1}{X}\frac{d^2 X}{dx^2} - \alpha_x^2 x^2\right) + \left(\frac{1}{Y}\frac{d^2 Y}{dy^2} - \alpha_y^2 y^2\right) + \left(\frac{1}{Z}\frac{d^2 Z}{dz^2} - \alpha_z^2 z^2\right) + \lambda = 0 \tag{4.159}$$

Equation (4.159) has been separated into terms each of which depend upon one variable only. Each term is therefore equal to a constant. We obtain three differential equations

$$\frac{d^2 X}{dx^2} + (\lambda_x - \alpha_x^2 x^2)X = 0 \tag{4.160}$$

$$\frac{d^2 Y}{dy^2} + (\lambda_y - \alpha_y^2 y^2)Y = 0 \tag{4.161}$$

$$\frac{d^2 Z}{dz^2} + (\lambda_z - \alpha_z^2 z^2)Z = 0 \tag{4.162}$$

where λ_x, λ_y and λ_z are separation constant such that

$$\lambda_x + \lambda_y + \lambda_z = \lambda \tag{4.163}$$

Equation (4.160) is same as the wave equation for the one dimensional harmonic oscillator and the wave function is given by

$$X(x) = \left[\left(\frac{\alpha_x}{\pi}\right)^{\frac{1}{2}} \frac{1}{2^{n_x} n_x!}\right]^{\frac{1}{2}} \exp\left(-\frac{\alpha_x x^2}{2}\right) H_{n_x}(\sqrt{\alpha}x) \tag{4.164}$$

and λ_x is restricted by the relation

$$\lambda_x = (2n_x + 1)\alpha_x \tag{4.165}$$

The quantum number n_x can assume the values 0, 1, 2, 3, ... The similar expression holds for $Y(y)$ and $Z(z)$. λ_y and λ_z are given by

$$\lambda_y = (2n_y + 1)\alpha_y \tag{4.166}$$

$$\lambda_z = (2n_z + 1)\alpha_z \tag{4.167}$$

where, $n_y = 0, 1, 2, ...$ and $n_z = 0, 1, 2, ...$ Using Eqs. (4.163) and (4.165)–(4.167)

$$\lambda = (2n_x + 1)\alpha_x + (2n_y + 1)\alpha_y + (2n_z + 1)\alpha_z$$
$$E = \left(n_x + \frac{1}{2}\right)\hbar\omega_x + \left(n_y + \frac{1}{2}\right)\hbar\omega_y + \left(n_z + \frac{1}{2}\right)\hbar\omega_z \tag{4.168}$$

The complete wave function is

$$\Psi(X, Y, Z) = \left[\frac{(\alpha_x \alpha_y \alpha_z)^{\frac{1}{2}}}{\pi^{\frac{3}{2}} 2^{n_x+n_y+n_z} n_x! n_y! n_z!}\right]^{\frac{1}{2}} \exp\left[-\frac{1}{2}\left(\alpha_x x^2 + \alpha_y y^2 + \alpha_z z^2\right)\right]$$
$$\times H_{n_x}(\sqrt{\alpha_x}x) H_{n_y}(\sqrt{\alpha_y}y) H_{n_z}(\sqrt{\alpha_z}z) \tag{4.169}$$

For the special case of isotropic oscillator in which

$$\omega_x = \omega_y = \omega_z; \alpha_x = \alpha_y = \alpha_z \tag{4.170}$$

The energy eigenvalue (4.168) is given by

$$E = \left(n_x + n_y + n_z + \frac{3}{2}\right)\hbar\omega = \left(n + \frac{3}{2}\right)\hbar\omega \tag{4.171}$$

where $n = n_x + n_y + n_z$ is called total quantum number. Since the energy for the system deepens only on the sum of the quantum numbers, all the energy levels for the isotropic oscillator except the lowest one are degenerate.

Degeneracy

The n-th energy level of the symmetric three dimensional harmonic oscillator has energy given by Eq. (4.171) for which $n = n_x + n_y + n_z$. Consider n_x fixed, n_y can take on the values from 0 to $n - n_x$, and n_z is then determined. There are now $(n - n_x + 1)$ possibilities for the pair n_y and n_z which are

$$(0, n - n_x), [1, (n - n_x - 1)], \ldots, (n - n_x, 0) \tag{4.172}$$

The degeneracy of the n-th energy level is therefore

$$\sum_{n=0}^{n}(n - n_x + 1) = (n + 1)\sum_{n=0}^{n}1 - \sum_{n=0}^{n}n_x$$

$$= (n + 1)(n + 1) - \frac{n(n + 1)}{2} = \frac{(n + 1)(n + 2)}{2} \tag{4.173}$$

4.9 Pauli Exclusion Principle

In 1925, W. Pauli discovered the fundamental principle that governs the electron configurations of multielectron atoms. His exclusion principle states that 'in a multielectron atom there can never be more than one electron in the same quantum state. Each electron must have a different set of quantum numbers n, l, m_l and m_s'. He established from the analysis of experimental data that the exclusion principle represents a property of electrons and not, particularly, of atoms. The exclusion principle operates in any system containing electrons.

It is seen that the complete wave function ψ of the hydrogen atom can be expressed as the product of three separate wave functions, each describing that part of ψ, which is a function of one of the three, coordinates r, θ and φ. A multielectron system consisting of n non-interacting electrons can be expressed as the product of wave functions $\psi(1)$, $\psi(2)$, ... $\psi(n)$ of the individual electrons, that is

$$\Psi(1, 2, \ldots, n) = \psi(1)\psi(2)\ldots\psi(n) \tag{4.174}$$

Each of the eigenfunction describing the electron require quantum numbers n, l, m_l to specify the mathematical form of its dependence on the three coordinates. In addition each require one more quantum number m_s to specify the orientation of the spin of the electron. To designate a particular set of four quantum numbers, the symbols such as a, b, c,. etc. are used. Let us consider a wave function used to describe a system of two electrons. Suppose electron number 1 is in quantum state a and electron number 2 is in state b. The wave function is

$$\Psi_I = \psi_a(1)\psi_b(2) \tag{4.175}$$

Because the electrons are identical, there is no physical way to distinguish the electron wave function given by Eq. (4.175) from the wave function

$$\Psi_{II} = \psi_a(2)\psi_b(1) \tag{4.176}$$

in which electron number 2 now has quantum state a, etc. Similarly no conceivable physical experiment could distinguish the six three electron wave functions:

$$\psi_a(1)\psi_b(2)\psi_c(3); \ \psi_a(3)\psi_b(1)\psi_c(2); \ \psi_a(2)\psi_b(3)\psi_c(1);$$
$$\psi_a(2)\psi_b(1)\psi_c(3); \ \psi_a(3)\psi_b(2)\psi_c(1); \ \psi_a(1)\psi_b(3)\psi_c(2) \tag{4.177}$$

Although we might determine that an atom had one electron in state a, another in state b, and a third in state c and so on, but we could not say that a particular electron (say the first electron) was in state a etc.

For two electron system consider linear combinations of Eqs. (4.175) and (4.176)

$$\Psi_S = \sqrt{\frac{1}{2!}}[\psi_a(1)\psi_b(2) + \psi_a(2)\psi_b(1)] \tag{4.178}$$

$$\Psi_A = \sqrt{\frac{1}{2!}}[\psi_a(1)\psi_b(2) - \psi_a(2)\psi_b(1)] \tag{4.179}$$

the factor $(1/2!)^{1/2}$ is needed to normalize Ψ_S and Ψ_A. The function Ψ_S is symmetric with respect to interchange of electron labels, i.e. $(1 \to 2$, and $2 \to 1)$ in Eq. (4.178) gives

$$\Psi_S \to \sqrt{\frac{1}{2!}}[\psi_a(1)\psi_b(2) + \psi_a(2)\psi_b(1)] = \Psi_S \tag{4.180}$$

The function Ψ_A is antisymmetric with respect to interchange of electron labels, i.e. $(1 \to 2$, and $2 \to 1)$ in Eq. (4.179) gives

$$\Psi_A \to \sqrt{\frac{1}{2!}}[\psi_a(1)\psi_b(2) - \psi_a(2)\psi_b(1)] = -\Psi_A \tag{4.181}$$

When the state a is identical to state b, then the antisymmetric wavefuction given by Eq. (4.179) is identically zero i.e.

$$\Psi_A \rightarrow \sqrt{\frac{1}{2!}} [\psi_a(1)\psi_b(2) - \psi_a(2)\psi_b(1)] = 0 \qquad (4.182)$$

and symmetric wave function given by Eq. (4.178) becomes

$$\Psi_S \rightarrow \sqrt{\frac{1}{2!}} [\psi_a(1)\psi_b(2) + \psi_a(2)\psi_b(1)] = \sqrt{2}\psi_a(1)\psi_a(2) \qquad (4.183)$$

Thus antisymmetric wave function automatically satisfies Pauli exclusion principle. The Pauli exclusion principle can also be stated as a system containing several electrons must be described by an antisymmetric total wave function.

The antisymmetric wave function (Eq. 4.179) can also be written in Slater determinant form

$$\Psi_A = \sqrt{\frac{1}{2!}} \begin{vmatrix} \psi_a(1) & \psi_a(2) \\ \psi_b(1) & \psi_b(2) \end{vmatrix} \qquad (4.184)$$

Similarly, antisymmetric wave function for a system of three electrons can be written in Slater determinant form

$$\Psi_A = \sqrt{\frac{1}{3!}} \begin{vmatrix} \psi_a(1) & \psi_a(2) & \psi_a(3) \\ \psi_b(1) & \psi_b(2) & \psi_b(3) \\ \psi_c(1) & \psi_c(2) & \psi_c(3) \end{vmatrix} \qquad (4.185)$$

With the help of Pauli's exclusion principle we can assign different quantum states to the electron in a given atom. For an atom, the electrons that have the same principal quantum number n are said to be in the same shell. For a given n, the electrons having the same value of l are said to be in the same sub-shell. Now we can calculate the maximum number of electrons belonging to the same shell or sub-shell.

(i) The case of a sub-shell

For electrons in a sub-shell we have same value of quantum numbers n and l. These electrons must differ either by the value of quantum number m_l (which can be one of the $2l + 1$ integral values between $- l$ and $+ l$) or by quantum number m_s (which can take values $+ 1/2$ or $- 1/2$). There exists, therefore, $2(2l + 1)$ distinct quantum states corresponding to the same value of n and l and therefore, there can be $2(2l + 1)$ electrons in a sub-shell of quantum number l. The maximum number of electrons in s, p, d and

f sub-shell are therefore 2, 6, 10 and 14, respectively. A sub-shell containing $2(2l + 1)$ electrons is said to be complete.

(ii) The case of a shell

For electrons in a shell we have same values of quantum number n but different quantum numbers l, m_l and m_s. The quantum numbers l can have all values from 1 to $n - 1$. The maximum number of electrons can be obtained by adding the maximum number of electrons in each sub-shell.

s electron ($l = 0$) +	p electron ($l = 1$)	$+ \cdots +$	Electron with $l = n - 1$
Maximum number	Maximum number		Maximum number
2	6		$2(2n - 1)$

The series may be written as

$$2(2l + 1) = 2 + 6 + 10 + \cdots + 2(2n - 1) = 2n^2 \tag{4.186}$$

The maximum number of electrons of quantum number n is therefore $2n^2$. The maximum numbers of electrons in various shells are given in Table 4.6

A shell containing the maximum number of electrons is called a complete shell. We may be tempted to say that electron configuration of any atom follows the general rule.

$$1s^2 2s^2 2p^6 3s^2 3p^6 3d^{10} 4s^2 4p^6 4d^{10} 4f^{14} \tag{4.187}$$

But this is not true. The actual order in which the levels must be filled so that the resulting energy is minimum (corresponding to a stable atom) is as follows:

$$1s^2 2s^2 2p^6 3s^2 3p^6 4s^2 3d^{10} 4p^6 5s^2 4d^{10} 5p^6 6s^2 4f^{14} 5d^{10} 6p^6 7s^2 6d^{10} \tag{4.188}$$

A given electron configuration is specified by quantum numbers n and l for each electron but not by m_l and m_s. Therefore, to each single configuration these will correspond a certain number of descriptions, differing in the values of m_l and m_s of each electron. A configuration will therefore have certain degeneracy.

Table 4.6 Number of electrons in various shells

Shell	n	Maximum number of electrons
K	1	2
L	2	8
M	3	18
N	4	32

Let us consider a configuration in which a single electron is in each sub-shell. No electron is characterized by the same quantum number n and l. Thus an electron i may have X_i states according to the values of m_l and m_s. X_i represents the number of places in the sub-shell with $X_i = 2(2l + 1)$. The number of different states corresponding to a configuration will therefore be

$$g = \prod_i X_i \qquad (4.189)$$

Now consider a configuration in which q electrons have the same quantum numbers n and l. For this value of l an electron can have $r = 2(2l + 1)$ different states characterized by different values of m_l and m_s. Now we have to put q electrons in r places each containing a maximum of one electron. The number of distinct combinations is

$$g = \frac{r!}{q!(r - q)!} \qquad (4.190)$$

The number g represents the number of different states corresponding to q electrons. The total degeneracy of the configuration is obtained from the product of g and the degeneracy due to electrons in other subshell and is

$$G = \prod_i g_i \qquad (4.191)$$

Solved Examples

Example 1 An electron moves in the x-direction with a speed of 3.6×10^6 m/s. The speed can be measured to a precision of 1%. With what precision can we simultaneously measure the position. What can we say about its motion in the y-direction.

Solution
The electron momentum is

$$p_x = mv_x = \left(9.11 \times 10^{-31}\, \text{kg}\right) \times \left(3.6 \times 10^6\, \text{m/s}\right)$$
$$= 3.3 \times 10^{-24}\, \text{kg m/s}$$

The uncertainty Δp_x is 1% of this value or $3 \cdot 3 \times 10^{-26}$ kg $\frac{\text{m}}{\text{s}}$. The uncertainty in position is then

$$\Delta x \sim \frac{\hbar}{\Delta p_x} = \frac{1.05 \times 10^{-34}\, \text{J s}}{3.3 \times 10^{-26}\, \text{kg m/s}} = 3.2\, \text{nm}$$

If the electron is moving in the x-direction, then we know that the speed in the y-direction precisely. Thus $p_y = 0$. The uncertainty relation

$$\Delta y \Delta p_y \sim \hbar$$

then requires that Δy be infinite. We can therefore know nothing at all about the electron's coordinate.

Example 2 A charged *pi* meson has a rest energy of 140 meV and a life time of 26 ns. Find the energy uncertainty of the *pi* meson.

Solution
We have

$$\Delta E \Delta t \sim \hbar$$

$$\Delta E \sim \frac{\hbar}{\Delta t} = \frac{6.58 \times 10^{16} \text{ eV s}}{26 \times 10^{-9} \text{ s}} = 2.5 \times 10^{-8} \text{ eV}$$

$$= 2.5 \times 10^{-14} \text{ MeV}$$

Example 3 An electron is moving with a speed 1455 m/s. What is de Brogli wavelength.

Solution
Putting $p_{\text{particle}} = \text{mv}$ in Eq. (4.1) and solving for λ gives

$$\lambda = \frac{h}{mv}$$

$$\lambda = \frac{6.626 \times 10^{-34} \text{ Js}}{9.1 \times 10^{-31} \text{ kg} \times 1455 \text{ m/s}} = 500 \times 10^{-9} \text{ m}$$

Example 4 What is the low lying energy spectrum for a particle of mass m in one dimension harmonic oscillator potential with a high, thin barrier at $x = 0$ i.e.

$$V(x) = \frac{1}{2}m\omega x^2 \quad |x| < b$$

$$= V_0 \qquad |x| > b$$

assume that the barrier is completely impenetrable.

Solution
The potential is of harmonic oscillator type for $x > b$ and $x < b$. Thus the low lying eigen-function must satisfy the condition that $\psi(x) = 0$ at $x = 0$. From Table 4.5 it is seen that this condition is satisfied for states of harmonic oscillator with odd quantum number ($2n + 1$). The corresponding energy by Eq. (4.143) will be

$$\left(2n + \frac{3}{2}\right)\hbar\omega$$

where $n = 0, 1, 2, 3, \ldots$, with a degeneracy of 2. Thus only the odd parity wave function is allowed for low lying states.

Example 5 Consider a one-dimensional particle which is confined within the region $0 \le x \le a$ and whose wave function is

$$\Psi(x, t) = sin(\pi x / a)exp(-i\omega t)$$

Find the potential $V(x)$.

Solution
Differentiating the given wave function

$$\frac{d\psi}{dx} = \frac{\pi}{a}\cos(\pi x/a)\exp(-i\omega t)$$

$$\frac{d^2\psi}{dx^2} = -\frac{\pi^2}{a^2}\sin(\pi x/a)\exp(-i\omega t) = -\frac{\pi^2}{a^2}\psi$$

$$\frac{d\psi}{dt} = -i\omega\sin(\pi x/a)\exp(-i\omega t) = -i\omega\psi$$

Substituting the value of $\frac{d^2\psi}{dx^2}$ and $\frac{d\psi}{dt}$ in Eq. (4.25)

$$i\hbar\frac{\partial\psi}{\partial t} = -\frac{\hbar^2}{2m}\nabla^2\psi + V(x, t)\psi$$

$$i\hbar(-i\omega)\psi = \left[\frac{\hbar^2}{2m}\frac{\pi^2}{a^2} + V\right]\psi$$

$$V = \hbar\omega - \frac{\hbar^2}{2m}\frac{\pi^2}{a^2}$$

Objective Type Questions

1. For which of the following three dimensional potentials would Schrödinger equation be separable ?
   (a) $V = x^2 y + \sin z$ (b) $V = x^2 + y + \tan^{-1}\sqrt{z}$ (c) $V = y\sin x + z\cos y + y\tan z$
   (d) $V = e^x y^{\frac{7}{2}} z^2$
2. Which among the following functions represent physically acceptable wave functions?
   (a) A sec x (b) 3 $\sin\pi\ x$ (c) $A\ \exp(x^2)$ (d) x
3. The magnitude of square of angular momentum is found to be $6\hbar^2$. What is the value of l?
   (a) 0 (b) 1 (c) 2 (d) 3.

4. Which of the following functions is antisymmetric ?
   (a) $\psi_a(1)\psi_a(2)[\alpha(1)\beta(2) - \beta(1)\alpha(2)]$
   (b) $[\psi_a(1)\psi_b(2) - \psi_b(1)\psi_a(2)][\alpha(1)\beta(2) - \beta(1)\alpha(2)]$
   (c) $\psi_a(1)\psi_a(2)[\alpha(1)\beta(2) + \beta(1)\alpha(2)]$
   (d) none

5. Operator corresponding to momentum p_x is
   (a) $-i\hbar\frac{\partial}{\partial x}$ (b) $i\hbar\frac{\partial}{\partial t}$ (c) $i\hbar\frac{\partial}{\partial x}$ (d) $-i\hbar\frac{\partial}{\partial t}$

6. Operator corresponding to energy is
   (a) $(a) - \frac{\hbar^2}{2m}\nabla^2$ (b) $\frac{\hbar^2}{2m}\nabla^2$ (c) $-\frac{2m}{\hbar^2}\nabla^2$ (d) $-\frac{1}{2m}\nabla^2$

7. The de Broglie wavelength of an electron whose speed is one half that of light is
   (a) 3.6 pm (b) 4.2 pm (c) 8.4 pm (d) none of these

8. If a charged particle of mass m and charge q is accelerated through a potential difference V, then its de Broglie wavelength is
   (a) $\lambda = \frac{h}{\sqrt{2qV}}$ (b) $\lambda = \frac{h}{\sqrt{2qm}}$ (c) $\lambda = \frac{h}{\sqrt{2mqV}}$ (d) $\lambda = \frac{h}{\sqrt{mqV}}$

9. In the first excited state of a three-dimensional harmonic oscillator, the energy is
   (a) $\hbar\omega$ (b) $2\hbar\omega$ (c) $3\hbar\omega/2$ (d) $5\hbar\omega/2$

10. When a particle is bounded to a limited space, then the probability of finding the particle at a finite distance will be
    (a) 0 (b) 1/2 (c) 1 (d) none of these

11. The wave function is
    (a) complex (b) real (c) complex but real and imaginary part can be separated (d) none

12. Which of the following is essential for a wave function to be acceptable?
    (a) it must be finite (b) it must be single valued (c) it must be continuous (d) all of the above are required for a function to be acceptable

13. The state is said to be stationary if its probability density is
    (a) function of time (b) does not depend on time (c) independent of space coordinates (d) none of these

14. The probability density of a state is always
    (a) Real (b) imaginary (c) always has probability equal to half (d) none of these

15. The Schrödinger equation for a particle
    (a) contain third order time derivative (b) contain second order time derivative (c) contain first order time derivative (d) none of these

16. The zero point energy of a linear harmonic oscillator is 1.66×10^{-32} J what would be its frequency ?
    (a) 10 Hz (b) 50 Hz (c) 250 Hz (d) 500 Hz

17. The total degeneracy for $n = 5$ energy level of hydrogen atom, without considering the spin is
    (a) 5 (b) 25 (c) 30 (d) 1

18. The degeneracy of the $n = 4$ level of three dimensional harmonic oscillator is
    (a) 15 (b) 16 (c) 8 (d) 12

Problems

1. For a particle in one dimension is the wave function $\psi(x) = x$ acceptable?
2. Obtain the expectation value of r and $(1/r)$ in the ground state of hydrogen atom.
3. A beam of electron bombards a sample of hydrogen. Through what potential difference must the electron be accelerated if the first line of the Balmer series is to be emitted.
4. The largest wavelength in the Lyman series is 121.5 nm and the shortest wave- length in the Balmer series is 364.6 nm. Find the longest wavelength of light that could ionize hydrogen.
5. What is the degeneracy of the ground state of carbon $1s^2 2s^2 2p^2$
6. What are the configurations of the electrons of atoms of the following elements (a) Si $(Z = 14)$ (b) Br $(Z = 35)$.
7. Show that $\Delta x \Delta \lambda \geq \frac{\lambda^2}{4\pi}$
8. Consider an electron moving freely along the x-axis. At time $t = 0$, the position of the particle is measured and is uncertain by the amount Δx_0. Obtain the uncertainty in the measured position of the electron at time
9. An electron is confined to a region of space of the size of an atom (0.1 nm). What is the uncertainty in the momentum of the electron.
10. A particle of mass m moves in one dimension under the influence of $V(x)$. Suppose it is in an energy eigenstate $\psi(x) = Ax \exp(-kx)$ $(0 \leq x \leq \infty, k > 0)$. Obtain A and $V(x)$.
11. A certain one particle one dimensional system has the potential energy $V = 2a^2\hbar^2 x^2/m$ and is in the stationary state $\psi(x) = bx \exp(-ax^2)$ where a and b are constant. Find the particle's energy.
12. Calculate the de Broglie wavelength for (a) a proton of kinetic energy 70 mMeV kinetic energy and (b) a 100 g bullet moving at 900 ms^{-1}.
13. Estimate the uncertainty in the position of (a) a neutron moving at 5×10^6 ms^{-1} and (b) a 50 kg person moving at 2 ms^{-1} (mass of neutron = 1.65×10^{-27} kg).
14. Consider a mass spring system where a 4 kg mass is attached to a mass less spring of constant $k = 196$ Nm^{-1}, the system is set to oscillate on a frictionless horizontal table. The mass is pulled 25 cm away from equilibrium position and then released. Find the energy spacing between two consecutive energy levels and total number of quanta involved.
15. A proton is subject to a harmonic oscillator potential $V(x) = (1/2)m\omega^2 x^2$, $\omega = 5.34 \times 10^{21}$ s^{-1}. Find the energies of five lowest states (express them in MeV).

Answers
Objective Type Questions

1. (b)	2. (b)	3. (c)	4. (a)	5. (a)	6. (a)	7. (d)	8. (c)
9. (c)	10. (d)	11. (a)	12. (d)	13. (b)	14. (b)	15. (c)	16. (a)

17. (b)   18. (a)

X-Ray Diffraction

5

5.1   X-Rays

X-rays are electromagnetic radiations of wavelength between ~10 pm and ~10 nm. X-rays are characterized by index of refraction very close to unity for all materials. X-rays are produced when a beam of highly accelerated particles such as electrons are allowed to strike a metal target. In the process, electrons suffer energy loss and this loss is emitted in the form of electromagnetic radiation. The X-rays are produced both by deceleration of electrons in the metal target and by the excitation of the core electrons in the atom of the target. The first process gives a broad continuous spectrum and the second gives sharp lines. When a moving electron is stopped suddenly, all its energy appears as photon of frequency ν of X-rays. The energy of an electron of charge e in dropping through a potential difference V is eV and

$$E = h\nu = \frac{hc}{\lambda} = eV$$

$$\lambda = \frac{hc}{E} = \frac{hc}{eV} \tag{5.1}$$

An electron will not lose all its energy in this way; it will have a number of glancing collisions with the atoms that it collides and causing them to vibrate. As a result of this, the temperature of the target increases. Equation (5.1), therefore, gives the minimum value λ can possibly have and accounts for the short wavelength cut-off. Larger wavelengths are more probable and so the rapid increase in the intensity. The intensity falls off gradually indicating that there is no upper limit. Figure 5.1 shows the X-ray spectrum that results when molybdenum target is bombarded by electron at 35 keV. The electron beam on striking the target not only gets decelerated but also a small fraction of electrons of the

© The Author(s) 2022
V. K. Jain, *Solid State Physics*, https://doi.org/10.1007/978-3-030-96017-9_5

Fig. 5.1 X-ray spectra of
molybdenum at 35 kV
accelerating potential

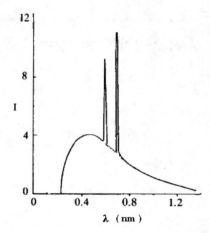

beam strikes the target and ejects the inner shell's electrons. The atom is then unstable,
and outer shell electrons in the same atom will drop into the hole (vacancy) caused by
the ejection of the electron. In doing so, it loses energy and a photon is emitted. If E is
the energy lost, we have

$$\lambda = \frac{hc}{E} \tag{5.2}$$

E is a definite quantity associated with the electron energy change in the atom. Therefore, the wavelength concerned is specific. Several wavelengths are possible, and they
constitute the characteristic X-ray line spectrum shown as peaks in Fig. 5.1. The energy
of the characteristic X-ray produced is very weakly dependent on the chemical structure in
which the atom is bound indicating that non-bonding shells of atoms are the characteristic X-ray source. The resulting characteristic spectrum is superimposed on the continuum.
An atom remains ionized for a very short time ($\sim10^{-14}$ s), and thus, the incident electrons
that arrive about every $\sim10^{-17}$ s can repeatedly ionize an atom. However, not all outer
electrons can fall into holes to provide X-rays.

Suppose an electron from K shell is ejected, and a hole is created in the shell. Subse-
quently, an electron from a higher shell L, M, N, ... will make a radiative transition filling
the hole. The energy of the photon emitted will be in the range from a few keV to a few
hundred keV and thus lie in the X-ray region. The emission spectrum, which results in
a line spectrum, forms a simple series. The lines originating from a K shell vacancy are
called K_α, K_β, K_γ, ... lines and correspond to $L \rightarrow K$, $M \rightarrow K$, $N \rightarrow K$, ... transition. K_α
is the strongest. The emission of K series is accompanied by other series. As a result of
lines of K series, the vacancies are created in L, M, N, ... shells which are filled by M,
N, ... shell electrons. Thus, K series may be accompanied by L, M, N, ... and so on.

According to the wavelength, X-rays can be classified as hard or soft. The K X-rays
are called hard while L radiation is soft and M, N, O X-rays are very soft. M and L

radiations arising from heavy elements are harder (i.e. of shorter wavelengths) than K radiation of lighter elements. K radiation of an element is more penetrating than that of $L, M, N, O,$ etc.

In most X-ray-related research work, only one well-defined wavelength of radiation is needed and for this X-ray filters are used. Filtering X-rays is probably the simplest way of approximately obtaining the monochromatic X-rays. The suitable filters are Ni, Zr, Co, Fe, Mn for Cu, Mo, Ni, Co, Fe radiation, respectively.

In 1913, Moseley observed that the wavelength of characteristic X-ray lines shift continuously with the change in the element atomic number Z. He observed that with increasing Z, the wavelength of X-ray decreases, and hence, the frequency of X-ray increases. A plot of square root of frequency versus Z gives a straight line. This increase in frequency with increase of Z is due to the increasing binding energy of the electron with increase of number of protons in the nucleus. Moseley law is given by

$$\sqrt{\frac{\bar{v}}{R}} = Z^*\sqrt{a} = \sqrt{a}(Z - \sigma_{nl}) \tag{5.3}$$

where $Z^* = Z - \sigma_{nl}$ is the effective charge of the nucleus, σ_{nl} is value of screening constant, R is Rydberg constant, $\bar{v}$ is frequency in cm^{-1} and

$$a = \left(\frac{1}{n_f^2} - \frac{1}{n_i^2}\right)$$

a is same for similar lines of all the elements, e.g. $K_{\alpha 1}$ line. σ is approximately equal for all the lines of a given series, e.g. for all the K lines of all the elements. Moseley law provides a simple test for the order of elements according to Z. It showed where elements were missing from the periodic table and led to the discovery of some of the elements.

The interatomic spacing in crystals is of the order of 10^{-8} cm and, therefore, if a wave having wavelength of the similar dimensions strike on it, the diffraction phenomena can be observed. For these X-rays, de Broglie waves due to electrons neutrons are suitable. For X-ray to have a wavelength of the order of 0.1 nm, the energy required is from Eq. (5.1)

$$E = \frac{hc}{\lambda} = \frac{6.62 \times 10^{-27} \times 3 \times 10^{10}}{10^{-8}} \text{ erg} = 19.86 \times 10^{-9} \text{ erg} = 12.4 \times 10^3 \text{ eV}.$$

According to de Broglie hypothesis, the wavelength associated with a particle is given by

$$\lambda = \frac{h}{mv} = \frac{h}{\sqrt{2mE}} \tag{5.4}$$

For neutrons, substituting the value of mass of neutron and Plank constant, we have

$$\lambda = \frac{0.28}{\sqrt{E}} \tag{5.5}$$

where λ is in Å and E in eV. For $\lambda = 1$ Å, the value of E is ~0.08 eV. This energy is of the same order as the energy of thermal neutrons (~0.025 eV).

For electrons, the relation (5.4) takes the form

$$\lambda = \frac{h}{\sqrt{2meV}} = \left(\frac{150}{V}\right)^{\frac{1}{2}} = \frac{12.24}{\sqrt{V}} \tag{5.6}$$

for $\lambda = 1$ Å, the value of E is ~0.08.

It is found that there are experimental difficulties with electrons and neutrons in this energy range as compared to X-rays. Neutron diffraction fundamentally differs from X-ray diffraction in the sense that X-rays are scattered by the electron cloud associated with the atom while neutron are scattered by the nuclei of the atom as neutrons are electrically neutral. This property makes it a useful tool to locate the position of light atom like hydrogen. As light atoms have only a few electrons to scatter the X-ray beam, therefore they do not contribute significantly to the X-ray diffraction pattern. Neutron diffraction can distinguish between different isotopes, whereas X-ray pattern does not. Since neutron has a magnetic moment, therefore neutron diffraction can be used for magnetic crystals. However, the disadvantage of neutron diffraction is that nuclear reactor is needed and being electrically neutral, it is difficult to detect than the ionizing X-ray.

The mechanism responsible for the electron scattering is the electric field associated with the atoms in the solid. This field is produced both by the nucleus and by the orbital electrons. The scattering length associated with the scattering of the electrons from the atom is large. Therefore, the electron beam has a short stopping distance. Hence, electron diffraction is particular sensitive to the physical properties of the surface and thus useful for studies of surface of solids and thin films.

5.2   Reciprocal Lattice

Consider a lattice whose vectors are a, b, and c forms a primitive basis. We define a new set of basis vectors a^*, b^* and c^* according to the relations

$$a^* = \frac{2\pi (b \times c)}{V} \tag{5.7}$$

$$b^* = \frac{2\pi (c \times a)}{V} \tag{5.8}$$

$$c^* = \frac{2\pi (a \times b)}{V} \tag{5.9}$$

where V is the volume of a unit cell given by

$$V = a \cdot (b \times c) = c \cdot (a \times b) = b \cdot (c \times a) \tag{5.10}$$

The vectors a^*, b^* and c^* can be used as a basis for a new lattice whose vectors are given by

$$G = ua^* + vb^* + wc^* \tag{5.11}$$

where u, v and w are any set of integers. The lattice defined by G is known as reciprocal lattice, and a^*, b^* and c^* are called reciprocal basis vectors. The relations (5.7)–(5.9) are still valid if we replace lattice vectors by reciprocal lattice vectors and inversely reciprocal lattice vectors by lattice vectors. If V^* is the volume of the reciprocal unit cell, then $V^*V = 1$ is also valid.

The vector a^* is normal to the plane defined by the vectors b and c; b^* is normal to the plane defined by the vectors c and a, and c^* is normal to the plane defined by a and b. The relation between reciprocal lattice vectors and primitive lattice vectors may be obtained as follows.

Let us consider

$$a^* \cdot a = \frac{2\pi (b \times c)}{V} \cdot a = 2\pi \frac{a \cdot (b \times c)}{a \cdot (b \times c)} = 2\pi \tag{5.12}$$

$$a^* \cdot b = \frac{2\pi (b \times c)}{V} \cdot b = 2\pi \frac{b \cdot (b \times c)}{a \cdot (b \times c)} = 0 \tag{5.13}$$

Similarly

$$b^* \cdot b = c^* \cdot c = 2\pi \tag{5.14}$$

$$a^* \cdot c = b^* \cdot a = b^* \cdot c = c^* \cdot a = c^* \cdot b = 0 \tag{5.15}$$

The reciprocal lattice possesses the same rotational symmetry as the direct lattice. The reciprocal lattice always falls in the same crystal system as its direct lattice. The reciprocal lattice for hexagonal, monoclinic, …, triclinic lattices is also hexagonal, monoclinic,…. triclinic, respectively.

5.3 BRAGG's Law

Consider two parallel planes of atoms AB and CD (Fig. 5.2). The Miller indices of the two planes are h, k and l, and they are separated by a distance d. Suppose that a parallel monochromatic and coherent (in phase) beam of X-rays of wavelength λ falls on these two planes at an angle θ. Let the two rays 1 and 2 are scattered by atoms P and Q. The

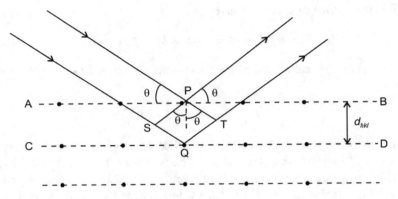

Fig. 5.2 X-ray diffraction from a crystal

path difference between 1-P-3 and 2-Q-R is

$$\text{Path diference} = SQ + QT \tag{5.16}$$

For diffraction, this path difference must be equal to an integral multiple of wavelength, that is,

$$SQ + QT = n\lambda \tag{5.17}$$

From Fig. 5.2

$$SQ = QT = d \sin \theta \tag{5.18}$$

Substituting Eqs. (5.18) in (5.17)

$$2d \sin \theta = n\lambda \tag{5.19}$$

This relation is known as Bragg's law, n is order of reflection which may be integer ($n = 1, 2, 3, \ldots$) consistent with $\sin\theta$ not exceeding unity. From Eq. (5.19), it is seen that diffraction intensities can be built only at certain values of θ, corresponding to a specific value of λ and d. From Eq. (5.19), we have

$$\theta = \sin^{-1} \frac{n\lambda}{2d} \tag{5.20}$$

From this, it is seen that rays diffracted by a crystal are given off in different directions corresponding to different values of the interplanar spacing d. From the experimentally observed diffraction angles, it is possible to determine the d of a crystal. From a list of such spacing, it is then possible to determine the lattice of the crystal.

The highest possible order can be determined by the condition

$$\sin \theta_{\max.} = 1 \text{ or } \frac{n\lambda}{2d} \leq 1 \tag{5.21}$$

This indicates that λ must not be greater than twice the interplanar spacing otherwise no diffraction will occur. Since each plane reflects 10^{-3} to 10^{-5} of the incident radiation so that 10^3 to 10^5 planes may contribute to the formation of Bragg reflected beam in a perfect crystal. Bragg's law is a consequence of the periodicity of the lattice.

5.4 Diffraction Condition

Let us consider the diffraction condition in terms of the reciprocal lattice. Assume that the incident X-ray is represented by a wave vector k and reflected ray by the wave vector k' as shown in Fig. 5.3.

From Fig. 5.3

$$k + \Delta k = k' \tag{5.22}$$

where the change Δk in k is only in the direction perpendicular to (hkl) plane. Assuming that scattering is elastic

$$k = k' = \frac{2\pi}{\lambda} \tag{5.23}$$

From Fig. 5.3

$$\Delta k = k' - k = 2\sin\theta|k|\hat{n} \tag{5.24}$$

From Eqs. (5.23) and (5.24)

Fig. 5.3 Definition of the scattering vector Δk

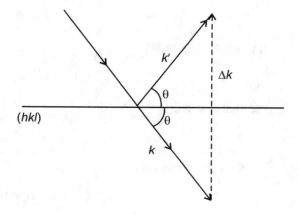

$$\Delta \mathbf{k} = \frac{4\pi}{\lambda}\sin\theta\,\hat{\mathbf{n}}$$

(5.25)

Let

$$\hat{\mathbf{n}} = \frac{\mathbf{G}_{hkl}}{|\mathbf{G}_{hkl}|}$$

(5.26)

From Eqs. (5.25) and (5.26)

$$\Delta \mathbf{k} = \frac{4\pi}{\lambda}\sin\theta\,\frac{\mathbf{G}_{hkl}}{|\mathbf{G}_{hkl}|}$$

(5.27)

Now

$$d_{hkl} = \frac{2\pi}{|\mathbf{G}_{hkl}|}$$

(5.28)

Substituting Eq. (5.28) in (5.27)

$$\Delta \mathbf{k} = \frac{2d_{hkl}}{\lambda}\sin\theta\,\mathbf{G}_{hkl}$$

(5.29)

When the combination of λ, θ and d_{hkl} is appropriate for the satisfaction of Bragg condition, we have

$$\Delta \mathbf{k} = \mathbf{G}_{hkl} = \mathbf{G}$$

(5.30)

Thus, the set of points generated by $\mathbf{G}$ is the array of diffraction spots by a crystal. From Eqs. (5.24) and (5.30)

$$\mathbf{G} = \mathbf{k}' - \mathbf{k}$$

$$\mathbf{k}' = \mathbf{k} + \mathbf{G}$$

(5.31)

On squaring Eq. (5.31)

$$k'^2 = k^2 + G^2 + 2\mathbf{k} \cdot \mathbf{G}$$

(5.32)

For elastic scattering

$$k' = k$$

Equation (5.32) takes the form

$$G^2 + 2\mathbf{k} \cdot \mathbf{G} = 0$$

(5.33)

This is Bragg's law in vector form. If $\mathbf{G}$ is reciprocal lattice vector so is $-\mathbf{G}$. Putting $\mathbf{G} = -\mathbf{G}$ in Eq. (5.33)

$$G^2 = 2\mathbf{k} \cdot \mathbf{G} \qquad (5.34)$$

The spacing between parallel lattice planes that is normal to the direction. Let

$$\mathbf{G} = h\mathbf{a}^* + k\mathbf{b}^* + l\mathbf{c}^* \qquad (5.35)$$

is

$$d_{hkl} = \frac{2\pi}{|\mathbf{G}|} \qquad (5.36)$$

Substituting Eq. (5.36) in Eq. (5.34)

$$\left(\frac{2\pi}{d_{hkl}}\right)^2 = 2\frac{2\pi}{\lambda} \times \frac{2\pi}{d_{hkl}}\sin\theta$$

$$2d_{hkl}\sin\theta = \lambda \qquad (5.37)$$

Here θ is the angle between the incident beam and the crystal plane. The integers hkl that define $\mathbf{G}$ are not necessarily identical with the indices of an actual crystal, because hkl may contain a common factor n, whereas in the definition of the indices the common factor has been eliminated. We thus obtain the Bragg result

$$2d_{hkl}\sin\theta = n\lambda \qquad (5.38)$$

where d_{hkl} is the spacing between the adjacent parallel planes with indices h/n, k/n, l/n.

5.5 Ewald Construction

Ewald in 1921 suggested a geometrical construction to represent diffraction condition as shown in Fig. 5.4. A vector $\mathbf{k}$ is drawn in the direction of the incident X-ray, and origin A is chosen such that k terminates at any reciprocal point B. We draw a sphere of radius $k = \frac{2\pi}{\lambda}$ about the origin A. If this sphere does not pass through any reciprocal lattice point, then that particular wavelength would not be diffracted by the crystal. If the sphere passes through any point C of the reciprocal lattice, then the diffraction will take place corresponding to this wavelength as the diffraction condition

$$\Delta\mathbf{k} = \mathbf{G}_{hkl} = \mathbf{G}$$

is satisfied.

Fig. 5.4 Ewald construction in
the reciprocal lattice

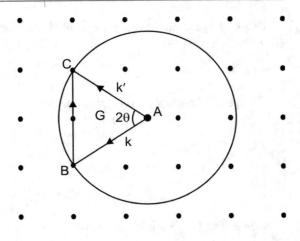

5.6   Laue Equations

Let the X-rays of wavelength λ and wave vector k $(2\pi/\lambda)$ be incident on the identical
atoms at a separation d (Fig. 5.5). The two atoms being identical would scatter X-rays in
the direction with a common wave vector k'. For elastic scattering $|k| = |k'|$. The path
difference between the X-ray scattered at P and that scattered at O is

$$PA - OB = d\cos\theta - d\cos\theta' \tag{5.39}$$

On assuming unit vector n and n' along k and k'

$$PA - OB = d\cos\theta - d\cos\theta' = d \cdot n - d \cdot n' = d \cdot (n - n') \tag{5.40}$$

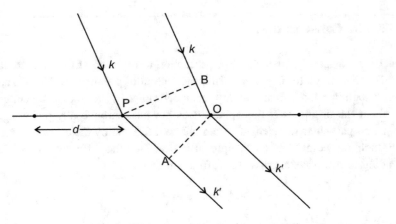

Fig. 5.5   Scattering of X-rays by a row of identical atoms separated by a distance d

The path difference at a point of constructive interference would meet the condition

$$d \cdot (n - n') = m\lambda \tag{5.41}$$

where m is an integer. The diffraction condition (5.41) will be satisfied by all the diffracted beams lying on the concentric cone with respect to d-row of atoms and has the semi-apex angle θ. Thus for any given angle of incidence, there will be a series of concentric cones surrounding the d-row of atoms, where each cone represents various orders of diffraction.

If a, b and c are primitive translation vector, then Eq. (5.41) would be written as

$$a \cdot (n - n') = a(\cos\alpha - \cos\alpha_0) = e\lambda \tag{5.42}$$

$$b \cdot (n - n') = b(\cos\beta - \cos\beta) = f\lambda \tag{5.43}$$

$$c \cdot (n - n') = c(\cos\gamma - \cos\gamma_0) = g\lambda \tag{5.44}$$

where e, f and g are integers, α_0, β_0 and γ_0 and α, β and γ respectively are the angles between the incident and scattered beam and the axes a, b and c. Equations (5.42)–(5.44) are known as Laue equations. In this case, there are three sets of cones, one each around a, b and c rows of atoms and the most intense diffracted beam will be directed along intersection of all three sets of cones. This reduces the number of possible diffracted beams.

On squaring and adding Eqs. (5.42)–(5.44)

$$a^2 \left(\cos^2\alpha + \cos^2\alpha_0 - 2\cos\alpha\cos\alpha_0\right) + b^2 \left(\cos^2\beta + \cos^2\beta_0 - 2\cos\beta\cos\beta_0\right)$$
$$+ c^2 \left(\cos^2\gamma + \cos^2\gamma_0 - 2\cos\gamma\cos\gamma_0\right) = \left(e^2 + f^2 + g^2\right)\lambda^2 \tag{5.45}$$

for simple cubic lattice $a = b = c$ and Eq. (5.45) is

$$a^2 \left(\cos^2\alpha + \cos^2\alpha_0 + \cos^2\beta + \cos^2\beta_0 + \cos^2\gamma + \cos^2\gamma_0 \right.$$
$$\left. -2\cos\alpha\cos\alpha_0 - 2\cos\beta\cos\beta_0 - 2\cos\gamma\cos\gamma_0\right) = \left(e^2 + f^2 + g^2\right)\lambda^2 \tag{5.46}$$

From solid geometry, we know that in an orthogonal coordinate system, α, β, and γ satisfy the following conditions

$$\cos^2\alpha + \cos^2\beta + \cos^2\gamma = 1 \tag{5.47}$$

$$\cos^2\alpha_0 + \cos^2\beta_0 + \cos^2\gamma_0 = 1 \tag{5.48}$$

$$\cos\alpha\cos\alpha_0 + \cos\beta\cos\beta_0 + \cos\gamma\cos\gamma_0 = \cos\varphi \tag{5.49}$$

where φ is the angle between incident and scattered beam. Substituting Eqs. (5.47)–(5.49), in Eq. (5.46) is

$$2a^2(1 - \cos\varphi) = (e^2 + f^2 + g^2)\lambda^2 \qquad (5.50)$$

Putting

$$d = \frac{a}{\sqrt{e^2 + f^2 + g^2}} \qquad (5.51)$$

in Eq. (5.50)

$$2(1 - \cos\varphi) = 4\sin^2\varphi = \frac{a^2\lambda^2}{a^2d^2} = \frac{\lambda^2}{d^2}$$

$$2\sin\frac{\varphi}{2} = \frac{\lambda}{d}$$

$$2d\sin\frac{\varphi}{2} = 2d\sin\theta = \lambda \qquad (5.52)$$

where $\varphi/2 = \theta$. Equation (5.52) is known as Bragg's law.

5.7   Brillouin Zones

A Brillouin zone is defined as a Wigner–Seitz primitive cell in the reciprocal lattice space. Consider the Bragg diffraction condition (5.34)

$$G^2 = 2\boldsymbol{k} \cdot \boldsymbol{G}$$

On dividing both sides by 4

$$\left(\frac{\boldsymbol{G}}{2}\right)^2 = \boldsymbol{k} \cdot \frac{\boldsymbol{G}}{2} \qquad (5.53)$$

Now consider reciprocal lattice as shown in Fig. 5.6. Consider a point O in the lattice and draw vectors $\boldsymbol{G}_A$ connecting O to reciprocal lattice point A. Similarly consider another vector $\boldsymbol{G}_B$ connecting O to reciprocal lattice point B. Now draw two planes 1 and 2 which are perpendicular bisector of $\boldsymbol{G}_A$ and $\boldsymbol{G}_B$, respectively. Any vector from origin to the plane 1 such as $\boldsymbol{k}_1$ will satisfy the diffraction condition (5.53)

$$\left(\frac{\boldsymbol{G}_A}{2}\right)^2 = \boldsymbol{k}_1 \cdot \frac{\boldsymbol{G}_A}{2}$$

Similarly, any vector $\boldsymbol{k}_2$ from origin to the plane 2 will satisfy the diffraction condition

Fig. 5.6 Reciprocal lattice
points near the point O

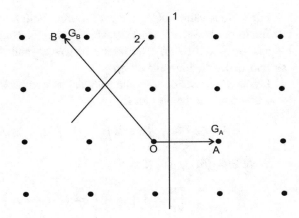

$$\left(\frac{G_B}{2}\right)^2 = k_2 \cdot \frac{G_B}{2}$$

Thus, planes 1, 2, etc., form the part of zone boundary. An X-ray beam in the crystal will be diffracted if its wave vector k has the magnitude and direction satisfying the relation (5.53).

Let us now consider one-dimensional monoatomic lattice in the reciprocal space as shown in Fig. 5.7. To construct Brillouin zone draw perpendicular plane to the point joining vector **OA** and **OB**. We write

$$G = \frac{2\pi}{a} n_x \hat{i}, \, k = k_x \hat{i} \quad (n_x = \pm 1, \pm 2, \pm 3, \ldots.) \tag{5.54}$$

Substituting Eq. (5.54) in Eq. (5.53)

$$\left(\frac{\pi}{a}\hat{i}n_x\right)^2 = \hat{i}k_x \cdot \left(\frac{\pi}{a}\hat{i}n_x\right)$$

$$k_x = \frac{\pi}{a} n_x \tag{5.55}$$

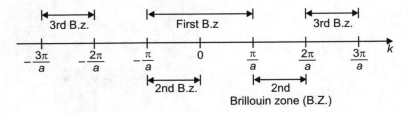

Fig. 5.7 Brillouin zones for a one-dimensional lattice

For various values of n_x, we get different Brillouin zones. For $n_x = \pm 1$, we have first Brillouin zone whose boundaries are $\frac{\pi}{a}$ and $-\frac{\pi}{a}$. For $n_x = \pm 2$, we have zone boundaries $\frac{2\pi}{a}$ and $-\frac{2\pi}{a}$ and so on. The region between $\frac{\pi}{a}$ and $\frac{2\pi}{a}$ and between $-\frac{\pi}{a}$ and $-\frac{2\pi}{a}$ forms second Brillouin zone and so on.

Let us now consider two-dimensional square lattice. The reciprocal lattice vector and $\boldsymbol{k}$ in this case can be written as

$$\boldsymbol{G} = \frac{2\pi}{a}(n_x \hat{\boldsymbol{i}} + n_y \hat{\boldsymbol{j}}), \boldsymbol{k} = k_x \hat{\boldsymbol{i}} + k_y \hat{\boldsymbol{j}} \quad (n_x, n_y = 0, \pm 1, \pm 2, \pm 3, \ldots.) \tag{5.56}$$

From Eqs. (5.53) and (5.56)

$$\left[\frac{\pi}{a}(n_x \hat{\boldsymbol{i}} + n_y \hat{\boldsymbol{j}})\right]^2 = \left(k_x \hat{\boldsymbol{i}} + k_y \hat{\boldsymbol{j}}\right) \cdot \frac{\pi}{a}(n_x \hat{\boldsymbol{i}} + n_y \hat{\boldsymbol{j}})$$

$$k_x n_x + k_y n_y = \frac{\pi}{a}\left(n_x^2 + n_y^2\right) \tag{5.57}$$

For $n_x = \pm 1, n_y = 0$, we have from Eq. (5.57)

$$k_x = \pm \frac{\pi}{a} \tag{5.58}$$

For $n_x = 0, n_y = \pm 1$, we have from Eq. (5.57)

$$k_y = \pm \frac{\pi}{a} \tag{5.59}$$

The relation given by Eqs. (5.58) and (5.59) forms the boundary of the first Brillouin zone of the square lattice as shown in Fig. 5.8.

Now consider $n_x = \pm 1, n_y = \pm 1$, we have from Eq. (5.56)

Fig. 5.8 First, second and third Brillouin zones are marked as I, II, III, respectively, for a two-dimensional square lattice

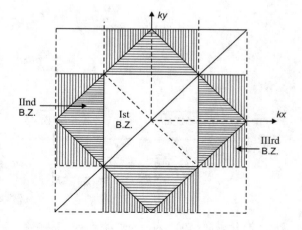

$$G = \pm \frac{2\pi}{a} \left(\hat{i} + \hat{j} \right), k = k_x \hat{i} + k_y \hat{j} \tag{5.60}$$

From Eqs. (5.53) and (5.60)

$$\left[\frac{\pi}{a} \left(\hat{i} + \hat{j} \right) \right]^2 = \pm \left(k_x \hat{i} + k_y \hat{j} \right) \cdot \frac{\pi}{a} \left(\hat{i} + \hat{j} \right)$$

$$\frac{2\pi}{a} = \pm (k_x + k_y) \tag{5.61}$$

dimensional square lattice.

From Eq. (5.61)

$$(k_x + k_y) = \frac{2\pi}{a}, -\frac{2\pi}{a} \tag{5.62}$$

$$k_x - k_y = \frac{2\pi}{a}, -k_x + k_y = \frac{2\pi}{a} \tag{5.63}$$

The area enclosed by these four lines given by Eqs. (5.62) and (5.63) minus the first zone is the second Brillouin zone.

The Brillouin zone for a three-dimensional lattice can be constructed in the same manner. In this case, the lattice vectors are bisected by perpendicular planes, the first Brillouin zone is now the smallest volume enclosed by three planes.

The first Brillouin zone is a unit because it satisfies all the necessary requirements. It has the property that its corresponding lattice point falls at the cell centre unlike the case of direct lattice, in which the lattice points usually lie at the corners of the cell. If first Brillouin zone is translated by the reciprocal lattice vector G, then whole reciprocal lattice space is covered.

5.8 Experimental Methods

There are essentially three experimental techniques used for X-ray diffraction:

(a) The rotating crystal method
(b) The Laue method and
(c) The powder method.

(a) *The Rotating crystal Method*

In this method, crystal whose diameter is about 1 mm is mounted on a spindly which can be rotated. An incident beam of monochromatic X-rays of wavelength λ falls on the crystal. The crystal is then rotated until a diffraction condition is achieved; that is, Bragg condition is satisfied. The diffracted beam that emerges from the crystal is recorded as a spot on the photographic plate. The shape and size of the unit cell as well as arrangement of atoms inside the cell are determined from the diffraction angles and intensities for various orientations.

(b) *The Laue Method*

In this method, a continuous X-ray beam falls on the crystal which have a fixed orientation relative to the incident beam. Flat photographic films are placed in front of and behind the crystal. Since X-rays have a continuous range of wavelengths, the crystal selects that particular wavelength which satisfy Bragg's law, and a diffraction beam emerges at the corresponding angle. The diffraction beam is recorded as a spot on the film. Since the wavelength corresponding to the spot is not measured, therefore only ratio of interplanar spacing can be determined.

(iii) *The Powder Method*

In this method, the monochromatic X-rays are falling on a finely powdered specimen. In a powdered sample, there are a large number of crystallites which are randomly oriented. Any crystallite giving reflection will produce a diffracted beam making an angle 2θ with the incident beam and, therefore, the locus of all such diffracted beams is a cone of semi-angle 2θ with the apex at the sample. Since λ and θ are measured, one can determine lattice parameters quite accurately.

5.9   Structure Factor

The atomic scattering factor or form factor f_s is a measure of the efficiency of an atom in scattering X-rays. It is defined as ratio of the amplitude of electromagnetic wave scattered by an atom and that of a wave scattered by a free electron. Consider an atom whose centre is at O. Let incident and scattered beams make an angle of 2θ with each other as shown in Fig. 5.9.

Consider spherical polar coordinate system (r, θ', φ') such that z-direction is along OC, which is perpendicular to BOA plane. Let us assume that electronic charge distribution is spherically symmetric and density of electrons at a distance r is $\rho(r)$ from O. The number of electrons in a volume element $\mathrm{d}V$ at a distance r from O is

$$\rho(r)\mathrm{d}V = \rho(r)r^2\mathrm{d}r \sin\theta'\mathrm{d}\theta'\mathrm{d}\varphi' \tag{5.64}$$

Since atoms have a finite size they can cause phase difference between the scattered waves due to various electrons. The phase difference between the X-rays scattered by the element of charge and the ray that would be scattered if the same charge were located at

Fig. 5.9 Calculation of the atomic factor

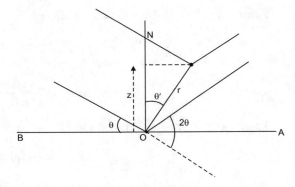

O. The phase difference φ is

$$\varphi = \frac{2\pi}{\lambda}(\text{Path difference}) = \frac{2\pi}{\lambda}\left(\hat{\boldsymbol{k}}' \cdot \boldsymbol{r} - \hat{\boldsymbol{k}} \cdot \boldsymbol{r}\right)$$

$$\varphi = \frac{2\pi}{\lambda}\Delta\hat{\boldsymbol{k}} \cdot \boldsymbol{r} = \Delta\boldsymbol{k} \cdot \boldsymbol{r} \tag{5.65}$$

where $\frac{2\pi}{\lambda}\Delta\hat{\boldsymbol{k}} = \Delta\boldsymbol{k}$.

The scattering amplitude from element dV will be $\rho(r)dV$ times as strong and out of phase by an amount φ with the scattering amplitude from the point electron. The ratio of the complex amplitude of the wave scattered by the element and the amplitude of the wave that would be scattered by the same charge in O is thus simply $\exp(i\varphi)$.

The atomic scattering factor

$$f_s = \int\limits_{r=0}^{\infty}\int\limits_{\theta'=0}^{\pi}\int\limits_{\varphi'=0}^{2\pi} \exp(i\varphi)\rho(r)r^2 \sin\theta' dr d\theta' d\varphi'$$

Substituting the value of φ

$$f_s = \int\limits_{r=0}^{\infty}\int\limits_{\theta'=0}^{\pi}\int\limits_{\varphi'=0}^{2\pi} \exp(i\Delta k \cdot r)\rho(r)r^2 \sin\theta' dr d\theta' d\varphi' \tag{5.66}$$

From the scattering theory

$$\exp(i\Delta k \cdot r) = \sum_{l=0}^{\infty}(2l+1)i^l j_l(\Delta kr)P_l(\cos\theta) \tag{5.67}$$

where P_l are Legendre polynomials, j_l are spherical Bessel function. Assuming that only $l = 0$ component will contribute towards the integral we have

$$P_0 = 1,\ j_0(\Delta kr) = \frac{\sin \Delta kr}{\Delta kr} \tag{5.68}$$

Substituting Eqs. (5.67) and (5.68) in (5.66)

$$f_s = \int\limits_{r=0}^{\infty} \int\limits_{\theta'=0}^{\pi} \int\limits_{\varphi'=0}^{2\pi} \frac{\sin \Delta kr}{\Delta kr} \rho(r) r^2 \sin \theta' dr d\theta' d\varphi'$$

$$f_s = 4\pi \int \rho(r) \frac{\sin \Delta kr}{\Delta kr} r^2 dr \tag{5.69}$$

For $\theta \to 0$, $\Delta k \to 0$ and hence $\frac{\sin \Delta kr}{\Delta kr} \to 1$ and Eq. (5.69) is then

$$f_s = 4\pi \int \rho(r) r^2 dr = Z \tag{5.70}$$

The atomic scattering factor is equal to Z only for $\theta = 0$ and decreases for all other values of θ. Let us consider, the intensity of an X-ray beam diffracted by a crystal with a unit cell of primitive translation vectors a, b, c. Suppose that waves emitted by the atoms at the corners of unit cell satisfied the diffraction conditions (5.42)–(5.44). Let the position of atoms is defined relative to corner atom as origin and is given by

$$r_k = u_k a + v_k b + w_k c \tag{5.71}$$

The phase difference between the beam scattered by atom k and the one scattered by the atom at the origin is given by

$$\varphi_k = \frac{2\pi}{\lambda} r_k (n - n') \tag{5.72}$$

where n and n' are unit vectors, respectively, in the direction of the incident and scattered beam. From Eqs. (5.71) and (5.72)

$$\phi_k = 2\pi (u_k e + v_k f + w_k g) \tag{5.73}$$

Let us now define the structural factor F as the ratio of amplitude of the wave scattered by all atoms in a unit cell and that scattered by a free electron for the same incident beam. The complex amplitude produced by atom k is

$$f_{sk} \exp(i\varphi_k) \tag{5.74}$$

where f_{sk} is the atomic scattering factor for atom k. The structure factor is then

$$F = \sum_k f_{sk} \exp(i\varphi_k) = \sum_k f_{sk} \exp[2\pi i (u_k e + v_k f + w_k g)] \tag{5.75}$$

where the summation is over all atoms in the unit cell.

Total diffracted intensity of the beam is due to the combination of the waves scattered by all the atoms in the unit cell. Each atom in the cell scatters with an amplitude proportional to its value of f and with a phase which is dependent on its position in the cell. Consider the sum of waves that arise from all the atoms in the unit cell, of the same wavelength but different amplitudes and phases. Each amplitude has two components $f_{sk} \cos \varphi_k$ and $f_{sk} \sin \varphi_k$, and all the intensity which is proportional to the square of the amplitude then becomes proportional to

$$F^2 = \left(\sum_k f_{sk} \cos \varphi_k \right)^2 + \left(\sum_k f_{sk} \sin \varphi_k \right)^2 \tag{5.76}$$

Solved Examples

Example 1 For *fcc* iron obtains the diffraction angle for the (220) set of planes. The lattice parameter for Fe is 0.2866 nm. The wavelength of X-ray is 0.1790 nm. The order of diffraction is one.

Solution
We have $h = 2, k = 2, l = 0$ and

$$d_{hkl} = \frac{a}{\sqrt{h^2 + k^2 + l^2}} = \frac{0.2866 \, \text{nm}}{\sqrt{2^2 + 2^2 + 0^2}} = \frac{0.2866 \, \text{nm}}{\sqrt{8}} = 0.1013 \, \text{nm}$$

From Bragg's law (for $n = 1$)

$$2d_{hkl} \sin \theta = \lambda$$

$$\sin \theta = \frac{\lambda}{2d_{hkl}} = \frac{0.1790 \, \text{nm}}{2 \times 0.1013 \, \text{nm}} = 0.884$$

$$\theta = \sin^{-1}(0.884) = 62.13°$$

The diffraction angle is 2θ

$$2\theta = 124.26°.$$

Example 2 Show that the reciprocal of simple cubic lattice of edge a is also a simple cubic lattice with a cube of edge $2\pi/a$.

Solution
The volume of cubic cell of edge a is a^3. The primitive lattice vectors of a simple cubic cell are

$$\boldsymbol{a} = a\hat{x}, \boldsymbol{b} = a\hat{y}, \boldsymbol{c} = a\hat{z}$$

Substituting the values of volume, a, b and c in Eqs. (5.7)–(5.9)

$$a^* = \frac{2\pi a^2 (\hat{y} \times \hat{z})}{a^3} = \frac{2\pi}{a}\hat{x}$$

$$b^* = \frac{2\pi a^2 (\hat{z} \times x)}{a^3} = \frac{2\pi}{a}\hat{y}$$

$$c^* = \frac{2\pi a^2 (\hat{x} \times \hat{y})}{a^3} = \frac{2\pi}{a}\hat{z}$$

The a^*, b^* and c^* are again vectors of simple cubic lattice with cell edge $2\pi/a$.

Example 3 Prove that the reciprocal lattice of body-centred cubic lattice is face-centred cubic lattice.

Solution
The primitive lattice vectors of *bcc* lattice (Example 23, Chap. 1) are

$$a' = \frac{a}{2}[\hat{x} + \hat{y} - \hat{z}]$$

$$b' = \frac{a}{2}[-\hat{x} + \hat{y} + \hat{z}]$$

$$c' = \frac{a}{2}[\hat{x} - \hat{y} + \hat{z}]$$

We have

$$b' \times c' = \frac{a^2}{4}[-\hat{x} + \hat{y} + \hat{z}] \times [\hat{x} - \hat{y} + \hat{z}] = \frac{a^2}{4}[\hat{z} + \hat{y} - \hat{z} + \hat{x} + \hat{y} + \hat{x}] = \frac{a^2}{2}[\hat{x} + \hat{y}]$$

Volume V is

$$V = a' \cdot (b' \times c') = \frac{a^3}{4}[\hat{x} + \hat{y} - \hat{z}] \cdot [\hat{x} + \hat{y}] = \frac{a^3}{2}$$

Substituting the values of volume $V = a^3/2$, a', b' and c' in Eqs. (5.7)–(5.9)

$$a^* = \frac{4\pi a^2 [(-\hat{x} + \hat{y} + \hat{z}) \times (\hat{x} - \hat{y} + \hat{z})]}{4a^3} = \frac{\pi}{a}[\hat{x} + \hat{y}]$$

Similarly

$$b^* = \frac{\pi}{a}[\hat{y} + \hat{z}]$$

$$c^* = \frac{\pi}{a}[\hat{x} + \hat{z}]$$

The a^*, b^* and c^* are vectors of face-centred cubic lattice with cell edge π/a.

Example 4 Show that the diffraction condition $\Delta k = G$ leads to Laue equations.

Solution
We have

$$\Delta k = G$$

Suppose

$$G = ha^* + kb^* + lc^*$$

where h, k, l are integers and a^*, b^* and c^* are reciprocal lattice vectors. Taking the scalar product of both Δk and G with a, b and c, we have

$$a \cdot \Delta k = a \cdot \left(ha^* + kb^* + lc^*\right) = 2\pi h \tag{5.77}$$

Similarly

$$b \cdot \Delta k = b \cdot \left(ha^* + kb^* + lc^*\right) = 2\pi k \tag{5.78}$$

$$c \cdot \Delta k = c \cdot \left(ha^* + kb^* + lc^*\right) = 2\pi l \tag{5.79}$$

Equations (5.77)–(5.79) are Laue equations. Equation (5.77) tells that Δk lies on a certain cone about the direction of a. Equation (5.78) tells that Δk lies on a cone about b as well, and Eq. (203) requires that Δk lies on a cone about c. Thus, a reflection Δk must satisfy all three equations; it must lie at the common line of intersection of three. Refletion from a single plane of atoms takes place in the direction of the line of intersection of two cones. The two cones will in general intercept each other provided the wave vector of the particle in the incident beam exceeds some threshold value determined by the first two Laue equations.

Example 5 The primitive translation vectors of the hexagonal space lattice may be taken as

$$a = \left(\frac{a\sqrt{3}}{2}\right)\hat{x} + \frac{a}{2}\hat{y}, \; b = -\left(\frac{a\sqrt{3}}{2}\right)\hat{x} + \frac{a}{2}\hat{y}, \; c = c\hat{z}$$

Find the primitive translations of the reciprocal lattice.

Solution
Volume of the cell is

$$a \cdot (b \times c) = \left[\left(\frac{a\sqrt{3}}{2}\right)\hat{x} + \frac{a}{2}\hat{y}\right] \cdot \left[\left(-\left(\frac{a\sqrt{3}}{2}\right)\hat{x} + \frac{a}{2}\hat{y}\right) \times c\hat{z}\right]$$

$$a \cdot (b \times c) = \left[\left(\frac{a\sqrt{3}}{2}\right)\hat{x} + \frac{a}{2}\hat{y}\right] \cdot \left[\frac{ac}{2}\hat{x} + ac\frac{\sqrt{3}}{2}\hat{y}\right] = \frac{a^2 c\sqrt{3}}{2}$$

The primitive reciprocal lattice vectors are

$$a^* = \frac{2\pi(b \times c)}{a \cdot (b \times c)} = 4\pi\frac{\left[\frac{ac}{2}\hat{x} + ac\frac{\sqrt{3}}{2}\hat{y}\right]}{a^2 c\sqrt{3}} = \frac{2\pi}{a}\left[\frac{\hat{x}}{\sqrt{3}} + \hat{y}\right]$$

Similarly,

$$b^* = \frac{2\pi}{a}\left[-\frac{\hat{x}}{\sqrt{3}} + \hat{y}\right], c^* = \frac{2\pi}{c}\hat{z}$$

Thus lattice is its own reciprocal, but with a rotation of axes.

Example 6 Show that $\boldsymbol{G}_{hkl}$ is normal to (hkl) crystal plane.

Solution
Let PQR is (hkl) plane (Fig. 5.10). The Miller indices are hkl. From Fig. 5.10.

$$\frac{a}{h} + u = \frac{b}{k}$$

$$u = \frac{b}{k} - \frac{a}{h}$$

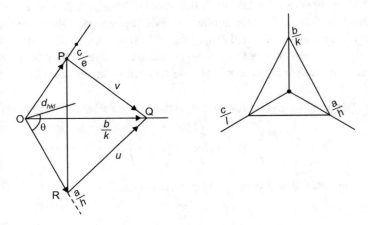

Fig. 5.10 Reciprocal lattice vector $\boldsymbol{G}_{hkl}$ is normal to the plane (hkl)

Similarly

$$v = \frac{b}{k} - \frac{c}{l}$$

Taking dot product of reciprocal lattice vector $G(=G_{hkl})$ with u

$$G \cdot u = \left(ha^* + kb^* + lc^*\right) \cdot \left(\frac{b}{k} - \frac{a}{h}\right) = b^*b - a^*a = 2\pi - 2\pi = 0$$

Similarly

$$G \cdot v = \left(ha^* + kb^* + lc^*\right) \cdot \left(\frac{b}{k} - \frac{c}{l}\right) = b^*b - c^*c = 2\pi - 2\pi = 0$$

Thus

$$G \cdot u = 0, \, G \cdot v = 0$$

G is perpendicular to u and v. U and v are different directions but are in the same (hkl) plane. Therefore, G is perpendicular to (hkl) plane.

Example 7 Show that

$$d_{hkl} = \frac{2\pi}{|G_{hkl}|}$$

Solution
Let $\hat{n}$ is unit vector normal to hkl plane. Therefore, $\hat{n}$ would be parallel to G $(=G_{hkl})$. Thus

$$G = G\hat{n}$$

$$\hat{n} = \frac{G}{|G|} = \frac{ha^* + kb^* + lc^*}{|G|}$$

Let n makes an angle φ with (a/h) vector then projection of (a/h) along n is

$$\frac{a}{h} \cdot \hat{n}$$

This distance is equal to distance d_{hkl} from O as shown in Fig. 5.10.

$$d_{hkl} = \frac{a}{h} \cdot \hat{n} = \frac{a}{h} \cdot \frac{G}{|G|} = \frac{a}{h} \cdot \frac{ha^* + kb^* + lc^*}{|G|} = \frac{aa^*}{|G|} = \frac{2\pi}{|G|} = \frac{2\pi}{|G_{hkl}|}$$

Example 8 Find the structure factor for the *bcc* lattice.

Solution

In *bcc* lattice, the atoms are at (0,0,0) and (1/2,1/2,1/2). Substituting these values of atomic coordinates in Eq. (5.75), we have

$$F = f \exp[2\pi i (0 + 0 + 0)] + f \exp[i\pi(e + f + g)]$$

$$F = f + f \exp[i\pi(e + f + g)]$$

For even values of $(e + f + g)$

$$F = f(1 + 1) = 2f$$

$$F^2 = FF^* = 4f^2$$

For odd values of $(e + f + g)$, $\exp[i\pi(e + f + g)] = -1$

$$F = f + f \exp[i\pi(\text{odd value})] = f(1 - 1) = 0$$

$$F^2 = FF^* = 0$$

Thus, the diffraction pattern does not contain lines such as (100), (300) or (111) but lines such as (200), (110), (222) will be present.

Example 9 Find the structure factor for the *fcc* lattice.

Solution

In *fcc* lattice, the atoms are at (0,0,0) and (1/2,1/2,0). (1/2,0,1/2) and (0,1/2,1/2). Substituting these values of atomic coordinates in Eq. (5.75), we have

$$F = f \exp[2\pi i (0 + 0 + 0)] + f \exp[i\pi(e + f)] + f \exp[i\pi(e + g)] + f \exp[i\pi(f + g)]$$

If all the integers are even, then

$$\exp[i\pi(e + f)] = \exp[i\pi(e + g)] = \exp[i\pi(f + g)] = 1$$

and

$$F = f + f + f + f = 4f$$

$$F^2 = FF^* = 4f4f^* = |4f|^2$$

If all the integers are odd, then

$$\exp[i\pi(e+f)] = \exp[i\pi(e+g)] = \exp[i\pi(f+g)] = 1$$

and

$$F = f + f + f + f = 4f$$

$$F^2 = FF^* = 4f4f^* = |4f|^2$$

If one integer is odd and two even (e is odd and f and g are even), then

$$\exp[i\pi(e+f)] = \exp[i\pi(e+g)] = -1$$

$$\exp[i\pi(f+g)] = 1$$

and

$$F = f\exp[2\pi i(0+0+0)] + f\exp[i\pi(e+f)] + f\exp[i\pi(e+g)] + f\exp[i\pi(f+g)]$$

is

$$F = f - f - f + f = 0.$$

$$F^2 = FF^* = 0$$

Similarly if two integers are odd and one even, we have $F = 0$ and

$$F^2 = FF^* = 0.$$

Thus, the diffraction pattern does not contain lines such as (100), (300) or (211) but lines such as (200), (111), (222) will be present.

Example 10   Find the structure factor for the CsCl crystal structure.

Solution
In CsCl crystal structure, the Cs ions are at $(0,0,0)$ and Cl ions are at $(1/2,1/2,1/2)$. Substituting these values of atomic coordinates in Eq. (5.75), we have

$$F = f_{Cs}\exp[2\pi i(0+0+0)] + f_{Cl}\exp[i\pi(e+f+g)]$$

For $e + f + g =$ even integer, we have

$$F = f_{Cs} + f_{Cl}$$

$$F^2 = FF^* = |f_{Cs} + f_{Cl}|^2$$

For $e + f + g = $ odd integer, we have

$$F = f_{Cs} - f_{Cl}$$

$$F^2 = FF^* = |f_{Cs} - f_{Cl}|^2.$$

Objective Type Questions

1. Which cannot be used for determining the crystal structure?
    (a) X-rays (b) electrons (c) neutrons (d) visible light
2. The wavelength of the X-rays for structural studies should be
    (a) ~1 Å (b) 100 Å (c) 1000 Å (d) 2000 Å
3. X-rays in the crystal interact with
    (a) Nucleus (b) orbital electrons (c) nucleus and orbital electrons (d) magnetic field produced by electrons
4. Neutron diffraction is useful in diffraction studies for
    (a) Light atoms (b) heavy atoms (c) compounds having non-magnetic atoms (d) none of these
5. In crystal structure determination with electron beam, the electron interacts with
    (a) Nucleus (b) orbital electrons (c) electric field produced by atoms (d) none of these
6. Which diffraction method is most useful for surface and thin film studies?
    (a) X-rays diffraction (b) electron diffraction (c) neutron diffraction (d) none of these
7. Magnetic crystals can be studied by neutron diffraction because neutron is
    (a) Electrically neutral (b) have a positive charge (c) have a magnetic moment (d) none of these
8. X-rays are scattered from
    (a) Atomic nuclei (b) phonons (c) electrons (d) electrostatic potential
9. What is kept stationary/constant with Laue method ?
    (a) Crystal (b) wavelength (c) both (d) nothing
10. What is the behaviour of f for X-rays when the angle between the incoming and outgoing beam increases? The scattering factor
    (a) Decreases (b) increases (c) stays constant
11. The reciprocal lattice of fcc is
    (a) fcc (b) bcc (c) sc (d) none of these
12. The reciprocal lattice of sc is
    (a) fcc (b) bcc (c) sc (d) none of these
13. The X-ray K_α lines come from the transition between
    (a) L → K (b) M → K (c) N → K (d) none of these

14. The shortest wavelength present in the radiation from X-ray tube whose accelerating potential is 50 keV is
    (a) 0.248 nm (b) 0.0248 nm (c) 2.48 nm (d) 2.48 mm

15. X-rays are produced when an element of high atomic weight is bombarded by high energy
    (a) Electrons (b) photons (c) neutrons (d) protons

16. Which is true?
    (a) The scattering power of an atom increases as θ increases
    (b) The geometrical factor is not a function of position of each atom in a unit cell.
    (c) The scattering power of an atom decreases as θ increases
    (d) The atomic scattering factor is ratio of the amplitude scattered by an electron to that scattered by the atom

17. The structure factor of the *fcc* lattice vanishes if
    (a) All the indices are even
    (b) All the indices or odd
    (c) One of the indices is odd and other two are even
    (d) None of these

18. The structure factor of the bcc lattice does not vanish if
    (a) All the indices are even
    (b) All the indices are odd
    (c) One of the indices is odd and other two are even
    (d) None of these

19. The structure factor for CsCl vanishes if
    (a) All the indices are even
    (b) All the indices are odd
    (c) One of the indices is odd and other two are even
    (d) None of these

20. Which is not true?
    (a) A diffraction pattern of a crystal is a map of the reciprocal lattice
    (b) Vectors in the reciprocal lattice have dimensions of length while that of direct lattice has dimensions of [1/length]
    (c) A Brillouin zone is defined as a Wigner–Seitz primitive cell in the reciprocal lattice
    (d) The volume of the primitive cell of reciprocal of *fcc* lattice is $4(2\pi/a)^3$

21. Suppose an X-ray tube with a *Cu* anode is used as a source of radiation. A suitable filter material would be
    (a) Cu (b) Co (c) Ni (d) Zr

22. Suppose an X-ray tube with a *Mo* anode is used as a source of radiation. A suitable filter material would be
    (a) Cu (b) Co (c) Ni (d) Zr

23. The reciprocal lattice of monoclinic is
    (a) Monoclinic (b) hexagonal (c) triclinic (d) cubic

24. Which of the following reflection would be missing in a *bcc* lattice
    (a) (211) (b) (110) (c) (111) (d) (200)

Problems

1. Prove that the reciprocal lattice of face-centred cubic lattice is body-centred cubic lattice.
2. Explain the concept of reciprocal lattice vector. Show that the reciprocal lattice for body-centred cubic lattice is a face-centred cubic lattice.
3. What are Laue equations for diffraction of X-rays by a crystalline solid? Show that Bragg's equation is a special case of Laue equation.
4. The (111) reflection line in powder diffraction pattern of KCl is absent, whereas it is present for NaCl lattice. Explain.
5. Discuss X-ray diffraction and obtain Bragg's law ($2\ k.G + G^2 = 0$) and show that its geometrical interpretation leads to the concept of Brillouin zones in crystals.
6. Find the geometrical structure factor for *fcc* structure in which all atoms are identical. Hence show that for *fcc* lattice no reflections can occur for which the indices are partly even and partly odd.
7. Calculate the geometrical structure factor for a *bcc* lattice. Name some important planes which will be missing from the X-ray diffraction pattern.
8. Name the three most important probes used in diffraction experiments on crystals. What is the one essential condition that they must all satisfy? Describe briefly what each probe is suitable for.

Answers
Objective Type Questions

1. (d)	2. (a)	3. (b)	4. (a)	5. (c)	6. (b)	7. (c)	8. (c)
9. (a)	10. (a)	11. (a)	12. (c)	13. (a)	14. (b)	15. (a)	16. (c)
17. (c)	18. (c)	19. (d)	20. (b)	21. (c)	22. (d)	23. (a)	24. (c)

Lattice Vibrations

6

6.1   Phonons

At all temperatures above 0 K, the atoms vibrate about their mean equilibrium position by absorption of thermal energy. In harmonic approximation, it is assumed that amplitude of these vibrations is small compared to interatomic spacing and interatomic forces obey Hook's law. Under these conditions, the vibrations of atoms in a solid are analogous to those of harmonic oscillator. In Chap. 4, it is shown that the energy of the simple harmonic oscillator is quantized and is given by

$$E = \left(n + \frac{1}{2}\right)\hbar\omega \quad (n = 0, 1, 2, \ldots)$$

Due to the analogy between harmonic oscillators and lattice vibrations, the name phonon has been given for the quanta of lattice vibration or quanta of thermal energy absorbed or emitted by the solid. The concept of phonon is analogous with photon, the quanta of electromagnetic radiation. Phonons have dual nature like photons and are indistinguishable particles. The phonons are classified as acoustical or optical phonons. They are associated with acoustical and optical modes of vibrations, respectively. At low temperature, acoustical modes are more populated than optical phonons. The relationship between the energy E of a phonon, the angular frequency ω and wave vector q (or k) is

$$E = \hbar\omega$$
$$\omega = v_s|q|$$

where v_s is the velocity of the sound wave. The phonon is a boson and obeys Bose–Einstein statistics. The occupation number of bosons in a state with energy E is

© The Author(s) 2022
V. K. Jain, *Solid State Physics*, https://doi.org/10.1007/978-3-030-96017-9_6

$$n(E) = \frac{1}{\exp\left(\frac{E}{k_B T}\right) - 1} = \frac{1}{\exp\left(\frac{\hbar\omega}{k_B T}\right) - 1}$$

The number depends on the temperature. At $T = 0$ K, $n(E) = 0$, but as T increases n (E) also increases, eventually reaching the value $k_B T/\hbar\omega$ at high temperatures. Phonons are created by raising the temperature, and therefore, their number in the system is not conserved. The phonon interacts with other particles and has a momentum $\hbar\boldsymbol{q}$. The functional dependence of the energy $\hbar\omega$ on $\hbar\boldsymbol{q}$ (or $\hbar\boldsymbol{k}$) or the angular frequency ω on the phonon wavevector $\boldsymbol{q}$ (or $\overrightarrow{k}$) is called phonon dispersion relation.

The phonons are mainly responsible for the thermal properties of solids like thermal conductivity, heat capacity, and thermal expansion. Phonons interact with each other. Phonons also interact with defects like vacancies, dislocation as well as with conduction electrons. The energy and momentum are conserved in phonon collision. The primary mechanism responsible for electrical resistance of metal and semiconductors is interaction of electrons with phonons. The emission of phonons by excited atoms and molecules in solids makes it possible for non-radiative electronic transition to occur. Phonons, in a relaxation process in solids, usually serve as a sink for energy stored by the other degrees of freedom. The phonons play an important role in the phenomenon of superconductivity.

6.2   The One-Dimensional Monoatomic Lattice

Consider a linear chain of identical atoms, of mass M spaced at a distance a, the lattice constant, connected by ideal Hook's law spring. When the lattice is equilibrium, each atom is positioned exactly at its lattice site. Suppose the lattice begins to vibrate, so that each atom is displaced from its site by a small amount. Because the atoms interact with each other, the various atoms move simultaneously (Fig. 6.1).

Let

U_n = displacement of atom n from its equilibrium position
U_{n-1} = displacement of atom $n - 1$ from its equilibrium position
U_{n+1} = displacement of atom $n + 1$ from its equilibrium position
The force exerted on nth atom as a result of its interaction with $(n + 1)$th atom is

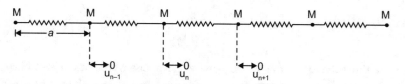

Fig. 6.1   One-dimensional monoatomic lattice

$$C(U_{n+1} - U_n)$$

where $(U_{n-1} - U_n)$ is relative displacement, C is interatomic force constant (the assumption that the force is proportional to relative displacement is known as the harmonic approximation and it holds provided the displacement is small).

The force exerted on nth atom by $(n-1)$th atom is

$$(U_{n-1} - U_n)$$

Applying Newton's second law to nth atom

$$M\frac{d^2U_n}{dt^2} = C(U_{n+1} - U_n) + C(U_{n-1} - U_n)$$
$$= C(U_{n+1} - 2U_n + U_{n-1}) \tag{6.1}$$

We have neglected the interaction of nth atom with all but its nearest neighbours. Let us consider the solution of the form

$$U_n = A\exp[i(kx - \omega t)] \tag{6.2}$$

where x is equilibrium position of the nth atom; that is, $x = na$

From Eq. (6.2)

$$U_n = A\exp[i(kna - \omega t)] \tag{6.3}$$

$$U_n = A\exp[i(k(n+1)a - \omega t)] \tag{6.4}$$

$$U_n = A\exp[i(k(n-1)a - \omega t)] \tag{6.5}$$

Substituting Eqs. (6.3)-(6.5) in Eq. (6.1)

$$-M\omega^2 A\exp[i(kna - \omega t)] = AC\exp(ikna)[\exp(i(ka - \omega t))$$
$$- 2\exp(-i\omega t) + \exp(-i(ka + \omega t)] \tag{6.6}$$

$$-M\omega^2 A\exp[i(kna - \omega t)] = AC\exp(i(kna - \omega t)[\exp(ika) - 2 + \exp(-ika)]$$
$$- M\omega^2 = C[(\exp](ika) + \exp(-ika)) - 2]$$
$$= C\left[\exp\left(\frac{ika}{2}\right) - \exp\left(-\frac{ika}{2}\right)\right]^2 \tag{6.7}$$

But

$$\sin x = \frac{\exp(ix) - \exp(-ix)}{2i}$$

$$2i \sin x = \exp(ix) - \exp(-ix) \tag{6.8}$$

Substituting Eq. (6.8) in Eq. (6.7)

$$-M\omega^2 = C\left[2i \sin \frac{ka}{2}\right]^2 = -4C \sin^2 \frac{ka}{2}$$

$$\omega^2 = \frac{4C}{M} \sin^2 \frac{ka}{2}$$

$$\omega = \pm\sqrt{\frac{4C}{M}} \sin \frac{ka}{2} = \sqrt{\frac{4C}{M}}\left|\frac{\sin ka}{2}\right| \tag{6.9}$$

The dispersion relation (ω versus k) is shown in Fig. 6.2.
Let us now consider

(a) *Long wavelength limit ($k \to 0$) or at low frequencies*

From Eq. (6.9)

$$\omega = \sqrt{\frac{4C}{M}}\left|\sin \frac{ka}{2}\right|$$

$$= \omega_m\left|\sin \frac{ka}{2}\right| \tag{6.10}$$

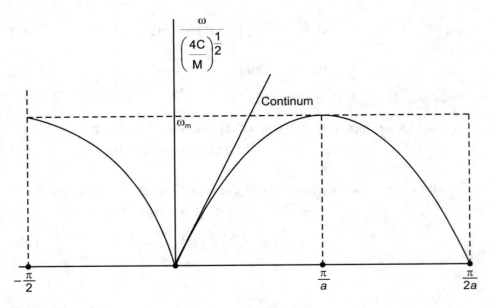

Fig. 6.2 Dispersion curve

where $\omega_m = \sqrt{\frac{4C}{M}}$. In the long wave limit $k \to 0$, hence

$$\sin \frac{ka}{2} \to \frac{ka}{2}$$

Equation (6.10) is

$$\omega = \omega_m \frac{ka}{2} \qquad (6.11)$$

The relation between ω and k is linear. This is expected because in this limit, the lattice behaves as an elastic continuum. From Eq. (6.10), the values of ω are between 0 and ω_m corresponding to k values between 0 and π/a. Thus, $0 < k < \pi/a$ and we have $0 < \omega < \omega_m$. Thus, these and only these frequencies will be transmitted by the lattice, while other frequencies will be strongly attenuated. The lattice therefore acts as a low-pass mechanical filter.

From Eq. (6.11)

$$\omega = \omega_m \frac{ka}{2} = v_s k \qquad (6.12)$$

The velocity of sound wave is given by

$$v_s = \omega_m \frac{a}{2} \qquad (6.13)$$

As k increases, the dispersion curve begins to deviate from the straight line and bends downward. Eventually, the curve saturates at $k = \pi/a$ with a maximum frequency equal to ω_m.

(b) *Phase and Group Velocity*

For an arbitrary dispersion relation, phase velocity is given by

$$v_p = \frac{\omega}{k} \qquad (6.14)$$

and group velocity

$$v_g = \frac{d\omega}{dk} \qquad (6.15)$$

v_p is the velocity of propagation of a pure wave of an exactly specified frequency ω and wave vector k while v_g describes the velocity of a wave pulse whose average frequency and wave vector are specified by ω and k. At higher frequencies, the group velocity and phase velocity are no longer equal. From Eqs. (6.10) and (6.15)

Fig. 6.2 **a** Variation of $\left(v_g/\sqrt{\frac{Ca^2}{M}}\right)$ versus k

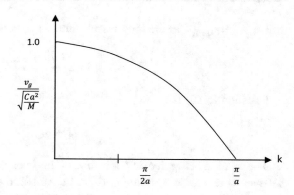

$$v_g = \frac{d\omega}{dk} = \frac{\omega_m a}{2}\cos\frac{ka}{2} = v_s \cos\frac{ka}{2} \tag{6.16}$$

Figure 6.2a shows the plot of $\left(v_g/\sqrt{\frac{Ca^2}{M}}\right)$ versus k

If $k \to 0$, $\cos\frac{ka}{2} \to 1$ and Eq. (6.16) is

$$v_g \to v_s \tag{6.17}$$

If $k = \pi/a$

$$\cos\frac{ka}{2} \to 0$$

$$v_g \to 0 \tag{6.18}$$

From Eqs. (6.10) and (6.14)

$$v_p = \frac{\omega}{k} = \frac{\omega_m}{k}\sin\frac{ka}{2} \tag{6.19}$$

If $k \to 0$, then from Eq. (6.19)

$$v_p \to \frac{\omega_m}{k}\frac{ka}{2} = \frac{\omega_m}{2}a \to v_s \tag{6.20}$$

If $k = \pi/a$, Eq. (6.19) is

$$v_p = \frac{\omega_m a}{\pi}\sin\frac{\pi a}{2a}$$

$$= \frac{\omega_m}{\pi}a = \frac{2\omega_m}{2\pi}a = \frac{2v_s}{\pi} \tag{6.21}$$

v_p represents the transfer of a signal or of the energy. Thus, at $k = \pi/a$ no signal or energy is being propagated and wave is a standing wave. At $k = \pi/a$, the group velocity vanishes. Thus,

$$\frac{2\pi}{\lambda} = \frac{\pi}{a}$$

$$\lambda = 2a \tag{6.22}$$

If $\lambda = 2a$, the adjacent atoms (which are at a distance a apart, that is, $a = \lambda/2$) are always 180° out of phase. When the wavelets reflected from B reaches that reflected from A, the two are in phase (Fig. 6.3). It follows that at $k = \pi/a$ all the scattered wavelets interfere constructively, and consequently, the reflection is at its maximum. Thus, $v_g = 0$ at $k = \pi/a$, it is because the reflected wave is so strong that, when combined with the incident wave, it leads to standing wave, which of course, has a vanishing group velocity. For a wavelength less than $\lambda = 2a$, for example, a the adjacent atoms would be in phase; this would not be a wave motion at all but a translation of the entire chain. Wavelength between a and $2a$ and less than a is not represented at all, since there is nothing to crest at these separations. Such wavelengths are actually identical with wavelengths greater than $2a$ as wave displacements are defined only at lattice points.

(iii) *First Brillouin Zone*

If we substitute $k \rightarrow k + \frac{2\pi n}{a}$ ($n = 1, 2, \ldots$) in Eq. (6.10), the frequencies corresponding to wave vector k_n are

$$\omega = \omega_m \left| \sin\left(k + \frac{2\pi n}{a} \right) \frac{a}{2} \right|$$

$$= \omega_m \left| \sin\left(\frac{ka}{2} + \pi n \right) \right|$$

$$= \omega_m \left| \sin\left(\frac{ka}{2} \right) \right| \tag{6.23}$$

This is identical with that of k. Therefore to obtain unique relationship between the wave vector k and the state of vibration of lattice, the wave vector must be confined to a range of values $2\pi/a$. Generally, one selects the range such that

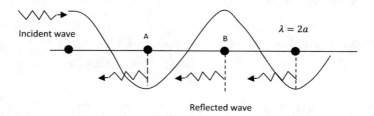

Fig. 6.3   Bragg reflection of lattice waves. The group velocity is zero at $= 2a$

$$-\frac{\pi}{a} \le k \le \frac{\pi}{a} \tag{6.24}$$

The region of k values defined by Eq. (6.24) is the first Brillouin zone of the monoatomic linear chain.

6.3   The Linear (One Dimensional) Diatomic Lattice

Consider a diatomic lattice in one dimension (Fig. 6.4). The distance between nearest neighbours is denoted by a. The particles are numbered in such a way that even numbers have a mass M_1 and odd ones have M_2.

Suppose that each atom interacts only with its nearest neighbours and force constants are identical between all pairs of nearest neighbour atoms. The equation of motion is

$$M_1 \frac{d^2 U_{2n}}{dt^2} = C(U_{2n+1} + U_{2n-1} - 2U_{2n}) \tag{6.25}$$

$$M_2 \frac{d^2 U_{2n+1}}{dt^2} - C(U_{2n} + U_{2n+2} - 2U_{2n+1}) \tag{6.26}$$

Let the solution is in the form of a travelling wave given by

$$U_{2n} = A \exp[i(2nka - \omega t)] \tag{6.27}$$

$$U_{2n+1} = B \exp[i(2n+1)ka - \omega t)] \tag{6.28}$$

$$U_{2n-1} = B \exp[i(2n-1)ka - \omega t)] \tag{6.29}$$

where k is the wave vector of a particular mode of vibration; A and B are the amplitude corresponding to particles of mass M_1 and M_2, respectively. Substituting Eqs. (6.27)–(6.29) in Eqs. (6.25) and (6.26)

$$-M_1 A\omega^2 \exp[i(2nka - \omega t)] = C \exp[i(2nka - \omega t)][B \exp(ika) + B \exp(-ika) - 2A] \tag{6.30}$$

$$-M_2 B\omega^2 \exp[i(2n+1)ka - \omega t] = C \exp[i(2n+1)ka - \omega t)]$$
$$[A \exp(-ika) + A \exp(ika) - 2B] \tag{6.31}$$

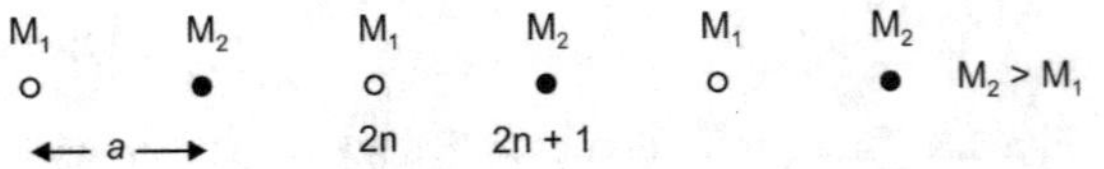

Fig. 6.4 One-dimensional diatomic lattice

$$M_1 \quad M_2 \quad M_1 \quad M_2 \quad M_1 \quad M_2$$
$$\circ \quad \bullet \quad \circ \quad \bullet \quad \circ \quad \bullet \quad M_2 > M_1$$
$$\leftarrow a \rightarrow \quad 2n \quad 2n+1$$

or

$$-M_1 A \omega^2 = C[2B \cos ka - 2A] \tag{6.32}$$

$$-M_2 B \omega^2 = C[2A \cos ka - 2B] \tag{6.33}$$

or

$$(M_1 \omega^2 - 2C)A + 2BC \cos ka = 0 \tag{6.34}$$

$$(M_2 \omega^2 - 2C)B + 2AC \cos ka = 0 \tag{6.35}$$

The system has non-vanishing solution for A and B only if the determinant of the coefficients A and B vanishes

$$\begin{vmatrix} M_1 \omega^2 - 2C & 2C \cos ka \\ 2C \cos ka & M_2 \omega^2 - 2C \end{vmatrix} = 0 \tag{6.36}$$

or

$$(M_1 \omega^2 - 2C)(M_2 \omega^2 - 2C) - 4C^2 \cos^2 ka = 0 \tag{6.37}$$

$$M_1 M_2 \omega^4 - 2C \omega^2 (M_1 + M_2) + 4C^2 (1 - \cos^2 ka) = 0 \tag{6.38}$$

$$M_1 M_2 \omega^4 - 2C \omega^2 (M_1 + M_2) + 4C^2 \sin^2 ka = 0 \tag{6.39}$$

$$\omega^4 - 2C \left(\frac{M_1 + M_2}{M_1 M_2} \right) \omega^2 + \frac{4C^2}{M_1 M_2} \sin^2 ka = 0 \tag{6.40}$$

$$\omega^2 = C \left(\frac{M_1 + M_2}{M_1 M_2} \right) \pm C \sqrt{ \left(\frac{M_1 + M_2}{M_1 M_2} \right)^2 - \frac{4 \sin^2 ka}{M_1 M_2} } \tag{6.41}$$

For small value of ka, that is, $ka \ll 1$, $\sin ka = ka$

$$\omega^2 = C \left(\frac{M_1 + M_2}{M_1 M_2} \right) \pm C \sqrt{ \left(\frac{M_1 + M_2}{M_1 M_2} \right)^2 - \frac{4(ka)^2}{M_1 M_2} } \tag{6.42}$$

$$\omega_+^2 = C \left(\frac{1}{M_1} + \frac{1}{M_2} \right) + C \left(\frac{1}{M_1} + \frac{1}{M_2} \right) \sqrt{ 1 - \frac{4(ka)^2}{M_1 M_2 \left(\frac{1}{M_1} + \frac{1}{M_2} \right)^2} } \tag{6.43}$$

$$\omega_+^2 \approx C\left(\frac{1}{M_1} + \frac{1}{M_2}\right) + C\left(\frac{1}{M_1} + \frac{1}{M_2}\right)\left[1 - \frac{4(ka)^2}{2M_1 M_2\left(\frac{1}{M_1} + \frac{1}{M_2}\right)^2}\right] \qquad (6.44)$$

$$\omega_+^2 \approx 2C\left(\frac{1}{M_1} + \frac{1}{M_2}\right) - \frac{2C(ka)^2}{M_1 M_2\left(\frac{1}{M_1} + \frac{1}{M_2}\right)} \qquad (6.45)$$

$$= 2C\left(\frac{1}{M_1} + \frac{1}{M_2}\right)$$

$$\omega_+ = \sqrt{2C\left(\frac{1}{M_1} + \frac{1}{M_2}\right)} \qquad (6.46)$$

$$\omega_-^2 \approx C\left(\frac{1}{M_1} + \frac{1}{M_2}\right) - C\left(\frac{1}{M_1} + \frac{1}{M_2}\right)\left[1 - \frac{4(ka)^2}{2M_1 M_2\left(\frac{1}{M_1} + \frac{1}{M_2}\right)^2}\right] \qquad (6.47)$$

$$\omega_- = ka\sqrt{\frac{2C}{M_1 + M_2}} \qquad (6.48)$$

The frequencies given by Eqs. (6.45) and (6.47) are referred to as two branches of the dispersion curve. Figure 6.5 shows the variation of ω versus k.

6.4   Acoustical and Optical Branches

Equation (6.45) referred as optical branch while Eq. (6.47) referred as acoustical branch. As $k \to 0$

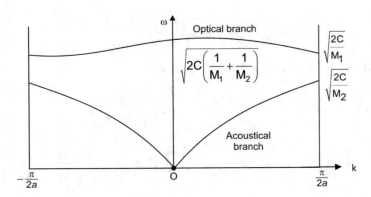

Fig. 6.5 Optical and acoustical branches of the dispersion relation for a diatomic linear lattice

$$\omega_+ \to \sqrt{2C\left(\frac{1}{M_1} + \frac{1}{M_2}\right)}$$

and

$$\omega_- \to 0$$

At $k = 2\pi/a$

$$\omega_+^2 = C\left(\frac{M_1 + M_2}{M_1 M_2}\right) + C\sqrt{\left(\frac{M_1 + M_2}{M_1 M_2}\right)^2 - \frac{4\sin^2\frac{\pi}{2a}a}{M_1 M_2}}$$

$$\omega_+^2 = C\left(\frac{M_1 + M_2}{M_1 M_2}\right) + C\sqrt{\left(\frac{M_1 + M_2}{M_1 M_2}\right)^2 - \frac{4}{M_1 M_2}}$$

$$\omega_+^2 = C\left(\frac{1}{M_1} + \frac{1}{M_2}\right) + C\sqrt{\left(\frac{1}{M_1} - \frac{1}{M_2}\right)^2} = C\left(\frac{1}{M_1} + \frac{1}{M_2}\right) + C\left(\frac{1}{M_1} - \frac{1}{M_2}\right)$$

$$\omega_+^2 = \frac{2C}{M_1}$$

$$\omega_+ = \sqrt{\frac{2C}{M_1}} \tag{6.49}$$

and

$$\omega_-^2 = C\left(\frac{M_1 + M_2}{M_1 M_2}\right) - C\sqrt{\left(\frac{M_1 + M_2}{M_1 M_2}\right)^2 - \frac{4\sin^2\frac{\pi}{2a}a}{M_1 M_2}}$$

$$\omega_-^2 = C\left(\frac{M_1 + M_2}{M_1 M_2}\right) - C\sqrt{\left(\frac{M_1 + M_2}{M_1 M_2}\right)^2 - \frac{4}{M_1 M_2}}$$

$$\omega_-^2 = C\left(\frac{1}{M_1} + \frac{1}{M_2}\right) - \sqrt{\left(\frac{1}{M_1} - \frac{1}{M_2}\right)^2}$$

$$= C\left(\frac{1}{M_1} + \frac{1}{M_2}\right) - C\left(\frac{1}{M_1} - \frac{1}{M_2}\right) = \frac{2C}{M_2}$$

$$\omega_- = \sqrt{\frac{2C}{M_2}} \tag{6.50}$$

The allowed frequencies of propagation are split into an upper branch known as the optical branch and a lower branch called acoustical branch. There is a band of frequencies between the two branches that cannot propagate. The width of this band depends on the difference of masses; if the two masses are equal, the two branches join (become degenerate) at $k = 2\pi/a$. One speaks here of a frequency gap. Therefore, the diatomic lattice acts as a band pass mechanical filter.

Ratio of Amplitudes for Two Branches
For optical branch:
   From Eq. (6.34)

$$\left(M_1\omega^2 - 2C\right)A + 2BC = 0$$

$$\frac{\omega^2}{2C} = \frac{B - A}{M_1 A} \tag{6.51}$$

From Eq. (6.35)

$$\left(M_2\omega^2 - 2C\right)B + 2AC = 0$$

$$\frac{\omega^2}{2C} = \frac{A - B}{M_2 B} \tag{6.52}$$

From Eqs. (6.51) and (6.52)

$$\frac{A - B}{M_2 B} = \frac{B - A}{M_1 A}$$

$$\frac{A}{B} = -\frac{M_2}{M_1} \tag{6.53}$$

The masses move opposite to each other; the light mass amplitude is greater, and the centre of mass remains stationary. If $M_1 = M_2$, $A/B = -1$ for the optical branch. For acoustic branch, as $k \rightarrow 0$

$$\cos ka \approx 1 - \frac{(ka)^2}{2} \tag{6.54}$$

We cannot neglect second-order term in this case. From Eqs. (6.34) and (6.35)

$$\left(M_1\omega^2 - 2C\right)A + 2BC\left(1 - \frac{(ka)^2}{2}\right) = 0 \tag{6.55}$$

and

$$\left(M_2\omega^2 - 2C\right)B + 2AC\left(1 - \frac{(ka)^2}{2}\right) = 0 \tag{6.56}$$

Equations (6.55) and (6.56) take the form

$$M_1\omega^2 A = 2AC - 2BC\left(1 - \frac{(ka)^2}{2}\right) \tag{6.57}$$

and

$$M_2\omega^2 B = 2BC - 2AC\left(1 - \frac{(ka)^2}{2}\right) \tag{6.58}$$

From Eqs. (6.48) and (6.57)

$$\frac{2(ka)^2 CAM_1}{(M_1 + M_2)}A = 2AC - 2BC\left(1 - \frac{(ka)^2}{2}\right) \tag{6.59}$$

From Eqs. (6.48) and (6.58)

$$\frac{2(ka)^2 CBM_2}{(M_1 + M_2)} = 2BC - 2AC\left(1 - \frac{(ka)^2}{2}\right) \tag{6.60}$$

On adding Eqs. (6.59) and (6.60)

$$\frac{2(ka)^2 C}{(M_1 + M_2)}(M_1 A + M_2 B) = \frac{2(ka)^2}{2}(A + B)C$$

$$\frac{(M_1 A + M_2 B)}{(M_1 + M_2)} = \frac{(A + B)}{2}$$

$$2(M_1 A + M_2 B) = (M_1 + M_2)(A + B)$$

$$(M_1 - M_2)A = (M_1 - M_2)B$$

$$\frac{A}{B} = 1 \tag{6.61}$$

Even if $M_1 \neq M_2$, both masses have the same amplitude, direction and phase in acoustic mode. In optical mode, the neighbouring atoms vibrate oppositely to each other; in acoustic mode, they go in the same direction (Fig. 6.6).

The acoustical branch can be stimulated by some kind of force that makes all the atoms to go in the same direction such as compressional wave or sound wave; this is why it is called acoustical branch. The optical branch, on the other hand, can be excited by an excitation that has an opposite effect on the two atoms. It is called the optical branch because the most common example of this is the response of an ionic crystal, for example NaCl to optical radiation. The frequency of the optical branch lies in the infrared region. The existence of optical vibration does not necessarily depend on a mass difference, or even on oppositely charged ions. It depends on there being two atoms per primitive cell so that they can vibrate in and out from the centre of mass (such structure as the diamond lattice and the hexagonal closed packed lattices, the basis of which consists of two identical atoms at inequivalent positions, has an optical branch).

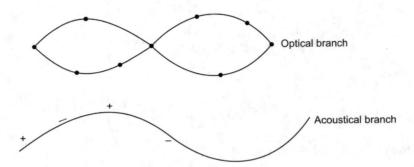

Fig. 6.6 Transverse optical, transverse acoustical waves in a diatomic lattice

1. If $M_1 = M_2$, the frequency range is the same in the monoatomic and diatomic lattices, and there is no forbidden gap.
2. If the heavy mass $M_2 \rightarrow \infty$, the acoustic branch disappears and the optical branch flattens in such a way that all k values have the same frequency (Fig. 6.7). Physically, this means that each atom is completely independent of the neighbours and oscillates at its own natural uncoupled frequency. If $M_2 \rightarrow \infty$, midpoint between each atom is tied down, isolating the atom from one another.
3. If $M_1 \rightarrow 0$, the optical branch disappears upward. The acoustic branch is unchanged (Fig. 6.8)
4. The optical branch is essentially the natural frequency of light atoms M_1, perturbed by the heavy sublattice. At $k = 2\pi/a$, the heavy sublattice is stationary and the light atom vibrates at their natural frequency $\sqrt{\frac{2C}{M_1}}$
5. At $k = \pi/a_0 (a_0 = 2a)$, the Brillouin zone edge, only one of the sublattice is vibrating. In the optical branch, it is the light atom; in the acoustical branch, it is heavy atom.

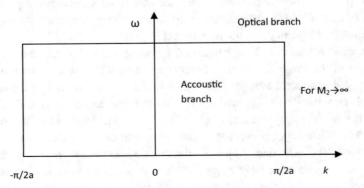

Fig. 6.7 The optical branch flattens for $M_2 = \infty$. The vertical axis indicates frequency. The acoustic branch disappears

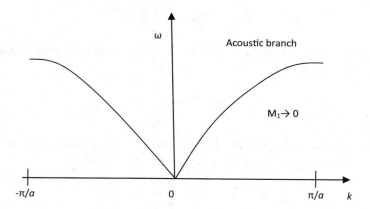

Fig. 6.8 The acoustic branch for $M_1 = 0$. The vertical axis indicates frequency. The optical branch disappears upward

Both branches represent standing waves, which have a phase difference of $\pi/2$. The frequency difference between the two modes arises entirely from the masses of the two atoms.

6. If there are n atoms in a primitive cell, there are $3n$ branches to the dispersion relation: 3 acoustical and $3n - 3$ optical branches; for example, in Ge or KBr there are two atoms in a primitive cell and thus have six branches. One longitudinal acoustic, one longitudinal optical, two transverse acoustical and two transverse optical branches.

6.5   Density of States

Consider the vibrational modes of a continuous elastic medium. Each normal mode of vibration of the medium has a characteristic wavelength. The medium is assumed to be a continuum if the wavelength λ is much larger than the interatomic separation a, that is, $\lambda \gg a$.

Consider the case in one dimension. For simplicity consider the vibrational modes of a string of length L whose both ends are fixed. When a continuous succession of waves, such as sinusoidal waves arrive at the fixed end of the string, a corresponding continuous successive wave is generated which are reflected back. Thus, the wave is reflected and re-reflected. Since both ends of the string are fixed, therefore, there will be nodes at the fixed ends. The different nodes are separated by $\frac{\lambda}{2}$. Suppose $u(z, t)$ represents the deflection of string at point z at any instant t. The wave on the string is then given by one-dimensional wave equation

$$\frac{\partial^2 u(z, t)}{\partial t^2} = v_s^2 \frac{\partial^2 u(z, t)}{\partial z^2} \quad (6.62)$$

where v_s is the velocity of propagation of the wave on the string. Since the string is fixed at both the ends, therefore

$$u(0, t) = u(L, t) = 0 \quad (6.63)$$

for all values of t.

Since both the ends of the string are fixed, therefore the solution of one-dimensional wave Eq. (6.62) corresponds to a stationary wave and is given by

$$u(z, t) = A \sin \frac{n\pi z}{L} \cos 2\pi v_n t \quad (6.64)$$

where n is a positive integer such that $n \geq 1$ and

$$n\frac{\lambda_n}{2} = L$$

$$\lambda_n = \frac{2L}{n} \quad (6.65)$$

and frequency is

$$v_n = \frac{v_s}{\lambda_n} = v_s \frac{n}{2L} \quad (6.66)$$

From Eq. (6.66), it is seen that frequency spectrum corresponds to an infinite number of equidistant lines

$$n = \frac{2L v_n}{v_s} \quad (6.67)$$

The possible modes of vibration in the frequency interval between v and $v + dv$ from Eq. (6.67) are

$$dn = \frac{2L}{v_s} dv_n \quad (6.68)$$

In three-dimensional case, the wave equation in analogy with Eq. (6.62) becomes

$$\frac{\partial^2 u(x, y, z, t)}{\partial t^2} = v_s^2 \left[\frac{\partial^2 u(x, y, z, t)}{\partial x^2} + \frac{\partial^2 u(x, y, z, t)}{\partial y^2} + \frac{\partial^2 u(x, y, z, t)}{\partial z^2} \right] \quad (6.69)$$

It is assumed that the faces of cubes are fixed, and each side of the cube is of length L. Extending the solution given by Eq. (6.64) for one dimensional to three-dimensional case, the solution takes the form

$$u(x, y, z, t) = u = A \sin \frac{n_x \pi x}{L} \sin \frac{n_y \pi y}{L} \sin \frac{n_z \pi z}{L} \cos 2\pi \nu t \qquad (6.70)$$

Differentiating Eq. (6.70) with respect to x twice

$$\frac{\partial u}{\partial x} = \frac{n_x \pi}{L} A \cos \frac{n_x \pi x}{L} \sin \frac{n_y \pi y}{L} \sin \frac{n_z \pi z}{L} \cos 2\pi \nu t$$

$$\frac{\partial^2 u}{\partial x^2} = -\left(\frac{n_x \pi}{L}\right)^2 A \sin \frac{n_x \pi x}{L} \sin \frac{n_y \pi y}{L} \sin \frac{n_z \pi z}{L} \cos 2\pi \nu t = -\left(\frac{n_x \pi}{L}\right)^2 u \qquad (6.71)$$

Similarly,

$$\frac{\partial^2 u}{\partial y^2} = -\left(\frac{n_y \pi}{L}\right)^2 u \qquad (6.72)$$

$$\frac{\partial^2 u}{\partial z^2} = -\left(\frac{n_z \pi}{L}\right)^2 u \qquad (6.73)$$

Differentiating Eq. (6.70) with respect to t twice

$$\frac{\partial^2 u}{\partial t^2} = -(2\pi \nu)^2 u \qquad (6.74)$$

Substituting Eqs. (6.66), (6.71)–(6.74) in Eq. (6.69)

$$(2\pi \nu)^2 u = v_s^2 \left[\left(\frac{n_x \pi}{L}\right)^2 + \left(\frac{n_y \pi}{L}\right)^2 + \left(\frac{n_z \pi}{L}\right)^2\right] u$$

$$4\pi^2 \nu^2 = v_s^2 \left(\frac{\pi}{\lambda}\right)^2 \left[n_x^2 + n_y^2 + n_z^2\right]$$

$$\left(\frac{\pi}{L}\right)^2 \left[n_x^2 + n_y^2 + n_z^2\right] = 4\left(\frac{\pi \nu}{v_s}\right)^2 = 4\left(\frac{\pi}{\lambda}\right)^2 \qquad (6.75)$$

Thus, the frequency or wavelength is determined by $n_x, n_y n_z$.

Consider a network of points where each point is determined by different combination of n_x, n_y and n_z. Let

$$R^2 = n_x^2 + n_y^2 + n_z^2 \qquad (6.76)$$

From Eqs. (6.75) and (6.76)

$$R^2 = \frac{4L^2 \nu^2}{v_s^2} \qquad (6.77)$$

Consider a shell between R and $R + dR$. The number of points in the shell is

$$\frac{1}{8}4\pi R^2 dR = \frac{2\pi L^2}{v_s^2} v^2 dR \tag{6.78}$$

where the factor (1/8) arises from the fact that integers are positive. Therefore, the number of modes lying between v and $v + dv$. will be given by the positive octant of the sphere. For each combination of integers n_x, n_y and n_z, there is a possible mode of vibration. Therefore, Eq. (6.78) gives the number of possible modes of vibration in the given range. From Eq. (6.77)

$$2R dR = \frac{4L^2}{v_s^2} 2v dv$$

$$dR = \frac{4L^2}{v_s^2}\frac{v dv}{R} = \frac{4L^2}{v_s^2}\frac{v dv}{2Lv}v_s = \frac{2L}{v_s}dv \tag{6.79}$$

The possible modes of vibration that is density of modes, from Eqs. (6.78) and (6.79)

$$D(v)dv = \frac{4\pi L^3}{v_s^3}v^2 dv = \frac{4\pi V}{v_s^3}v^2 dv \tag{6.80}$$

where $V = L^3$ is the volume of the solid. From Eq. (6.80), it is seen that $D(v)$ increases as square of frequency v. Further, the frequencies vary between 0 and ∞.

In the case of elastic waves, there are transverse as well as longitudinal waves. The velocity v_l of longitudinal wave may not be equal to velocity v_t of transverse wave. Since for each frequency or wavelength there are one longitudinal and two transverse modes, therefore Eq. (6.80) is written as

$$D(v)dv = 4\pi V\left(\frac{1}{v_l^3} + \frac{2}{v_t^3}\right)v^2 dv \tag{6.81}$$

Writing $\omega = 2\pi v$, Eq. (6.81) becomes

$$D(\omega)d\omega = 4\pi V\left(\frac{1}{v_l^3} + \frac{2}{v_t^3}\right)\left(\frac{\omega}{2\pi}\right)^2\frac{d\omega}{2\pi} = \frac{V}{2\pi^2}\left(\frac{1}{v_l^3} + \frac{2}{v_t^3}\right)\omega^2 d\omega \tag{6.82}$$

$$D(\omega) = \frac{V}{2\pi^2}\left(\frac{1}{v_l^3} + \frac{2}{v_t^3}\right)\omega^2 \tag{6.83}$$

If the velocity of modes are independent of polarization, that is,

$$v_l = v_t = v \tag{6.84}$$

Equations (6.82) and (6.83) take the form

$$D(\omega)d\omega = \frac{3V}{2\pi^2 v^3}\omega^2 d\omega \tag{6.85}$$

$$D(\omega) = \frac{3V}{2\pi^2 v^3}\omega^2 \tag{6.86}$$

The above equation gives density of vibrational modes of phonons or density of states.

Solved Examples

Example 1 Show that for monoatomic one-dimensional lattice frequencies corresponding to k and $k + (2\pi n/a)$ are identical.

Solution
The frequency corresponding to k is given by Eq. (6.9)

$$\omega = \sqrt{\frac{4C}{M}}\left|\sin\frac{ka}{2}\right|$$

Substituting $k + \frac{2\pi n}{a}$ for k. in Eq. (6.9)

$$\omega_n = \sqrt{\frac{4C}{M}}\left|\sin\left(k + \frac{2\pi n}{a}\right)\frac{a}{2}\right|$$

$$= \sqrt{\frac{4C}{M}}\left|\sin k\frac{a}{2}\right| = \omega.$$

Example 2 Obtain density of state for one-dimensional monoatomic linear lattice.

Solution
The frequency ω is a periodic function giving several k values of each ω less than ω_{max}. All of these k values are not allowed. Let the end of linear chain (lattice) is held fixed, then there would be an integral multiple of half wavelengths in the length $L = Na$ of the chain so that

$$L = \frac{n\lambda}{2}(n = 1, 2, \ldots, N).$$

and

$$k = \frac{2\pi}{\lambda} = \frac{n\pi}{L} = \frac{n\pi}{Na}$$

$$k = \frac{\pi}{L}, \frac{2\pi}{L}, \ldots, \frac{N\pi}{L}$$

The density of state $g(k)$ is defined as the number of states per unit interval of k

$$g(k) = \frac{1}{\left(\frac{\pi}{L}\right)} = \frac{L}{\pi}$$

Number of states in the interval k and $k + dk$ is

$$g(k)dk = \frac{L}{\pi}dk$$

We require the density of state per unit frequency range $g(\omega)$. If dn is the number of states in the interval dk centred at k and an interval $d\omega$ at ω, then

$$dn = g(k)dk = g(\omega)d\omega$$

$$g(\omega) = \frac{g(k)}{(d\omega/dk)} = \frac{L}{\pi v_g}$$

Differentiating Eq. (6.10)

$$\frac{d\omega}{dk} = \frac{a}{2}\omega_m \cos ka = \frac{a}{2}\omega_m \sqrt{1 - \sin^2 ka} = \frac{a}{2}\sqrt{\omega_m^2 - \omega^2}$$

$$g(\omega) = \frac{2L}{\pi a\sqrt{\omega_m^2 - \omega^2}}$$

$g(\omega)$ tends to infinity as $\omega_m \to \omega$.

Objective Type Questions

1. A phonon is
   (a) Quantized spin wave
   (b) Quanta of electromagnetic radiation
   (c) Quanta of lattice vibrations
   (d) None of these
2. The speed of the phonon is
   (a) Equal to the velocity of light in vacuum
   (b) Equal to velocity of light in the medium
   (c) Equal to velocity of sound in the medium
   (d) None of these
3. Which is not true?
   (a) The average number of phonons depends on temperature
   (b) The average number of phonons is equal to zero at 0 K
   (c) The average number of phonons in a system are conserved
   (d) The average number of phonons is equal to $\frac{k_B T}{\hbar \omega}$ at very high temperature
4. Phonons obey
   (a) The Bose–Einstein statistics

(b) The Fermi–Dirac statistics

(c) The Maxwell–Boltzmann statistics

(d) None of these

5. For one-dimensional monoatomic linear lattice, which is not true in the long wavelength limit?

  (a) The velocity of sound wave is given by $\frac{\omega_m}{2a}$

  (b) The dispersion ration is $\omega = \left(\frac{\omega_m a}{2}\right)k$

  (c) $\omega_m = \sqrt{\frac{4C}{M}}$.

  (d) Group velocity is equal to phase velocity

6. Which is not true for one-dimensional monoatomic linear lattice?

  (a) At higher frequencies, the group velocity and phase velocity are equal

  (b) At $k = \frac{\pi}{a}$. no energy is being propagated

  (c) The first Brillouin zone is specified by $-\frac{\pi}{a} \le k \le \frac{\pi}{a}$.

  (d) At higher frequency, the group velocity is given by $v_g = v_s \sin\frac{ka}{2}$.

7. The frequency corresponding to optical branch is given by

  (a) $\sqrt{2C\left(\frac{1}{M_1} + \frac{1}{M_2}\right)}$ (b) $2C\sqrt{\left(\frac{1}{M_1} + \frac{1}{M_2}\right)}$ (c) $ka\sqrt{\frac{2C}{M_1+M_2}}$ (d) $\frac{2C}{M_1+M_2}$

8. The frequency corresponding to acoustic branch is given by

  (a) $\sqrt{2C\left(\frac{1}{M_1} + \frac{1}{M_2}\right)}$ (b) $2C\sqrt{\left(\frac{1}{M_1} + \frac{1}{M_2}\right)}$ (c) $ka\sqrt{\frac{2C}{M_1+M_2}}$ (d) $\frac{2C}{M_1+M_2}$.

9. For $M_1 = M_2$, the amplitude A/B for optical branch is equal to

  (a) 1 (b) -1 (c) 0 (d) ∞

10. For $M_1 = M_2$, the amplitude A/B for acoustic branch is equal to

  (a) 1 (b) -1 (c) 0 (d) ∞

11. At the Brillouin zone edge of one-dimensional linear lattice, the optical and acoustic branches have a phase difference of

  (a) $\frac{\pi}{2}$ (b) π (c).2π. (d)$\frac{3\pi}{2}$

12. For three atoms in a primitive cell of one-dimensional linear lattice, the number of acoustical and optical branches to the dispersion curve is

  (a) 6 and 3 (b) 3 and 6 (c) 9 and 0 (d) 0 and 9

13. The frequency of optical branch lies in

  (a) Radiofrequency region

  (b) Infrared region

  (c) Optical region

  (d) Ultraviolet region

14. Which is not true

  (a) In the long wavelength limit, the lattice behaves as an elastic continuum

  (b) Number of acoustical branches for a one-dimensional linear lattice is always three.

  (c) If $M_1 = M_2$, the optical and acoustical branches appear for a one-dimensional diatomic linear chain, if the atoms are at inequivalent positions of the cell

(d) In the analysis of monoatomic and diatomic linear chains in one dimension only nearest neighbour interaction is taken into consideration.

Problems

1. What are phonons? Derive the dispersion relation for a linear monoatomic lattice. Discuss the condition when the lattice behaves as a continuum medium. How would the group and phase velocity vary in the first Brillouin zone?
2. Show that the number of normal modes of vibration in a monoatomic lattice of finite length is equal to the number of mobile atoms.
3. Derive the vibrational modes of a diatomic linear chain of atoms. What is the difference between the two branches? Why are they named so?
4. Why is the linear diatomic chain of lattice called a band-pass filter?
5. Obtain an expression for density of vibrational modes in three dimensions.

Answers
Objective Type Questions

1. (c)   2. (b)   3. (c)   4. (a)   5. (a)   6. (d)   7. (a)   8. (c)
9. (b)   10. (a)   11. (a)   12. (b)   13. (b)   14. (b)

Thermal Properties of Solids

7

7.1 Classical Lattice Heat Capacity

The first law of thermodynamics states that

$$Q = dU - pdV$$

dQ is the amount of heat added to the system and is equal to increase in energy dU of the system plus the amount of work done by the system and T is temperature. The specific heat at constant volume is

$$C_V = \left(\frac{\partial W}{\partial t}\right)_V = \left(\frac{\partial U}{\partial t}\right)_V \tag{7.1}$$

where partial derivatives are used because the energy may be a function of other variables in addition to temperature and the subscript denotes that volume is kept constant.

Thus to calculate the specific heat, one should have an expression for the internal energy, that is, total energy of all the particles. The contribution of the phonons to the heat capacity of a crystal is called the lattice heat capacity and is denoted by C_{lat}.

Let us evaluate the value of specific heat. For this consider, a model in which it is assumed that each atom of solid is bound to its site by a harmonic force. When the solid is heated, the atom vibrates about their equilibrium position like a set of harmonic oscillators. The average energy, $\overline{E}$, for a one-dimensional oscillator is equal to $k_B T$, where k_B is Boltzmann constant. Therefore, the average energy per atom, regarded as a three-dimensional harmonic oscillator is $3k_B T$. The energy per mole is

$$\overline{E} = 3N_A k_B T = 3RT \tag{7.2}$$

where is Avogadro number and $R = N_A k_B$.

© The Author(s) 2022
V. K. Jain, *Solid State Physics*, https://doi.org/10.1007/978-3-030-96017-9_7

Fig. 7.1 Dependence of specific heat of solids on temperature

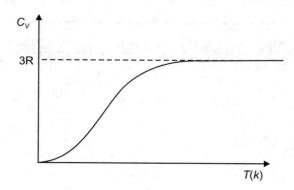

Differentiating Eq. (7.2)

$$C_V = \left(\frac{\partial \overline{E}}{\partial t}\right)_V = 3R \qquad (7.3)$$

It predicts a constant value for the specific heat at all temperatures. This is known as Dulong and Petit's law. This result is in agreement with experiment at high temperatures, but it fails completely at low temperatures. Experimental observations about specific heat C_V show that: (a) At high temperatures, C_V is independent of temperature and equals to $3R$ per mole of atoms for all solids. It holds to just below the melting point, at which C_V starts to rise (b) At low temperatures, C_V begins to fall off and approaches zero as $T \to 0$ (c) The approach to zero near $T = 0$ is by a T^3 law, that is $C_V \propto (\text{constant})T^3$. The behaviour of C_V versus temperature is shown in Fig. 7.1.

The discrepancy between the experimental results and theory is resolved first by Einstein and then by Debye using quantum theory.

7.2   Einstein Model of Specific Heat

In this model, the atoms are treated as independent oscillator. Einstein assumed a rigid lattice of identical oscillators, all having the same resonance frequencies. From the discussion of lattice vibrations, it is clear that the optical branch has a single resonant frequency when one of the masses becomes very large. The light masses then vibrate entirely independent of each other with their own resonant frequency. This is essentially the Einstein model.

According to quantum mechanics, the energy of an isolated oscillator is restricted to the value

$$E_n = \left(n + \frac{1}{2}\right)\hbar\omega \qquad (7.4)$$

where n is positive integer or zero. That is $n = 0, 1, 2, 3, \ldots$ The constant ω is the frequency of the oscillator. Thus, the energy of the oscillator is quantized. Equation (7.4) refers to an isolated oscillator but the atomic oscillators in a solid are not isolated. They are continually exchanging energy with the thermal bath surrounding the solid. The energy of the oscillator is therefore continually changing, but the average value at thermal equilibrium is given by

$$\overline{E} = \frac{\sum_{n=0}^{\infty} \left(n + \frac{1}{2}\right) \hbar\omega \exp\left[-\frac{\left(n+\frac{1}{2}\right)\hbar\omega}{k_B T}\right]}{\sum_{n=0}^{\infty} \exp\left[-\frac{\left(n+\frac{1}{2}\right)\hbar\omega}{k_B^T}\right]} \tag{7.5}$$

The $\exp\left[-\frac{\left(n+\frac{1}{2}\right)\hbar\omega}{k_B T}\right] = \exp\left[-\frac{E_n}{k_B T}\right]$ is Boltzmann factor, which gives the probability that the energy state E_n is occupied and the sum in the denominator is inserted for correct normalization. $n = 0$ is included in the summation to get the right average we need to include those oscillators that are not oscillating at all. Let

$$x = -\frac{\hbar\omega}{k_B T} \tag{7.6}$$

$$\overline{E_n} = \frac{1}{2}\hbar\omega + \frac{\sum\limits_{n=0}^{\infty} n\hbar\omega \exp\left[-\frac{n\hbar\omega}{k_B T}\right]}{\sum\limits_{n=0}^{\infty} \exp\left[-\frac{n\hbar\omega}{k_B T}\right]}$$

$$\overline{E} = \frac{1}{2}\hbar\omega + \hbar\omega\left[\frac{\exp(x) + 2\exp(2x) + 3\exp(3x) + \cdots}{1 + \exp(x) + \exp(2x) + \cdots}\right] \tag{7.7}$$

Let

$$u = 1 + \exp(x) + \exp(2x) + \cdots \tag{7.8}$$

$$\frac{du}{dx} = \exp(x) + 2\exp(2x) + 3\exp(3x) + \cdots \tag{7.9}$$

From Eqs. (7.7)–(7.9)

$$\overline{E} = \frac{1}{2}\hbar\omega + \hbar\omega\frac{1}{u}\frac{du}{dx} = \frac{1}{2}\hbar\omega + \hbar\omega\frac{d}{dx}(\log u) \tag{7.10}$$

$$\overline{E} = \frac{1}{2}\hbar\omega + \hbar\omega\frac{d}{dx}[\log(1 + \exp(x) + \exp(2x) + \cdots] \tag{7.11}$$

For any quantity y, we have

$$\frac{1}{1-y} = 1 + y + y^2 + y^3 + \cdots \tag{7.12}$$

From Eqs. (7.5) and (7.6)

$$\overline{E} = \frac{1}{2}\hbar\omega + \hbar\omega\frac{d}{dx}\log\frac{1}{[1-\exp(x)]} = \hbar\omega\frac{d}{dx}[\log 1 - \log(1 - \exp(x))]$$

$$\overline{E} = \frac{1}{2}\hbar\omega + \hbar\omega\frac{d}{dx}[-\log(1 - \exp(x))] = \hbar\omega\frac{\exp(x)}{1 - \exp(x)} = \hbar\omega\frac{\exp\left(-\frac{\hbar\omega}{k_BT}\right)}{1 - \exp\left(-\frac{\hbar\omega}{k_BT}\right)}$$

$$\overline{E} = \frac{1}{2}\hbar\omega + \frac{\hbar\omega}{\exp\left(\frac{\hbar\omega}{k_BT}\right) - 1} \tag{7.13}$$

This is average energy per oscillator. The energy U is given by

$$U = 3N\overline{E} = \frac{3N}{2}\hbar\omega + 3N\frac{\hbar\omega}{\exp\left(\frac{\hbar\omega}{k_BT}\right) - 1} \tag{7.14}$$

For $k_BT \gg \hbar\omega$ (high temperature limit)

$$\exp\left[\frac{\hbar\omega}{k_BT}\right] = 1 + \frac{\hbar\omega}{k_BT} + \cdots \tag{7.15}$$

From Eqs. (7.14) and (7.15)

$$U = \frac{3N}{2}\hbar\omega + 3N\frac{\hbar\omega}{\left(1 + \frac{\hbar\omega}{k_BT} + \cdots\right) - 1} = \frac{3N}{2}\hbar\omega + 3Nk_BT \tag{7.16}$$

$$C_V = \frac{\partial U}{\partial T} = 3Nk_B = 3R \tag{7.17}$$

This is Dulong and Petit law. Thus, the model is satisfactory at high temperature limit. At low temperature, that is, $k_BT \ll \hbar\omega$

$$\exp\left[\frac{\hbar\omega}{k_BT}\right] \gg 1$$

and Eq. (7.14) is then

$$U = 3N\overline{E} = \frac{3N}{2}\hbar\omega + 3N\frac{\hbar\omega}{\exp\left(\frac{\hbar\omega}{k_BT}\right)} = \frac{3N}{2}\hbar\omega + 3N\hbar\omega\exp\left(-\frac{\hbar\omega}{k_BT}\right) \tag{7.18}$$

The specific heat is

$$C_V = \frac{\partial U}{\partial T} = 3N\hbar\omega\left(\frac{\hbar\omega}{k_BT^2}\right)\exp\left[-\frac{\hbar\omega}{k_BT}\right] = 3Nk_B\left(\frac{\hbar\omega}{k_BT}\right)^2\exp\left[-\frac{\hbar\omega}{k_BT}\right] \quad (7.19)$$

The exponential term in Eq. (7.19) is controlling the specific heat, and therefore, $C_V \to 0$ as $T \to 0\mathrm{K}$.

Differentiating Eq. (7.14) with respect to T

$$C_V = \frac{\partial U}{\partial T} = 3N\hbar\omega\left[-\frac{\left(-\frac{\hbar\omega}{k_BT^2}\right)\exp\left(\frac{\hbar\omega}{k_BT}\right)}{\left[\exp\left(\frac{\hbar\omega}{k_BT}\right)-1\right]^2}\right] = 3Nk_B\left[\left(\frac{\hbar\omega}{k_BT}\right)^2\frac{\exp\left(\frac{\hbar\omega}{k_BT}\right)}{\left[\exp\left(\frac{\hbar\omega}{k_BT}\right)-1\right]^2}\right]$$

$$(7.20)$$

Introducing Einstein temperature θ_E by

$$k_B\theta_E = \hbar\omega \quad (7.21)$$

Equation (7.20) is

$$C_V = 3Nk_B\left[\left(\frac{\theta_E}{T}\right)^2\frac{exp\left(\frac{\theta_E}{T}\right)}{\left[\exp\left(\frac{\theta_E}{T}\right)-1\right]^2}\right] \quad (7.22)$$

A plot of C_V versus T (Fig. 7.2) indicates that the theory is in agreement with experiment, at least qualitatively, over the entire temperature range. From Eq. (7.22), we have $C_V \to 0$ as $T \to 0\mathrm{K}$. The temperature θ_E is adjustable parameter chosen to produce the best fit to the measured values over the whole temperature range.

In the high temperature limit, $T \gg \theta_E$

$$exp\left(\frac{\theta_E}{T}\right) = 1 + \frac{\theta_E}{T} + \quad (7.23)$$

Substituting Eqs. (7.23) in (7.22)

Fig. 7.2 Einstein specific heat

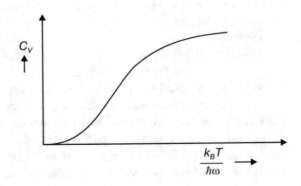

$$C_V = 3R\left[\left(\frac{\theta_E}{T}\right)^2 \frac{\left(1+\frac{\theta_E}{T}\right)}{\left[1+\frac{\theta_E}{T}-1\right]^2}\right] = 3R\left[\left(\frac{\theta_E}{T}\right)^2 \frac{\left(1+\frac{\theta_E}{T}\right)}{\left[\frac{\theta_E}{T}\right]^2}\right] = 3R \qquad (7.24)$$

as $\frac{\theta_E}{T}$ can be ignored in comparison with 1, in the numerator.

Low Temperature Range $T \ll \theta_E$.

Under this condition, $exp\left(\frac{\theta_E}{T}\right)$ is much larger than unity and Eq. (7.22) is

$$C_V = 3R\left[\left(\frac{\theta_E}{T}\right)^2 \frac{\exp\left(\frac{\theta_E}{T}\right)}{\exp\left(\frac{2\theta_E}{T}\right)}\right] = 3R\left[\left(\frac{\theta_E}{T}\right)^2 \exp\left(-\frac{\theta_E}{T}\right)\right] = B(T)\exp\left(-\frac{\theta_E}{T}\right)$$

$$(7.25)$$

where $B(T)$ is a function relatively insensitive to temperature. Because of exponential factor $\exp\left(-\frac{\theta_E}{T}\right)$ specific heat approaches zero very rapidly and vanishes at $T = 0$ K. Although the fact that $C_V \rightarrow 0 \,as\, T \rightarrow 0 K$ agrees with experiment, the manner in which this is approached is not. Equation (7.25) indicates that $C_V \rightarrow 0$ exponentially, while experiment shows that C_V approaches zero as T^3. This is the basic weakness of Einstein model.

7.3   Debye Model of Specific Heat

Debye assumed that a crystalline solid can be represented by an isotropic elastic continuum. Atoms in a solid do not vibrate with the same frequency but have the same dispersion. Because there is finite number of atoms, however, not all values of the frequency are allowed. There is a maximum allowed frequency that any atom can have. Thus, it is possible to propagate wave through solids covering a wavelength region extending from low frequencies (sound waves) up to short waves (infrared absorption). The essential difference between the Debye model and the Einstein model is that Debye considers the vibrational modes of a crystal as a whole, whereas Einstein's starting point was to consider the vibration of a single atom, assuming the atomic vibrations to be independent of each other.

The solids are by no means continuous but are built up of atoms, that is, discrete mass points. However, the Debye continuum is justified on the following basis. Consider an elastic wave propagated in a crystal of volume V. As long as the wavelength of the wave is large compared with interatomic distances, the crystal looks like a continuum from the point of view of the wave. The essential assumption of Debye is now that this continuum model may be employed for all possible vibrational modes of the crystal. Further, the fact that the crystal actually consists of atoms is taken into account by limiting the total number of vibrational modes to $3 N$, N being the total number of atoms. In other words, the frequency spectrum corresponding to a perfect continuum is cut off so as to comply with a total of $3 N$ modes.

In the Debye model, the velocity of sound is taken as constant as it would be for a classical elastic continuum. The dispersion relation is then

$$\omega = vk \tag{7.26}$$

$v =$ constant velocity of sound.

The average energy of the oscillator of frequency ω is

$$\overline{E} = \frac{1}{2}\hbar\omega + \frac{\hbar\omega}{\exp\left[\frac{\hbar\omega}{k_B T}\right] - 1}$$

The first term is called zero-point energy of oscillator. This has effect of shifting all energy levels by the constant amount and therefore can be ignored in obtaining specific heat. Because zero-point energy term disappears when the energy term is differentiated to obtain the specific heat. Thus, it can be written as

$$\overline{E} = \frac{\hbar\omega}{\exp\left[\frac{\hbar\omega}{k_B T}\right] - 1} \tag{7.27}$$

All oscillators having frequency ω will have an average energy given by Eq. (7.27). However, we have a whole spectrum of frequencies; there will be different average energy for each frequency, that is,

$$U = \sum_{n=1}^{3N} \overline{E_n} \tag{7.28}$$

From Eqs. (7.27) and (7.28)

$$U = \sum_{n=1}^{3N} \frac{\hbar\omega_n}{\exp\left[\frac{\hbar\omega_n}{k_B T}\right] - 1} \tag{7.29}$$

The total energy of the vibration for the entire lattice is

$$U = \int \overline{E} g(\omega) d\omega \tag{7.30}$$

where the integration is over all the allowed frequencies. $g(\omega)$ is density of state, $g(\omega) d\omega$ is the number of modes in the range ω and $\omega + d\omega$, and energy of each of these modes is equal to $\overline{E_n}$.

The density of states is given by

$$g(\omega) = \frac{3V}{2\pi^2} \frac{\omega^2}{v^3} \tag{7.31}$$

Substituting Eqs. (7.27) and (7.31) in Eq. (7.30)

$$U = \int \frac{\hbar\omega}{\exp\left[\frac{\hbar\omega}{k_B T}\right] - 1} \frac{3V}{2\pi^2} \frac{\omega^2}{v^3} d\omega$$

$$U = \frac{3V}{2\pi^2 v^3} \int \frac{\hbar\omega^3}{\exp\left[\frac{\hbar\omega}{k_B T}\right] - 1} d\omega \tag{7.32}$$

The lower limit of the integral is $\omega = 0$. The upper cut-off frequency is obtained by requiring that the total number of modes included must be equal to the number of degrees of freedom of the entire solid. Since this number is equal to $3N$, because each atom has three degrees of freedom. Thus,

$$\int_0^{\omega_D} g(\omega)d\omega = 3N \tag{7.33}$$

where the cut-off frequency is denoted by ω_D, called Debye cut-off frequency (Fig. 7.3). In fact out of three modes, there is one longitudinal mode with $v = v_L$ and two transverse modes with $v = v_T$ so that Eq. (7.31) is

$$g(\omega) = \frac{3V\omega^2}{2\pi^2}\left(\frac{1}{v_L^2} + \frac{2}{v_T^2}\right)$$

Substituting this value of $g(\omega)$ in Eq. (7.33)

$$\int_0^{\omega_D} \frac{3V}{2\pi^2}\omega^2\left(\frac{1}{v_L^2} + \frac{2}{v_T^2}\right)d\omega = 3N\left(\frac{1}{v_L^2} + \frac{2}{v_T^2}\right)\int_0^{\omega_D} \omega^2 d\omega$$

$$= \frac{2\pi^2 N}{V}\left(\frac{1}{v_L^2} + \frac{2}{v_T^2}\right)\frac{\omega_D^3}{3} = 2\pi^2 n$$

If the velocities $v = v_L = v_T$, that is phonon velocity is independent of polarization

Fig. 7.3 Debye cut-off frequency

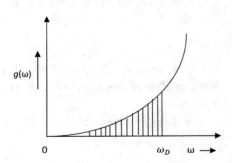

$$\omega_D = v\left(6\pi^2 n\right)^{\frac{1}{3}} \tag{7.34}$$

where $n = N/V$ is concentration of atoms. From Eqs. (7.26) and (7.34)

$$k_D = \left(6\pi^2 n\right)^{\frac{1}{3}} \tag{7.35}$$

Specific heat at constant volume from Eq. (7.32) is

$$C_V = \frac{3V}{2\pi^2 V^3} \int_0^{\omega_D} \frac{\hbar\omega^3 \left(\frac{\hbar\omega}{k_B T^2}\right) \exp\left(\frac{\hbar\omega}{k_B T}\right)}{\left[\exp\left(\frac{\hbar\omega}{k_B T}\right) - 1\right]^2} d\omega$$

$$C_V = \frac{3V}{2\pi^2 V^3} \frac{\hbar^2}{k_B T^2} \int_0^{\omega_D} \frac{\omega^4 \exp\left(\frac{\hbar\omega}{k_B T}\right)}{\left[\exp\left(\frac{\hbar\omega}{k_B T}\right) - 1\right]^2} d\omega \tag{7.36}$$

Let

$$x = \frac{\hbar\omega}{k_B T} \quad \text{and} \quad k_B \theta_D = \hbar\omega_D \tag{7.37}$$

From Eqs. (7.36) and (7.37)

$$C_V = \frac{3V}{2\pi^2 V^3} \frac{\hbar^2}{k_B T^2} \int_0^{\theta_D/T} \frac{\left(\frac{k_B T}{\hbar}\right)^5 x^4 \exp(x)}{[\exp(x) - 1]^2} dx$$

$$C_V = \frac{3V}{2\pi^2 V^3} \left(\frac{k_B T}{\hbar}\right)^5 \frac{\hbar^2}{k_B T^2} \int_0^{\theta_D/T} \frac{x^4 \exp(x)}{[\exp(x) - 1]^2} dx \tag{7.38}$$

Substituting the value of v from Eq. (7.34) in Eq. (7.38) and using $n = N/V$

$$C_V = \frac{3V}{2\pi^2 \omega_D^3} 6\pi^2 \frac{N}{V} \left(\frac{k_B T}{\hbar}\right)^5 \frac{\hbar^2}{k_B T^2} \int_0^{\theta_D/T} \frac{x^4 \exp(x)}{[\exp(x) - 1]^2} dx$$

$$C_V = 9Nk \left(\frac{k_B T}{\hbar\omega_D}\right)^3 \int_0^{\theta_D/T} \frac{x^4 \exp(x)}{[\exp(x) - 1]^2} dx$$

$$C_V = 9R \left(\frac{T}{\theta_D}\right)^3 \int_0^{\theta_D/T} \frac{x^4 \exp(x)}{[\exp(x) - 1]^2} dx \tag{7.39}$$

The value of θ_D depends on the given substance and a typical value is 300 K. The corresponding Debye frequency lies in the infrared region of the spectrum. From Eq. (7.34), we have

$$\theta_D \sim \omega_D \sim vn^{\frac{1}{3}}$$

and consequently

$$\theta_D \sim \sqrt{\frac{Y}{\rho}} n^{\frac{1}{3}} \sim \sqrt{\frac{Y}{M}}$$

where Y is Young's modulus and M the atomic mass (using $\rho = nM$). Therefore, Debye temperature depends on the elastic constant of the substance Y and on the atomic mass M. For stiffer crystal and small M, the Debye temperature is high. The Debye temperature for, carbon, which is stiff and light is 1860 K, while the Debye temperature of lead which is soft and heavy is 102 K.

Let us consider high temperature limit, that is, $T \gg \theta_D$. For this case, x is quite small and expanding the exponential in Eq. (7.39)

$$C_V = 9R \left(\frac{T}{\theta_D}\right)^3 \int_0^{\theta_D/T} \frac{x^4(1 + x + \cdots)}{[1 + x + \cdots - 1]^2} dx$$

$$C_V \approx 9R \left(\frac{T}{\theta_D}\right)^3 \int_0^{\theta_D/T} \frac{x^4}{x^2} dx \text{ for } x \ll 1$$

$$C_V \approx 9R \left(\frac{T}{\theta_D}\right)^3 \int_0^{\theta_D/T} x^2 dx = 9R \left(\frac{T}{\theta_D}\right)^3 \frac{x^3}{3}\Bigg|_0^{\frac{\theta_D}{T}}$$

$$C_V = \frac{9R}{3} \left(\frac{T}{\theta_D}\right)^3 \left(\frac{\theta_D}{T}\right)^3 = 3R \tag{7.40}$$

Let us consider the low temperature limit. At low temperature, $T \to 0$ implies $x_{max} \to \infty$; therefore, upper limit of the integral goes to infinity instead cutting it off at ω_D. We have Riemann zeta function

$$\int_0^\infty \frac{x^{s-1}}{\exp(x) - 1} dx = (s - 1)! \sum_{n=1}^\infty \frac{1}{n^s} \tag{7.41}$$

and

$$\int_0^\infty \frac{x^3}{\exp(x) - 1} dx = \frac{\pi^4}{15} \tag{7.42}$$

From Eqs. (7.32), (7.37), (7.41) and (7.42)

$$U = \frac{3V}{2\pi^2} \frac{(k_B T)^4}{(\hbar v)^3} \int_0^\infty \frac{x^3}{\exp(x) - 1} dx = \frac{3V}{2\pi^2} \frac{(k_B T)^4}{(\hbar v)^3} \frac{\pi^4}{15} \tag{7.43}$$

Substituting the value of v from Eq. (7.34) and using Eq. (7.37) in Eq. (7.43)

$$U = \frac{3}{5} \pi^4 k_B N T \left(\frac{T}{\theta_D}\right)^3 \tag{7.44}$$

Fig. 7.4 Temperature
variation of the specific heat
per gm atom according to
Debye theory

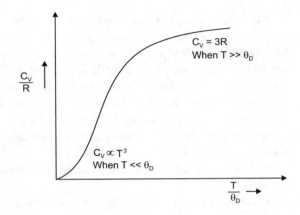

$$C_V = \frac{\partial U}{\partial T}\bigg|_V = \frac{12}{5}\pi^4 N k_B \left(\frac{T}{\theta_D}\right)^3 = 234R\left(\frac{T}{\theta_D}\right)^3 \qquad (7.45)$$

The T^3 dependence is in agreement with the experimental results (Fig. 7.4). At sufficiently low temperatures, the T^3 approximation is quite good; that is when only long wavelength acoustic modes are thermally excited. These are just the modes that may be treated as an elastic continuum. The energy of the short wavelength modes is too high for them to be populated significantly at low temperature. For actual crystals, the temperature at which the T^3 approximation holds is quite low. It may be necessary to be below $T = \theta/50$ to get reasonable pure T^3 behaviour.

The Einstein model ignores the presence of the very low frequency, long wavelength modes which can absorb heat even at very low temperature, because their energies of quantization are very small. The exponential freezing of the modes does not actually occur, and the specific heat has a finite, though small value. In spite of its successes, the Debye model also remains only an approximation. According to the theory, the Debye temperature is independent of temperature, whereas actually it has been found to vary with temperature. The theory cannot be applied to crystal composed of more than one type of atoms. Further, total numbers of vibrational modes is assumed to be $3N$, but the solid is an elastic continuum and should possess infinite frequencies. In order to improve on the Debye model, one needs to remove the long wavelength approximation and use, instead of the correct dispersion relation and the corresponding density of states.

7.4 Thermal Expansion

Classical physics regarded a solid crystal as an assembly of atoms held together in a periodic array by certain attractive forces. The atoms were assumed to be free to vibrate about their equilibrium positions under the constraints of the resulting forces and to a first approximation, forces and atomic displacements would be related by Hook's law. The effect of thermal energy than would be to set these atoms into vibration as harmonic oscillators about their equilibrium positions. Since the oscillators are purely harmonic, therefore the potential is a parabolic function of position. The minimum of the potential energy curve is the classical equilibrium position of the atom, if it is at rest. If the inter-atomic forces were such that the atom if set in motion thermally would vibrate about its equilibrium position as an ideal classical harmonic oscillator. The atom would execute vibrations about the equilibrium position, the maximum displacement from the equilibrium position in either direction being equal, and the average distance $<x>$ would be equal to value of lattice constant a at zero temperature. There would thus be no thermal expansion.

Let the potential well in which the atoms vibrate have approximately the appearance as shown in Fig. 7.5.

In this case, although nearly parabolic about the minimum point A, the actual will deviate from the parabolic form more and more as the distance from the minimum point increases. If the atom has energy U_0, it should according to the classical picture vibrate between the extreme amplitude limits B and C, the vibrations are somewhat anharmonic in character. But the distance DC between the equilibrium positions is now greater than the distance BD between the equilibrium position and the maximum compression position. The average interatomic distance $<x>$ is thus greater than the zero temperature lattice constant, and thermal expansion is observed. Therefore, in order to account for the thermal expansion, it is necessary to consider anharmonic terms in the potential. We approximate the true potential more accurately by adding high-order (anharmonic) terms as follows

$$U(x) = cx^2 - gx^3 - fx^4 \tag{7.46}$$

where c, g and f all positive. The first term is harmonic while second and third terms are anharmonic terms. The origin of energy and position is taken at a, the lattice constant

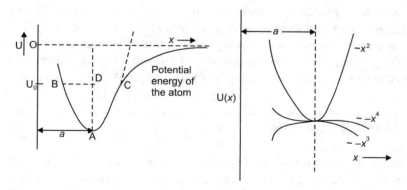

Fig. 7.5 Variation of potential energy with interatomic separation

at 0 K. Figure 7.5 shows the plot of various terms of $U(x)$ versus displacement x. x represents the increase in the lattice constant or the thermal expansion. As shown in Fig. 7.5, the cubic term steepens the left side and flattens the right; it therefore represents the asymmetry of the mutual repulsion of the atom. The quadratic term flattens the bottom of the curve. It represents a softening effect since it reduces the energy necessary to produce a displacement.

The average value of x is calculated using the Boltzmann distribution function, which weights all the possible values of x according to their thermodynamical probabilities

$$< x >= \frac{\int_{-\infty}^{\infty} x \exp\left[-\frac{U(x)}{k_B T}\right] dx}{\int_{-\infty}^{\infty} \exp\left[-\frac{U(x)}{k_B T}\right] dx} \tag{7.47}$$

Substituting the value of $U(x)$

$$< x >= \frac{\int_{-\infty}^{\infty} x \exp\left[-\frac{cx^2 - gx^3 - fx^4}{k_B T}\right] dx}{\int_{-\infty}^{\infty} \exp\left[-\frac{cx^2 - gx^3 - fx^4}{k_B T}\right] dx} \tag{7.48}$$

If the anharmonic terms are small in comparison with $k_B T$

$$\exp\left[-\frac{cx^2 - gx^3 - fx^4}{k_B T}\right] \rightarrow \exp\left[-\frac{cx^2}{k_B T}\right] \exp\left[\frac{gx^3 + fx^4}{k_B T}\right]$$

$$= \exp\left[-\frac{cx^2}{k_B T}\right]\left[1 + g\frac{x^3}{k_B T} + f\frac{x^4}{k_B T} + \cdots\right] \tag{7.49}$$

Substituting Eq. (7.49) in Eq. (7.48)

$$< x >= \frac{\int_{-\infty}^{\infty} \exp\left[-\frac{cx^2}{k_B T}\right]\left[x + g\frac{x^4}{k_B T} + f\frac{x^5}{k_B T} + \cdots\right] dx}{\int_{-\infty}^{\infty} \exp\left[-\frac{cx^2}{k_B T}\right] dx} \tag{7.50}$$

In the denominator, the small anharmonic terms are neglected. From Eq. (7.50)

$$< x > = \frac{\int_{-\infty}^{\infty} x \exp\left[-\frac{cx^2}{k_B T}\right] dx}{\int_{-\infty}^{\infty} \exp\left[-\frac{cx^2}{k_B T}\right] dx}$$

$$+ \frac{\int_{-\infty}^{\infty} \exp\left[-\frac{cx^2}{k_B T}\right]\left[g\frac{x^4}{k_B T}\right] dx}{\int_{-\infty}^{\infty} \exp\left[-\frac{cx^2}{k_B T}\right] dx}$$

$$+ \frac{\int_{-\infty}^{\infty} \exp\left[-\frac{cx^2}{k_B T}\right]\left[f\frac{x^5}{k_B T}\right] dx}{\int_{-\infty}^{\infty} \exp\left[-\frac{cx^2}{k_B T}\right] dx} \tag{7.51}$$

The terms containing odd x have zero value. Using

$$\int_0^\infty x^{2n} \exp(-ax^2)dx = \frac{1\cdot 3\cdot 5\ldots\ldots(2n-1)}{2^{n+1}a^n}\sqrt{\frac{\pi}{a}} \tag{7.52}$$

Using Eq. with $n=2$, $a = \frac{c}{k_BT}$ in Eq. (7.51)

$$<x> = +\frac{\int_{-\infty}^\infty \exp\left[-\frac{cx^2}{k_BT}\right]\left[g\frac{x^4}{k_BT}\right]dx}{\int_{-\infty}^\infty \exp\left[-\frac{cx^2}{k_BT}\right]dx} = \frac{g}{k_BT}\frac{1\cdot 3}{4c^2}(k_BT)^2 = \frac{3gk_BT}{4c^2} \tag{7.53}$$

in the classical region. The expansion coefficient

$$\alpha = \frac{\partial\langle x\rangle}{\partial T} = \frac{3gk_B}{4c^2} \tag{7.54}$$

It appears to be independent of temperature. This result was obtained, however, without taking the quantization of the energy into consideration. At low temperature, the integral would have to be replaced by the summation because of discrete nature of the energy and this may be difficult to evaluate. However, we can write Eq. (7.53) as

$$<x> = \frac{3g}{4c^2}<E> \tag{7.55}$$

The quantum expression of <E> is

$$<E> = \frac{\hbar\omega}{\exp\left(\frac{\hbar\omega}{k_BT}\right)-1} \tag{7.56}$$

Using Eq. (7.56) in Eq. (7.55)

$$<x> = \frac{3g}{4c^2}\frac{\hbar\omega}{\exp\left(\frac{\hbar\omega}{k_BT}\right)-1}$$

$$<x> = \frac{3g}{4c^2}\hbar\omega\exp\left(-\frac{\hbar\omega}{k_BT}\right) \tag{7.58}$$

and we have

$$\int_0^\infty x^n \exp(-ax^p)dx = \frac{k!}{pa^k}\left(n > -1, p > 0, a > 0, k = \frac{n+1}{p}\right) \tag{7.59}$$

Applying Eq. (7.58) in Eq. (7.54).

$$\alpha = \frac{\partial\langle x\rangle}{\partial T} = \frac{3gk_B}{4c^2}\left(\frac{\hbar\omega}{k_BT}\right)^2\exp\left(-\frac{\hbar\omega}{k_BT}\right) \tag{7.60}$$

this vanishes as $T \to 0$ K.

7.5 Thermal Conductivity

The transfer of energy arising from the temperature difference between adjacent parts of a body is called thermal conduction. Heat energy can be transmitted through a crystal via the motion of phonons, photons, free electrons or holes, electron–hole pairs or excitons (bound electron–hole pair). The electronic components of heat conduction are usually the largest component in a metal, but almost all of the thermal current in a non-metal is carried by the lattice vibrations except at the highest temperature where photon may become dominant.

The thermal conductivity coefficient K of a solid is defined with respect to steady-state flow of heat down a long rod with a temperature gradient dT/dx. The j_U flux of thermal energy or the energy transmitted across unit is per unit time which is

$$j_U = -K \frac{dT}{dx} \tag{7.61}$$

where K is called thermal conductivity. We choose the direction of heat flow to be the direction of increasing x, since heat flows in the direction of decreasing temperature. We introduce a minus sign, that is, we wish j_U to be positive when dT/dx is negative. Equation (7.61) implies that the process of thermal energy transfer is a random process. The energy does not simply enter one end of the specimen and proceed directly in a straight path to the other end, but rather the energy diffuses through the specimen, suffering frequent collisions. If the energy were propagated directly through the specimen without deflection, then the expression for the thermal flux would not depend on the temperature gradient, but only on the difference in temperature ΔT between the ends of the specimen, regardless of the length of the specimen. The random nature of the conductivity process brings the temperature gradient and mean free path into the expression for the thermal flux.

Consider an ideal gas in which a temperature gradient exists. In the gas, we construct three parallel planes separated by one mean free path length l as shown in Fig. 7.6. If the gas molecules are moving randomly, on the average one-sixth of them move downward through the plane P and one-sixth move upward. If there are N molecules with average velocity v, then $(Nv/6)$ is the particle current density up or down. The energy per particle is $\left(\frac{3k_B T}{2} \right)$ so the energy current j_U up or down is

$$(j_U)_{\text{down}} = \frac{1}{6} N v \left(\frac{3}{2} k_B T_1 \right) \tag{7.62}$$

$$(j_U)_{\text{up}} = \frac{1}{6} N v \left(\frac{3}{2} k_B T_2 \right) \tag{7.63}$$

Fig. 7.6 Three imaginary planes separated by mean free path length in an ideal gas having a temperature gradient

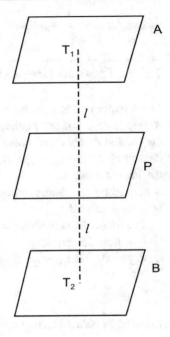

The net energy current density

$$j_U = (j_U)_{\text{up}} - (j_U)_{\text{down}} = -\frac{1}{4} N v k_B (T_1 - T_2) \tag{7.64}$$

We assume that energy is transferred by means of particle colliding with each other. Thus, a particle passing downward through P with temperature T_1 acquired its temperature at the location of its last collision, namely at a distance l above the plane P, or is in the plane A. It suffers another collision in P and another at a distance l below, in B, where it acquires the temperature T_2. The temperature gradient is then

$$\frac{dT}{dx} = \frac{T_1 - T_2}{2l} \tag{7.65}$$

From Eqs. (7.61), (7.64) and (7.65)

$$j_U = -\frac{1}{4} N v k_B (T_1 - T_2) = -K \frac{T_1 - T_2}{2l} \tag{7.66}$$

$$K = \frac{1}{2} N k_B v l = \frac{1}{3} C_V l v \tag{7.67}$$

where $C_V = \frac{3}{2} N k_B$ is specific heat of an ideal gas. Let us now apply this to the case of solids.

Let n is the concentration of molecules. The flux of particles in the x-direction is $\left(\frac{n}{2} < |v_x| >\right)$. In equilibrium, there is a flux of equal magnitude in the $-x$-direction. The $< >$ denotes average value. Let c is heat capacity of a particle, then in moving from a region at local temperature $T + \Delta T$ to a region at local temperature T a particle will give up energy $c\Delta T$. Now ΔT between the ends of a free path of the particle is given by

$$\Delta T = \frac{dT}{dx}l = \frac{dT}{dx}v_x\tau \tag{7.68}$$

where τ is the average time between collisions, l is the mean free path of phonon.

The net flux of energy (from both senses of particle flux) is therefore using Eq. (7.68)

$$j_U = -n < v_x > c\Delta T = -n < v_x^2 > c\tau\frac{dT}{dx} = -\frac{1}{3}n < v^2 > c\tau\frac{dT}{dx} \tag{7.69}$$

For phonons, v is constant and writing $l = v\tau$, $C = nc$. Equation (7.69) reduces to

$$j_U = -\frac{1}{3}Clv\frac{dT}{dx} \tag{7.70}$$

Comparing Eqs. (7.70) and (7.61)

$$K = \frac{1}{3}Cvl \tag{7.71}$$

7.6 Lattice Thermal Resistivity

Heat can be transported in a solid by two different mechanisms. In all solids, there is a possibility of heat being conducted by the lattice vibrations; however, if the solid also has a reasonable electrical conductivity and hence has free electrons, the electrons can also act as carriers of heat. In the dielectric crystals, all the heat is transported by the lattice vibrations. Equation (7.71) gives an expression for thermal conductivity. We assume that v is constant for phonon. The absolute magnitude and temperature dependence of conductivity are determined by variation of C and l. The temperature variation of C according to Debye model is proportional to T^3 at low temperatures and is constant at higher temperatures. The main problem is the evaluation of l. There are several scattering mechanisms which are effective at the same time.

The most important scattering mechanisms for phonons are (i) scattering of phonon by interaction with one another (ii) geometrical scattering (collision of a phonon with crystal boundary, point defects, dislocation, etc.). These mechanisms determine the phonon mean free path. If the forces between atoms were purely harmonic, there would be no mechanism for collision between phonons and the mean free path should be limited solely by collisions of a phonon with crystal boundary and by lattice imperfections. With anharmonic lattice interaction, there is a coupling between different phonons which limits the value of the mean free path.

7.7   Normal and Umklapp Processes

The lattice thermal conductivity of a solid is determined by two contributions (a) specific heat and (b) mean free paths of the phonons. The phonons are major heat carriers in solids. In harmonic approximation, the phonons travel freely without attenuation. As a result, they have unlimited free path resulting in an infinite thermal conductivity. However, thermal equilibrium is attained in a solid as mean free path is restricted by (a) the anharmonic terms of Eq. (7.46) (b) imperfections and impurities in the crystals and (c) finite size of the lattice.

The anharmonicity results in scattering or collisions between phonons. Scattering from other phonons can be classified into (a) Normal process and (b) Umklapp process depending on the energies involved.

Normal Process
Consider a phonon of wave vector k_1 which collides with another phonon of wave vector k_2. As a result of this collision, a wave vector k_3 is formed. The probability of such a collision is determined by the magnitude of the anharmonic terms. The property of the resulting phonon with wave vector k_3 is determined by laws of momentum and energy conservation.

The energy conservation gives

$$\hbar\omega_1 + \hbar\omega_2 = \hbar\omega_3 \tag{7.72}$$

Momentum conservation gives

$$k_1 + k_2 = k_3 \tag{7.73}$$

Such a process is known as Normal process. This process is shown in Fig. 7.7 for two-dimensional square lattices. The square with dotted lines represents the first Brillouin zone in the reciprocal space.

The Brillouin zone contains all the possible independent values of vector. The vector k with head towards the centre of Brillouin zone indicates that the phonon is absorbed in the scattering process while vector k with arrow head away from the centre of Brillouin zone indicates that the phonon is emitted in the scattering process. In the scattering of two phonons with wave vector k_1 and wave vector k_2, another phonon having wave vector k_3 is emitted. In the Normal process, the direction of energy flow is not changed. The Normal process does not make contribution towards the thermal resistance. In this process,

Fig. 7.7 Normal process

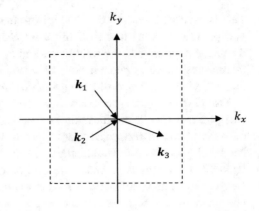

(a) energy is conserved (b) momentum is conserved and (c) initial energy of phonons involved in the scattering is small.

Umklapp Process

Umklapp process is a scattering process of phonons in which the resultant phonon's wave vector falls outside the first Brillouin zone. Consider the scattering process in which the initial phonons momenta are not small. In this case, the third phonon wave vector falls outside the first Brillouin zone. This phonon can be brought back into the first Brillouin zone by adding the reciprocal vector $\boldsymbol{G}$. The process can be represented by the relation

$$\boldsymbol{k}_1 + \boldsymbol{k}_2 = \boldsymbol{k}_3 + \boldsymbol{G} \tag{7.74}$$

This process is shown in Fig. 7.8. The scattering of two phonons with wave vectors $\boldsymbol{k}_1$ and $\boldsymbol{k}_2$ results in a phonon such that the resulting $\boldsymbol{k}_1 + \boldsymbol{k}_2$ lies outside the Brillouin zone.

Fig. 7.8 Umklapp process

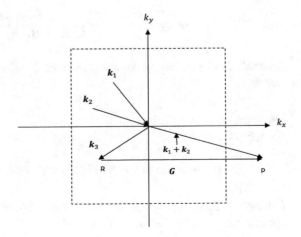

This is brought back into the Brillouin zone by reciprocal lattice vector G such that the new point R is in the first Brillouin zone. P and R are equivalent points. In this process, the energy is carried to the right by k_1 and k_2 but to the left by k_3. This process provides thermal resistance to phonon flow. Since $G (= \frac{2\pi}{a} n)$ may accept a number of direction, the scattering may be considered as approximately random.

From Eq. (7.74), it is seen that Umklapp process can occur only when the phonons have energy larger than a certain minimum value. At sufficiently low temperature $(T \ll \theta_D)$ only the region of reciprocal space close to the centre of Brillouin zone remains populated with phonons. But Umklapp process cannot occur unless $k_1 + k_2$ extends beyond the Brillouin zone boundary. Thus Umklapp process is unimportant. At temperatures above Debye temperature, all phonon modes are excited. A major proportion of all phonons will undergo Umklapp process and large momentum change with the scattering process. The energy of phonons k_1 and k_2 for Umklapp process to occur should be of the order of $\frac{1}{2} k_B \theta_D$ (where k_B is the Boltzmann constant) because both phonons must have wave vectors of the order of $\frac{G}{2}$ for scattering to occur. If both k_1 and k_2 are small and therefore have small energy, there is no way to get from their scattering a phonon k_3 outside the first zone. The Umklapp process must conserve energy. At low temperature, the number of suitable phonons of energy $\frac{1}{2} k_B \theta_D$ may be expected to vary approximately as $\exp(-\theta_D/2T)$.

Solved Examples

Example 1 The Debye temperature of a compound is 300 K. Determine the corresponding Debye frequency.

Solution
From Eq. (7.37).

$$\omega_D = \frac{k_B \theta_D}{\hbar} = \frac{1.38 \times 10^{-23} \times 300}{1.055 \times 10^{-34}} = 3 \times 10^{13} \text{ s}^{-1}$$

Example 2 Determine the Debye frequency if velocity is 5×10^5 cm/s and $n = 10^{22}$ atoms/cm^3.

Solution
From Eq. (7.34)

$$\omega_D = v \left(6\pi^2 n\right)^{\frac{1}{3}} = 5 \times 10^5 \left(6 \times (3.14)^2 \times 10^{22}\right) = 4 \times 10^{13} \text{ s}^{-1}$$

Example 3 The Debye temperature of carbon (diamond structure) is 1850 K. Calculate the specific heat per kmol for diamond at 20 K.

Solution

From Eq. (7.45)

$$C_V = \frac{\partial U}{\partial T}\bigg|_V = \frac{12}{5}\pi^4 N k_B \left(\frac{T}{\theta_D}\right)^3 = 234R\left(\frac{T}{\theta_D}\right)^3$$

$$R = N k_B = 6.026 \times 10^{26} \times 1.38 \times 10^{-23} = 8.31 \times 10^3 \, JK^{-1}kmol^{-1}$$

Substituting the value of R in Eq. (7.45)

$$C_V = 234R\left(\frac{T}{\theta_D}\right)^3 = 234 \times 8.31 \times 10^3 \left(\frac{20}{1850}\right)^3 = 2.45 \, JK^{-1}kmol^{-1}$$

Example 4 The specific heat of NaCl at 20 K is $C_V = T^3 \times 10^{-4}$. Determine the Debye temperature.

Solution

From Eq. (7.45)

$$C_V = 234R\left(\frac{T}{\theta_D}\right)^3$$

$$\theta_D^3 = \frac{234RT^3}{C_V} = \frac{234 \times 8.31 \times 10^3 \times T^3}{T^3 \times 10^{-4}} = \frac{234 \times 8.31 \times 10^3}{10^{-4}}$$

$$\theta_D = \left(\frac{234 \times 8.31 \times 10^3}{10^{-4}}\right)^{\frac{1}{3}} = 288 \text{ K.}$$

Objective Type Questions

1. Which is not true?
   (a) The contribution of phonons to the heat capacity of solids is called the lattice heat capacity
   (b) Dulong and Petit's law predicts a constant value of specific heat at all temperatures
   (c) According to Einstein model $C_V \propto \exp\left[\frac{\theta_E}{T}\right]$
   (d) The Einstein model assumes atoms as independent oscillator in the calculation of specific heat

2. The Einstein temperature is defined as
   (a) $\theta_E = \frac{k_B}{\hbar\omega}$ (b) $\theta_E = \frac{\hbar\omega}{k_B}$ (c) $\theta_E = \frac{\hbar\omega}{k_B T}$ (d) $\theta_E = \frac{\hbar\omega_D}{k_B}$

3. The Debye frequency is given by
   (a) $\left(6\pi^2 n v^3\right)^{\frac{1}{3}}$ (b) $\left(6\pi^2 n v\right)^{\frac{1}{3}}$ (c) $\left(6\pi^2 v^3\right)^{\frac{1}{3}}$ (d) $6\pi^2 n v^3$

4. The density of modes according to Debye model is
   (a) $\frac{3V}{2\pi}\frac{\omega^2}{v^3}$ (b) $\frac{3V}{2\pi^3}\frac{\omega^2}{v^3}$ (c) $\frac{3V}{2\pi^2}\frac{\omega^2}{v^3}$ (d) none of these

5. According to the Debye model, the specific heat of one-dimensional lattice at low temperature is proportional to
   (a) T (b) T^2 (c) T^3 (d) none of these
6. According to the Debye model, the specific heat of two-dimensional lattice at low temperature is proportional to
   (a) T (b) T^2 (c) T^3 (d) none of these
7. Anharmonic terms in the expression

$$U(x) = cx^2 - gx^3 - fx^4$$

   are
   (a) cx^2, gx^3 (b) cx^2, fx^4 (c) gx^3, fx^4 (d) cx^2, gx^3, fx^4
8. Thermal expansion coefficient is proportional to
   (a) $\left(\frac{\omega}{T}\right)^2 \exp\left(-\frac{\hbar\omega}{k_BT}\right)$ (b) $\left(\frac{\omega}{T}\right) \exp\left(-\frac{\hbar\omega}{k_BT}\right)$ (c) $\left(\frac{\omega}{T}\right)^3 \exp\left(-\frac{\hbar\omega}{k_BT}\right)$ (d) $\omega^2 \exp\left(-\frac{\hbar\omega}{k_BT}\right)$
9. Debye frequency lies in
   (a) Ultraviolet region (b) visible region (c) infrared region (d) radiofrequency region
10. Which is true?
    (a) Debye temperature is higher for small Y and large atomic mass
    (b) Debye temperature is higher for large Y and large atomic mass
    (c) Debye temperature is higher for small Y and small atomic mass
    (d) Debye temperature is higher for larger Y and small atomic mass
11. Einstein frequency lies in
    (a) Ultraviolet region (b) visible region (c) infrared region (d) radiofrequency region

Problems

1. State Dulong and Petit's law and show how the departure from the law at lower temperatures has been explained by Einstein's theory?
2. Derive an expression for the specific heat of solids following the Einstein model. How does the specific heat depend on the temperature and on what extent does the model agree with the experimental results?
3. Derive an expression for specific heat of solids on the basis of Debye model. How does the Debye model differ from the Einstein model? Discuss the variation of specific heat with temperature.
4. Obtain the density of states for vibration of a one-dimensional lattice of length L carrying $N + 1$ particles at a separation a.
5. The Einstein frequency for copper is 4.8×10^{12} Hz. Calculate its Einstein temperature.
6. According to Debye's theory, calculate the atomic specific heat of carbon in form of diamond at a temperature 20 K. The Debye temperature of the diamond crystal is 2230 K. What will be the minimum frequency of modes of vibrations?

7. Describe Normal process and Umklapp process. What is the difference between the two processes?

Answers
Objective Type Questions

1. (c)   2. (b)   3. (a)   4. (c)   5. (a)   6. (b)   7. (c)   8. (a)
9. (c)   10. (d)   11. (c)

Free Electron Theory of Metal

8

8.1 Drude–Lorentz Model

P. Drude in 1900 proposed that physical properties of metals can be explained in terms of free electron model. According to this model, a metal consists of stationary ions and valence electrons. The valence electrons form the free electron gas. These electrons move in the volume of the metal. Drude applied kinetic theory of gases to the free electron gas. The basic assumption of Drude theory of metal is

(1) A metal consists of positive metal ions whose valence electrons are free to move between the ions as if they constituted an electron gas.
(2) The metal is held together by electrostatic force of attraction between the positively charged ions and negatively charged electron gas. The repulsion between the electrons is ignored, and the potential field due to positive ions is assumed to be uniform.
(3) The electrons are free to move about the whole volume of the metal like the molecules of a perfect gas in a container.
(4) In the absence of electric or magnetic field, each electron of free electron gas moves uniformly in a straight line, and in the presence of field, they move according to Newton law's of motion.
(5) The electrons move from one place to another in the metal without any change in energy. During the movement, they occasionally collide elastically with the ions (which are fixed in the lattice) and other free electrons. Between collisions, the interactions of electrons with the others and with the ions are ignored. They have velocities determined at a constant temperature according to Maxwell–Boltzmann distribution law.

© The Author(s) 2022
V. K. Jain, *Solid State Physics*, https://doi.org/10.1007/978-3-030-96017-9_8

(6) Collisions are instantaneous which suddenly change the velocity of electron with a probability per unit time $(1/\tau)$ where τ is known as relaxation time, collision time or mean free time.

(7) Electrons are assumed to reach thermal equilibrium with their surrounding through collisions. They emerge after collision at a random direction with speed appropriate to the temperature of the region where collision has taken place.

8.1.1   Electrical Conductivity

In the absence of an electric field, the free electron moves randomly in the metal. During the motion, they collide with fixed positive ions as well as with other electrons. Since the motion is completely random, therefore the average velocity of electrons in any direction is zero. If a constant electric field E is applied inside a metal, the electrons experience a force $F = -eE$. As a result, they move in a direction opposite to the direction of the electric field. This electron undergoes frequent collisions, and it is assumed that immediately after collision the electron velocities are completely random; that is, electron from a collision does not remember if it had been previously accelerated or not. Thus, the momentum gained under the influence of electric field is lost. As a result of the application of electric field and random motion, the electrons are subjected to a very slow directional motion. This motion is called drift, and the average velocity of this motion is called drift velocity v_d.

When an electric field E is applied to the metal, the electrons are accelerated in the direction of the field and acquire an average drift velocity and momentum p parallel to E. In time dt, an electron of charge $-e$ acquires an additional momentum $-eE dt$ through the acceleration by the field E. Let in time dt, a fraction dn of the total number of electrons n per unit volume makes collision where

$$\frac{dn}{n} = \frac{dt}{\tau} \tag{8.1}$$

where τ is mean time between collisions. Immediately after the collision, the electron velocities are completely random, and hence, the momentum gained under the influence of the electric field is lost. The momentum gained in time dt is

$$-neE dt \tag{8.2}$$

The momentum destroyed in collision from Eq. (8.1) is

$$p dn = n p \frac{dt}{\tau} \tag{8.3}$$

For equilibrium, there must be balance. From Eqs. (8.2) and (8.3)

$$n p \frac{dt}{\tau} = -n e E \, dt$$

$$p = -e E \tau \tag{8.4}$$

The current density J is

$$J = -n e v_d = -\frac{ne}{m} p = \frac{ne^2}{m} E \tau \tag{8.5}$$

The electrical conductivity σ is

$$\sigma = \frac{J}{E} = \frac{ne^2 \tau}{m} \tag{8.6}$$

Thus conductivity is independent of the sign of the charge of the carriers, since reversing this sign changes the direction of their drift motion but not the direction of current flow. From Eq. (8.4), it is seen that mean drift velocity

$$v_d = \frac{p}{m} = -\frac{e}{m} E \tau = \left| \frac{e}{m} \right| E \tau \tag{8.7}$$

The mobility μ is defined as

$$\mu = \frac{v_d}{E} = \left| \frac{e}{m} \right| \tau \tag{8.8}$$

The mobility is thus proportional to τ. The unit of μ is $\frac{m^2}{V\,s}$ or $\frac{cm^2}{V\,s}$.

8.1.2 Wiedemann and Franz Law

Since metals are much better conductors of heat than electrical insulators, therefore it is assumed that the thermal conduction in a metal is also mainly due to free electrons. From Eq. (7.71), the thermal conductivity is

$$K = \frac{1}{3} C_e v l \tag{8.9}$$

The electronic specific heat C_e for electron gas

$$C_e = \frac{1}{3} N v l \frac{dE}{dT} \tag{8.10}$$

The kinetic energy E is

$$E = \frac{1}{2} m v^2 = k_B T \tag{8.11}$$

From Eqs. (8.6) and (8.9)

$$\frac{K}{\sigma} = \frac{(1/3)Nvl(dE/dT)}{N(e^2/m)\tau} = \frac{mv^2}{3e^2}\frac{dE}{dT} \qquad (8.12)$$

If the electron obeys classical statistics

$$E = \frac{1}{2}mv^2 = \frac{3}{2}k_BT \qquad (8.13)$$

$$mv^2 = 3k_BT \qquad (8.14)$$

From Eqs. (8.12)–(8.14)

$$\frac{K}{\sigma} = \frac{3k_BT}{3e^2}\frac{d}{dT}\left(\frac{3}{2}k_BT\right) = \frac{3}{2}\left(\frac{k_B}{e^2}\right)^2 T \qquad (8.15)$$

This equation shows that the ratio of the thermal and electrical conductivity should be proportional to absolute temperature for a given metal and it should be the same for all metals at a given temperature. This is Wiedemann and Franz law. The numerical value of $(K/\sigma T)$ given by Eq. (8.15) is in good agreement with the experimental values for copper, silver and gold over the limited temperature range of the experiments. The experimental values of K and σ themselves and their variation with temperature do not, however, fit the theory.

8.1.3   Specific Heat of Electrons

According to the kinetic theory of gases, free particle has an average energy of $\frac{1}{2}k_BT$ at equilibrium temperature T. The average energy per mole is

$$< E > = N_A\left(\frac{3}{2}k_BT\right) = \frac{3}{2}N_Ak_BT = \frac{3}{2}RT \qquad (8.16)$$

where N_A is Avogadro number and $R = N_Ak_B$. The electronic specific heat is

$$C_e = \frac{\partial <\overline{E}>}{\partial T} = \frac{\partial}{\partial T}\left(\frac{3}{2}RT\right) = \frac{3}{2}R \qquad (8.17)$$

The total heat capacity C in metals including phonons should then

$$C = C_{\text{ph}} + C_e \qquad (8.18)$$

At high temperatures

$$C = 3R + \frac{3}{2}R = 4.5R \tag{8.19}$$

The total heat capacity indicates that for metals C is nearly equal to $3R$ at high temperatures.

8.1.4 Shortcoming of the Model

1. The model does not explain temperature dependence of electrical conductivity and thermal conductivity
2. According to the model, the specific heat of metals should be greater than that of insulators by $3R/2$. In practice, the specific heat of metal at ordinary temperature is not significantly greater than those of insulators, showing that the contribution from electrons is much smaller than $3R/2$.
3. It does not take into account collisions between electrons themselves.
4. The de Broglie wavelength of electrons is on the nanometre scale. Therefore, they cannot be treated as classical particles.

8.2 Free Electron Gas in an Infinite Square Well Potential

Consider a particle of mass m confined in a one-dimensional box of length L. The potential is zero inside the box and infinite elsewhere, that is,

$$V = 0 \quad 0 < x < L$$

$$V = \infty \quad x \leq 0, x \geq L \tag{8.20}$$

The time independent Schrödinger equation in one dimension is

$$\frac{d^2\psi}{dx^2} + \frac{2m}{\hbar^2}[E - V]\psi = 0 \tag{8.21}$$

For $0 < x < L$, $V = 0$, and Eq. (8.21) reduces to

$$\frac{d^2\psi}{dx^2} + \frac{2m}{\hbar^2}E\psi = 0$$

$$\frac{d^2\psi}{dx^2} + k^2\psi = 0 \tag{8.22}$$

where

$$k^2 = \frac{2m}{\hbar^2} E \tag{8.23}$$

The wave number k is real. The general solution of Eq. (8.22) is

$$\psi(x) = A \sin kx + B \cos kx \tag{8.24}$$

where A and B can be any complex number. Since the particle is confined within the box, therefore the probability $|\psi(x)|^2$ of finding the particle outside the box must be infinitesimal or zero. Thus, $\psi(x)$ must also be zero. As $\psi(x) = 0$ for $x \le 0, x \ge L$, we have on the boundary; that is, at $x = 0$ and L.

$$\psi(x = 0) = \psi(x = L) = 0$$

The boundary conditions are $\psi(x) = 0$ at $x = 0$ and L. The derivative of the wave function $d\psi/dx$, cannot vanish at $x = 0, L$, since then the wave function $\psi(x)$ would be trivial solution $\psi(x) = 0$. Thus, $d\psi/dx$ will be discontinuous at the points $x = 0$ and L where the potential makes infinite jumps. Using the boundary condition $\psi(x)$ at $x = 0$, Eq. (8.22) can be written as

$$\psi(0) = 0 = A \sin 0 + B \cos 0 = B \tag{8.25}$$

that is

$$B = 0 \tag{8.26}$$

At $x = L$

$$\psi(L) = 0 = A \sin kL + 0 \cos kL = A \sin kL \tag{8.27}$$

Equation (8.27) implies either $A = 0$ or $\sin kL = 0$, that is

$$kL = n\pi \qquad (n = 0, \pm1, \pm2, \ldots) \tag{8.28}$$

If $A = 0$, then $\psi = 0$ everywhere implying that particle is not in the box. Thus

$$kL = n\pi \qquad (n = 0, \pm1, \pm2, \ldots) \tag{8.29}$$

$$k^2 L^2 = n^2 \pi^2 \qquad (n = 0, \pm1, \pm2, \ldots) \tag{8.30}$$

Substituting the value of k^2 from Eq. (8.23) in Eq. (8.30)

$$\frac{2mE}{\hbar^2} L^2 = n^2 \pi^2$$

$$E = \frac{n^2\pi^2\hbar^2}{2mL^2} \tag{8.31}$$

where

$$n = 0, \pm 1, \pm 2, \ldots \tag{8.32}$$

It is evident that $n = 0$ gives the result that $\psi = 0$ everywhere, corresponding to the case where the particle is not in the box. With $n = -1, -2$, we have the same wave function with change in sign, which gives the same probability when squared and therefore can be neglected. Thus, the allowed values of n are 1, 2, 3,.... and Eq. (8.31) is

$$E = \frac{n^2\pi^2\hbar^2}{2mL^2} \quad (n = 1, 2, \ldots) \tag{8.33}$$

We have labelled the energy with a subscript corresponding to the value of n on the right hand side.

Thus, there is an infinite sequence of discrete energy levels that correspond to all positive integer values of the quantum number n. This is called energy quantization. The quantization entered through the confinement of the particle in the box.

The wave function (eigen function) corresponding to quantum number n using Eq. (8.29) is

$$\psi_n(x) = A \sin kx = A \sin \frac{n\pi}{L}x \tag{8.34}$$

The particle must exist somewhere in space, and the probability $|\psi(x)|^2$ of finding the particle is

$$\int_{-\infty}^{\infty} \psi_n^*(x)\psi_n(x)dx = 1 \tag{8.35}$$

But $\psi_n(x) = 0$ outside $0 < x < L$. Therefore, the condition (8.35) becomes

$$\int_0^L \psi_n^*(x)\psi_n(x)dx = 1 \tag{8.36}$$

Substituting the value of $\psi_n(x)$ from Eq. (8.34) in Eq. (8.36)

$$A^*A \int_0^L \sin^2 \frac{n\pi x}{L}dx = 1$$

$$A^*A \int_0^L \frac{1}{2}\left[1 - \cos \frac{2n\pi x}{L}\right]dx = 1$$

$$A^*A \int_0^L \frac{1}{2}dx - A^*A \int_0^L \frac{1}{2}\left[\cos \frac{2n\pi x}{L}\right]dx = 1$$

$$A^*A \left[\frac{x}{2}\Big|_0^L - \frac{\sin \frac{2n\pi x}{L}}{\frac{2n\pi}{L}}\Big|_0^L\right] = A^*A\left[\frac{L}{2} - 0\right] = 1$$

$$A^*A = \frac{2}{L}$$

$$A = \exp(i\delta)\sqrt{\frac{2}{L}} \qquad (8.37)$$

where δ is a phase factor whose value lies between 0 and 2π. Thus

$$\psi_n(x) = \sqrt{\frac{2}{L}} \exp(i\delta) \sin \frac{n\pi}{L} x \qquad (8.38)$$

$\psi_n^*(x)\psi_n(x) = |\psi_n(x)|^2$ is independent of time. This ensures that the probability of finding the particle somewhere in the box remains constant with time. We choose $\delta = 0$ by convention and thus

$$\psi_n(x) = \sqrt{\frac{2}{L}} \sin \frac{n\pi}{L} x \qquad (8.39)$$

The particle in a box cannot have any arbitrary levels as in classical physics. Since $n = 0$ is not allowed, the lowest possible energy of the particle called zero point energy is nonzero. Further, the separation between energy level decreases as the box length increases and zero point energy decreases.

Now the N electrons are to be accommodated on the line or one-dimensional box. According to the Pauli exclusion principle, no two electrons can have all their quantum number identical. In one-dimensional solid, the quantum number of conduction electrons is n (the principal quantum number), m_s (magnetic quantum number), $m_s = \pm 1/2$, according to spin orientation. A state designated by n can accommodate two electrons, one with spin up and one with spin down.

Let n_F represents the value of n for the topmost filled energy level. Since a level can accommodate only two electrons, therefore the number of electrons $N = 2n_F$. We define Fermi energy of the ground state of N electron system as the energy of the topmost filled level. From Eq. (8.31)

$$E_F = \frac{\hbar^2}{2m} \left(\frac{n_F \pi}{L}\right)^2 = \frac{\hbar^2}{2m} \left(\frac{N\pi}{2L}\right)^2 \qquad (8.40)$$

Let us now consider a particle of mass m inside a cubic box of side L. The potential V is

$$V(x, y, z) = 0 \quad \text{for} \quad 0 \le x \le L; \quad 0 \le y \le L; \quad 0 \le z \le L \qquad (8.41)$$

For the region inside box where the potential is zero, the time independent Schrödinger equation in three dimensions is

$$V(x, y, z) = \infty \quad \text{elsewhere}$$

$$\frac{\partial^2 \psi}{\partial x^2} + \frac{\partial^2 \psi}{\partial y^2} + \frac{\partial^2 \psi}{\partial z^2} + \frac{2mE}{\hbar^2} \psi = 0 \qquad (8.42)$$

where $\psi = \psi(x, y, z)$. Using the method of separation of variables

$$\psi(x, y, z) = X(x)Y(y)Z(z) \tag{8.43}$$

Substituting Eq. (8.43) in Eq. (8.42)

$$YZ\frac{d^2 X}{dx^2} + XZ\frac{d^2 Y}{dy^2} + XY\frac{d^2 Z}{dz^2} + \frac{2mE}{\hbar^2}XYZ = 0 \tag{8.44}$$

On dividing by $\psi(x, y, z)$

$$\frac{1}{X}\frac{d^2 X}{dx^2} + \frac{1}{Y}\frac{d^2 Y}{dy^2} + \frac{1}{Z}\frac{d^2 Z}{dz^2} = -\frac{2mE}{\hbar^2} \tag{8.45}$$

Since the first term of Eq. (8.45) is independent of y and z, therefore the variation of y and z can only change second and third terms, respectively, without changing the first term on the left hand side of Eq. (8.45). But E is constant therefore the second and third terms of the left hand side of Eq. (8.45) also be constant, independent of y, z.

Since the terms on the left hand side of Eq. (8.45) are constant, we can write

$$\frac{1}{X}\frac{d^2 X}{dx^2} = -\frac{2mE_x}{\hbar^2} \tag{8.46}$$

$$\frac{1}{Y}\frac{d^2 Y}{dy^2} = --\frac{2mE_y}{\hbar^2} \tag{8.47}$$

$$\frac{1}{Z}\frac{d^2 Z}{dz^2} = -\frac{2mE_z}{\hbar^2} \tag{8.48}$$

where

$$E = E_x + E_y + E_z \tag{8.49}$$

writing

$$k_x^2 = \frac{2mE_x}{\hbar^2} \tag{8.50}$$

$$k_y^2 = \frac{2mE_y}{\hbar^2} \tag{8.51}$$

$$k_z^2 = \frac{2mE_z}{\hbar^2} \tag{8.52}$$

Using Eqs. (8.50–8.52), Eqs. (8.46–8.48) become

$$\frac{d^2 X}{dx^2} + k_x^2 X = 0 \tag{8.53}$$

$$\frac{d^2Y}{dy^2} + k_y^2 Y = 0 \tag{8.54}$$

$$\frac{1}{Z}\frac{d^2Z}{dz^2} + k_z^2 = 0 \tag{8.55}$$

Equations (8.53)–(8.55) are similar to Eqs. (8.33)–(8.34), therefore

$$E_x = \frac{\pi^2\hbar^2}{2mL^2}n_x^2 \tag{8.56}$$

$$E_y = \frac{\pi^2\hbar^2}{2mL^2}n_y^2 \tag{8.57}$$

$$E_z = \frac{\pi^2\hbar^2}{2mL^2}n_z^2 \tag{8.58}$$

$$X = \sqrt{\frac{2}{L}}\sin\frac{n_x\pi x}{L} \tag{8.59}$$

$$Y = \sqrt{\frac{2}{L}}\sin\frac{n_y\pi y}{L} \tag{8.60}$$

$$Z = \sqrt{\frac{2}{L}}\sin\frac{n_z\pi z}{L} \tag{8.61}$$

where $n_i(i = x, y, z) = 1, 2, 3, \ldots$. The possibility that any one of the n_i could be zero is excluded because this would result in the wave function vanishing everywhere. If there is a particle in the box, the wave function cannot be zero.

From Eqs. (8.49), (8.56)–(8.58)

$$E = \frac{p^2}{2m} = \frac{\pi^2\hbar^2}{2m}\left[\left(\frac{n_x}{L}\right)^2 + \left(\frac{n_y}{L}\right)^2 + \left(\frac{n_z}{L}\right)^2\right] = \frac{\pi^2\hbar^2}{2mL^2}\left(n_x^2 + n_y^2 + n_z^2\right) \tag{8.62}$$

$$\psi = XYZ\sqrt{\frac{8}{L^3}}\sin\frac{n_x\pi x}{L}\sin\frac{n_y\pi y}{L}\sin\frac{n_z\pi z}{L}$$

8.3   Density of States

From Eq. (8.62)

$$\frac{p^2L^2}{\pi^2\hbar^2} = \left(n_x^2 + n_y^2 + n_z^2\right) = R^2 \tag{8.63}$$

This equation represents a sphere of radius R. The number of different sets of integers corresponding to the range between R and $R + dR$ is $4\pi R^2 dR$. Differentiating Eq. (8.63)

$$2RdR = \frac{L^2}{\pi^2\hbar^2}2pdp = \frac{8L^2 pdp}{h^2} \qquad (8.64)$$

Multiplying Eq. (8.64) by $2\pi R$ and using Eq. (8.63)

$$4\pi R^2 dR = \frac{16\pi L^2 pdp}{h^2}\frac{pL}{\pi\hbar} = \frac{32\pi L^3 p^2 dp}{h^3} \qquad (8.65)$$

or

$$\frac{1}{8} \times 4\pi R^2 dR = \frac{1}{8} \times \frac{32\pi L^3 p^2 dp}{h^3} = 4\pi p^2 dp \frac{L^3}{h^3} \qquad (8.66)$$

where the factor 1/8 arises from the fact that the integers asssre positive, and therefore, the number of states of energy/momentum lies between $p(E)$ and $p(E) + dp(E)$ will be given by the positive octant of the sphere. For each set of integers n_x, n_y, n_z, there is one wave function, that is, one state, the spin is not included in this case. Taking into account the fact that electron has a spin which can accept two possible values, the number of possible states corresponding to the momentum range p and $p + dp$ is

$$D(p)dp = \frac{8\pi p^2 dp L^3}{h^3} = \frac{8\pi p^2 dp V}{h^3} \qquad (8.67)$$

where $V = L^3$. From Eq. (8.62)

$$\frac{2pdp}{2m} = EE \quad \text{or} \quad pdp = mdE \qquad (8.68)$$

Substituting Eqs. (8.62) and (8.68) in Eq. (8.67)

$$D(p)dp = D(E)dE = \frac{8\pi\sqrt{2mE}mdEV}{h^3}$$

$$D(E)dE = \frac{4\pi V}{h^3}(2m)^{\frac{3}{2}}E^{\frac{1}{2}}dE \qquad (8.69)$$

This gives the density of states $D(E)$ for a particle of mass m with energy between E and $E + dE$.

8.4 Fermi–Dirac Distribution and Fermi Level

In order to investigate the distribution of electrons above $0\,K$, the Fermi–Dirac distribution function $F(E)$ is used (Fig. 8.1). It gives the probability that an electron has an energy E at a temperature T and is given by

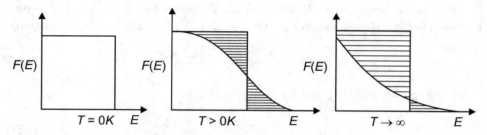

Fig. 8.1 Fermi–Dirac distribution function

$$F(E) = \frac{1}{\exp\left[\frac{E-E_F}{k_B T}\right] + 1}$$

(8.70)

At $T = 0$ K, the $F(E)$ has the property

$$F(E) = \frac{1}{\exp[-\infty] + 1} = 1 \ \text{ for } \ E < E_F$$

(8.71)

and

$$F(E) = \frac{1}{\exp[\infty] + 1} = \frac{1}{\infty} = 0 \ \text{ for } \ E > E_F$$

(8.72)

That is, all energy levels up to $E = E_F$ will be occupied and none will be occupied above E_F. E_F is called Fermi level, and the corresponding energy is called Fermi energy. The physical meaning of E_F is, therefore, that it represents the highest occupied energy level at $T = 0$.

8.5   The Population Density, Fermi Energy, Fermi Wave Vector, Fermi Velocity and Fermi Temperature for Metals

The population density $N(E)$ defined as the number of electrons per unit volume having an energy within a band is

$$N(E) = D(E)F(E)$$

(8.73)

where

$$D(E) = \frac{4\pi}{h^3}(2m)^{\frac{3}{2}} E^{\frac{1}{2}}$$

(8.74)

The total number of electrons is

$$N = \int_0^\infty N(E)dE = \int_0^\infty D(E)F(E)dE \tag{8.75}$$

At $T = 0$ K

$$N = \frac{4\pi}{h^3}(2m)^{\frac{3}{2}} \int_0^{E_F} E^{\frac{1}{2}}dE = \frac{4\pi}{h^3}(2m)^{\frac{3}{2}}\frac{2}{3}E_F^{\frac{3}{2}} = \frac{8\pi}{3h^3}(2mE_F)^{\frac{3}{2}} \tag{8.76}$$

At $T = 0$ K

$$E_F = \left(\frac{3Nh^3}{8\pi}\right)^{\frac{2}{3}}\frac{1}{2m} \tag{8.77}$$

E_F is determined essentially by the number of electrons per unit volume. The Fermi wave vector is given by

$$E_F = \frac{p^2}{2m} = \frac{(\hbar k_F)^2}{2m} \tag{8.78}$$

$$k_F^2 = \frac{8m\pi^2 E_F}{h^2} = \frac{8m\pi^2}{h^2}\left(\frac{3Nh^3}{8\pi}\right)^{\frac{2}{3}}\frac{1}{2m} = 4\pi^2\left(\frac{3N}{8\pi}\right)^{\frac{2}{3}} \tag{8.79}$$

$$k_F = 2\pi\left(\frac{3N}{8\pi}\right)^{\frac{1}{3}} = \left(3\pi^2 N\right)^{\frac{1}{3}} \tag{8.80}$$

The electron velocity v_F at the Fermi surface is

$$v_F = \frac{\hbar k_F}{m} = \frac{\hbar}{m}\left(3\pi^2 N\right)^{\frac{1}{3}} \tag{8.81}$$

The Fermi temperature is

$$T_F = \frac{E_F}{k_B} = \frac{\hbar^2}{2mk_B}\left(3\pi^2 N\right)^{\frac{2}{3}} \tag{8.82}$$

The Fermi temperature T_F is not an actual temperature, but only a convenient reference notation. It has nothing to do with the temperature of the electron gas.

8.6   Variation of Fermi Energy with Temperature

For $k_B T \ll E_F$

For all measurements below melting point of metals $k_B T$ is small compared with E_F ($k_B T \sim 0.025$ eV at 300 K). For $E = E_F$, $F(E) = 1/2$. Hence, the physical meaning of E_F may be stated: at the Fermi level, the probability for occupation is $1/2$. The equation for occupied state is now

$$N(E) = D(E)F(E) = \frac{1}{2\pi^2}\left(\frac{2m}{\hbar^2}\right)^{\frac{3}{2}} \frac{E^{\frac{1}{2}}}{\exp\left[\frac{E-E_F}{k_BT}\right]+1} \tag{8.83}$$

Total number of free electrons will be

$$N = \int_0^\infty N(E)dE = \frac{1}{2\pi^2}\left(\frac{2m}{\hbar^2}\right)^{\frac{3}{2}} \int_0^\infty \frac{E^{\frac{1}{2}}}{\exp\left[\frac{E-E_F}{k_BT}\right]+1}dE \tag{8.84}$$

Consider the integral

$$I = \int_0^\infty \frac{E^{\frac{1}{2}}}{\exp\left[\frac{E-E_F}{k_BT}\right]+1}dE = \frac{2}{3}\int_0^\infty \frac{d\left(E^{\frac{3}{2}}\right)}{\exp\left[\frac{E-E_F}{k_BT}\right]+1} \tag{8.85}$$

Integrating by parts

$$I = \frac{2}{3}E^{\frac{3}{2}} \frac{1}{\exp\left[\frac{E-E_F}{k_BT}\right]+1}\Bigg|_0^\infty + \frac{2}{3k_BT}dE \int_0^\infty \frac{E^{\frac{3}{2}}\exp\left[\frac{E-E_F}{k_BT}\right]}{\left[\exp\left[\frac{E-E_F}{k_BT}\right]+1\right]^2} \tag{8.86}$$

The first term on the right hand side is zero for both the limit. Any function f (E) in the neighbourhood of $E = E_F$ by Taylor's series in power of E-E_F is

$$f(E) = f(E_F) + (E - E_F)f'(E_F) + \frac{(E-E_F)^2}{2!}f''(E_F) + \dots \tag{8.87}$$

If $f(E)$ is $E^{3/2}$, then from Eq. (8.87)

$$E^{\frac{3}{2}} = E_F^{\frac{3}{2}} + \frac{3}{2}(E - E_F)E_F^{\frac{1}{2}} + \frac{3}{4}\frac{(E-E_F)^2}{2!}E_F^{-\frac{1}{2}} + \dots \tag{8.88}$$

The integral I is therefore

$$I$$
$$= \frac{2}{3k_BT}\int_0^\infty \frac{\exp\left[\frac{E-E_F}{k_BT}\right]}{\left[\exp\left[\frac{E-E_F}{k_BT}\right]+1\right]^2}\left[E_F^{\frac{3}{2}} + \frac{3}{2}(E - E_F)E_F^{\frac{1}{2}} + \frac{3}{4}\frac{(E-E_F)^2}{2!}E_F^{-\frac{1}{2}} + \dots\right]dE \tag{8.89}$$

Let

$$\frac{E - E_F}{k_BT} = y, \quad dE = k_BTdy \tag{8.90}$$

Also, the lower limit is taken $(-\infty)$ instead of $\left(-\frac{E}{k_BT}\right)$ because of the E_F in the neighbourhood of E. From Eqs. (8.89) and (8.90)

$$I = \frac{2}{3} \int_{-\infty}^{\infty} \frac{\exp(y)}{\left[\exp(y) + 1\right]^2} E_F^{\frac{3}{2}} dy + k_B T \int_{-\infty}^{\infty} \frac{y\exp(y)}{\left[\exp(y) + 1\right]^2} E_F^{\frac{1}{2}} dy$$

$$+ \frac{(k_B T)^2}{4} \int_{-\infty}^{\infty} \frac{\exp(y)y^2}{\left[\exp(y) + 1\right]^2} E_F^{-\frac{1}{2}} dy \qquad (8.91)$$

Using the standard integrals

$$\int_{-\infty}^{\infty} \frac{\exp(y)dy}{\left[\exp(y) + 1\right]^2} = 1, \quad \int_{-\infty}^{\infty} \frac{y\exp(y)dy}{\left[\exp(y) + 1\right]^2} = 0, \quad \int_{-\infty}^{\infty} \frac{y^2 \exp(y)dy}{\left[\exp(y) + 1\right]^2} = \frac{\pi^2}{3},$$
$$(8.92)$$

We have for Eq. (8.91)

$$I = \frac{2}{3} E_F^{\frac{3}{2}} + 0 + \frac{(k_B T)^2}{4} \frac{\pi^2}{3} E_F^{-\frac{1}{2}} + \ldots = \frac{2}{3} E_F^{\frac{3}{2}} \left[1 + \frac{\pi^2}{8} \left(\frac{k_B T}{E_F} \right)^2 + \ldots \right] \qquad (8.93)$$

Substituting Eq. (8.93) in Eq. (8.84)

$$N = \frac{1}{3\pi^2} \left(\frac{2m E_F}{\hbar^2} \right)^{\frac{3}{2}} \left[1 + \frac{\pi^2}{8} \left(\frac{k_B T}{E_F} \right)^2 \right] \qquad (8.94)$$

At $T = 0$

$$N = \frac{1}{3\pi^2} \left(\frac{2m E_{F_0}}{\hbar^2} \right)^{\frac{3}{2}} \qquad (8.95)$$

where the subscript 0 refers to quantity at $T = 0$

$$E_{F_0} = \frac{\hbar^2}{2m} (3\pi^2 N)^{\frac{2}{3}} \qquad (8.96)$$

From Eqs. (8.94) and (8.96)

$$\frac{1}{3\pi^2} \left(\frac{2m E_{F_0}}{\hbar^2} \right)^{\frac{3}{2}} = \frac{1}{3\pi^2} \left(\frac{2m E_F}{\hbar^2} \right)^{\frac{3}{2}} \left[1 + \frac{\pi^2}{8} \left(\frac{k_B T}{E_{F_0}} \right)^2 \right]$$

$$E_{F_0}^{\frac{3}{2}} = E_F^{\frac{3}{2}} \left[1 + \frac{\pi^2}{8} \left(\frac{k_B T}{E_{F_0}} \right)^2 \right] \qquad (8.97)$$

$$E_F = E_{F_0} \left[1 + \frac{\pi^2}{8} \left(\frac{k_B T}{E_{F_0}} \right)^2 \right]^{-\frac{2}{3}} = E_{F_0} \left[1 - \frac{2}{3} \frac{\pi^2}{8} \left(\frac{k_B T}{E_{F_0}} \right)^2 \right]$$

$$= E_{F_0} \left[1 - \frac{\pi^2}{12} \left(\frac{k_B T}{E_{F_0}} \right)^2 \right] \qquad (8.98)$$

It is observed that E_F decreases as T increases.

8.7   Heat Capacity of Free Electrons

The average energy of an electron in a metal at temperature T is

$$< E >_T = \frac{1}{N} \int_0^\infty EN(E)dE \tag{8.99}$$

Using Eq. (8.83) in Eq. (8.99)

$$< E >_T = \frac{V}{2\pi^2 N} \left(\frac{2m}{\hbar^2}\right)^{\frac{3}{2}} \int_0^\infty \frac{E^{\frac{3}{2}}}{\exp\left[\frac{E-E_F}{k_B T}\right] + 1} dE \tag{8.100}$$

Consider the integral I

$$I = \int_0^\infty \frac{E^{\frac{3}{2}}}{\exp\left[\frac{E-E_F}{k_B T}\right] + 1} dE = \frac{2}{5} \int_0^\infty \frac{d\left(E^{\frac{5}{2}}\right)}{\exp\left[\frac{E-E_F}{k_B T}\right] + 1} dE \tag{8.101}$$

Integrating I by parts

$$I = \frac{2}{5} \frac{d\left(E^{\frac{5}{2}}\right)}{\exp\left[\frac{E-E_F}{k_B T}\right] + 1} \Bigg|_0^\infty + \frac{2}{5} \frac{1}{k_B T} \int_0^\infty \frac{E^{\frac{5}{2}} \exp\left[\frac{E-E_F}{k_B T}\right]}{\left(\exp\left[\frac{E-E_F}{k_B T}\right] + 1\right)^2} dE \tag{8.102}$$

The first term on the right hand side of Eq. (8.102) is zero. The function $E^{5/2}$ in the neighbourhood of $E = E_F$, by Taylor's series, in power of $E - E_F$ is by Eq. (8.87) is

$$E^{\frac{5}{2}} = E_F^{\frac{5}{2}} + \frac{5}{2}(E - E_F)E_F^{\frac{3}{2}} + \frac{15}{4}\frac{(E - E_F)^2}{2!}E_F^{\frac{1}{2}} + \dots \tag{8.103}$$

Substituting this value of $E^{5/2}$ in Eq. (8.102)

I

$$= \frac{2}{5}\frac{1}{k_B T} \int_0^\infty \frac{\exp\left[\frac{E-E_F}{k_B T}\right]}{\left(\exp\left[\frac{E-E_F}{k_B T}\right] + 1\right)^2} \left[E_F^{\frac{5}{2}} + \frac{5}{2}(E - E_F)E_F^{\frac{3}{2}} + \frac{15}{4}\frac{(E - E_F)^2}{2!}E_F^{\frac{1}{2}} + \dots\right] dE \tag{8.104}$$

Assuming $\frac{E-E_F}{k_B T} \gg 1$ and putting

$$\frac{E - E_F}{k_B T} = y, dE = k_B T dy \tag{8.105}$$

Also, the lower limit is taken ($-\infty$) instead of $\left(-\frac{E}{k_BT}\right)$ because of the E_F in the neighbourhood of E.

Equation (8.105) is

$$I = \frac{2}{5}\int_{-\infty}^{\infty} \frac{\exp(y)}{\left[\exp(y)+1\right]^2}\left[E_F^{\frac{5}{2}} + \frac{5}{2}yk_BTE_F^{\frac{3}{2}} + \frac{15}{8}y^2(k_BT)^2E_F^{\frac{1}{2}} + \ldots\right]dy \qquad (8.106)$$

Using the standard integrals [Eq. (8.92)] in Eq. (8.106), we have

$$I = \frac{2}{5}E_F^{\frac{5}{2}} + 0 + \frac{3}{4}(k_BT)^2\frac{\pi^2}{3}E_F^{\frac{1}{2}} + \ldots. \qquad (8.107)$$

Retaining up to second term only and then substituting this integral value in Eq. (8.100)

$$<E>_T = \frac{V}{2\pi^2N}\left(\frac{2m}{\hbar^2}\right)^{\frac{3}{2}}\frac{2}{5}E_F^{\frac{5}{2}}\left[1 + \frac{5\pi^2}{8}\left(\frac{k_BT}{E_F}\right)^2\right] \qquad (8.108)$$

At $T = 0$ K, the average energy is

$$<E>_0 = \frac{V}{2\pi^2N}\left(\frac{2m}{\hbar^2}\right)^{\frac{3}{2}}\frac{2}{5}E_{F_0}^{\frac{5}{2}} = \frac{3}{5}\frac{V}{3\pi^2}\left(\frac{2m}{\hbar^2}\right)^{\frac{3}{2}}E_{F_0}^{\frac{3}{2}}\frac{1}{N}E_{F_0} \qquad (8.109)$$

Using Eq. (8.96)

$$<E>_0 = \frac{3}{5}E_{F_0} \qquad (8.110)$$

Substituting Eq. (8.110) in Eq. (8.108)

$$<E>_T = \frac{<E>_0}{E_{F_0}^{\frac{5}{2}}}E_F^{\frac{5}{2}}\left[1 + \frac{5\pi^2}{8}\left(\frac{k_BT}{E_F}\right)^2\right] \qquad (8.111)$$

Using Eq. (8.98)

$$E_F^{\frac{5}{2}} = E_{F_0}^{\frac{5}{2}}\left[1 - \frac{\pi^2}{12}\left(\frac{k_BT}{E_{F_0}}\right)^2\right]^{\frac{5}{2}} \qquad (8.112)$$

$$E_F^{\frac{5}{2}} = E_{F_0}^{\frac{5}{2}}\left[1 - \frac{5\pi^2}{24}\left(\frac{k_BT}{E_{F_0}}\right)^2\right] \qquad (8.113)$$

From Eqs. (8.111) and (8.113)

$$<E>_T = <E>_0\left[1 - \frac{5\pi^2}{24}\left(\frac{k_BT}{E_{F_0}}\right)^2\right]\left[1 + \frac{5\pi^2}{8}\left(\frac{k_BT}{E_{F_0}}\right)^2\right] \qquad (8.114)$$

$$< E >_T =< E >_0 \left[1 - \frac{5\pi^2}{24}\left(\frac{k_B T}{E_{F_0}}\right)^2 + \frac{5\pi^2}{8}\left(\frac{k_B T}{E_{F_0}}\right)^2 - \frac{5\pi^2}{24}\left(\frac{k_B T}{E_{F_0}}\right)^4 \times \frac{5\pi^2}{8} \right]$$

(8.115)

Ignoring the last term on the right hand side of Eq. (8.115) and rearranging the terms

$$< E >_T =< E >_0 \left[1 + \frac{5\pi^2}{12}\left(\frac{k_B T}{E_{F_0}}\right)^2 \right]$$

(8.116)

Electronic specific heat at constant volume is

$$C_e = \frac{\partial \langle E \rangle_T}{\partial T} = \frac{5\pi^2}{6}\langle E \rangle_0 \left(\frac{k_B}{E_{F_0}}\right)^2 T = \frac{5\pi^2}{6}\frac{3}{5}E_{F_0}\left(\frac{k_B}{E_{F_0}}\right)^2 T = \frac{\pi^2 k_B^2}{2E_{F_0}}T$$

(8.117)

$$C_e = \frac{\pi^2 k_B}{2}\frac{T}{T_F}$$

(8.118)

As $T \ll T_F$ (at normal temperature) is much less than $3/2 k_B$ suggested by a classical free electron model and agrees much better with experiments. At low temperatures, electronic specific heat varies with T at low temperatures.

The electronic specific heat behaviour can be understood qualitatively as follows. Since all electrons within $k_B T$ of the Fermi level are excited, therefore only a fraction $(k_B T/E_F)$ of the electrons is affected. Therefore, the number of electrons excited per mol is about $(N_A k_B T/E_F)$. Now each of these electrons absorbs an energy $k_B T$, on the average. The thermal energy per mole is therefore

$$< E >= N_A \frac{k_B T}{E_F} \times k_B T$$

(8.119)

The electronic specific heat is therefore

$$C_e = \frac{\partial < E >}{\partial T} = \frac{2N_A k_B^2 T}{E_F} = \frac{2R k_B T}{E_F}$$

(8.120)

The specific heat of the electrons is reduced from the classical value by a fraction of $\frac{k_B T}{E_F}$. A typical value for T_F corresponding to $E_F = 5$ eV is 60,000 K. Thus in order for the specific heat of the electron in a solid to reach its classical value, the solid must be heated to a temperature corresponding to T_F. But this is not possible as the solid would have melted and evaporated. At all practical temperatures, therefore, the specific heat of electrons is far below its classical value.

8.8   Thermionic Emission

An electron in a metal or in any solid finds itself in the field of all nuclei and all other electrons. The potential energy for such an electron may therefore is expected to be periodic. According to the free electron model, the potential energy of an electron inside the metal is zero while outside it is taken as infinite. If the potential energy outside is infinitely large, then no electron could ever set out of metal. Therefore, we consider the potential energy outside is E_s $(E_F + e)$ because all the states up to Fermi level E_F are filled. $e\,\varphi$ is work function. That is energy required to remove the electron from the metal (Fig. 8.2).

Since the electrons are not observed to be emitted from the metal at room temperature, therefore it is assumed that the potential energy of an electron at rest inside the metal must be lower than the electron at rest outside the metal. Let us define the energy of a free electron at rest inside the metal as zero. In order to escape from the metal, an electron must have an energy perpendicular to the surface of at least $E_{min}(E_s)$. Thus, if x is the coordinate perpendicular to the surface, an electron must have a momentum $p_x \geq p_{min}$ in order to escape where

$$\frac{(p_x)_{min}}{2m} = E_{min}$$

However, even if an electron at the surface has a momentum $p_x \geq p_{min}$, it does not necessarily escape; but may be reflected by the potential barrier, thus reducing the emission of number of electrons.

The differential current density dJ_x due to these electrons in the direction normal to the emitting surface which is along the x-direction is given by considering a metal block at a temperature $T(K)$. Let the emitting surface of the metal be parallel to the yz plane (Fig. 8.3). Let E be the energy of the electrons in the metal having the momentum components p_x, p_y and p_z and velocity components v_x, v_y and v_z along the x-, y- and z-directions, respectively. The number of electrons occupying the quantum states in the momentum component interval between p_x and $p_x + dp_x$, p_y and $p_y + dp_y$ and p_z and

Fig. 8.2 Free electron model of a metal surface suitable for thermionic emission

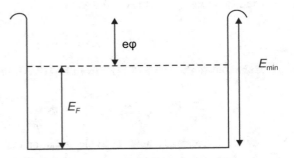

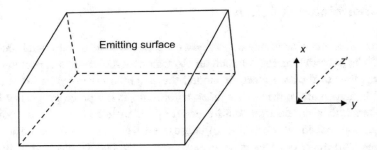

Fig. 8.3 Metal block illustrating the electron emission

$p_z + dp_z$ is given by

$$dn = \frac{2}{h^3} f(E) dp_x dp_y dp_z \qquad (8.121)$$

where $f(E)$ is Fermi–Dirac distribution function which is given by Eq. (8.70)

$$dJ_x = ev_x dn \qquad (8.122)$$

Substituting Eq. (8.121) in Eq. (8.122)

$$dJ_x = \frac{2e}{h^3} f(E) v_x dp_x dp_y dp_z \qquad (8.123)$$

Putting $v_x = p_x/m$ in Eq. (8.123)

$$dJ_x = \frac{2e}{mh^3} f(E) p_x dp_x dp_y dp_z \qquad (8.124)$$

The total emission current density J_x in the direction of x-axis is obtained by integrating Eq. (8.124) for all values of momenta between $-\infty$ and ∞ along the directions of y- and z-axes and between $(p_x)_{min}$ and ∞ along the x-direction. $(p_x)_{min}$ is the minimum momentum in the x-direction that is required to overcome the surface potential energy barrier. Equation (8.124) takes the form

$$J_x = \frac{2e}{mh^3} \int_{(p_x)_{min}}^{\infty} \int_{-\infty}^{\infty} \int_{-\infty}^{\infty} f(E) p_x dp_x dp_y dp_z \qquad (8.125)$$

Substituting the value of $f(E)$ from Eq. (8.70) in Eq. (8.125)

$$J_x = \frac{2e}{mh^3} \int_{(p_x)_{min}}^{\infty} \int_{-\infty}^{\infty} \int_{-\infty}^{\infty} \left[1 + \exp\left(\frac{E - E_F}{k_B T} \right) \right]^{-1} p_x dp_x dp_y dp_z \qquad (8.126)$$

We are interested only in those electrons for which $p_x \geq p_{min}$, which is the total energy of the electrons of interest is at least equal to E_{min}. On the other hand

$$E_{\min} - E_F = \varphi \gg k_B T$$

For all metals at temperature below melting point. Hence, the term of unity in the distribution function may be neglected, that is,

$$1 + \exp\left(\frac{E - E_F}{k_B T}\right) \cong \exp\left(\frac{E - E_F}{k_B T}\right) \tag{8.127}$$

Substituting Eq. (8.127) in Eq. (8.126)

$$J_x = \frac{2e}{mh^3} \int_{(p_x)_{\min}}^{\infty} \int_{-\infty}^{\infty} \int_{-\infty}^{\infty} \exp\left(\frac{E_F - E}{k_B T}\right) p_x \mathrm{d}p_x \mathrm{d}p_y \mathrm{d}p_z$$

$$J_x = \frac{2e}{mh^3} \exp\left(\frac{E_F}{k_B T}\right) \int_{(p_x)_{\min}}^{\infty} \int_{-\infty}^{\infty} \int_{-\infty}^{\infty} \exp\left(\frac{-E}{k_B T}\right) p_x \mathrm{d}p_x \mathrm{d}p_y \mathrm{d}p_z \tag{8.128}$$

The energy E is

$$E = \frac{p_x^2 + p_y^2 + p_z^2}{2m} \tag{8.129}$$

Substituting Eq. (8.129) in Eq. (8.128)

$$J_x = \frac{2e}{mh^3} \exp\left(\frac{E_F}{k_B T}\right) \int_{(p_x)_{\min}}^{\infty} \exp\left(-\frac{p_x^2}{2mk_B T}\right) p_x \mathrm{d}p_x \int_{-\infty}^{\infty} \exp\left(-\frac{p_y^2}{2mk_B T}\right) \mathrm{d}p_y$$

$$\int_{-\infty}^{\infty} \exp\left(-\frac{p_z^2}{2mk_B T}\right) \mathrm{d}p_z \tag{8.130}$$

Using the integral

$$\int_{-\infty}^{\infty} \exp(-ax^2)\mathrm{d}x = \sqrt{\frac{\pi}{a}}$$

We have

$$\int_{-\infty}^{\infty} \exp\left(-\frac{p_y^2}{2mk_B T}\right) \mathrm{d}p_y = \sqrt{2\pi mk_B T} \tag{8.131}$$

$$\int_{-\infty}^{\infty} \exp\left(-\frac{p_z^2}{2mk_B T}\right) \mathrm{d}p_z = \sqrt{2\pi mk_B T} \tag{8.132}$$

Substituting Eqs. (8.131) and (8.132) in Eq. (8.130)

$$J_x = \frac{2e}{mh^3} (2\pi mk_B T) \exp\left(\frac{E_F}{k_B T}\right) \int_{(p_x)_{\min}}^{\infty} \exp\left(-\frac{p_x^2}{2mk_B T}\right) p_x \mathrm{d}p_x \tag{8.133}$$

$$J_x = \frac{2e}{mh^3}(2\pi m k_B T) \exp\left(\frac{E_F}{k_B T}\right) \int_{\frac{(p_x)^2_{min}}{2mk_B T}}^{\infty} \exp\left(-\frac{p_x^2}{2mk_B T}\right)(-mk_B T)\mathrm{d}\left(-\frac{p_x^2}{2mk_B T}\right)$$

$$J_x = -\frac{2e}{h^3}(2\pi m)(k_B T)^2 \exp\left(\frac{E_F}{k_B T}\right)\left[\exp\left(-\frac{p_x^2}{2mk_B T}\right)\right]\bigg|_{\frac{(p_x)^2_{min}}{2mk_B T}}^{\infty}$$

$$J_x = \frac{4\pi m e}{h^3}(k_B T)^2 \exp\left(\frac{E_F}{k_B T}\right)\left[\exp\left(-\frac{(p_x)^2_{min}}{2mk_B T}\right)\right] \tag{8.134}$$

Putting

$$E_{min} = \frac{(p_x)^2_{min}}{2m} \tag{8.135}$$

In Eq. (8.134)

$$J_x = \frac{4\pi m e}{h^3}(k_B T)^2 \exp\left(-\frac{E_{min} - E_F}{k_B T}\right) \tag{8.136}$$

$$J_x = \frac{4\pi m e}{h^3}(k_B T)^2 \exp\left(-\frac{e\varphi}{k_B T}\right) \tag{8.137}$$

where

$$E_{min} - E_F = e\varphi \tag{8.138}$$

Putting

$$A = -\frac{4\pi m e k_B^2}{h^3}, \quad \frac{e\varphi}{k_B} = b \tag{8.139}$$

In Eq. (8.137)

$$J_x = AT^2 \exp\left(-\frac{b}{T}\right) \tag{8.140}$$

Equation (8.140) is known as Richardson–Dushman equation. Equation (8.140) can be written as

$$\frac{J_x}{T^2} = A \exp\left(-\frac{b}{T}\right) \tag{8.141}$$

Taking logarithm

$$\ln \frac{J_x}{T^2} = \ln A - \frac{b}{T} \tag{8.142}$$

$$\log_{10} \frac{J_x}{T^2} = \log_{10} A - \frac{0.434b}{T} \tag{8.143}$$

If A and b are assumed constants, then a plot of $\log_{10} \frac{J_x}{T^2}$ versus $1/T$ should be straight line with a slope $-0.434b$ and an intercept $\log_{10} A$ on the $\log_{10} \frac{J_x}{T^2}$ axis (Fig. 8.4).

Experimentally it is found that plots of $\log_{10} \frac{J_x}{T^2}$ versus $1/T$ are straight lines for most of the emitting surfaces. From the plot, A and b can be determined. The value of b leads to the determination of the work function. The fact that the experimental values of $\log_{10} \frac{J_x}{T^2}$ versus $1/T$ give a straight line shows that Richardson's equation correctly gives the nature of variation of the emission current with temperature. For most of the metals experimentally determined value of A is found to be 60 amp cm^{-2} K^{-2} in sharp contrast to the theoretical value of 120.4 ampcm^{-2} K^{-2}. The important factors contributing to the disparity between the theoretical and experimental values of A are the following.

(i) In obtaining Richardson's equation, it is assumed that all the electrons having kinetic energy associated with the velocity component normal to the emitting surface equal to, or greater than the surface potential energy barrier escape from the surface. According to quantum mechanics, some electrons are reflected back from the potential energy barrier back into the surface. If r represents the fraction of the electrons reflected from the barrier, Eq. (8.140) takes the form

$$J_x = A(1-r)T^2 \exp\left(-\frac{b}{T}\right) \qquad (8.144)$$

(ii) As a result of thermal expansion, the work function may vary with temperature. Further, work function is very sensitive to the impurities on the metal surface.
(iii) The apparent work function increases if a negative space charge exists in the vicinity of the emitter.
(iv) The number of electrons emitted may depend on the nature of surface of the metal.

Fig. 8.4 A plot of $\log_{10}\left(\frac{J_x}{T^2}\right)$ versus (1/T) for an emitter

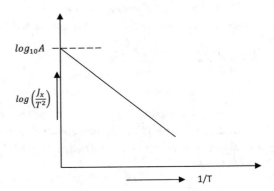

(v) In the derivation of Eq. (8.140), it has been assumed that the work function is same over the whole area of the emitter; this is valid only if the emitter is a single crystal, because φ varies from one crystallographic plane to another.

In view of the discrepancy in the value of A, it is appropriate to consider Richardson's equation as an empirical relation with A and b treated as numerical constants which are adjusted to fit the experimental data.

8.9   Boltzmann Equation

In the steady state, flow of electricity or heat, the distribution function for momentum component or space coordinates of the electron will be different from that in thermal equilibrium in the absence of flow.

Consider the electron distribution function $f(r, p, t)$. This is the number of electrons at point r with momentum p in phase space at any moment t. The function $f(r, p, t)$ can be changed because of

1. Motion of electrons; that is, the electrons are moving into the volume element around $\vec{r}$.
2. External forces which are acting on the electrons can cause a change in the momentum of the electrons.
3. The scattering or collisions of electron with impurities, defects, electrons, phonons etc.

In the steady state, the function $f(r, p, t)$ should not change, that is,

$$\frac{\mathrm{d} f(r, p, t)}{\mathrm{d} t} = 0 \tag{8.145}$$

Since change in distribution function can be caused by diffusion, external forces or collision, therefore, the Eq. (8.145) can be written as

$$\frac{\mathrm{d} f(r, p, t)}{\mathrm{d} t} = \frac{\partial f(r, p, t)}{\partial t}\bigg|_{\text{Diffusion}} + \frac{\partial f(r, p, t)}{\partial t}\bigg|_{\text{External field}} + \frac{\partial f(r, p, t)}{\partial t}\bigg|_{\text{collision}} = 0 \tag{8.146}$$

Let x, y, z represent the components of the position r of electron, and p_x, p_y, p_z represent the components of the momentum $\vec{p}$ of the electron. Then the number of electrons in volume element $\mathrm{d}x\mathrm{d}y\mathrm{d}z\mathrm{d}p_x\mathrm{d}p_y\mathrm{d}p_z$ is given by

$$f(x, y, z; p_x, p_y, p_z; t)\mathrm{d}x\mathrm{d}y\mathrm{d}z\mathrm{d}p_x\mathrm{d}p_y\mathrm{d}p_z \tag{8.147}$$

or

$$f(\mathbf{r}, \mathbf{p}, t)d^3 r d^3 p$$

At any instant $t + \delta t$ (δt is very small), the position coordinates can be written as

$$x + dx, y + dy, z + dz$$

or

$$x + p_x \frac{\delta t}{m}; y + p_y \frac{\delta t}{m}; z + p_z + \frac{\delta t}{m} \tag{8.148}$$

where m is the mass of the electron. The momentum coordinates are

$$p_x + dp_x, p_y + dp_y, p_z + dp_z$$

or

$$p_x + F_x \delta t; p_y + F_y \delta t; p_z + F_z \delta t \tag{8.149}$$

where F_x, F_y, F_z are the components of the external force $\vec{F}$.

As the time passes, the volume element (in the absence of collisions) in such a way that the distribution function does not change, that is,

$$f\left(x, y, z; p_x, p_y, p_z; t\right) = f\left(x + p_x \frac{\delta t}{m}, y + p_y \frac{\delta t}{m}, z + p_z \right.$$
$$\left. + \frac{\delta t}{m}; p_x + F_x \delta t, p_y + F_y \delta t, p_z + F_z \delta t; t + \delta t\right)$$

or

$$f(\mathbf{r}, \mathbf{p}, t) = f(\mathbf{r} + d\mathbf{r}; \mathbf{p} + d\mathbf{p}, t + dt) \tag{8.150}$$

However, collision does occur, thus

$$f(\mathbf{r} + d\mathbf{r}; \mathbf{p} + d\mathbf{p}, t + dt) - f(\mathbf{r}, \mathbf{p}, t) = \left. \frac{\partial f(\mathbf{r}, \mathbf{p}, t)}{\partial t} \right|_{\text{collision}} dt \tag{8.151}$$

Expanding the left hand side of Eq. (8.151)

$$f(\mathbf{r}, \mathbf{p}, t) + d\mathbf{r} \cdot \nabla_r f(\mathbf{r}, \mathbf{p}, t) + d\mathbf{p} \cdot \nabla_p f(\mathbf{r}, \mathbf{p}, t)$$
$$+ dt \frac{\partial f(\mathbf{r}, \mathbf{p}, t)}{\partial t} - f(\mathbf{r}, \mathbf{p}, t) = \left. \frac{\partial f(\mathbf{r}, \mathbf{p}, t)}{\partial t} \right|_{\text{collision}} dt$$

or

$$\frac{d\mathbf{r}}{dt} \cdot \nabla_r f + \frac{d\mathbf{p}}{dt} \cdot \nabla_p f + \frac{\partial f}{\partial t} = \left. \frac{\partial f}{\partial t} \right|_{\text{collision}}$$

$$\frac{\partial f}{\partial t} + \boldsymbol{v} \cdot \nabla_r f + \boldsymbol{F} \cdot \nabla_p f = \left.\frac{\partial f}{\partial t}\right|_{\text{collision}} \tag{8.152}$$

where $\boldsymbol{F}$ is external force. Equation (8.152) is Boltzmann equation. In the steady state, $\frac{\partial f}{\partial t} = 0$. Therefore, Eq. (8.152) in the steady state is

$$\boldsymbol{v} \cdot \nabla_r f + \boldsymbol{F} \cdot \nabla_p f = \left.\frac{\partial f}{\partial t}\right|_{\text{collision}} \tag{8.153}$$

The collision or scattering restores equilibrium which was disturbed by the external forces. Let $f(\boldsymbol{r}, \boldsymbol{p}, t)$ does change only slightly from its equilibrium value f_0 and in time τ called relaxation time, the system attains equilibrium. The relaxation time τ generally depends on $\boldsymbol{r}$ and v. The change in f with time due to collision can be written as

$$\left.\frac{\partial f}{\partial t}\right|_{\text{collision}} = -\frac{(f - f_0)}{\tau} \tag{8.154}$$

$$\frac{\partial f}{(f - f_0)} = -\frac{\partial t}{\tau}$$

On integration

$$f(t) = f_0 + [f(0) - f_0]\exp\left(-\frac{t}{\tau}\right)$$

From Eqs. (8.152) and (8.154)

$$\frac{\partial f}{\partial t} + \boldsymbol{v} \cdot \nabla_r f + \boldsymbol{F} \cdot \nabla_p f = -\frac{(f - f_0)}{\tau} \tag{8.155}$$

$$\frac{\partial f}{\partial t} + \boldsymbol{v} \cdot \nabla_r f + \boldsymbol{a} \cdot \nabla_v f = -\frac{(f - f_0)}{\tau} \tag{8.156}$$

where $\boldsymbol{F} = m\boldsymbol{a}$, $\boldsymbol{a}$ is acceleration. Equation (8.155) or (8.156) is Boltzmann transport equation in relaxation time approximation.

8.10 Electrical Conductivity

Consider an electric field $\boldsymbol{E}$ applied to a metal in the y-direction. As a result of this field, the electrons will be moving with a velocity v_y in the y-direction. The force on the electron is

$$F_y = -eE_y$$

The current density is given by

$$j_y = -nev_y \tag{8.157}$$

where n is the average number of electrons per unit volume. To determine the value of n, consider the number of possible states per unit volume with momentum between $\mathbf{p}$ and $\mathbf{p} + d\mathbf{p}$. The number of states is given by Eq. (8.67) and is

$$D(p)dp = \frac{8\pi}{h^3} p^2 dp \tag{8.158}$$

Consider the number of electrons which have momentum in the range dp_x at p_x, dp_y at p_y and dp_z at p_z. Now construct a plot in momentum space such that each point represents a particular component p_x, p_y and p_z with

$$p_x^2 + p_y^2 + p_z^2 = p^2$$

This will be spherical shell with radius between $\mathbf{p}$ and thickness $d\mathbf{p}$. This spherical shell in momentum space contains all the states with momentum between $\mathbf{p}$ and $\mathbf{p} + d\mathbf{p}$. Therefore, the number of states in the volume element $dp_x dp_y dp_z$ would be

$$D(p_x, p_y, p_z)dp_x dp_y dp_z = \frac{D(p)dp}{4\pi\, p^2 dp} dp_x dp_y dp_z \tag{8.159}$$

From Eqs. (8.158) and (8.159)

$$D(p_x, p_y, p_z)dp_x dp_y dp_z = \frac{8\pi}{h^3} \frac{p^2 dp}{4\pi\, p^2 dp} dp_x dp_y dp_z$$

$$D(p_x, p_y, p_z)dp_x dp_y dp_z = \frac{2}{h^3} dp_x dp_y dp_z \tag{8.160}$$

In thermal equilibrium and in the absence of fields, the average number of occupied states would be

$$\frac{2}{h^3} f_0(p)dp_x dp_y dp_z \tag{8.161}$$

where $f_0(p)$ is the Fermi–Dirac distribution function in terms of total momentum p. In the steady state current, the average number of electrons per unit volume in the range of $dp_x dp_y dp_z$ from Eq. (8.161) would be

$$n = \iiint \frac{2}{h^3} f_0(p)dp_x dp_y dp_z \tag{8.162}$$

From Eqs. (8.157) and (8.162)

$$j_y = -\frac{2e}{h^3} \iiint v_y(f - f_0)dp_x dp_y dp_z \tag{8.163}$$

The term f_0 is added to indicate that the current is essentially determined by the deviation $(f - f_0)$ from the Fermi–Dirac distribution, f_0 is spherically symmetric. From Eqs. (8.155) and (8.157)

$$-eE_y \frac{\partial f}{\partial p_y} = -\frac{f - f_0}{\tau}$$

Since deviation of f from f_0 is very small, therefore f is replaced by f_0 and the above equation can be written as

$$eE_y \frac{\partial f_0}{\partial p_y} = \frac{f - f_0}{\tau} \tag{8.164}$$

Since

$$E = \frac{p_x^2}{2m} + \frac{p_y^2}{2m} + \frac{p_z^2}{2m}$$

Therefore

$$\frac{\partial f_0}{\partial p_y} = \frac{\partial f_0}{\partial E} \frac{\partial E}{\partial p_y} = \frac{\partial f_0}{\partial E} \frac{2p_y}{2m} = v_y \frac{\partial f_0}{\partial E} \tag{8.165}$$

From Eqs. (8.164) and (8.165)

$$\frac{f - f_0}{\tau} = eE_y v_y \frac{\partial f_0}{\partial E} \tag{8.166}$$

From Eqs. (8.163) and (8.166)

$$j_y = -\frac{2e^2}{h^3} E_y \iiint v_y^2 \tau \frac{\partial f_0}{\partial E} dp_x dp_y dp_z \tag{8.167}$$

Further, τ and $\frac{\partial f_0}{\partial E}$ are functions of energy only. Replacing v_y^2 by $\frac{v^2}{3}$ and $dp_x dp_y dp_z$ by $4\pi p^2 dp$ in Eq. (8.167)

$$j_y = -\frac{16\pi e^2 \sqrt{2m}}{3h^3} E_y \int_0^\infty E^{\frac{3}{2}} \tau(E) \frac{\partial f_0}{\partial E} dE \tag{8.168}$$

All the electrons take part in conduction mechanism. However, only the relaxation time of electrons at the Fermi surface occurs in the conductivity. Thus, the distribution function is changed only near the Fermi surface. To a good approximation $E^{\frac{3}{2}} \tau(E)$ under the integral sign may thus be replaced by the quantity $E_F^{\frac{3}{2}} \tau_F$ in front of the integer. Since $\frac{\partial f_0}{\partial E}$ is only nonzero over a small range of E at the Fermi surface and treating $-\frac{\partial f_0}{\partial E}$ as a delta function, that is,

$$\int_0^\infty \frac{\partial f_0}{\partial E} dE = -1 \qquad (8.169)$$

From Eqs. (8.77), (8.168) and (8.169)

$$j_y = \frac{16\pi e^2 \sqrt{2m}}{3h^3} E_y \frac{3nh^3}{8\pi 2m\sqrt{2m}} \tau_F = \frac{ne^2 \tau_F}{m} E_y$$

The electric conductivity σ is

$$\sigma = \frac{j_y}{E_y} = \frac{ne^2 \tau_F}{m} \qquad (8.170)$$

where n is the number of electrons per unit volume.

Solved Examples

Example 1 An electron is confined in the ground state of a one-dimensional box of width 10^{-10} m. Its energy is 38 eV. Calculate the energy of the electron in its first excited state.

Solution
From Eq. (8.33)

$$E_n = \frac{n^2 \pi^2 \hbar^2}{2mL^2} \quad (n = 1, 2, 3, \cdots)$$

$$E_1 = \frac{\pi^2 \hbar^2}{2mL^2}$$

The first excited state ($n = 2$) is

$$E_2 = \frac{4\pi^2 \hbar^2}{2mL^2} = 4E_1 = 4 \times 38\,\text{eV} = 152\,\text{eV}$$

Example 2 For a particle in a box, show that the fractional difference in the energy between adjacent eigenvalues is

$$\frac{\Delta E_n}{E_n} = \frac{2n + 1}{n^2}$$

Solution
From Eq. (8.33)
$$E_n = \frac{n^2 \pi^2 \hbar^2}{2mL^2} \text{ and } E_{n+1} = \frac{(n+1)^2 \pi^2 \hbar^2}{2mL^2}$$

$$\frac{E_{n+1} - E_n}{E_n} = \frac{\left[(n^2 + 2n + 1) - n^2\right]\pi^2\hbar^2}{n^2\pi^2\hbar^2} = \frac{2n + 1}{n^2}$$

Example 3 Calculate the magnitude of the Fermi wave vector for 4.2×10^{21} electrons confined in a box of volume 1 cm^3.

Solution
From Eq. (8.80)

$$k_F = \left(3\pi^2 N\right)^{\frac{1}{3}} = \left(3 \times (3.14)^2 \times 4.2 \times 10^{21}\right)^{\frac{1}{3}} = 4.989 \times 10^7 \text{cm}^{-1}$$

Example 4 Calculate the magnitude of the Fermi energy in eV for 4.2×10^{21} electrons confined in a box of volume 1 cm^3.

Solution
From Eq. (8.78)

$$E_F = \frac{(\hbar k_F)^2}{2m}$$

Substituting the value of k_F from Example 3

$$E_F = \frac{\left(6.62 \times 10^{-34} \text{ Js} \times 4.989 \times 10^9 \text{ m}^{-1}\right)^2}{8 \times (3.14)^2 \times 9.11 \times 10^{-31} \text{ Kg}} = 1.51 \times 10^{-19} \text{ J}$$

$$E_F = \frac{1.51 \times 10^{-19}}{1.602 \times 10^{-19}} \text{ eV} = 0.9425 \text{ eV}$$

Example 5 Calculate the Fermi energy (in eV) and pressure of electrons in silver at 0 K, assuming that it has one free electron per atom. The number density of electron $= 5.9 \times 10^{28}$ m^{-3}, mass of the electron $= 9.11 \times 10^{-31}$ kg and $h = 6.62 \times 10^{-34}$ Js.

Solution
We have from Eq. (8.77)

$$E_F = \frac{\hbar^2}{2m}\left(3\pi^2 N\right)^{\frac{2}{3}} = \frac{\left(6.62 \times 10^{-34}\right)^2}{8 \times (3.14)^2 \times 9.11 \times 10^{-31} \text{ Kg}}\left[3 \times (3.14)^2 \times 5.9 \times 10^{28}\right]^{\frac{2}{3}}$$

$$E_F = 8.83596 \times 10^{-19} \text{ J} = \frac{8.83596 \times 10^{-19}}{1.602 \times 10^{-19}} = 5.52 \text{ eV}$$

Now

$$\frac{1}{2}m\bar{v}^2 = \frac{3}{5}E_F(0)$$

$$\text{Pressure} = \frac{2}{5}\left(\frac{N}{V}\right)E_F = \frac{2}{5}(5.9 \times 10^{28}) \times 8.83596 \times 10^{-19} = 20.85 \times 10^9 \text{ J/m}^3$$

$$\text{Pressure} = \frac{20.85 \times 10^9 \text{ J/m}^3}{1.013 \times 10^5 \text{ Nm}^{-2}} = 2.058 \times 10^5 \text{ atmosphere}$$

Example 6 For a system of non-interacting electrons, show that the probability of finding an electron in a state with energy Δ above E_F is the same as the probability of finding an electron absent from a state with energy Δ below E_F at any given temperature.

Solution
According to the Fermi–Dirac distribution, the probability for a level E to be occupied is given by Eq. (8.70)

$$F(E) = \frac{1}{\exp\left[\frac{E-E_F}{k_BT}\right] + 1}$$

The probability for finding an electron at $E = E_F + \Delta$ is

$$F(E) = \frac{1}{\exp\left[\frac{E_F+\Delta-E_F}{k_BT}\right] + 1} = \frac{1}{\exp\left[\frac{\Delta}{k_BT}\right] + 1}$$

The probability for not finding electrons at $E = E_F - \Delta$ is

$$1 - F(E_F - \Delta) = 1 - \frac{1}{\exp\left[\frac{E_F-\Delta-E_F}{k_BT}\right] + 1}$$

$$= \frac{\exp\left[-\frac{\Delta}{k_BT}\right]}{\exp\left[-\frac{\Delta}{k_BT}\right] + 1} = \frac{1}{1 + \exp\left[\frac{\Delta}{k_BT}\right]}$$

The two probabilities have the same value.

Example 7 Evaluate the temperature at which there is one per cent probability that state with an energy 0.5 eV above the Fermi energy will be occupied by an electron.

Solution
From Eq. (8.70), we have

$$F(E) = \frac{1}{\exp\left[\frac{E-E_F}{k_BT}\right] + 1}$$

$$0.01 = \frac{1}{\exp[x] + 1}$$

where $x = \frac{E - E_F}{k_B T}$

$$1 + \exp[x] = 100$$

$$\exp[x] = 99$$

$$x = 2.303 \log 99$$

Putting
$E - E_F = 0.5 \, eV$ and $x = \frac{0.5 \, eV}{k_B T}$.
We have

$$x = \frac{0.5 \, eV}{k_B T} = 2.303 \log 99$$

$$k_B T = \frac{0.5 \text{ eV}}{2.303 \times \log 99} = 0.109 \text{ eV} = 0.109 \times 1.6 \times 10^{-19} \text{ J}$$

$$T = \frac{(0.109 \times 1.6 \times 10^{(-19)})}{(1.38 \times 10^{(-23)})} K = 1264 \text{ K}$$

Example 8 Obtain the expression of wavelength associated with an electron having an energy equal to the Fermi energy.

Solution
We have from Eq. (8.77)

$$E_F = \left(\frac{3Nh^3}{8\pi}\right)^{\frac{2}{3}} \frac{1}{2m}$$

The Fermi wave vector is given by

$$E_F = \frac{p^2}{2m} = \frac{(\hbar k_F)^2}{2m}$$

$$k_F^2 = \frac{8m\pi^2 E_F}{h^2} = \frac{8m\pi^2}{h^2}\left(\frac{3Nh^3}{8\pi}\right)^{\frac{2}{3}} \frac{1}{2m} = 4\pi^2\left(\frac{3N}{8\pi}\right)^{\frac{2}{3}}$$

$$k_F = \frac{2\pi}{\lambda_F} = 2\pi\left(\frac{3N}{8\pi}\right)^{\frac{1}{3}} = (3\pi^2 N)^{\frac{1}{3}}$$

$$\lambda_F = \left(\frac{8\pi}{3N}\right)^{\frac{1}{3}}$$

Example 9 Obtain an expression for thermal conductivity of the metal using Drude model.

Solution

Consider a long rod as shown in Fig. 8.5. Temperature at left end is T_1 and at the right end is T_2 such that $T_2 > T_1$. The temperature gradient $\frac{dT}{dz}$ is positive. Heat flows from high temperature end to low temperature end. The flow of energy per unit area per unit time j_U is related to ΔT as

$$j_U = -K\Delta T \tag{8.171}$$

where K is the thermal conductivity of the metal. The direction of j_U is always in a direction opposite to direction of temperature gradient. j_U is directed towards low temperature side while temperature gradient is directed towards the high temperature. K is always positive.

The energy of an electron depends on the temperature of the place where it had its last collision, the higher the temperature the more the energy it comes with. The energy per electron at equilibrium at z is $E[T(z)]$. If there are n electrons with velocity v_z, then $\frac{n}{2}v_z$ is the number density towards $+z$-direction or $-z$-direction. At point z, the most nearby collision occurs at $z - v_z\tau$ and $z + v_z\tau$ in the $-z$- and $+z$-directions, respectively. The energy acquired by the electrons due to collisions at z is $E[T(z - v_z\tau)]$ for electrons coming from $-z$-direction and $E[T(z + v_z\tau)]$ for electrons coming from $+z$-direction. The thermal current density is

$$j_U = \frac{nv_z}{2}\{E[T(z - v_z\tau)] - E[T(z + v_z\tau)]\} \tag{8.172}$$

This is the energy arriving at z per unit time per unit area. If $v_z\tau$ is small, we may expand right hand side of Eq. (8.172) about z and obtain

Fig. 8.5 Thermal conduction through a metal rod

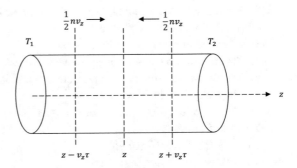

$$j_U = -n v_z^2 \tau \left(\frac{dE}{dT}\right)\left(\frac{dT}{dz}\right) \tag{8.173}$$

In three dimensions, z component of the average velocities of electron can have components along x and y. Since

$$\langle v_x^2 \rangle = \langle v_y^2 \rangle = \langle v_z^2 \rangle = \frac{v^2}{3} \tag{8.174}$$

From Eqs. (8.173) and (8.174)

$$j_U = -\frac{1}{3} n v^2 \tau \left(\frac{dE}{dT}\right)\left(\frac{dT}{dz}\right) \tag{8.175}$$

Putting $n\frac{dE}{dT} = C_e$ electronic specific heat in Eq. (8.175)

$$j_U = -\frac{1}{3} v^2 \tau C_e \left(\frac{dT}{dz}\right) \tag{8.176}$$

Comparing Eq. (8.171) and (8.176), the thermal conductivity is

$$K = \frac{1}{3} v^2 \tau C_e = \frac{1}{3} l v C_e \tag{8.177}$$

where l is mean free path.

Objective Type Questions

1. Which is not true?
   (a) Drude–Lorentz model does not explain the temperature dependence of electrical conductivity
   (b) Drude–Lorentz model does not explain the temperature dependence of thermal conductivity
   (c) According to Drude–Lorentz model, the total heat capacity in metals is 3R
   (d) Wiedemann and Franz law states that ratio of thermal and electrical conductivity is proportional to temperature
2. For a particle in a box in one dimension
   (a) The separation between energy levels decreases as the box length decreases
   (b) The zero-point energy for a particle in a box is not equal to zero
   (c) The energy is given by $E_n = \frac{n^2 \pi^2 \hbar^2}{2mL}$
   (d) The wave function vanishes at $x = 0$ and L
3. The unit of mobility is
   (a) $\frac{m^2}{V s}$ (b) $\frac{m}{V}$ (c) $\frac{V s}{m}$ (d) $\frac{m^2}{(V s)^2}$
4. The Fermi wave vector is given by
   (a) $(3\pi N)^{\frac{1}{3}}$ (b) $(3\pi N)^{\frac{2}{3}}$ (c) $(3\pi N^2)^{\frac{1}{3}}$ (d) $(3\pi^2 N)^{\frac{1}{3}}$

5. The Fermi energy at 0 K is

(a) $\left(\frac{3Nh^3}{8\pi}\right)^{\frac{2}{3}}$ (b) $\frac{1}{2m}\left(\frac{3Nh^3}{8\pi}\right)^{\frac{2}{3}}$ (c) $\frac{1}{2m}\left(\frac{3Nh^3}{8\pi}\right)^{\frac{3}{2}}$ (d) $\left(\frac{3Nh^3}{8\pi}\right)^{\frac{3}{2}}$

6. The electron velocity at the Fermi surface is

(a) $\left(3\pi^2 N\right)^{\frac{1}{3}}$ (b) $\frac{\hbar^2}{2mk_B}\left(3\pi^2 N\right)^{\frac{1}{3}}$ (c) $\frac{\hbar}{m}\left(3\pi^2 N\right)^{\frac{1}{3}}$ (d) $\frac{(\hbar k_F)^2}{2m}$

7. The Fermi temperature is

(a) $\left(3\pi^2 N\right)^{\frac{1}{3}}$ (b) $\frac{\hbar^2}{2mk_B}\left(3\pi^2 N\right)^{\frac{2}{3}}$ (c) $\frac{\hbar}{m}\left(3\pi^2 N\right)^{\frac{1}{3}}$ (d) $\frac{(\hbar k_F)^2}{2m}$

8. Which is not true ?

(a) Fermi temperature is not an actual temperature
(b) Fermi energy decreases as temperature increases
(c) Fermi energy is determined by the number of electrons per unit volume
(d) Electronic specific heat is proportional to T^2

Problems

1. Calculate the extent of energy range between $F(E) = 0.9$ and 0.1 at 200 K and express it as a function of E_F which is 3 eV.
2. The entropy S of a Fermi–Dirac gas at a finite temperature is given by the relation. Show that entropy is proportional to temperature.
3. Calculate the electronic specific heat at 300 K for silver. The Fermi temperature of silver is 6.37×10^4 K.
4. Calculate the number of states lying in an energy interval of 0.02 eV above the Fermi energy for sodium of unit volume. For sodium $E_F = 3.22$ eV.
5. Calculate the mean free path of sodium, if its Fermi energy is 2.1 eV and the electrical conductivity is 1.5×10^7 Ω^{-1} m^{-1}.
6. Consider a one-dimensional box of width 1 mm. Show that $n = 1.63 \times 10^4$ corresponds to a state of energy 0.01 eV.
7. Calculate the number of energy states available for the electrons in a cubical box of side 1 cm lying below an energy of 1 eV.
8. Estimate the relative contribution of electrons and lattice to the specific heat of sodium at constant volume at 20 K. The Fermi temperature of sodium is 3.8×10^4 K, and its Debye temperature is 150 K.

Answers
Objective Type Questions

1. (c)   2. (b)   3. (a)   4. (d)   5. (b)   6. (c)   7. (b)   8. (d)

Band Theory

9

9.1 Electronic Energy Levels of a Free Atom

The state of an electron in an atom is determined by four quantum numbers: the principal n, the orbital l, the magnetic m_l and the spin m_s numbers. In a hydrogen atom the principal quantum number, n, describes the steady state energy of the electron

$$E = -\frac{R}{n^2}$$

where R (=13.6 eV) is called Rydberg constant and $n = 1, 2, 3, \ldots$.

The orbital quantum number, l, describes the orbital angular momentum L

$$L = \hbar\sqrt{l(l+1)}$$

where $l = 0, 1, 2, \ldots, n-1$, a total of n values.

The magnetic quantum number, m_l, describes the orientation of the orbital angular momentum with respect to some specified direction. The orientation of L with respect to this direction is such that its projection on to this direction is multiple of $\hbar$

$$L_z = m_l \hbar$$

The number m_l may assume the following set of integral values

$$m_l = -l, -(l+1), \ldots, 0, 1, 2, \ldots, l$$

$(2l + 1)$ values in all.

The spin quantum number m_s describes the orientation of the spin-angular momentum with respect to the specified direction, the vector s_z may only with respect to this direction so that its projection along this direction is

V. K. Jain, *Solid State Physics*, https://doi.org/10.1007/978-3-030-96017-9_9

$$s_z = m_s \hbar$$

where m_s can take values 1/2 and $-1/2$.

The states with orbital quantum number $l = 0$ are termed as s-states; those with $l = 1$ are termed as p states; $l = 2$, d state; $l = 3$, f states, etc. Electrons in those states are termed s–, p–, d–, f–, etc. electrons.

In contrast to the hydrogen atom the energy of an electron in many electron atom (e.g. Li, Na, Ge, Si) depends not only on n but on l as well. Only discrete values of n and l being allowed, the energy spectrum of electrons in atom may assume only discrete values too. All the s levels are non-degenerate. The p levels are threefold degenerate. The degeneracy of d levels is fivefold and so on. Generally, a level with orbital quantum number l is a $(2l + 1)$ fold degenerate and can accommodate $2(2l + 1)$ electrons.

9.2   Origin of Energy Bands

To understand the concept of energy bands, let us start from isolated atoms and then bring them together to form a solid. Isolated atoms have discrete energy levels. When atoms come closer, the electrons respond to the influence of the other nuclei and electrons. When the distance between the atoms becomes comparable to or less than the spatial extension of the electronic wave function associated with a particular atom, the consequence is that the electrons are no longer identifiable with specific atom, but belong to the crystal as a whole. According to the Pauli exclusion principle no two electrons could have the same identical set of quantum numbers (i.e. being in the same identical energy state). Similarly in the large environment of the crystal, no two electrons can have the same identical energy state. Therefore, each atomic level must split into multiplet levels to allow all the electrons which formerly were in similar single atomic levels. What was a single level for the isolated atoms broadened out into a large number of closely spaced levels of the solid. Such continuous energy levels having finite energy width are called energy band. In general, each energy band is separated from neighbouring energy band by a forbidden energy gap.

Consider the case of Na metal. The electronic configuration of Na atom is $1s^2 2s^2 2p^6 3s^1$. When the atoms of Na are far apart, there is no interaction between them and the allowed states of the system are just the states of a single atom repeated N times in space (N is number of atoms). As the interatomic spacing decreases, the valence orbitals of the atom begin to overlap and the energy of $3s$ level of Na splits into levels which are so close together that they form an energy band (since there are 2.7×10^{22} atoms in 1 cm^3, there must be 2.7×10^{22} different energy levels, thus there must exists a band of energy levels with very small difference between the individual levels). Similarly, the $3p$ (empty) level also spread into a band. As the interatomic spacing is still reduced, the $2s$ and $2p$ levels also start to split, a further reduction causes $1s$ level to split. The valence

Fig. 9.1 Energy bands in a
sodium crystal

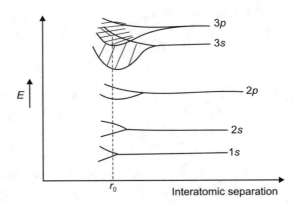

level begins to split at a larger separation than all the others (1s, 2s, 2p) because these
electrons being farthest away from the parent atom as less tightly bound and more likely
to be affected by external influence (Fig. 9.1).

What is true of 3s electrons in Na is also true for the 1s, 2s and 2p electrons. They also
belong to the whole crystal and forms energy band as a result of the interaction of Na
atom forming a solid. The difference is that there are more tightly bound electrons and
therefore give rise to much narrower band then associated with valence electron. They
interact less with each other since their wave functions do not overlap so much. A further
reduction in the interatomic spacing causes broadening of the bands formed by 3s and
3p levels. The energy passes through a minimum at some specific interatomic distance
r_0 corresponding to equilibrium position. This is to be expected since all physical system
leads to a minimum energy state corresponding to stable equilibrium. A forbidden gap
remains between the bands associated with 2p and 3s electrons.

For atomic spacing somewhat larger than the equilibrium spacing, overlap of bands
formed from the 3s electrons and those formed from the 3p electrons occur. Because of
the overlapping of allowed bands, all energy above the bottom of 3s bands is allowed
at the equilibrium Na atom spacing. Because of the overlapping of the bands we can no
longer properly refer to them as 3s or 3p band but must recognize that there is a mixing
of s and p-like wave function. The energy curves bend upwards again for small spacing
between Na atoms because of growing interaction between closely spaced atoms.

Let us consider a bivalent metal, for example, Mg. Its electronic configuration is
$1s^2 2s^2 2p^6 3s^2$. In magnesium 3p and 3s bands overlap. A p atomic orbital can hold six
electrons, therefore a p band formed by N atoms can hold 6N electrons. Similarly, the
3s band can hold 2N electrons. Consequently, (3s + 3p) in Mg can hold 8N electrons
in all. With only 2N electrons in the band, it is only one-fourth filled. Therefore it is a
conductor.

Consider example of As having electronic configuration (Z = 33)
$1s^2 2s^2 2p^6 3s^2 3p^6 3d^{10} 4s^2 4p^5$. It has five valence electrons and crystallizes into a
structure with two atoms per unit cell. Thus ten electrons almost exactly fill five bands.

Fig. 9.2 Formation of energy band in Si

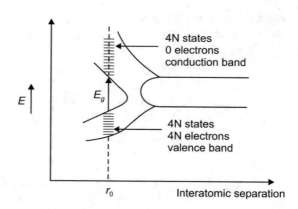

The fifth band however being not quite fills because of a little overlap into the sixth band. Therefore *As* is a semimetal.

Now consider the example of Si (Fig. 9.2). The electronic configuration of Si ($Z = 14$) is $1s^2 2s^2 2p^6 3s^2 3p^2$. The outer two subshells ($3s$ and $3p$) contain two *s* electrons and two *p* electrons. The s atomic subshell has two possible states; therefore, there will be $2N$ possible *s* states in a solid of N atoms and the number of *s* electrons will be $2N$. Thus, $2N$ electrons completely fill the $2N$-s subshell. In case of *p* atomic subshell, there are six possible states; therefore, there will be $2N$ electrons filling $2N$-p sub shell out of possible $6N$ states. Now if the spacing between the atoms decreases, they start interacting and their atomic wave function overlap. As a result, the solid becomes a system obeying Pauli exclusion principle and an electron does not belong to a particular atom. If the interatomic distance is decreased sufficiently, the $2N$-s states and $6N$-p states spread out. If the interatomic distance further decreases to the limit of crystal lattice spacing, these bands overlap and $6N$-p states and $2N$-s states give $8N$ states. As the available electrons are $4N$, half of the $8N$ states are filled. The band occupied by these electrons is called the valence band. Hence at $r = r_0$, there is a valence band filled with $4N$ electrons separated by a forbidden band of energy E_g from an empty band having $4N$ states. This empty band is known as conduction band.

9.3 Bloch Theorem

Consider a simple model of a solid (e.g. a metal) in which the positive ions comprise the uniform array of fixed sites. The valence electrons are assumed to be free. They are conduction electrons. For sodium, for instance, there is one free electron per ion. Each such electron finds itself in a periodic potential supported by the ions (Fig. 9.3). If the distance between sites is a, then inside the metal the potential is periodic in the distance a

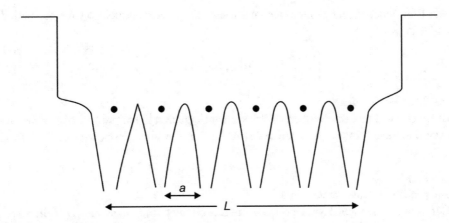

Fig. 9.3   Periodic potential that an electron sees in a one dimensional crystalline solid

$$V(x) = V(x + a) \tag{9.1}$$

The periodic property of $V(x)$ as given in Eq. (9.1) fails at the ends of the lattice. To remove this difficulty it is assumed that there are very large numbers of ion sites in the length of the sample. The change in the character of the potential at the ends of the sample is therefore relatively unimportant. It is assumed that when an electron leaves the end of the sample, it re-enters the front of the sample. The idea is best realized if the one dimensional potential function is assumed to lie on a circle of radius r (Fig. 9.4) which is very large compared to distance between the ion sites. The ends of the chain are joined together so that $n = 0$ and $n = N$ particle coincide.

Fig. 9.4   Boundary condition

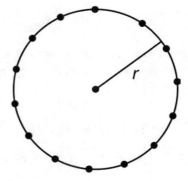

The time independent Schrödinger wave equation in one dimension for the periodic potential is

$$\left[-\frac{\hbar^2}{2m}\nabla^2 + V(x)\right]\psi(x) = E\psi(x) \tag{9.2}$$

where $V(x) = V(x+a)$.

Bloch showed that the solution of Schrodinger equation is the product of a plane wave function modulated by the function $u_k(x)$ which has the same periodicity as the lattice.

$$\psi(x) = \exp(ikx)u_k(x) \tag{9.3}$$

where k is Bloch wave vector.

Consider N identical lattice points around a circular ring separated by a distance a. The boundary condition is

$$\psi(x + Na) = \psi(x) \tag{9.4}$$

where N is number of ion sites. Changing the origin of the coordinate in Schrödinger equation by a and using $V(x) = V(x+a)$ we have $\psi(x+a)$ also the solution. We have

$$\frac{d}{d(x+a)} = \frac{d}{dx}$$

and Eq. (9.2) is then

$$\left[-\frac{\hbar^2}{2m}\nabla^2 + V(x+a)\right]\psi(x+a) = E\psi(x+a)$$

or

$$\left[-\frac{\hbar^2}{2m}\nabla^2 + V(x)\right]\psi(x+a) = E\psi(x+a) \tag{9.5}$$

Thus $\psi(x+a)$ is also the solution of Eq. (9.2). Assuming, energy eigenvalues are non-degenerate for each energy level and hence $\psi(x+a)$ must be multiple of $\psi(x)$. Thus

$$\psi(x+a) = c\psi(x) \tag{9.6}$$

On applying the translation a, we have

$$\psi(x+2a) = c\psi(x+a) = c^2\psi(x)$$

Again applying the translation a

$$\psi(x+3a) = c\psi(x+2a) = c^2\psi(x+a) = c^3\psi(x)$$

Applying the translation N times

$$\psi(x + Na) = c^N \psi(x) = \psi(x) \tag{9.7}$$

Because $n = 0$ and $n = N$ are same points. From Eq. (9.7)

$$c^N = 1$$

or

$$c^N = \exp[i2\pi n]; \quad n = 0, \pm 1, \pm 2, \ldots \tag{9.8}$$

From Eq. (9.8)

$$c = \exp\left[i\frac{2\pi n}{N}\right] = \exp\left[i\frac{2\pi na}{Na}\right] = \exp[ika] \tag{9.9}$$

where $k = \frac{2\pi n}{Na}$ is Bloch wave vector. Equation (9.6) is then

$$\psi(x + a) = c\psi(x) = \exp[ika]\psi(x) \tag{9.10}$$

The property (9.10) may also, following Bloch be expressed in a different form. If we rewrite Eq. (9.10) using Eq. (9.3)

$$\exp[ik(x + a)]u_k(x + a) = \exp[ika]\exp[ikx]u_k(x)$$

or

$$u_k(x + a) = u_k(x)$$

The function u_k has the periodicity of the lattice. The Bloch theorem is one of the most important results in all solid state physics because it tells us the mathematical form of an electron wave function in the presence of a periodic potential energy.

9.4 Kronig–Penney Model

In this model it is assumed that the periodic potential consists of an infinite number of potential barriers of width b spaced at interval $a + b$ leaving regions of zero potential of width a between the barriers. The height of each barrier to be V_0 (Fig. 9.5).

In the region

$$0 < x < a, \quad V = 0 \tag{9.11}$$

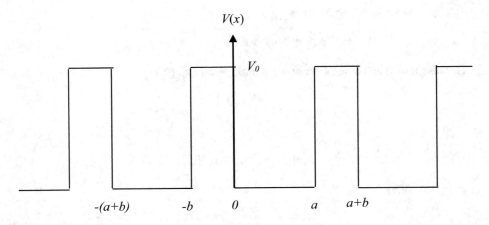

Fig. 9.5 Kronig–Penney model

and

$$-b < x < 0, \quad V = V_0 \tag{9.12}$$

The Schrödinger equations for the two regions are

$$\frac{d^2\psi}{dx^2} + \frac{2m}{\hbar^2}E\psi = 0 \quad \text{for } 0 < x < a \tag{9.13}$$

$$\frac{d^2\psi}{dx^2} + \frac{2m}{\hbar^2}(E - V_0)\psi = 0 \quad \text{for } -b < x < 0 \tag{9.14}$$

It is assumed that the energy of the electrons under consideration is smaller than V_0, that is, $E < V_0$. Let

$$q = \sqrt{\frac{2mE}{\hbar^2}} \text{ and } \beta = \sqrt{\frac{2m}{\hbar^2}(V_0 - E)} \tag{9.15}$$

The solution of Eqs. (9.13) and (9.14) must be of the form of Bloch function, that is,

$$\psi(x) = \exp[ikx]u_k(x) \tag{9.16}$$

Substituting Eqs. (9.16) in Eqs. (9.13) and (9.14) and omitting the subscript k

$$\frac{d^2u}{dx^2} + 2ik\frac{du}{dx} + (q^2 - k^2)u = 0 \text{ for } 0 < x < a \tag{9.17}$$

$$\frac{d^2u}{dx^2} + 2ik\frac{du}{dx} - (\beta^2 + k^2)u = 0 \text{ for } -b < x < 0 \tag{9.18}$$

The solutions of Eqs. (9.17) and (9.18) are

$$u_1 = A \exp[i(q - k)x] + B \exp[-i(q + k)x] \tag{9.19}$$

$$u_2 = C \exp[(\beta - ik)x] + D \exp[-(\beta + ik)x] \tag{9.20}$$

where A, B, C and D are constants. The solutions for the two regions must join smoothly at the boundary so that we must have from the continuity of wave function and their derivatives

$$u_1|_{x=0} = u_2|_{x=0} \tag{9.21}$$

$$\left.\frac{du_1}{dx}\right|_{x=0} = \left.\frac{du_2}{dx}\right|_{x=0} \tag{9.22}$$

boundary condition from the periodicity of u

$$u_1(a) = u_2(-b) \tag{9.23}$$

$$\left.\frac{du_1}{dx}\right|_{x=a} = \left.\frac{du_2}{dx}\right|_{x=-b} \tag{9.24}$$

The boundary conditions lead to

$$A + B = C + D \tag{9.25}$$

$$i(q - k)A - i(q + k)B = (\beta - ik)C - (\beta + ik)D \tag{9.26}$$

$$A \exp[i(q - k)a] + B \exp[-i(q + k)a]$$
$$= C \exp[-(\beta - ik)b] + D \exp[(\beta + ik)b] \tag{9.27}$$

$$i(q - k)A \exp[i(q - k)a] - i(q + k)B \exp[-i(q + k)a]$$
$$= (\beta - ik)C \exp[-(\beta - ik)b] - (\beta + ik)D \exp[(\beta + ik)b] \tag{9.28}$$

The four Eqs. (9.25)–(9.28) have solution only if the determinant of the coefficients A, B, C, D vanishes. This leads to complicated relation between k and E. Considerable simplification is obtained by allowing $b \to 0$ and $V_0 \to \infty$ in such a way that the product bV_0 remains finite. Then one obtains

$$P\frac{\sin qa}{qa} + \cos qa = \cos ka \tag{9.29}$$

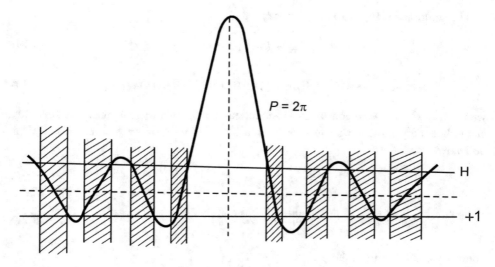

Fig. 9.6 Plot of the function given by Eq. (9.29) for $P = 2\pi$ as a function of qa

where $P = \frac{mV_0ab}{\hbar^2}$ is a measure for the area V_0b of the potential barrier. The increasing P has the physical meaning of binding a given electron more strongly to a particular potential well.

The left hand side of this equation is plotted as a function of qa in Fig. 9.6 for a value of $P = 2\pi$: allowed values of k are obtained only when the function lies between 1 and -1, and hence only certain range of values of q are allowed. Since $q = \sqrt{\frac{2mE}{\hbar^2}}$ this means that the energy E is restricted to lie within certain range, which forms the allowed energy bands. The allowed energy bands are narrowest for small values of q (low values of energy E) and become broader as E increases, the unallowed bands getting narrower.

P is characteristic of the strength of the potential barriers which separate regions of zero potential. As the energy E of the electron increases and P is kept constant, the electron can move readily surmount or penetrate the potential barrier becomes less and less important as far as the behavior of the electron is concerned. Ultimately, as energy approaches infinity, the electron becomes as though free. For $P \to 0$, Eq. (9.29) takes the form

$$qa = ka \text{ or } q^2 = k^2$$
$$\frac{2mE}{\hbar^2} = k^2$$

$$E = \frac{\hbar^2 k^2}{2m} \tag{9.30}$$

Fig. 9.7 Allowed (shaded) and forbidden (open) energy range as function of *P*

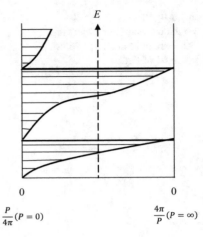

$$\frac{P}{4\pi}(P = 0) \qquad\qquad \frac{4\pi}{P}(P = \infty)$$

which is appropriate to the completely free particle. For the case when $P \to \infty$, the allowed solutions are those for which qa is multiple of π, that is

$$\sin qa = 0$$

$$qa = n\pi$$

$$\frac{2mE}{\hbar^2}a^2 = q^2a^2 = n^2\pi^2$$

$$E = \frac{n^2\pi^2\hbar^2}{2ma^2} \tag{9.31}$$

This gives the energy levels of a particle in a one dimensional box of width a (Fig. 9.7).

9.5   *E-k* Curves: Brillouin Zones

From Eq. (9.29) it is seen that an electron moving in a periodic potential can have energy values between allowed region for which left hand side lies between $+1$ and -1. A plot of energy E of electron versus wave number k is shown in Fig. 9.8. The right hand side of Eq. (9.29) is for values of k given by

$$\cos ka = \pm 1 \tag{9.32}$$

$$k = \frac{n\pi}{a} \quad (n = \pm 1, \pm 2, \pm 3, \ldots) \tag{9.33}$$

Thus the discontinuities in the E versus k curve occur for

Fig. 9.8 Plot of energy E versus wavenumber k showing band structure due to periodic potential of lattice

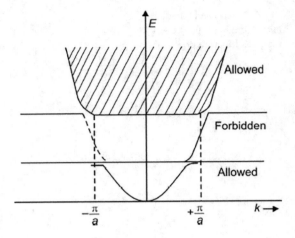

$$k = \frac{n\pi}{a} \qquad\qquad (9.34)$$

These k values define the boundaries of the first, second, third, etc. Brillouin zones. The region extending from $k = -\frac{\pi}{a}$ to $k = -\frac{\pi}{a}$ is called the first Brillouin zone while the second Brillouin zone extend from $k = -\frac{\pi}{a}$ to $k = -\frac{2\pi}{a}$ and $k = +\frac{\pi}{a}$ to $k = +\frac{2\pi}{a}$. Brillouin zones in one dimensional case are shown in Fig. 9.9.

Let us now consider a square lattice. The lattice points of the reciprocal lattice are separated by a distance $2\pi/a$ forming again a square lattice. The boundary conditions of Brillouin zone for a square lattice are given by Eq. (5.57).

$$n_x k_x + n_y k_y = \frac{\pi}{a}\left(n_x^2 + n_y^2\right)$$

where n_x and n_y are integers. The first Brillouin zone is enclosed by the four lines corresponding to $n_x = \pm 1$, $n_y = 0$ and $n_x = 0$, $n_y = \pm 1$ (Fig. 9.10). The first Brillouin zone is a square lattice of edge $2\pi/a$. The square PQRS is determined by four lines corresponding to the four sets of integers n_x, $n_y = \pm 1$, ± 1. The area between PQRS and

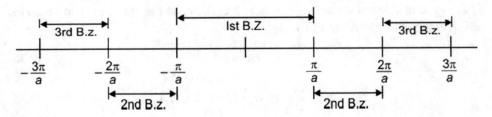

Fig. 9.9 First three Brillouin zones in one dimension

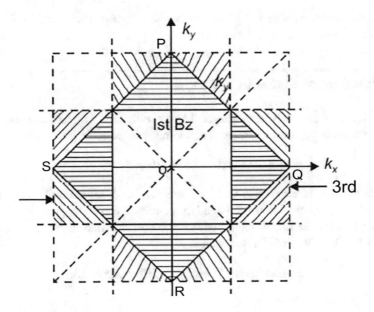

Fig. 9.10 First three Brillouin zones for a square lattice of edge *a*

the first Brillouin zone forms the second Brillouin zone (Fig. 9.10). The zone boundaries for simple cubic lattice are given by the solution of the equation.

$$n_x k_x + n_y k_y + n_z k_z = \frac{\pi}{a}\left(n_x^2 + n_y^2 + n_z^2\right)$$

From Fig. 9.10, it is seen that the second zone has the same area as the first, that is, $(2\pi/a)^2$. The same is true for third zone. In general all the zones have equal area.

Each energy band $E_n(\mathbf{k})$ satisfy the following symmetry properties

$$E_n(\mathbf{k} + \mathbf{G}) = E_n(\mathbf{k})$$
$$E_n(-\mathbf{k}) = E_n(\mathbf{k})$$

and $E_n(\mathbf{k})$ has the same rotational symmetry as the real lattice. The first property indicates $E_n(\mathbf{k})$ is periodic, with a period equal to the reciprocal lattice vector. Thus any two points in $\mathbf{k}$-space related to each other by a displacement equal to a reciprocal lattice vector have the same energy. The various pieces of the second zone may be translated by reciprocal lattice vector to fit precisely over the first zone. Similarly, higher-order zones can be appropriately translated to fit over the first zone. The second property shows that band is symmetric with respect to inversion around $\mathbf{k} = 0$.

The symmetry properties reduce the labour involved in determining the energy band. For a square lattice we need to know only half of the first zone because of inversion symmetry. The rotational symmetry reduces this even further. For square lattice only one eight of the zone be specified independently.

9.6   Number of States in a Band

Let us consider a linear crystal constructed by primitive cells of lattice constant a. The boundary condition requires that wave function must be periodic, that is,

$$\psi(x + L) = \psi(x)$$

Assuming that L is very large such that the number N of primitive cell satisfy the condition $N \gg 1$. In this case the boundary condition can be applied for finite linear lattice as an infinite one dimensional lattice can be divided into macroscopic sections of length $L = Na$. The Bloch function requires

$$\exp(ikx)u_k(x) = \exp[i(x + L)a]u_k(x + L)$$

Since $u_k(x)$ is periodic, therefore,

$$u_k(x + L) = u_k(x)$$

and hence

$$\exp(ikL) = 1$$

and

$$k = \frac{2\pi n}{L} \quad \left(n = 0, \pm 1, \pm 2, \ldots, \pm\frac{N}{2}\right)$$

The number of states or orbitals inside the first zone, whose length is $2\pi/a$ is therefore equal to

$$\left(\frac{2\pi}{a}\right)/\left(\frac{2\pi}{L}\right) = \frac{L}{a} = N$$

Thus each band has N states inside the first zone. According to Pauli exclusion principle, each such state can accommodate at most two electrons, of opposite spin. Therefore, the maximum number of electrons that may occupy a single band is $2N$.

9.7 Electrons and Holes

The band theory of solids indicates that the allowed energy bands are separated by forbidden gap as shown in Fig. 9.11. The magnitude of the forbidden gap depends from solid to solid. The last completely filled (at least at 0 K) is called valence band and the next band with higher energy is conduction band which may be empty or partially filled.

In the valence band, nearly all the energy levels are filled with electrons. The term electron is usually means conduction band electrons. The term hole is used for the absence of electron in a valence band. Electrons carry $-e$ charge while holes carry $+e$ charge. In an energy band diagram, higher position represents higher energy of electrons. The minimum energy of the electrons in the conduction band is E_C (energy of the bottom of the conduction band), any energy of electron above E_C is the kinetic energy of the electron. A lower position in energy diagram represents a higher hole energy. Therefore, to move a hole from E_V (energy of top of valence band) to bottom of valence band requires energy. The electrons and holes in a solid interact with Coulomb field in the solid. They move in periodic potential of the lattice, and therefore mass of electrons and holes may not be same as that of free electron. Each solid has unique mass of electrons and holes.

A full band (an allowed band where all the states are occupied) can carry no electric current, since for every electron with a positive value of k there is another with the value $-k$. Suppose we have a band which is full except for one state at the top of a band which has a negative value of k. If an electron occupies this state it would have negative charge and negative mass. Its momentum $p = \hbar k$ would be negative, but its velocity would be positive so that it would carry a negative current. However, the presence of such an electron would fill the band and the net momentum and current would be zero. Hence

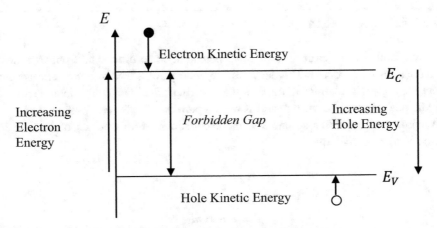

Fig. 9.11 Electron and hole levels tend to retain lowest energy state

the momentum and current due to all the other electrons must be equivalent to that of one particle with positive momentum giving a positive current, and the same value of $|m^*|$. Such a particle is called a positive hole, since the behaviour corresponds to that of a particle with a positive charge (and positive mass). The advantage of the concept of positive holes is that the momentum and current of a nearly-filled band with n empty states can be attributed to the presence of an equivalent number n of entities which behave like particles with positive charge and effective mass m^*.

Let us consider a single hole in a filled band of a one-dimensional lattice. Let the charge on the electron is $-e$ and the velocities of the electrons is denoted by v_i. In the absence of an electric field, the current due to all the electrons in a completely filled band is

$$\mathbf{I} = -e \sum_i \mathbf{v}_i = -e \left[\mathbf{v_j} + \sum_{i \neq j} \mathbf{v}_i \right] = 0 \tag{9.35}$$

If the electron j was missing, we should have

$$\mathbf{I}' = -e \sum_{i \neq j} \mathbf{v}_i = e\mathbf{v}_j \tag{9.36}$$

Rate of change of current $\mathbf{I}'$, on applying the electric field $\mathbf{F}$ is

$$\frac{d\mathbf{I}'}{dt} = e \frac{dv_j}{dt} = -\frac{e^2 \mathbf{F}}{m_j^*} = \frac{e^2 \mathbf{F}}{\left| m_j^* \right|} \tag{9.37}$$

Since holes tend to reside in the upper part of the nearly filled band, m_j^* is negative and the right hand side of Eq. (9.37) becomes positive. Thus a band in which an electron is missing behaves as a positive hole with an effective mass m^*.

9.8   Effective Mass

When we deal with electron in a periodic potential, some difference from free electron behavior is to be expected. This leads to the concept of effective mass. The advantage of the concept of the effective mass is that the dynamical behavior of an electron in a periodic potential can be treated as if it were a particle of mass m^*.

According to Newton's second law the acceleration of an electron of mass m in an electric field $\mathbf{F}$ is given by

$$\mathbf{a} = \frac{d\mathbf{v}}{dt} = \frac{F}{m}$$

$$\mathbf{F} = m\mathbf{a} = m \frac{d\mathbf{v}}{dt}$$

Inside the crystal, the Brillouin zone boundary gives rise to gap in the electron energies. Therefore, the velocity of the electron as represented by its group velocity

$$v_g = \frac{d\omega}{dk} = \frac{d}{dk}\left(\frac{E}{\hbar}\right) = \frac{1}{\hbar}\frac{dE}{dk} \tag{9.38}$$

where ω is the angular frequency of the de Broglie waves and is related to the energy of the particle by the relation

$$E = \hbar\omega$$

$$\frac{dv_g}{dt} = \frac{d}{dt}\left(\frac{1}{\hbar}\frac{dE}{dk}\right) = \frac{1}{\hbar}\frac{d^2E}{dk^2}\frac{dk}{dt} \tag{9.39}$$

Since $p = \hbar k$

$$\frac{dv_g}{dt} = \frac{1}{\hbar}\frac{d^2E}{dk^2}\frac{1}{\hbar}\frac{dp}{dt} = \frac{1}{\hbar^2}\frac{d^2E}{dk^2}F = \frac{F}{m^*} \tag{9.40}$$

where

$$\frac{1}{m^*} = \frac{1}{\hbar^2}\frac{d^2E}{dk^2} \tag{9.41}$$

m^* is known as the effective mass is thus determined by the curvature of the E versus k curve, that is, on $\frac{d^2E}{dk^2}$. The effective mass is a tensor. Figure 9.12 shows energy, velocity and effective mass and f_k as a function of k. From Fig. 9.12 it is seen that $v = 0$ at the top and bottom of the energy band as from the periodicity of the E-k curve it is seen that at these positions $dE/dk = 0$. The absolute value of the velocity reaches a maximum corresponding to inflection point of the E-k curve. Beyond this point the velocity decreases with increasing energy. This behaviour is different from that of free electron. For free electrons velocity increases with increase in E.

Suppose the electron is initially in the state k when the field has acted on the electron for a small time dt it has gained an energy. The value of E for a free particle is $E = \frac{\hbar^2 k^2}{2m}$ and $\frac{d^2E}{dk^2} = \frac{\hbar^2}{m}$ resulting in $m^* = m$, that is the effective mass is equal to the true mass for a free particle. For an electron in a periodic potential the effective mass may departs markedly from the true mass. The difference between m^* and the true mass m represents the effect of motion of the electron which results from the electric potential of the ions forming the crystal lattice; when a force is applied to the electron, its change in momentum is different from that of a free electron, and $p = \hbar k$ is often referred to as the crystal momentum.

The relation between E and k as derived from the Kronig–Penney model has the form as shown in Fig. 9.8. It does not differ greatly from that for a free electron except near the edge of the allowed band. At the points where $\cos ka = \pm 1$, there is a discontinuity

Fig. 9.12 Energy, velocity,
effective mass and f_k as a
function of k

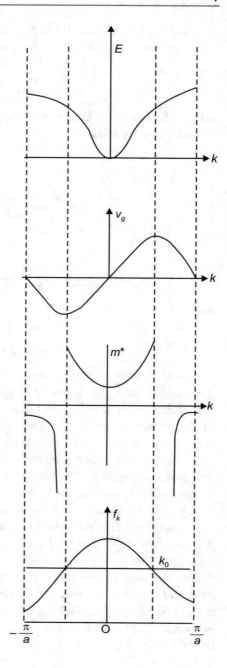

in the relation between E and k. From the shape of E versus k curve, it is seen that the effective mass m^* becomes negative near the top of an allowed band, because $d^2 E/dk^2$ becomes negative. At the bottom of the band $d^2 E/dk^2$ is positive and the effective mass is positive. At the inflection points in the E-k curve (curvature is zero), the effective mass is infinite. This means that in the upper half of the band the electron behaves as a positively charged particle.

The crystal does not weigh any less if the effective mass is less than the free electron mass, nor is Newton's second law is violated for the crystal taken as a whole. The electron in a periodic potential is accelerated relative to the lattice in an applied electric field or magnetic field as if the mass of the electron were equal to the effective mass.

Consider an electron in an external field F. Suppose the electron is initially in the state k when the field has acted on the electron for a small time dt it has gained an energy

$$dE = evF dt = \frac{eF}{\hbar}\left(\frac{dE}{dk}\right)dt$$

$$\frac{dE}{dk}dk = \frac{eF}{\hbar}\left(\frac{dE}{dk}\right)dt \tag{9.42}$$

$$\frac{dk}{dt} = \frac{eF}{\hbar} \tag{9.43}$$

$$a = \frac{dv}{dt} = \frac{1}{\hbar}\left(\frac{d^2 E}{dk^2}\right)dt \tag{9.44}$$

The degree of freedom of an electron in a crystal field is defined by

$$f_k = \frac{m}{m^*} = \frac{m}{\hbar^2}\left(\frac{d^2 E}{dk^2}\right) \tag{9.45}$$

it is a measure of the extent to which an electron in state k is free. If m^* is large f_k is small and particle behaves as a heavy particle. If $f_k = 1$ the electron behaves as a free particle.

9.9   Metals, Semiconductors and Insulators

The electronic band structure scheme consists of energy bands separated by band gap. The last completely filled band is called valence band. The next band with higher energy is called conduction band. The conduction band can be empty or partially filled. The energy separation between bottom of conduction band and top of valence band is called band gap or forbidden gap. The width of band gap affects the conductivity of the materials. On the basis of band theory of solids, the materials can be classified into three categories:

(i)   Metals
(ii) Semiconductors and
(iii) Insulators.

Metals

In a metal the valence band is full of electrons, while the conduction band is only partially filled at 0 K. In a metal there is no band gap between the valence and conduction band as shown in Fig. 9.13. The addition of very small amount of energy will allow electrons to

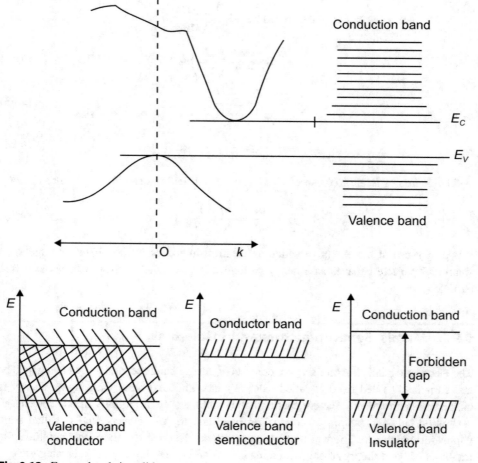

Fig. 9.13 Energy bands in solids

move within conduction band, some rising to a higher level and others returning to lower level. This movement of electrons constitutes electric current. In some metals the valence band and the conduction band actually overlap. This effectively gives a partially filled top band. The electric resistivity of conductor is $\sim 10^{-6}\,\Omega$ cm.

Semiconductor

In a semiconductor, at $0\,K$ the highest occupied band is completely filled with electrons and the next band is completely empty. The separation between the lowest point of the conduction band and the top of the valence band is the band gap E_g (Fig. 9.13). The lowest point in the conduction band is called the conduction band edge; the highest pint in the valence band is called the valence band edge. The band gap of semiconductors ordinarily varies from 0.2 to 2.5 eV. At room temperature the resistivity of semiconductor varies from 10^{-3} to 10^6 Ω cm. The electric resistivity of semiconductor is temperature dependent. The elements Ge and Si are regarded as basic semiconductors. The band gap of Ge is 0.67 eV while that of Si is 1.14 eV. The semiconductor compounds of chemical formula AB where A is a trivalent element and B is a pentavalent element are called III–V (three-five) compounds, for example InSb and GaAs. When A is divalent and B is hexavalent the compound is called a II–VI compound, for example, ZnS and CdS. Si and Ge are sometimes called diamond-type semiconductors, because they have the crystal structure of diamond. SiC is a IV–IV compound. As the two elements forming these compounds become more widely separated with respect to their relative positions in the periodic table, that is, electronegativities become more dissimilar, the atomic bonding becomes more ionic and the magnitude of the band gap energy increases. Further, the size of the energy gap commonly decreases as we move down in the periodic table. For example as we move $C \rightarrow Si \rightarrow Ge \rightarrow Sn$ the band gap is 6 eV $\rightarrow$ 1.1 eV $\rightarrow$ 0.7 eV $\rightarrow$ 0.1 eV; for $GaP \rightarrow GaAs \rightarrow GaSb$, the band gap is 2.3 eV $\rightarrow$ 1.4 eV $\rightarrow$ 0.7 eV; $AlSb \rightarrow GaSb \rightarrow InSb$, the band gap is 1.6 eV $\rightarrow$ 0.7 eV $\rightarrow$ 0.029 eV.

Semiconductors possess the following characteristic properties

(i)   A pure semiconductor has a negative temperature coefficient of resistance unlike a metal which has a positive temperature coefficient of resistance.

(ii) Semiconductors give high thermoelectric power with signs both positive and negative relative to a given metal.

(iii) A semiconductor is light sensitive, generating either a photovoltaic or a change in resistance upon irradiation by light.

(iv) The junction between a p- and n-type semiconductor having majority carriers holes and electrons, respectively shows rectifying property.

Insulators

In the insulator the valence band is full and the conduction band is empty. However, the energy gap between valence band and the conduction band is very large (Fig. 9.13). It would require a large amount of energy to make electron jump from valence band to conduction band. At very high temperature or under very large electric field this may happen. The insulators have negative temperature coefficients of resistance. The band gap of typical insulator like diamond is 6 eV. The electric resistivity of insulators is of the order of 10^{12} Ω cm.

The difference between semiconductors and insulators is related to the size of the energy gap.

Frequently, when making a simplified analysis of the energy band structure of semiconductor, instead of the actual dispersion curves which bound the valence and conduction band, use is made of two parallel lines, one drawn tangentially to the bottom of the conduction band and the other to the top of the valence band (Fig. 9.12). The first line is taken to represent the bottom of the conduction band and the second the top of the valence band. The separation between them is equal to the forbidden band width E_g.

9.10   Direct and Indirect Band Gap Semiconductors

The band gap represents the minimum energy difference between the top of the valence band and the bottom of the conduction band. However, the top of the valence band and the bottom of the conduction band are not generally at the same value of the wave vector (Fig. 9.14). In a direct band gap semiconductor the top of valence band and the bottom of the conduction band occur at the same value of wave vector, for example, in CdS, CdTe, alpha tin, InSb, etc. In an indirect band gap semiconductor, the maximum of the valence

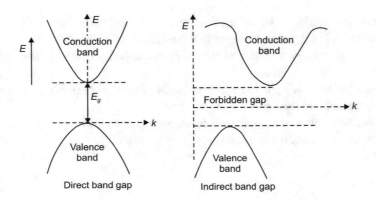

Fig. 9.14 Direct and indirect band gap semiconductors

band and the minimum of the conduction band occur at different values of wave vector, the examples are Si, Ge, PbSe, PbTe, etc.

The difference between the direct and indirect band gap semiconductors is most important in optical devices. A photon can provide the energy to produce electron–hole pair. Each photon of energy $E = h\nu$ has momentum $p = h\nu/c$ (c is velocity of light). An optical photon has an energy ~10^{-19} J. A typical photon has a very small momentum. A photon of an energy E_g can produce an electron–hole pair in a direct band gap semiconductor very easily, because the electron does not need to be given very much momentum. However, in indirect band gap semiconductors, for the conservation of momentum, an electron must also undergo a significant change in the momentum for a photon of energy E_g to produce an electron–hole pair. The optical measurements determine whether the band gap is direct or indirect.

9.11 Nearly Free Electron Model

Consider an electron of mass m confined in a one dimensional box of length L. The potential is zero in the box and infinite elsewhere. The energy eigenvalues of such a free electron are given by Eq. (8.23)

$$E = \frac{\hbar^2 k^2}{2m} \tag{9.46}$$

The eigenfunction can be written as

$$\psi = \frac{1}{\sqrt{L}} \exp(ikx) \tag{9.47}$$

The energy eigenvalues are doubly degenerate because substitution of k or $-k$ in Eq. (9.46) gives the same energy eigenvalues.

The electrons in a lattice are not free. They interact with electrons and potential produced by central metal ions. In nearly free electron model, it is assumed that electrons are perturbed only by the ionic potential V. Further, it is also assumed that the potential V is very weak, that is, $E \gg V$ and electron–electron interaction is ignored. The potential V is periodic with the same period as lattice. In one dimensional lattice of lattice constant a

$$V(x) = V(x + a) \tag{9.48}$$

Expanding periodic potential $V(x)$ in Fourier series

$$V(x) = \sum_{n=-\infty}^{\infty} V_n \exp\left(\frac{i2\pi nx}{a}\right) \tag{9.49}$$

and

$$V(x + a) = \sum_{n=-\infty}^{\infty} V_n \exp\left[\frac{i2\pi n(x + a)}{a}\right]$$

$$= \sum_{n=-\infty}^{\infty} V_n \exp\left[\frac{i2\pi nx}{a}\right] \exp(i2\pi n) = V(x) \tag{9.50}$$

The Hamiltonian for the free electron is

$$H_0 = \frac{p^2}{2m} = \frac{\hbar^2 k^2}{2m}$$

The total Hamiltonian including the perturbation is

$$H = H_0 + H' = \frac{\hbar^2 k^2}{2m} + V(x) \tag{9.51}$$

with

$$H' = V(x) = \sum_{n=-\infty}^{\infty} V_n \exp\left(\frac{i2\pi nx}{a}\right) \tag{9.52}$$

The zero energy line is set at the average of $V(x)$. Hence $V_0 = 0$. The unperturbed wave function corresponding to H_0 is

$$\psi(x) = \frac{1}{\sqrt{L}} \exp(ikx) \tag{9.53}$$

According to time independent non-degenerate first order perturbation theory, the energy shift is given by the matrix element $H_{kk'}$ with perturbation H', that is,

$$H_{kk'} = \int_{-\frac{L}{2}}^{\frac{L}{2}} \psi_{k'}^*(x) H' \psi_k(x) dx \tag{9.54}$$

where

$$\psi_{k'}^* = \frac{1}{\sqrt{L}} \exp(-ik'x) \tag{9.55}$$

$$\psi_k = \frac{1}{\sqrt{L}} \exp(ikx) \tag{9.56}$$

From Eqs. (9.52), (9.54)–(9.56)

$$H_{kk'} = \frac{1}{L} \sum_{n=-\infty}^{\infty} \int_{-\frac{L}{2}}^{\frac{L}{2}} \exp(-ik'x) V_n \exp\left(\frac{i2\pi nx}{a}\right) \exp(ikx) dx$$

$$H_{kk'} = \frac{1}{L} \sum_{n} V_n \int_{-\frac{L}{2}}^{\frac{L}{2}} \exp(-ik'x) \exp\left(\frac{i2\pi nx}{a} + ikx\right) dx \tag{9.57}$$

Let

$$q = k - k' + \frac{2\pi n}{a} \tag{9.58}$$

From Eqs. (9.57) and (9.58)

$$H_{kk'} = \frac{1}{L} \sum_{n} V_n \int_{-\frac{L}{2}}^{\frac{L}{2}} \exp(iqx) dx \tag{9.59}$$

Now

$$\frac{1}{L} \int_{-\frac{L}{2}}^{\frac{L}{2}} \exp(iqx) dx = \frac{\exp\left(\frac{iqL}{2}\right) - \exp\left(-\frac{iql}{2}\right)}{iqL} = \frac{2i \sin \frac{qL}{2}}{iqL} = \frac{\sin \frac{qL}{2}}{\frac{qL}{2}} \tag{9.60}$$

The function $\frac{\sin \frac{qL}{2}}{\frac{qL}{2}}$ is an even function. The graph of $\frac{\sin \frac{qL}{2}}{\frac{qL}{2}}$ would be damped oscillator function having value 1 at $\frac{qL}{2} = 0$ or at $q = 0$ or $L \to \infty$. The plot of $\frac{\sin \frac{qL}{2}}{\frac{qL}{2}}$ shows that two zero's are located at

$$\frac{qL}{2} = \pm\pi$$

or

$$q = \pm\frac{2\pi}{L} \tag{9.61}$$

Thus

$$\lim_{L\to\infty} \frac{1}{L} \int_{-\frac{L}{2}}^{\frac{L}{2}} \exp(iqx) dx = \delta_{q,0} \tag{9.62}$$

The matrix element given by Eq. (9.59) is nonzero only if the initial and final states are related by the reciprocal lattice vector, that is,

$$k' = k + \frac{2\pi n}{a} \tag{9.63}$$

The initial and final states must have the same energy, thus

$$\frac{\hbar^2 k'^2}{2m} = \frac{\hbar^2 k^2}{2m} \tag{9.64}$$

$$\frac{\hbar^2}{2m}\left(k + \frac{2\pi n}{a}\right)^2 = \frac{\hbar^2 k^2}{2m}$$

$$\frac{4\pi^2 n^2}{a^2} + \frac{4\pi n k}{a} = 0$$

$$k = -\frac{\pi n}{a} \tag{9.65}$$

Since n varies from $-\infty$ to $+\infty$, the sign of n can flip in Eq. (9.65), that is,

$$k = \frac{\pi n}{a} \tag{9.66}$$

The periodic potential will therefore have largest effect on the states with wave vector

$$k = \pm\frac{\pi}{a}, \pm\frac{2\pi}{a}, \pm\frac{3\pi}{a}, \cdots \tag{9.67}$$

The eigenstate of H_0 is doubly degenerate therefore, degenerate perturbation theory is to be used

$$H_0\psi_1 = E_0\psi_1$$
$$H_0\psi_2 = E_0\psi_2$$

where

$$\psi_1 = \frac{1}{\sqrt{L}}\exp(ikx) = \frac{1}{\sqrt{L}}\exp\left(i\frac{\pi x}{a}\right) \tag{9.68}$$

$$\psi_2 = \frac{1}{\sqrt{L}}\exp(-ikx) = \frac{1}{\sqrt{L}}\exp\left(-i\frac{\pi x}{a}\right) \tag{9.69}$$

Since ψ_1 and ψ_2 are eigenfunctions of H_0, their linear combination is also a solution of H_0, that is

$$\psi = c_1\psi_1 + c_2\psi_2 \tag{9.70}$$

Such that

$$H_0\psi = E_0\psi$$

Consider

$$(H_0 + V)\psi = (E_0 + \Delta E)\psi \tag{9.71}$$

From Eqs. (9.70) and (9.71)

$$(H_0 + V)(c_1\psi_1 + c_2\psi_2) = (E_0 + \Delta E)(c_1\psi_1 + c_2\psi_2)$$

$$c_1 H_0\psi_1 + c_2 H_0\psi_2 + c_1 V\psi_1 + c_2 V\psi_2$$

$$= c_1 E_0\psi_1 + c_2 E_0\psi_2 + c_1\Delta E\psi_1 + c_2\Delta E\psi_2$$

$$c_1 E_0\psi_1 + c_2 E_0\psi_2 + c_1 V\psi_1 + c_2 V\psi_2$$

$$= c_1 E_0\psi_1 + c_2 E_0\psi_2 + c_1\Delta E\psi_1 + c_2\Delta E\psi_2$$

$$c_1 V\psi_1 + c_2 V\psi_2 = c_1\Delta E\psi_1 + c_2\Delta E\psi_2 \tag{9.72}$$

Multiplying Eq. (9.72) by ψ_2^* from left and integrating

$$c_1 \int \psi_2^* V\psi_1 dx + c_2 \int \psi_2^* V\psi_2 dx = c_1\Delta E \int \psi_2^*\psi_1 dx + c_2\Delta E \int \psi_2^*\psi_2 dx$$

$$c_1 V_{12}^* + c_2 V_{22} = c_2\Delta E \tag{9.73}$$

where

$$V_{12}^* = \int \psi_2^* V\psi_1 dx; \quad V_{22} = \int \psi_2^* V\psi_2 dx$$

Multiplying Eq. (9.72) by ψ_1^* from left and integrating

$$c_1 \int \psi_1^* V\psi_1 dx + c_2 \int \psi_1^* V\psi_2 dx = c_1\Delta E \int \psi_1^*\psi_1 dx + c_2\Delta E \int \psi_1^*\psi_2 dx$$

$$c_1 V_{11} + c_2 V_{12} = c_1\Delta E \tag{9.74}$$

where

$$V_{11} = \int \psi_1^* V\psi_1 dx; \quad V_{12} = \int \psi_1^* V\psi_2 dx$$

A general solution exists if

$$\begin{vmatrix} V_{11} - \Delta E & V_{12} \\ V_{12}^* & V_{22} - \Delta E \end{vmatrix} \begin{pmatrix} c_1 \\ c_2 \end{pmatrix} = 0 \tag{9.75}$$

For Eq. (9.74) to exists the determinant should be zero, that is,

$$\begin{vmatrix} V_{11} - \Delta E & V_{12} \\ V_{12}^* & V_{22} - \Delta E \end{vmatrix} = 0$$

On solving the determinant

$$(V_{11} - \Delta E)(V_{22} - \Delta E) - |V_{12}|^2 = 0$$
$$\Delta E^2 - \Delta E(V_{11} + V_{22}) + (V_{11}V_{22} - |V_{12}|^2) = 0$$

The roots are

$$\Delta E = \frac{V_{11} + V_{22}}{2} \pm \sqrt{\left(\frac{V_{11} - V_{22}}{2}\right)^2 + |V_{12}|^2} \tag{9.76}$$

Therefore, perturbation not only shifts but also removes the degeneracy of the energy levels.

Using Eqs. (9.68) and (9.69) we obtain various matrix elements, that is,

$$V_{11} = \int \psi_1^* V \psi_1 dx = \frac{1}{L} \int \left[\exp\left(\frac{i\pi x}{a}\right)\right]^* V(x) \exp\left(\frac{i\pi x}{a}\right) dx = \frac{1}{L} \int V(x) dx = V_0$$

$$V_{22} = \int \psi_2^* V_2 dx = \frac{1}{L} \int \left[\exp\left(-\frac{i\pi x}{a}\right)\right]^* V(x) \exp\left(-\frac{i\pi x}{a}\right) dx = \frac{1}{L} \int V(x) dx = V_0$$

Thus

$$V_{11} = V_{22} = V_0 \tag{9.77}$$

$$V_{12} = \int \psi_1^* V \psi_2 dx = \frac{1}{L} \int \left[\exp\left(\frac{i\pi x}{a}\right)\right]^* V(x) \exp\left(-\frac{i\pi x}{a}\right) dx = V_{-1} = V_1^* \tag{9.78}$$

Since

$$V^*(x) = \sum_n \exp\left(-\frac{2i\pi nx}{a}\right) V_n^* = \sum_n \exp\left(\frac{2i\pi nx}{a}\right) V_{-n}^*$$

Relabeling the summation $n \rightarrow -n$

$$V(x) = \sum \exp\left(\frac{2i\pi nx}{a}\right) V_n = -V_n^* = V_n$$

At $= \pm\frac{\pi}{a}$, the energy from Eq. (9.60) is

$$\Delta E = V_0 \pm V_1$$

Fig. 9.15 Plot of energy E versus wave number k showing band structure in nearly free electron model (Only half of the picture)

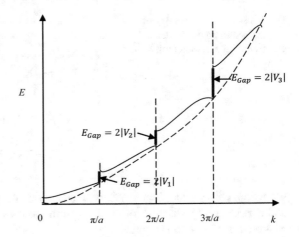

Total energy is

$$E = E_0 + \Delta E = \frac{\pi^2 \hbar^2}{2ma^2} + V_0 \pm |V_1| \tag{9.79}$$

Now V_0 can be dropped from Eq. (9.79), because it is just a shift in energy. The energy gap is

$$E_{\text{Gap}} = E_+ - E_- = \left(\frac{\pi^2 \hbar^2}{2ma^2} + V_0 + |V_1|\right) - \left(\frac{\pi^2 \hbar^2}{2ma^2} + V_0 - |V_1|\right) = 2|V_1| \tag{9.80}$$

Likewise the energy at $k = \pm 2\frac{\pi}{a}$ is

$$E_{\text{Gap}} = 2|V_2| \tag{9.81}$$

and so on. Figure 9.15 shows energy gap at the band edges in the nearly free electron model. The band gap $k = \pm\frac{\pi}{a}$ is $2|V_1|$ and the band gap $k = \pm\frac{2\pi}{a}$ is $2|V_2|$ and so on.

9.12 Tight-Binding Approximation

In tight-binding approximation it is assumed that the electrons are tightly bound to atoms and the effect of other neighbouring atoms is very little. The wave function of the electron overlaps very little with the wave function of the electrons of the neighbouring atoms. An electron in an atom has discrete energy levels. Atoms are brought close together, the wave functions for electrons in atoms overlap to a small extent. In crystals the discrete energy levels are broadened in to energy bands.

Consider an electron in a free atom. The potential energy of the electron, under consideration, in the free atom is due to field of the nucleus plus that of rest of the electrons of

Fig. 9.16 Potential energy of an electron in an atom (dashed curve) and in a crystal (fully drawn)

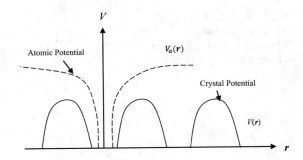

the atom. This potential energy is represented by $V_a(r)$ where subscript a indicate atom, and r is the vector distance of electron from the nucleus of the atom. The form of $V_a(r)$ is shown in Fig. 9.16 by dotted lines.

Let $\psi_k(r)$ is the wave function of electron in the free atom. The Hamiltonian operating on $\psi_k(r)$ gives energy E_0. From time independent Schrödinger equation

$$H\psi_k(r) = E_0\psi_k(r) \tag{9.82}$$

From Eqs. (9.81) and (9.82)

$$H\psi_k(r) = \left[-\frac{\hbar^2}{2m}\nabla^2 + V_a(r)\right]\psi_k(r) = E_0\psi_k(r) \tag{9.83}$$

It is assumed that E_0 is non-degenerate and wave function $\psi_k(r)$ is normalized.

The atoms are now brought close together to form a crystal. The potential energy of the electron is shown in Fig. 9.16 by solid lines. The potential energy of the electron in the crystal is represented by $V(r)$. The potential energy is periodic having periodicity of the formed crystal. Taking an atom as the origin, then position of any other atom n is represented by a vector r_n, where r_n is a lattice vector. The potential energy $V(r)$ consists of two parts

$$V(r) = V_a(r) + V'(r) = V_a(r) + V(r) - V_a(r)$$

where $V_a(r)$ is the atomic potential due to the atom at origin and $V'(r)$ is that due to all other atoms. $V'(r)$ is small in the neighbourhood of the origin. In tight-binding approximation, the electron wave function is only slightly effected by the other atoms when r is close to r_n. The wave function of the electron in this case can be represented by linear combination

$$\psi_k(r) = \sum_n \exp(i\mathbf{k} \cdot r_n)\psi(r - r_n) \tag{9.84}$$

The wave function given by Eq. (9.84) is periodic under lattice translation $\mathbf{R}$. Putting $r = r + R$ in Eq. (9.84)

$$\psi_k(\mathbf{r} + \mathbf{R}) = \sum_n \exp(i\mathbf{k} \cdot \mathbf{r}_n)\psi(\mathbf{r} + \mathbf{R} - \mathbf{r}_n)$$

$$\psi_k(\mathbf{r} + \mathbf{R}) = \sum_n \exp(i\mathbf{k} \cdot \mathbf{r}_n)\psi[\mathbf{r} - (\mathbf{r}_n - \mathbf{R})]$$

$$\psi_k(\mathbf{r} + \mathbf{R}) = \exp(i\mathbf{k} \cdot \mathbf{R}) \sum_n \exp[i\mathbf{k} \cdot (\mathbf{r}_n - \mathbf{R})]\psi[\mathbf{r} - (\mathbf{r}_n - \mathbf{R})]$$

$$\psi_k(\mathbf{r} + \mathbf{R}) = \sum_n \exp(i\mathbf{k} \cdot \mathbf{R})\psi_k(R)$$

Thus Eq. (9.84) satisfies the characteristic of a periodic function.
The Hamiltonian given by Eq. (9.81) can now be written as

$$H = \frac{p^2}{2m} + V(r) = \frac{p^2}{2m} + V_a(r) + V'(r) = \frac{p^2}{2m} + V_a(r) + V(r) - V_a(r) \qquad (9.85)$$

The energy of an electron with wave vector $\mathbf{k}$ in the crystal using the wave function given by Eq. (9.84) is

$$E(\mathbf{k}) = \frac{\int \psi_k^* H \psi_k \mathrm{d}r}{\int \psi_k^* H \psi_k \mathrm{d}r} \qquad (9.86)$$

Neglecting the small overlapping of atomic wave function centred at various lattice points, the denominator of Eq. (9.86) is

$$\int \psi_k^* H \psi_k \mathrm{d}r = \sum_n \int \exp[i(\mathbf{r_n} - \mathbf{r}_n)]\psi^*(\mathbf{r} - \mathbf{r}_n)\psi(\mathbf{r} - \mathbf{r}_n)\mathrm{d}r \qquad (9.87)$$

where summation over n only the term $n = m$ is retained. We may write Eq. (9.86) as

$$\int \psi_k^* H \psi_k \mathrm{d}r = N \qquad (9.88)$$

where N is the number of atoms in the crystal.
From Eqs. (9.84)–(9.86) and

$$E(\mathbf{k}) = \frac{1}{N}\left\{ \int \sum_m \exp(-i\mathbf{k} \cdot \mathbf{r}_m)\psi^*(\mathbf{r} - \mathbf{r}_m)\left[\frac{p^2}{2m} + V(r)\right] \right.$$

$$\left. \sum_n \exp(i\mathbf{k} \cdot \mathbf{r}_n)\psi(\mathbf{r} - \mathbf{r}_n)\mathrm{d}r \right\}$$

$$E(\mathbf{k}) = \frac{1}{N}\left\{ \int \sum_m exp(-i\mathbf{k} \cdot \mathbf{r}_m)\psi^*(\mathbf{r} - \mathbf{r}_m)\left[\frac{p^2}{2m} + V_a(r) + V'(r)\right] \right.$$

$$\left. \sum_n \exp(i\mathbf{k} \cdot \mathbf{r}_n)\psi(\mathbf{r} - \mathbf{r}_n)\mathrm{d}r \right\} \qquad (9.89)$$

From the above equation taking $\frac{p^2}{2m} + V_a(r)$ as Hamiltonian and solving

$$\frac{1}{N}\left\{ \int \sum_m \exp(-i\boldsymbol{k}\cdot\boldsymbol{r}_m)\psi^*(\boldsymbol{r}-\boldsymbol{r}_m)\left[\frac{p^2}{2m}+V_a(\boldsymbol{r})\right]\sum_n \exp(i\boldsymbol{k}\cdot\boldsymbol{r}_n)\psi(\boldsymbol{r}-\boldsymbol{r}_n)\mathrm{d}r\right\}$$

Using Eq. (9.83)

$$\frac{1}{N}\left\{ \int \sum_n \psi^*(\boldsymbol{r}-\boldsymbol{r}_n)[E_0]\psi(\boldsymbol{r}-\boldsymbol{r}_n)\mathrm{d}r\right\} = E_0$$

Here we have neglect in the summation over m all terms except $m = n$. From Eqs. (9.83) and (9.89)

$$E(\boldsymbol{k}) = E_0 + \frac{1}{N}\left\{ \int \sum_m \exp(-i\boldsymbol{k}\cdot\boldsymbol{r}_m)\psi^*(\boldsymbol{r}-\boldsymbol{r}_m)[V'(\boldsymbol{r})]\sum_n \exp(i\boldsymbol{k}\cdot\boldsymbol{r}_n)\psi(\boldsymbol{r}-\boldsymbol{r}_n)\mathrm{d}r\right\}$$
$$(9.90)$$

Now due to periodicity of the crystal and due to summation over all values of m, each term of the summation over n yields the same value. The summation over n is then just the value of any term of the summation times the number of terms in the sum which is N. Thus Eq. (9.90) can be written as

$$E(\boldsymbol{k}) = E_0 + \left\{ \int \sum_m \exp(-i\boldsymbol{k}\cdot\boldsymbol{r}_m)\psi^*(\boldsymbol{r}-\boldsymbol{r}_m)[V'(\boldsymbol{r})]\sum_n \exp(i\boldsymbol{k}\cdot\boldsymbol{r}_n)\psi(\boldsymbol{r}-\boldsymbol{r}_n)\mathrm{d}r\right\}$$

$$E(\boldsymbol{k}) = E_0 + \left\{ \int \sum_n \exp[i\boldsymbol{k}\cdot(\boldsymbol{r}_n-\boldsymbol{r}_m)]\psi^*(\boldsymbol{r}-\boldsymbol{r}_m)[V'(\boldsymbol{r})]\psi(\boldsymbol{r}-\boldsymbol{r}_n)\mathrm{d}r\right\} \qquad (9.91)$$

We define two positive quantities

$$\beta = -\int \psi^*(\boldsymbol{r}-\boldsymbol{r}_n)[V'(\boldsymbol{r})]\psi(\boldsymbol{r}-\boldsymbol{r}_n)\mathrm{d}r \qquad (9.92)$$

$$\gamma = -\int \psi^*(\boldsymbol{r}-\boldsymbol{r}_m)[V'(\boldsymbol{r})]\psi(\boldsymbol{r}-\boldsymbol{r}_n)\mathrm{d}r \qquad (9.93)$$

If the overlap of the atomic functions is very small then γ will be small and the electron are quite immobile. On the other hand if the overlap is very large then the electrons become quite mobile. The value of γ may be obtained from atomic wave functions the atomic potential and near neighbour distance. The integral for β and γ is independent of $\boldsymbol{k}$.

From Eqs. (9.91)–(9.93)

$$E(\boldsymbol{k}) = E_0 - \beta - \gamma \sum_n \exp[i\boldsymbol{k} \cdot (\boldsymbol{r}_n - \boldsymbol{r}_m)] \tag{9.94}$$

where summation extends over nearest neighbours of atom n only and β, γ are positive as V' is negative. From Eq. (9.94) it is seen that energy of the electron in tight-binding approximation differs from energy of electron in the free atom by a constant factor β plus a term which depends on the wave vector $\boldsymbol{k}$ and β is a small quantity since the function is appreciable only near the origin where V' is small. The term in γ transforms the discrete atomic levels into energy band in the solid.

As an example consider the case of simple cubic (s.c.) lattice having dimension $a = b = c$. In this lattice a given atom has nearest neighbours located at $(\pm a, 0, 0); (0, \pm a, 0)$ and $(0, 0, \pm a)$. Thus putting $\boldsymbol{r}_n - \boldsymbol{r}_m = \boldsymbol{\rho}$ we have

$$\boldsymbol{\rho} = (\pm a, 0, 0); (0, \pm a, 0) \text{ and } (0, 0, \pm a) \tag{9.95}$$

Using Eqs. (9.94) and (9.95)

$$E(\boldsymbol{k}) = E_0 - \beta - \gamma \sum_n \exp[i\boldsymbol{k} \cdot \boldsymbol{\rho}]$$

$$\sum_n \exp[i\boldsymbol{k} \cdot \boldsymbol{\rho}] = \{[\exp(ik_x a] + [\exp(-ik_x a] + [\exp(ik_y a] + [\exp(-ik_y a] + [\exp(ik_z a] + [\exp(-ik_z a]\}$$

$$\sum_n \exp[i\boldsymbol{k} \cdot \boldsymbol{\rho}] = [2\cos k_x a + 2\cos k_y a + 2\cos k_z a] \tag{9.96}$$

From Eqs. (9.95) and (9.96)

$$E(\boldsymbol{k}) = E_0 - \beta - 2\gamma[\cos k_x a + \cos_y a + \cos k_z a]$$

The part of E($\boldsymbol{k}$) which depends on $\boldsymbol{k}$ is periodic. We restrict to first Brilloiun zone, that is,

$$-\frac{\pi}{a} \le k_x \le \frac{\pi}{a}; -\frac{\pi}{a} \le k_y \le \frac{\pi}{a}; -\frac{\pi}{a} \le k_z \le \frac{\pi}{a}$$

The first Brillouin zone in this case is a cube of edge $\frac{2\pi}{a}$ in k space. The minimum value of energy corresponds to $k_x = k_y = k_z = 0$ and

$$E(\boldsymbol{k}) = E_0 - \beta - 2\gamma[1 + 1 + 1] = E_0 - \beta - 6\gamma \tag{9.97}$$

The maximum value occurs at $k_x = k_y = k_z = \frac{\pi}{a}$ and

$$E(k) = E_0 - \beta - 2\gamma[\cos \pi + \cos \pi + \cos \pi] = E_0 - \beta - 2\gamma[-1 - 1 - 1]$$
$$E(k) = E_0 - \beta + 6\gamma \tag{9.98}$$

Thus the energy bands are contained within band of width 12γ. Thus it is seen that width of the band increases as γ increases or the overlap of wave functions on the neighbouring atoms increases. Thus the inner electronic levels of the free atoms give rise to narrow bands in solids.

The bottom of the band corresponds to

$$\cos k_x a = \cos_y a = \cos_z a = 1$$

That is to $k = 0$ in this case. As long as k is small, the cosine term may be expanded. To the first approximation

$$\cos_x a = 1 - \frac{k_x^2 a^2}{2}$$

$$\cos_y a = 1 - \frac{k_y^2 a^2}{2}$$

$$\cos k_z a = 1 - \frac{k_z^2 a^2}{2}$$

$$E_k(r) = E_0 - \beta - 2\gamma\left[3 - \frac{\left(k_x^2 + k_y^2 + k_z^2\right)a^2}{2}\right]$$

$$E(k) = E_0 - \beta - 6\gamma + \gamma k^2 a^2 \tag{9.99}$$

Thus with reference to the bottom of the band the energy of the electron is proportional to k^2 similar to the case of free electron. Therefore in this region the effective mass can be evaluated by comparing the energy

$$\frac{\hbar^2 k^2}{2m^*} = \gamma k^2 a^2 \tag{9.100}$$

$$m^* = \frac{\hbar^2}{2\gamma a^2} \tag{9.101}$$

$$\frac{m^*}{m} = \frac{\hbar^2}{2\gamma m a^2}$$

If the overlap of atomic functions is very small then γ will be small and the effective mass will be large. If the overlap of the wave functions is very small, the electrons are quite immobile. If the overlap is very large then the mass becomes small and the electrons become quite mobile and the tight-binding approximation is not accurate.

Solved Examples

Example 1 The energy of a band is given by

$$E(k) = E_\alpha - \beta - 2\gamma \left(\cos ak_x + \cos ak_y + \cos ak_z\right)$$

where E_α, β and γ are constants and a is lattice constant. Obtain the effective mass when k_x, k_y and k_z are very small.

Solution

Under the assumption of k_x, k_y and k_z are very small, we have

$$\cos ak_x = 1 - \frac{(ak_x)^2}{2} + \cdots$$

$$\cos ak_y = 1 - \frac{(ak_y)^2}{2} + \cdots$$

$$\cos ak_z = 1 - \frac{(ak_z)^2}{2} + \cdots$$

$$E(k) = E_\alpha - \beta - 2\gamma \left[\frac{3}{2} - \frac{a^2}{2}\left(k_x^2 + k_y^2 + k_z^2\right)\right] = E_\alpha - \beta - 2\gamma \left[\frac{3}{2} - \frac{a^2 k^2}{2}\right]$$

where $k_x^2 + k_y^2 + k_z^2 = k^2$.

Differentiating E with respect to k

$$\frac{dE}{dk} = 2a^2\gamma k; \quad \frac{d^2E}{dk^2} = 2\gamma a^2$$

The effective mass is thus

$$m^* = \frac{\hbar^2}{d^2E/dk^2} = \frac{\hbar^2}{2\lambda a^2}$$

Example 2 In a crystal the energy is given by

$$E(k) = \text{Constant} + b(k - k_0)^2$$

where b and k_0 are constants. Obtain the effective mass.

Solution

We have

$$m^* = \frac{\hbar^2}{d^2 E/dk^2}$$

From expression of energy

$$\frac{dE}{dk} = 2b(k - k_0); \quad \frac{d^2 E}{dk^2} = 2b$$

The effective mass is

$$m^* = \frac{\hbar^2}{2b}$$

Example 3 The energy near a valence band edge is given by $E_k = -1 \times 10^{-26} \, k^2$ erg. An electron is removed from the state $k = 1 \times 10^7 k_x$ cm^{-1}. The band is otherwise full. Give the sign and magnitude of (i) effective mass of the hole (ii) direction and magnitude of the wave vector of the hole (iii) the crystal momentum of the hole (iv) the velocity of the hole.

Solution

(i) we have

$$m^* = \frac{\hbar^2}{d^2 E/dk^2}$$

$$E_k = -1 \times 10^{-26} k^2 \text{erg}$$

Thus

$$d^2 E/dk^2 = -2 \times 10^{-26} \text{erg}$$

$$m^* = \frac{(1.055 \times 10^{-27})^2}{-2 \times 10^{-26}} g \approx -5 \times 10^{-29} g$$

since $d^2 E/dk^2$ is negative at the top of the band, therefore m^* should be positive.

(ii) since kh $= -$ke, therefore

$$k_h = -10^7 k_x \text{cm}^{-1}$$

(iii) Crystal momentum is

$$\hbar k = 1.055 \times 10^{-27} \text{ erg.sec} \times \left(-10^7 k_x \text{cm}^{-1}\right)$$

(iv) The velocity of the hole is

$$v = \frac{1}{\hbar}\frac{dE}{dk} = \frac{1}{1.055 \times 10^{-27}} \times \left(-2 \times 10^{-26} \times 10^7\right)k_x \approx -2 \times 10^8 k_x \text{cm/s}$$

Example 4 Which of ZnS and CdSe will have the largest band gap energy?

Solution

As the two elements forming these compounds become more widely separated with respect to their relative positions in the periodic table, that is, the electronegativities become more dissimilar, the atomic bonding becomes more ionic and the magnitude of the band gap energy increases, the materials tend to become more insulators. The band gap of CdS is 2.4 eV while those of ZnTe is 2.26 eV.

Example 5 Prove that for the Kronig–Penney potential with $P \ll 1$, the energy of the lowest energy band at $k = 0$ is

$$E = \frac{\hbar^2 P}{ma^2}$$

Solution

From Eq. (9.29) for $k = 0$, we have

$$P\frac{\sin qa}{qa} + \cos qa = \cos 0 = 1$$

$$P\frac{\sin qa}{qa} = 1 - \cos qa$$

$$\frac{P}{qa} = \frac{1 - \cos qa}{\sin qa}$$

For small qa

$$\cos qa = 1 - \frac{q^2 a^2}{2} + \cdots$$

$$\sin qa = qa - \frac{q^3 a^3}{6} + \cdots$$

$$\frac{P}{qa} \approx \frac{1 - 1 + (q^2 a^2 / 2)}{qa} = \frac{qa}{2}$$

$$P = \frac{q^2 a^2}{2} = \frac{2mEa^2}{2\hbar^2}$$

$$E = \frac{\hbar^2 P}{ma^2}$$

Example 6 Show that in the tight-binding approximation, the energy $E(\mathbf{k})$ for one dimensional linear lattice of lattice constant a is given by.

$$E(k) = E_0 - \beta - 2\gamma \cos ka$$

The 2 nearest neighbour of the origin are at $\pm a$. Determine the band width and effective mass.

Solution

For one dimensional lattice, there are two nearest neighbours at $x = \pm a$. The value of $\exp(i\mathbf{k} \cdot \boldsymbol{\rho})$ is

$$\exp(i\mathbf{k} \cdot \boldsymbol{\rho}) = \exp(ikx) = \exp(ika) + \exp(-ika) = 2\cos ka$$

Substituting this value $\exp(i\mathbf{k} \cdot \boldsymbol{\rho})$ in Eq. (9.94)

$$E(k) = E_0 - \beta - 2\gamma \cos ka \qquad (9.102)$$

The minimum value of energy corresponds to $k = 0$ and is

$$E(k) = E_0 - \beta - 2\gamma$$

The maximum value of energy corresponds to $k = \pm\frac{\pi}{a}$ and is

$$E(k) = E_0 - \beta - 2\gamma \cos \pi = E_0 - \beta + 2\gamma$$

Thus the energy bands are contained within band of width 4γ. Thus it is seen that width of the band increases as γ increases or the overlap of wave functions on the neighbouring atoms increases.

For small values of k, that is, $k \ll \frac{\pi}{a}$, we can Taylor expand Eq. (9.102)

$$E(k) = E_0 - \beta - 2\gamma \left(1 - \frac{k^2 a^2}{2}\right)$$

$$E(k) = E_0 - \beta - 2\gamma - \gamma k^2 a^2 \qquad (9.103)$$

Thus with reference to the bottom of the band the energy of the electron is proportional to k^2 similar to the case of free electron. Therefore in this region the effective mass can be evaluated by comparing the energy

$$\frac{\hbar^2 k^2}{2m^*} = \gamma k^2 a^2$$

$$m^* = \frac{\hbar^2}{2\gamma a^2} \qquad (9.104)$$

Example 7 Show that in the tight-binding approximation, the energy $E(k)$ for body centre cubic lattice is given by

$$E(k) = E_0 - \beta - 8\gamma \cos\frac{k_x a}{2} \cos\frac{k_y a}{2} \cos\frac{k_z a}{2}$$

The 8 nearest neighbour of the origin are at

$$\frac{a}{2}(\pm1, 1, 1), \frac{a}{2}(\pm1, 1, -1), \frac{a}{2}(\pm1, -1, 1), \frac{a}{2}(\pm1, -1, -1)$$

Solution

The 8 nearest neighbour of the origin are at

$$r_n - r_m = \rho = \frac{a}{2}(\pm1, 1, 1), \frac{a}{2}(\pm1, 1, -1), \frac{a}{2}(\pm1, -1, 1), \frac{a}{2}(\pm1, -1, -1)$$

If $k = k_x i + k_y j + k_z k$ then corresponding 12 values of $\exp[i k \cdot \rho]$ are $\exp[i k \cdot \rho]$ for $\rho = \frac{a}{2}(\pm1, 1, 1)$ is

$$\exp\left[\frac{i}{2}(k_x a + k_y a + k_z a)\right] + \exp\left[\frac{i}{2}(-k_x a + k_y a + k_z a)\right]$$

$$\exp\left[\frac{i}{2}(k_y a + k_z a)\right]\left[\exp\left(i\frac{k_x a}{2}\right) + \exp\left(-i\frac{k_x a}{2}\right)\right]$$

$$\left[\exp\left(i\frac{k_y a}{2}\right)\exp\left(\frac{ik_z a}{2}\right)\right]\left(2\cos\frac{k_x a}{2}\right)$$

$$\left[\cos\frac{k_y a}{2} + i\sin\frac{k_y a}{2}\right]\left[\cos\frac{k_z a}{2} + i\sin\frac{k_z a}{2}\right]\left(2\cos\frac{k_x a}{2}\right)$$

$$2\cos\frac{k_x a}{2}\left[\cos\frac{k_y a}{2}\cos\frac{k_z a}{2} + i\cos\frac{k_y a}{2}\sin\frac{k_z a}{2} + i\sin\frac{k_y a}{2}\cos\frac{k_z a}{2} - \sin\frac{k_y a}{2}\sin\frac{k_z a}{2}\right]$$

$$(9.105)$$

Similarly
$\exp[i k \cdot \rho]$ for $\rho = \frac{a}{2}(\pm1, 1, -1)$ is

$$2\cos\frac{k_x a}{2}\left[\cos\frac{k_y a}{2}\cos\frac{k_z a}{2} - i\cos\frac{k_y a}{2}\sin\frac{k_z a}{2} + i\sin\frac{k_y a}{2}\cos\frac{k_z a}{2} + \sin\frac{k_y a}{2}\sin\frac{k_z a}{2}\right]$$

$$(9.106)$$

$\exp[i\mathbf{k} \cdot \boldsymbol{\rho}]$ for $\boldsymbol{\rho} = \frac{a}{2}(\pm 1, -1, 1)$ is

$$2\cos\frac{k_x a}{2}\left[\cos\frac{k_y a}{2}\cos\frac{k_z a}{2} + i\cos\frac{k_y a}{2}\sin\frac{k_z a}{2} - i\sin\frac{k_y a}{2}\cos\frac{k_z a}{2} + \sin\frac{k_y a}{2}\sin\frac{k_z a}{2}\right]$$

(9.107)

$\exp[i\mathbf{k} \cdot \boldsymbol{\rho}]$ for $\boldsymbol{\rho} = \frac{a}{2}(\pm 1, -1, -1)$ is

$$2\cos\frac{k_x a}{2}\left[\cos\frac{k_y a}{2}\cos\frac{k_z a}{2} - i\cos\frac{k_y a}{2}\sin\frac{k_z a}{2} - i\sin\frac{k_y a}{2}\cos\frac{k_z a}{2} - \sin\frac{k_y a}{2}\sin\frac{k_z a}{2}\right]$$

(9.108)

On adding Eqs. (9.105)–(9.108), $\exp[i\mathbf{k} \cdot \boldsymbol{\rho}]$for.
$\boldsymbol{\rho} = \frac{a}{2}(\pm 1, 1, 1), \frac{a}{2}(\pm 1, 1, -1), \frac{a}{2}(\pm 1, -1, 1), \frac{a}{2}(\pm 1, -1, -1)$ is

$$\exp[i\mathbf{k} \cdot \boldsymbol{\rho}] = 8\cos\frac{k_x a}{2}\cos\frac{k_y a}{2}\cos\frac{k_z a}{2}$$

(9.109)

From Eqs. (9.94) and (9.109)

$$E(\mathbf{k}) = E_0 - \beta - 8\gamma \cos\frac{k_x a}{2}\cos\frac{k_y a}{2}\cos\frac{k_z a}{2}$$

(9.110)

For small values of $k_x, k_y and k_z$

$$\cos\frac{k_x a}{2}\cos\frac{k_y a}{2}\cos\frac{k_z a}{2} = \left(1 - \frac{k_x^2 a^2}{2}\right)\left(1 - \frac{k_y^2 a^2}{2}\right)\left(1 - \frac{k_z^2 a^2}{2}\right)$$

$$\cos\frac{k_x a}{2}\cos\frac{k_y a}{2}\cos\frac{k_z a}{2} \approx 1 - \frac{a^2}{2}\left[k_x^2 + k_y^2 + k_z^2\right] = 1 - \frac{k^2 a^2}{2}$$

$$E(\mathbf{k}) = E_0 - \beta - 8\gamma\left(1 - \frac{k^2 a^2}{2}\right) = E_0 - \beta - 8\gamma + 4k^2 a^2$$

(9.111)

Objective Type Questions

1. The permissible energy levels taken in groups are called

   (a) permissible energy levels (b) energy bands (c) conduction bands (d) forbidden energy levels

2. The materials in which valence band and conduction band overlaps are called

   (a) insulators, (b) Conductors (c) Semiconductors (d) Superconductors

3. The material in which the highest occupied energy level is completely filled is called

   (a) insulator (b) conductor (c) semiconductor (d) superconductor

4. The material in which the gap between the filled energy band and next higher permitted energy band is small, is called

   (a) insulator (b) Conductor (c) semiconductor (d) superconductor
5. The substances with resistivity of order of 10^{-4} Ω-m are called

   (a) insulators (b) semiconductors (c) conductors (d) good conductors
6. The substances with resistivity of the order of 10^{-8} Ω-m are called

   (a) insulators (b) semiconductors (c) conductors (d) good conductors
7. At temperature near absolute zero, a pure semiconductor behaves like

   (a) an insulator (b) a conductor (c) a superconductor (d) none of these
8. Pure Si at 0 K is an

   (a) semiconductor (b) metal (c) insulator (d) none of these
9. $InSb$ is

   (a) II–VI type semiconductor (b) III–V type semiconductor (c) IV–IV type semiconductor (d) none of these
10. ZnS is

    (a) II–VI type semiconductor (b) III–V type semiconductor (c) IV–IV type semiconductor (d) none of these
11. The example of direct band gap semiconductor is

    (a) Ge (b) Si (c) $PbSe$ (d) $InSb$
12. Which of the following statement about semiconductor is false?
    (a) when heated they conduct by heavy electrons flow through the conduction band and negative holes flow through the valence band
    (b) they acts as insulator at 0 K
    (c) when heated they conduct electricity through both electrons in conduction band and holes in the valence band
    (d) resistance decreases with increased temperature
13. Which of the following material is not a semiconductor?

    (a) Silicone (b) Germanium (c) Gallium arsenide (d) Gallium nitride
14. Which has the least band gap?

    (a) C (b) Si (c) Ge (d) Sn
15. Which has the least band gap?

    (a) $AlSb$ (b) $GaSb$ (c) $InSb$
16. Which is not true?
    (a) the effective mass is a tensor quantity
    (b) the curvature d^2E/dk^2 is positive at the conduction band minima, however, it is negative at the valance band maxima

(c) the curvature d^2E/dk^2 is negative at the conduction band minima, however, it is positive at the valance band maxima

(d) valence band electrons with negative charge and negative mass move in an electric field in the same direction as holes with positive charge and positive mass.

17. Which is False?

(a) in indirect band gap semiconductors, the minima of conduction band and maxima of the valence band occur at the same value of k

(b) the probability of recombination of electrons and holes for direct band gap semiconductors is much higher than that for indirect band gap semiconductors

(c) direct band gap semiconductors give up the energy released during recombination in the form of light

(d) recombination in indirect band gap semiconductors occurs through some defect states within the band gap, and the energy is released in the form of heat given to the lattice

18. Which is not true?

(a) the electron energy increases as one moves up the conduction band, and electrons gravitate downward towards the bottom of the conduction band

(b) the hole energy increases as one moves down the valence band, and holes gravitate upwards towards the top of the valence band

(c) the wave particle motion of electrons in a lattice is not the same as that for a free electron, because of the interaction with the periodic potential of the lattice

(d) the effective mass of an electron in a band with a given (E, k) relation is given by

$$m^* = \frac{\hbar^2}{d^2 E/dk^2}$$

19. According to Kronig–Penney model, which is not correct?

(a) allowed energy bands are narrower for low value of energy

(b) allowed energy bands are broader for large value of energy

(c) P is a measure of the area of the potential barrier

(d) all values of q are allowed

Problems

1. Discuss the Kronig–Penney monoatomic linear lattice. Explain how does this lead to the formation of allowed and forbidden energy bands.

2. Explain the concept of effective mass and derive an expression for the same.

3. How does the band theory of solid lead to the classification of solids into conductors, semiconductors and insulators?

4. Describe what is meant by a hole, in the context of the theory of electrons in solids. Explain how a hole might be created or annihilated.
5. If the energy E and the wave vector k of electrons are related as

$$E = \frac{h^2}{ma^2}(1 - \cos ka), \text{ for } 0 \leq k \leq \frac{\pi}{a}$$

where a is the distance and m is the mass of a free electron, calculate the effective mass of the electrons.
6. State and prove Bloch's theorem.
7. Show that the number of orbitals in a band inside the first zone is equal to the number of unit cells in the crystal.
8. Draw the first two zones for a two dimensional rectangular lattice.
9. Show that in the tight-binding approximation, the energy $E(k)$ for face centre cubic lattice is given by

$$E(k) = E_0 - \beta - 4\gamma \left(\cos \frac{k_x a}{2} \cos \frac{k_y a}{2} + \cos \frac{k_y a}{2} \cos \frac{k_z a}{2} + \cos \frac{k_z a}{2} \cos \frac{k_x a}{2} \right)$$

The 12 nearest neighbour of the origin are at $\frac{a}{2}(\pm 1, \pm 1, 0)$, $\frac{a}{2}(\pm 1, 0, \pm 1)$, $\frac{a}{2}(0, \pm 1, \pm 1)$. In the limit of small ka

$$E(k) = E_0 - \beta - 12\gamma + 2\gamma k^2 a^2$$

Answers
Objective Type Questions

1. (b)	2. (b)	3. (a)	4. (c)	5. (b)	6. (c)	7. (a)	8. (c)
9. (b)	10. (a)	11. (d)	12. (a)	13. (a)	14. (d)	15. (c)	16. (c)
17. (a)	18. (a)						

Semiconductors

10

10.1　Doping and Defects in Semiconductors

The electrical and optical properties of a pure semiconductor can be modified by addition of a small amount of impurities called dopant. This process of introducing the dopant in a pure semiconductor is known as doping. The concentration of impurities is very low and is of the order of one part in a million or less. The doping material is added during the process of forming of a semiconductor. The atoms of impurities are of a slightly different valence number than the atoms of the semiconductor. The dopants are of two kinds (a) donors: they donate an electron to the conduction band and (b) acceptor: they accept an electron from the valence band and create a hole in the valence band.

The addition of donors in semiconductors forms a semiconductor known as *n*-type semiconductors. The example of *n*-type semiconductor is addition of V group element in the lattice formed by the IV group elements like Ge. The introduction of acceptor in the semiconductor is known as *p*-type semiconductor. The example of *p*-type semiconductor is addition of III group element in the lattice formed by the IV group elements like Si. The energy levels of donor lie in forbidden gap near the bottom of the conduction band, while the energy levels of acceptor lie near the top of valence band. The donors become positively charged when electron jumps from donor levels to conduction band, and acceptor becomes negatively charged when it accept an electron from valence band.

The compound semiconductors GaAs is formed by III–V group elements. The II group elements like Be or Zn act as acceptor when they substitute for group III element Ga in GaAs. The group VI element like Se can substitute for group V element and become donor. The dopant can also be amphoteric in compound semiconductors, for example Si acts as both acceptor and donor. The silicon acts as a donor if it replaces Ga in GaAs semiconductor and becomes acceptor if it substitutes for As.

© The Author(s) 2022
V. K. Jain, *Solid State Physics*, https://doi.org/10.1007/978-3-030-96017-9_10

The concentration of dopant is very low such that the dopants are far apart. In this case, there is no interaction between them, and the semiconductors are referred as non-degenerate semiconductors. If the dopant concentration is increases, the distance between dopant atoms decreases and a point will be reached when dopant charge carriers (for example, electrons of donors) will begin to interact with each other. The energy level of dopant splits into band of energies. With increasing concentration of the dopant, the band of dopant may overlap with the bottom of the conduction band or top of the valence band for n-type and p-type semiconductors, respectively. In such a case, the Fermi level approaches conduction band or valence band. For very large concentration of dopant, the Fermi level enters conduction band for n-type semiconductor and valence band for p-type semiconductor. In such a case the semiconductor is called degenerate n-type semiconductor or degenerate p-type semiconductor. Because of high level of doping, a semiconductor starts to act like a metal. In the next section semiconductors with and without doping are described.

Impurities are classified as shallow impurities and deep impurities. Shallow impurities are impurities which require very little energy of the order of thermal energy or less for ionization. Deep impurities require energies larger than the thermal energy to ionize. Therefore, only a fraction of impurities in the semiconductor contribute to the free charge carriers. Deep impurities which are quite far away from the top of valence band or bottom of conduction band are unlikely to ionize. Such impurities can act as a recombination centre in which electrons or holes fall and annihilate each other. Such deep impurities are called trap.

In a perfect crystal structure of a semiconductor, all the atoms are arranged at the designated position. In real crystals there is deviation from ideal structure. The real crystal contains defects or imperfections. The defects may create electronic energy states in the forbidden gap of the semiconductor. These electronic energy states can be categorized into two types (a) shallow levels and (b) deep levels. The shallow levels are very close to bottom of conduction band for donors or top of valence band for acceptor. Deep levels are quite deep in the forbidden gap. The shallow levels have small ionization energy, while deep levels have higher ionization energy. The deep levels contribute very little to free charge carriers. Defects with deep levels in the forbidden gap are often referred to as trap. The traps are described in Chap. 16. Deep levels modify the properties of semiconductors and therefore are important. Deep levels are desirable in some applications, for example, in switching devices.

Defects in semiconductors can be divided into (a) point defects, (ii) line defects, (iii) planar defects and (iv) volume defects. Various point defects are described in Chap. 3. Point defects are zero dimensional defects, and therefore, perturbation of lattice is localized about a lattice site and involves only a few nearest neighbours. In semiconductors, the point defects are of interest are vacancies or self-interstitial and impurity atoms substituting for the host atoms or interstitial sites. If an atom is removed from the regular lattice site, then the empty lattice site is called a vacancy defect as mentioned in Chap. 3. Vacancy in Ge or Si semiconductor can have various charge states like 2+, 1+, 0, etc. The vacancy formation energy in Ge is smaller than in Si. For Ge it is ~ 1.7 to $2.5 eV$ while for Si it is $\sim$4.0 eV. Interstitials are atoms which occupy a site which is not occupied by atoms in a perfect lattice. This defect has same atom as that of lattice of crystal or an impurity. In semiconductor, divacancy can be formed when two neighbouring atoms are removed. A defect is formed when a vacancy is trapped next to a substitutional donor atom. This is known as E centre. A vacancy trapped next to an oxygen atom in intestinal position is known as A centre (vacancy-oxygen).

Most of primary and secondary defects are electrically active and induce deep levels in the forbidden gap of the semiconductor. A deep level may act as a minority carrier trap, majority carrier trap or recombination centre depending on the position in the forbidden gap of the semiconductor. A majority carrier trap is an electron trap in n-type semiconductor or hole trap in p-type semiconductor. Conversely, a minority carrier is a hole trap for n-type semiconductor or an electron trap for p-type semiconductor. Recombination centres are deep levels with approximately equal capture cross section for both holes and electrons.

In contrast to point defects, line defects called dislocations involve a larger number of atomic sites that can be connected by a line. The dislocations are described in Chap. 3. Planar defects are important in polycrystalline materials. In polycrystalline materials, small regions of a few microns in diameter are crystalline. But these microcrystallites may have different orientations. A grain boundary separates regions of different crystalline orientations (i.e. grains). The atoms in the grain boundary will not be in perfect crystalline arrangement. Grain size varies from 1 μm to 1 mm. The grain boundaries are discussed in Chap. 3. The volume defects occur if the process of growth crystals of semiconductors is poor. This causes semiconductors to contain regions that are amorphous or may contain voids.

10.2   Intrinsic Semiconductors

Semiconductors which are highly purified are termed as intrinsic semiconductors, for example Ge, Si, Se, Te, GaAs, InAs, etc. Consider a pure crystal of Ge containing no impurities. The Ge atom in the crystal has four valence electrons, and each forms four electron–pair bond in the diamond structure. Figure 10.1 shows a two dimensional representation of such a crystal. At 0 K all the electrons are tightly bound in the covalent bonds, and the crystal is a perfect insulator. At higher temperatures, vibrations of lattice atoms occasionally impart sufficient thermal energy to some of the bound electrons enabling them to leave the bond and move freely inside the crystal. As electron leaves a bond, a hole is created. As a result of thermal motion, an electron in a covalent bond adjacent to the holes jumps into the original vacant sites, thus transferring the location of the hole to another site. In this way the motion of the hole can be regarded as the transfer of ionization from one atom to next by the jumping of the bound electron from a

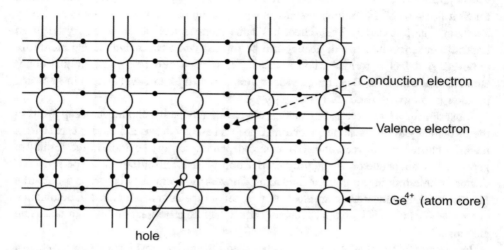

Fig. 10.1 A two dimensional schematic representation of Ge crystal showing the atom cores and valence electrons

covalent bond to a neighbouring vacant site. If n_e and n_h represent the electron and hole concentration, respectively, then $n_e = n_h = n_i$.

According to band theory of solid, in the intrinsic semiconductor the valence band is full but the conduction band is empty at 0 K (Fig. 10.2). Since there are no free charge carriers, the conductivity of intrinsic semiconductor is zero at 0 K. As the temperature increases, the electrons of the valence band become excited and some of them gain enough thermal energy to overcome the energy gap or forbidden gap and go over to the conduction band. When an electron jumps to the conduction band, it leaves behind a space or hole in the valence band. This hole is effectively positive, and an electron can jump into it from another part of the valence band leaving behind a hole. In this way hole moves in the valence band. The electrons in the conduction band are in excited states and have only a finite life time. An electron from the conduction band can drop into the top of valence band, recombining with a hole and releasing an energy E_g. Conversely an electron–hole pair can be created by lifting an electron from the valence band to the conduction band. Both processes occur repeatedly giving a dynamical equilibrium concentration which is a function of temperature.

The conduction can take place by negative electrons moving within the conduction band and by positive holes moving within the valence band. This gives a small conductivity whose magnitude depends on the temperature and on the width of energy gap between the valence and conduction band. The number of electrons excited into conduction band is proportional to $\exp(-E_g/2k_B T)$.

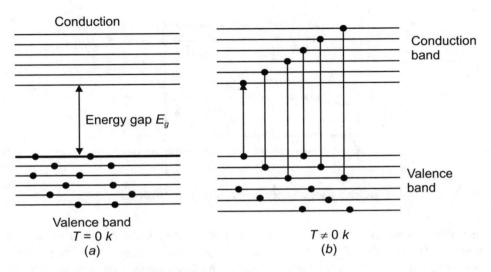

Fig. 10.2 Energy band diagram of an intrinsic semiconductor

For each electron in the conduction band, there will be a corresponding hole in the valence band. Both electron and hole contribute to the conductivity σ

$$\sigma = |e|(\mu_e + \mu_h)$$

where μ_e and μ_h are mobilities of electron and hole, respectively. For intrinsic semiconductor $n_e = n_h = n_i$ and

$$\sigma = n_i|e|(\mu_e + \mu_h)$$

10.3 Extrinsic Semiconductors

A pure semiconductor to which a very small amount of dopant has been added is called extrinsic semiconductor. Depending on the type of impurity or dopant present, an extrinsic semiconductor is further classified as n-type or p-type semiconductor.

(i) Donor or n-type semiconductor

Consider the Ge crystal. Ge has a diamond-type lattice in which every atom is surrounded by four nearest neighbours bound to it. The substitutional impurity atoms, having five or more valence electrons, donate the excess electrons to the crystal because only four of

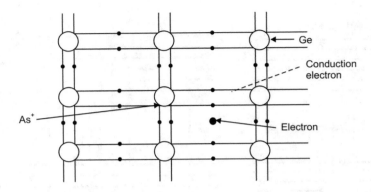

Fig. 10.3 A two dimensional schematic representation of Ge crystal doped with As

their electrons are required in bond formation. Suppose some Ge atoms of Ge crystal are substituted by pentavalent As atoms. Four of the valence electrons of the As (donor) atom behave as they would in a *Ge* atom. The remaining fifth electron now sees a positively charged As ion to which it is attracted as shown in Fig. 10.3.

The As$^+$ ion has a charge of $+e$ and the fifth electron is loosely bound it by coulombic attraction. The orbit of the electron will be quite large compared to the interatomic spacing so that many atoms will be encompassed by the path of the electron about the ion. The electrostatic force between the electron and the impurity ion (As$^+$) will be modified by the polarization of the intervening atoms. Thus it can be assumed that the electron has been immersed in a uniform polarizing medium of semiconductor crystal (Ge), and the electron plus impurity ion system forms hydrogen like atom in the uniform continuous medium of dielectric constant K. According to this model the energy required to remove the electron is

$$E_n = \frac{m^* e^4}{2K^2 n^2 \hbar^2} = \left(\frac{m^*}{m}\right) \frac{13.6 \text{ eV}}{n^2 K^2}$$

Substituting the values of dielectric constant, effective mass and $n = 1$, the energy is ~0.0054 eV for Ge and ~0.0198 eV for Si.

Thus the energy states of the donor atom lie within the forbidden energy gap, a few hundredth of an electron volt below the conduction band as shown in Fig. 10.4. In terms of band theory the process may be described as follows. The energy level of the fifth electron of the *As* atom occupies positions between the valence band and the conduction band. These positions are directly below the bottom of the conduction band at a distance of $E_d \sim 0.01$ eV. In the band picture, these levels are shown by discrete dotted lines as the impurity atoms are far apart in the host lattice. When an electron occupying such an impurity level receives additional energy greater than E_d, it goes to the conduction band. The remaining positive charge is localized on the immobile As atom and does not take

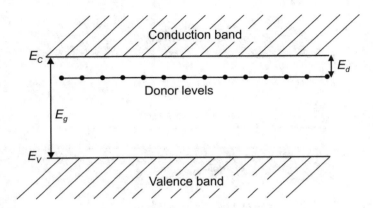

Fig. 10.4 Energy band diagram of a n-type semiconductor

part in electrical conductivity. The crystal as a whole remains neutral because the electron remains in the crystal.

(ii) **Acceptor or _p_-type semiconductor**

Suppose an element of group III such as Al is introduced in small amount into the crystal structure of Ge. For each atom of group III element introduced into the crystal structure there remains an unfilled bond indicated by broken line. The trivalent atom can accept an electron, indicated by the small open circle _b_ (Fig. 10.5). A valence electron from a nearby Ge atom such as at position a needs a small amount of energy to jump into the position b to complete the unfilled bond. This small energy is provided by the thermal

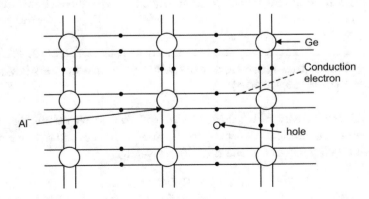

Fig. 10.5 A two dimensional schematic representation of Ge crystal doped with Al

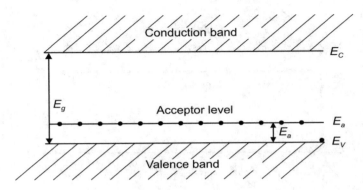

Fig. 10.6 Energy band diagram of a p-type semiconductor

motion of the crystal. In going from a to b the electron breaks its own covalent bond and creates a hole at a.

The unfilled energy state created by position like b is located just above the valence bond of the semiconductor is shown in Fig. 10.6. The energy difference between these states known as acceptor level, and top of the valence bond is much smaller than the band gap. Since acceptor levels are very close to valence band, electrons can easily jump into them from valence band. When an electron jump up to these levels it leaves behind a hole in the valence band; it is the movement of holes within the valence band that causes the greatest conduction in p-type materials.

Conduction occurs in both type of extrinsic semiconductor by the movement of both holes and electrons. The particles that contribute most of the conduction are called majority carriers and the other minority carriers. In a p-type material the majority carriers are holes, and in n-type semiconductors the majority carriers are electrons. In p-type thermally generated electrons are minority carriers while in n-type, thermally generated holes are minority carriers.

10.4   Electrons and Hole Concentration in Intrinsic Semiconductor (Law of Mass Action)

The electrical conductivity of a semiconductor will depend on the number of charge carrier in thermal equilibrium. Under the condition of thermal equilibrium the number of electrons with energy between E and $E + \mathrm{d}E$ is given by

$$\mathrm{d}n_e = f(E)D(E)\mathrm{d}E \qquad (10.1)$$

where $f(E)$ is the Fermi–Dirac function

$$f(E) = \frac{1}{\exp\left[\frac{E-E_F}{k_B T}\right] + 1} \tag{10.2}$$

and E_F is Fermi level defined as the energy at which function $f(E) = \frac{1}{2}$. $D(E)\,dE$ gives the density of states of electron of mass m_e^* between E and $E + dE$.

Consider the case of intrinsic semiconductor where at low temperatures only a few electrons are excited into the conduction band. In the limit of extremely few electrons, the chance of an electron occupying a given state is very low and the restriction imposed by the Pauli exclusion principle play little role. Therefore, one can use Maxwell–Boltzmann statistics or $E - E_F > > k_B T$ and hence

$$\exp\left[\frac{E - E_F}{k_B T}\right] \gg 1$$

and

$$f(E) = \exp\left[-\left(\frac{E - E_F}{k_B T}\right)\right] \tag{10.3}$$

This approximation is appropriate for electron in the conduction band which is empty at 0 K. At this temperature the valence band is full and this corresponds to energy well below the Fermi level. In this region $E - E_F \gg k_B T$ and we have

$$f_V = 1 - f(E) = 1 - \frac{1}{\exp\left[\frac{E-E_F}{k_B T}\right] + 1} = \frac{\exp\left[\frac{E-E_F}{k_B T}\right]}{1 + \exp\left[\frac{E-E_F}{k_B T}\right]} \tag{10.4}$$

Since $E \ll E_F$, the second term in the denominator can be ignored and hence

$$f_V = \exp\left[\frac{E - E_F}{k_B T}\right] \tag{10.5}$$

$1 - f(E)$ is the probability that an electron state is unoccupied, that is, the probability that the hole is present.

The density of states for electrons is given by

$$D(E)\,dE = \frac{4\pi}{h^3}(2m_e^*)^{\frac{3}{2}}(E - E_C)^{\frac{1}{2}}\,dE \quad \text{(for } E > E_C) \tag{10.6}$$

For holes

$$D(E)\,dE = \frac{4\pi}{h^3}(2m_h^*)^{\frac{3}{2}}(E_V - E)^{\frac{1}{2}}\,dE \quad \text{(for } E_V > E) \tag{10.7}$$

because the electron and holes energies are measured with reference to the conduction band edge E_C and valance band edge E_V, respectively.

The number of electrons in the conduction band is

$$n_e = \int_{E_C}^{E_{top}} f_C(E)D(E)dE \tag{10.8}$$

The function $f_C(E)$ decreases very rapidly as E grows; therefore, it is permissible to substitute infinity for upper limit and hence

$$n_e = \int_{E_C}^{\infty} f_C(E)D(E)dE \tag{10.9}$$

From Eqs. (10.3), (10.6) and (10.9)

$$n_e = \int_{E_C}^{\infty} f_C(E)D(E)dE = \frac{4\pi}{h^3}(2m_e^*)^{\frac{3}{2}} \int_{E_C}^{\infty} (E - E_C)^{\frac{1}{2}} \exp\left[-\left(\frac{E - E_F}{k_B T}\right)\right]dE$$

$$n_e == \frac{4\pi}{h^3}(2m_e^*)^{\frac{3}{2}}(k_B T)^{\frac{1}{2}}$$

$$\times \exp\left[-\frac{E_C - E_F}{k_B T}\right] \int_{E_C}^{\infty} \left(\frac{E - E_C}{k_B T}\right)^{\frac{1}{2}} \exp\left[-\left(\frac{E - E_C}{k_B T}\right)\right]d(E - E_C) \tag{10.10}$$

Putting $\frac{E - E_C}{k_B T} = y$; $dE = k_B T dy$ in Eq. (10.10)

$$n_e = \frac{4\pi}{h^3}(2m_e^* k_B T)^{\frac{3}{2}} \exp\left[-\frac{E_C - E_F}{k_B T}\right] \int_{E_C}^{\infty} y^{\frac{1}{2}} \exp[-y]dy \tag{10.11}$$

Now

$$\int_{E_C}^{\infty} y^{\frac{1}{2}} \exp[-y]dy = \frac{\sqrt{\pi}}{2} \tag{10.12}$$

From Eqs. (10.11) and (10.12)

$$n_e = 2\left(\frac{2\pi m_e^* k_B T}{h^2}\right)^{\frac{3}{2}} \exp\left[-\frac{E_C - E_F}{k_B T}\right] = N_C \exp\left[-\frac{E_C - E_F}{k_B T}\right] \tag{10.13}$$

where

$$N_C = 2\left(\frac{2\pi m_e^* k_B T}{h^2}\right)^{\frac{3}{2}} \tag{10.14}$$

Similarly the number of holes in the valence band

$$n_h = \int_{-\infty}^{E_V} [1 - f(E)]D(E)dE = N_V \exp\left[-\frac{E_F - E_V}{k_B T}\right] \tag{10.15}$$

where

$$N_V = 2\left(\frac{2\pi m_h^* k_B T}{h^2}\right)^{\frac{3}{2}} \tag{10.16}$$

The product $n_e n_h$ from Eqs. (10.13) and (10.15) is

$$n_e n_h = 4\left(\frac{2\pi k_B T}{h^2}\right)^3 (m_e^* m_h^*)^{\frac{3}{2}} \exp\left[\frac{-E_C - E_F + E_F + E_V}{k_B T}\right]$$

$$n_e n_h = 4\left(\frac{2\pi k_B T}{h^2}\right)^3 (m_e^* m_h^*)^{\frac{3}{2}} \exp\left[-\frac{E_C - E_V}{k_B T}\right]$$

$$n_e n_h = 4\left(\frac{2\pi k_B T}{h^2}\right)^3 (m_e^* m_h^*)^{\frac{3}{2}} \exp\left[-\frac{E_g}{k_B T}\right] \tag{10.17}$$

where $E_C - E_V = E_g$ is the energy gap. Equation (10.17) does not involve the Fermi level. It is an expression of the law of mass action.

In an intrinsic semiconductor $n_e = n_h$. We have from Eqs. (10.13), (10.15) and (10.16)

$$2\left(\frac{2\pi m_e^* k_B T}{h^2}\right)^{\frac{3}{2}} \exp\left[-\frac{E_C - E_F}{k_B T}\right] = 2\left(\frac{2\pi m_h^* k_B T}{h^2}\right)^{\frac{3}{2}} \exp\left[-\frac{E_F - E_V}{k_B T}\right]$$

$$\times \exp\left[\frac{E_F + E_F - E_C - E_V}{k_B T}\right] = \left(\frac{m_h^*}{m_e^*}\right)^{\frac{3}{2}}$$

$$E_F = \frac{1}{2}(E_C + E_V) + \frac{3}{4}k_B T \ln\left(\frac{m_h^*}{m_e^*}\right) \tag{10.18}$$

For most intrinsic semiconductors where $m_e^* = m_h^*$, Fermi level lies in the middle of the energy gap. In some case, such as InSb, $\frac{m_e^*}{m_h^*} = 1/20$, the level varies markedly in position with temperature and at room temperature is shifted well towards the bottom of the conduction band. If $m_e^* > m_h^*$, the Fermi level shifted downwards towards the top of the valence band.

Since number of electrons are equal to number of holes, $n_e = n_h = n_i$ and from Eq. (10.17)

$$n_e n_h = n_i^2 = 4\left(\frac{2\pi k_B T}{h^2}\right)^3 (m_e^* m_h^*)^{\frac{3}{2}} \exp\left[-\frac{E_g}{k_B T}\right]$$

$$n_i = 2\left(\frac{2\pi k_B T}{h^2}\right)^{\frac{3}{2}} (m_e^* m_h^*)^{\frac{3}{4}} \exp\left[-\frac{E_g}{2k_B T}\right] \tag{10.19}$$

Equation (10.19) gives an expression for intrinsic carrier concentration at any temperature T.

10.5   Concentration of Electrons and Holes in Extrinsic Semiconductors

Consider a n-type semiconductor consists of a conduction band below which there are n_d donor levels per cm^3 of energy E_d. The influence of the valence band is neglected, that is, the model may be applied only at relatively low temperatures. At 0 K all donor levels are filled with electrons. At low temperatures, only a small fraction of donor is ionized and Fermi level is expected to lie halfway between the donor levels and the bottom of the conduction band.

Let us assume that E_F lies more than a few $k_B T$ below the bottom of the conduction band. In such a case the density of conduction electrons n_e is given by

$$n_e = 2\left(\frac{2\pi m_e^* k_B T}{h^2}\right)^{\frac{3}{2}} \exp\left[\frac{E_F - E_C}{k_B T}\right]$$

Thus number must be equal to the density of ionized donors. If we assume that E_F lies more than a few $k_B T$ above the donor levels, the density of empty donors is equal to

$$n_d[1 - f(E_d)] = n_d\left[1 - \frac{1}{\exp\left[\frac{E_d - E_F}{k_B T}\right] + 1}\right] = n_d \frac{\exp\left[\frac{E_d - E_F}{k_B T}\right]}{\exp\left[\frac{E_d - E_F}{k_B T}\right] + 1}$$

$$= n_d \exp\left[\frac{E_d - E_F}{k_B T}\right] \qquad (10.20)$$

as $E_F > E_d$, exponential term in the denominator can be ignored in comparison with unity.

Since the number of electrons in the conduction band is equal to the number of vacancies in the donor levels, we have from Eqs. (10.13) and (10.20)

$$2\left(\frac{2\pi m_e^* k_B T}{h^2}\right)^{\frac{3}{2}} \exp\left[\frac{E_F - E_C}{k_B T}\right] = n_d \exp\left[\frac{E_d - E_F}{k_B T}\right] \qquad (10.21)$$

Taking log on both side

$$\ln\left[2\left(\frac{2\pi m_e^* k_B T}{h^2}\right)^{\frac{3}{2}}\right] + \left[\frac{E_F - E_C}{k_B T}\right] = \ln n_d + \left[\frac{E_d - E_F}{k_B T}\right]$$

$$\left[\frac{E_F - E_C}{k_B T}\right] - \left[\frac{E_d - E_F}{k_B T}\right] = \ln n_d - \ln\left[2\left(\frac{2\pi m_e^* k_B T}{h^2}\right)^{\frac{3}{2}}\right]$$

$$\frac{2E_F - E_C - E_d}{k_B T} = \ln \frac{n_d}{2\left(\frac{2\pi m_e^* k_B T}{h^2}\right)^{\frac{3}{2}}}$$

$$E_F = \frac{E_C + E_d}{2} + \frac{k_B T}{2} \ln \frac{n_d}{2\left(\frac{2\pi m_e^* k_B T}{h^2}\right)^{\frac{3}{2}}} \tag{10.22}$$

At $T = 0$ K

$$E_F = \frac{E_C + E_d}{2} \tag{10.23}$$

that is the Fermi level lies exactly half way between the donor levels and the bottom of the conduction band. Substituting the value of E_F from Eq. (10.22) in Eq. (10.13)

$$n_e = 2\left(\frac{2\pi m_e^* k_B T}{h^2}\right)^{\frac{3}{2}} \exp\left[\frac{E_d - E_C}{2k_B T} + \frac{1}{2} \ln \frac{n_d}{2\left(\frac{2\pi m_e^* k_B T}{h^2}\right)^{\frac{3}{2}}}\right]$$

$$n_e = 2\left(\frac{2\pi m_e^* k_B T}{h^2}\right)^{\frac{3}{2}} \exp\left(\frac{E_d - E_C}{2k_B T}\right) \times \frac{\sqrt{n_d}}{\sqrt{2}\left(\frac{2\pi m_e^* k_B T}{h^2}\right)^{\frac{3}{4}}}$$

$$n_e = \sqrt{2n_d}\left(\frac{2\pi m_e^* k_B T}{h^2}\right)^{\frac{3}{4}} \exp\left(\frac{E_d - E_C}{2k_B T}\right) \tag{10.24}$$

Let $E_C - E_d = \Delta E$

$$n_e = \sqrt{2n_d}\left(\frac{2\pi m_e^* k_B T}{h^2}\right)^{\frac{3}{4}} \exp\left(-\frac{\Delta E}{2k_B T}\right) \tag{10.25}$$

thus concentration of electrons in the conduction band is proportional to square root of the donor concentration and varies exponentially with temperature.

In p-type semiconductors, acceptor level lies above the valence band. If n_a is the number of acceptors per unit volume, then the number of electrons in the acceptor levels of energy E_a per unit volume is given by

$$n_a f(E) = \frac{n_a}{\exp\left(\frac{E_a - E_F}{k_B T}\right) + 1} \tag{10.26}$$

If $E_a - E_F \gg k_B T$

$$n_a f(E) = n_a \exp\left(-\frac{E_a - E_F}{k_B T}\right) \tag{10.27}$$

equating the total number of electrons in the acceptor levels with the number of holes in the valence band

$$2\left(\frac{2\pi m_h^* k_B T}{h^2}\right)^{\frac{3}{2}} \exp\left(\frac{E_V - E_F}{k_B T}\right) = n_a \exp\left(\frac{E_F - E_a}{k_B T}\right) \tag{10.28}$$

Taking log on both sides

$$\ln\left[2\left(\frac{2\pi m_h^* k_B T}{h^2}\right)^{\frac{3}{2}}\right] + \left(\frac{E_V - E_F}{k_B T}\right) = \ln n_a + \left(\frac{E_F - E_a}{k_B T}\right)$$

$$E_F = \frac{E_V + E_a}{2} - \frac{k_B T}{2} \ln \frac{n_a}{2\left(\frac{2\pi m_h^* k_B T}{h^2}\right)^{\frac{3}{2}}} \tag{10.29}$$

At $T = 0$ K

$$E_F = \frac{E_V + E_a}{2} \tag{10.30}$$

that is, Fermi level is half way between E_a and E_V.

Let us assume that E_F lies more than a few $k_B T$ above the top of the valence band. In such a case the density of holes n_h is given by

$$n_h = 2\left(\frac{2\pi m_h^* k_B T}{h^2}\right)^{\frac{3}{2}} \exp\left(\frac{E_V - E_F}{k_B T}\right)$$

Substituting the value of E_F from Eq. (10.29)

$$n_h = 2\left(\frac{2\pi m_h^* k_B T}{h^2}\right)^{\frac{3}{2}} \exp\left[\frac{E_V - E_a}{2k_B T} + \frac{1}{2}\ln\frac{n_a}{2\left(\frac{2\pi m_h^* k_B T}{h^2}\right)^{\frac{3}{2}}}\right]$$

$$n_h = 2\left(\frac{2\pi m_h^* k_B T}{h^2}\right)^{\frac{3}{2}} \exp\left(\frac{E_V - E_a}{2k_B T}\right) \times \exp\left[\ln\frac{\sqrt{n_a}}{\sqrt{2}\left(\frac{2\pi m_h^* k_B T}{h^2}\right)^{\frac{3}{4}}}\right]$$

$$n_h = 2\left(\frac{2\pi m_h^* k_B T}{h^2}\right)^{\frac{3}{2}} \exp\left(\frac{E_V - E_a}{2k_B T}\right) \times \left[\frac{\sqrt{n_a}}{\sqrt{2}\left(\frac{2\pi m_h^* k_B T}{h^2}\right)^{\frac{3}{4}}}\right]$$

$$n_h = \sqrt{2n_a}\left(\frac{2\pi m_h^* k_B T}{h^2}\right)^{\frac{3}{4}} \exp\left(\frac{E_V - E_a}{2k_B T}\right) \tag{10.31}$$

thus concentration of holes in the valence band is proportional to square root of the acceptor concentration and varies exponentially with temperature.

Suppose that a crystal is grown containing an excess of donor atoms; and smaller number of acceptor atoms. In this case the smaller number of acceptor states are first occupied by electrons from the donor atoms rather than by electrons transferring from states in the valence band. At 0 K, all the states in the valence band are filled. Similarly, all the acceptor states are filled by donor electrons, as far as possible. At a higher temperature, the donors become ionized and donor electrons occupies empty quantum states in the conduction band. The number of effective electrons at this temperature therefore is determined by the number of donor atoms less the number of acceptor atoms present. Such a crystal has a higher density of occupied states in the conduction band than of vacant states in the valence band. This excess is equal to the difference in the densities of donor and acceptor present. When a crystal contains more acceptor atoms, the reverse distribution is obtained.

10.6   Temperature Dependence of Electrons and Holes Concentration

Intrinsic Semiconductors

In an intrinsic semiconductor, the source of charge carriers, that is, electrons and holes are the valence and conduction band. The carrier concentration depends exponentially on the band gap given by Eq. (10.19). The important feature of this expression is that n_i increases very rapidly-exponentially with temperature by virtue of exponential factor. As the temperature is raised, a great number of electrons are excited across the energy gap. Figure 10.7 shows a plot of n_e versus $1/T$. The curve is a straight line of slope equal to $-E_g/2k_B$. The slope depends on the band gap. From this curve the band gap can be determined. The $T^{3/2}$ dependence is very weak in comparison with the exponential dependence. In intrinsic semiconductor, concentration of carriers is $\sim 10^{11}$ cm^{-3} while in metals it is 10^{21} cm^{-3}.

Fig. 10.7 Plot of $\ln n_e$ (n_e is in per cm^{-3}) versus $1/T$

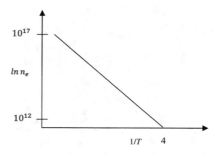

Extrinsic Semiconductors

In an extrinsic semiconductor, there are two sources of charge carriers. In n-type semiconductor the source of majority charge carriers, that is, electrons are:

(a) The donor energy levels close to the bottom of conduction band. The ionization energy of donors is of the order of few meV.
(b) The valence band of the semiconductor, with ionization energy equal to band gap between valence band and conduction band which is of the order of few eV.
    In p-type semiconductors the sources of majority charge carriers, that is, holes are:
(c) The acceptor energy levels close to the top of valence band. The ionization energy of acceptor is of the order of few meV.
(d) The valence band of the semiconductor, with ionization energy equal to band gap between valence band and conduction band which is of the order of few eV. The electrons are lifted from valence band to conduction band leaving behind the holes.

Because of the large difference in the ionization energies for the two processes to create charge carriers, they operate in different temperatures. Therefore, the variation of concentration of charge carriers with temperature is divided into different regions. Figure 10.8 shows the plot of logarithm of electron concentration versus temperature for an n-type semiconductor. The same is true for hole concentration a p-type semiconductor. The main features of this plot are:

(a) **Low Temperature Range**

At low temperature the average energy of lattice thermal vibrations $k_B T$ is much less than the width of the forbidden band E_g and because of that, the vibrations are incapable of providing sufficient excitation of the electrons of the valence band to shift them to the conduction band. But this energy is enough to excite and shift to conduction band

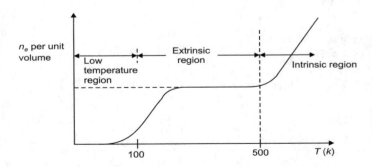

Fig. 10.8 Electron concentration versus temperature (in K)

the electrons occupying the donor level E_d and to the valence band the holes occupying the acceptor level E_a and this requires an energy 100 times less than E_g. Therefore, at low temperatures practically only the dopant charge carriers (i) electrons in the n-type semiconductors and (ii) holes in p-type semiconductors are excited. This range is also called ionization range.

(b) Extrinsic Range

As temperature rises, the electron concentration in the conduction band increases and that on donor levels decreases-the donor levels become exhausted. The behaviour of acceptor levels in p-type semiconductor is similar. In case of complete exhaustion the electron concentration in the conduction band of an n-type semiconductor becomes practically equal to concentration of donor impurity n_d

$$n_e \approx n_d$$

and the hole concentration in a p-type semiconductor to that of acceptor impurity n_a

$$n_h \approx n_a$$

The exhaustion or saturation temperature of the impurity levels T_s is higher if (i) the impurity activation energy E_d or E_a is higher (ii) the impurity concentration is higher. For Ge, $n_d = 10^{22}$ m^{-3}, $E_d = 0.01$ eV and $T_s = 30$ K.

(iii) High Temperature Range

As the temperature is raised still higher the excitation of intrinsic carriers becomes more intense. More and more electron–hole pairs are created. The semiconductor increasingly approaches the state of an intrinsic semiconductor. Until the concentration of intrinsic carriers remains much less than n_d ($n_i \ll n_d$) the total concentration $n = n_i + n_d$ remains practically constant and equal approximately to n_d.

However at sufficiently high temperatures, the intrinsic carrier concentration may not only become equal to n_d but may substantially exceeds it ($n_i \gg n_d$). In this case $n = n_i + n_d$ is approximately n_i and marks the transition to intrinsic conductivity. The temperature T_1 of such transition is higher the greater the width of the forbidden band and the impurity concentration. For *Ge*, with $n_d = 1022$ m^{-3}, $T_1 = 450$ K.

10.7 Fermi Level

The Fermi–Dirac distribution given by Eq. (8.70) determines the probability that an energy state E is filled with an electron under thermal equilibrium conditions. In the expression,

E_F is called Fermi level. It is the highest energy level that an electron can occupy at 0 K. It is defined as the energy at which the function $F(E)$ is $\frac{1}{2}$. It changes with temperature and as electrons are added to or removed from the solids. In a metal $E_F \gg k_B T$ at ordinary temperature, and the only electrons which can take part in the conduction process are close to the Fermi level. In thermal equilibrium, the Fermi level is constant and is not a function of position. The position of Fermi level relative to conduction band or valence band in a semiconductor is not obvious.

In the limit $T = 0$ K,

$$F(E) = 1 \quad \text{for} \quad E < E_F$$
$$F(E) = \frac{1}{2} \quad \text{for} \quad E = E_F$$
$$F(E) = 0 \quad \text{for} \quad E > E_F$$

Fermi level determines the distribution of electrons or holes in a solid. To a first approximation all the energy states above the Fermi levels are empty, and all the energy states below the Fermi level are filled with electrons in a metal.

Fermi Level in Semiconductors

In semiconductors the valence bands and conduction bands are separated by a band gap. In band diagram the position of Fermi level determines which of the two carriers holes or electrons are dominant. If the semiconductor has more electrons than holes, the Fermi level is positioned above the centre of the band gap. If holes are more than electrons, the Fermi level lies below the centre of the band gap. For equal number of holes and electrons, the Fermi level is at the centre of the band gap.

Fermi Level of an Intrinsic Semiconductor

The Fermi level in an intrinsic semiconductor is given by Eq. (10.18)

$$E_F = \frac{1}{2}(E_C + E_V) + \frac{3}{4}k_B T \ln\left(\frac{m_h^*}{m_e^*}\right) \tag{10.32}$$

For intrinsic semiconductor the effective mass of electrons and holes are usually equal, hence Eq. (10.32) takes the form

$$E_F = \frac{1}{2}(E_C + E_V) \tag{10.33}$$

The Fermi level lies in the middle of energy gap. However, in some case effective mass of electrons and holes may be different. In such a case, the relation (10.33) is not true. If the effective mass of electron is more than the effective mass of hole, then E_F value obtained from Eq. (10.32) is less than that given by Eq. (10.33). Therefore the Fermi level is shifted downwards towards the top of valence band. Further, if the effective mass of hole is larger than the effective mass of electron, the E_F obtained from Eq. (10.32) is larger than that given by Eq. (10.33). As a result the Fermi level is shifted upwards towards the bottom of the conduction band. The Fermi level also changes with the change of temperature. The Fermi level shifts upwards from the centre of the band gap with increase of temperature and with decreasing temperature it moves down from the middle position of the band gap.

Fermi Level in Extrinsic Semiconductors

In an intrinsic semiconductor, the Fermi level is positioned at the centre or near the centre of the band gap. In an intrinsic semiconductor, the number of holes and electrons is equal. In n-type or p-type semiconductors, the dopants are fully ionized and there is an imbalance in the electron and holes concentration. This is in turn reflected in the position of Fermi level. The Fermi level is shifted towards either conduction band or valence band depending on the type of dopant in the semiconductor.

n-Type Semiconductor

In n-type semiconductors, the Fermi level is given by Eq. (10.22) and is

$$E_F = \frac{E_C + E_d}{2} + \frac{k_B T}{2} \ln \frac{n_d}{2\left(\frac{2\pi m_e^* k_B T}{h^2}\right)^{\frac{3}{2}}}$$

The Fermi level depends on temperature T and donor concentration n_d. At 0 K from Eq. (10.22)

$$E_F = \frac{E_C + E_d}{2} \tag{10.34}$$

The Fermi level lies exactly half way between donor levels and the bottom of the conduction band.

Effect of Temperature on Fermi Level

With increase in temperature, the doping becomes less important as the entire donors have already donated their free electrons. The thermal generation of holes and electrons becomes important. With increasing temperature the number of electrons reaching conduction band from valence band increases, leaving behind holes in valence band. These electron–hole pairs are intrinsic carriers, and the number of pairs increases with increasing temperature. A temperature is ultimately attained when numbers of holes are approximately equal to the number of electrons. The n-type semiconductor then behaves like an intrinsic semiconductor. As a result the Fermi level of n-type semiconductor approaches towards the centre of band gap as shown in Fig. 10.9.

Fig. 10.9 Variation of Fermi
level with temperature

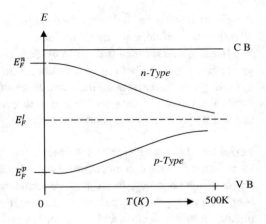

Effect of Concentration of Dopants

In n-type semiconductor, the density of conduction electrons n_e must be equal to density
of ionized donors n_d. Putting $n_e = n_d$ in Eq. (10.13)

$$n_d = n_e = 2\left(\frac{2\pi m_e^* k_{\mathrm{B}} T}{h^2}\right)^{\frac{3}{2}} \exp\left[\frac{(E_F - E_C)}{k_{\mathrm{B}} T}\right]$$

$$n_d = n_e = N_C \ \exp\left[\frac{(E_F - E_C)}{k_{\mathrm{B}} T}\right] \tag{10.35}$$

where $N_C = 2\left(\frac{2\pi m_e^* k_{\mathrm{B}} T}{h^2}\right)^{\frac{3}{2}}$

Taking log of Eq. (10.35)

$$\ln \frac{N_C}{n_d} = -\frac{(E_F - E_C)}{k_{\mathrm{B}} T}$$

$$E_F = E_C - k_{\mathrm{B}} T \ln \frac{N_C}{n_d} \tag{10.36}$$

From Eq. (10.36) it is seen that Fermi level of n-type semiconductor is slightly below
the bottom of conduction band. As the concentration of donor increases the Fermi level
increases and approaches the bottom of conduction band. It is shown in Fig. 10.10.

p-Type Semiconductors

In p-type semiconductor the Fermi level is given by Eq. (10.29)

$$E_F = \frac{E_V + E_a}{2} - \frac{k_{\mathrm{B}} T}{2} \ln \frac{n_a}{2\left(\frac{2\pi m_h^* k_{\mathrm{B}} T}{h^2}\right)^{\frac{3}{2}}}$$

Fig. 10.10 Variation of Fermi level with dopant concentration

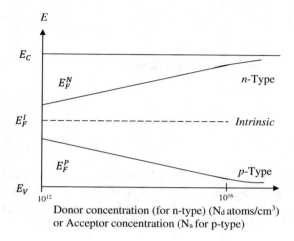

Donor concentration (for n-type) (N_d atoms/cm^3)
or Acceptor concentration (N_a for p-type)

The Fermi level depends on temperature T and acceptors concentration n_a. At 0 K from Eq. (10.22)

$$E_F = \frac{E_V + E_a}{2} \tag{10.37}$$

The Fermi level lies exactly halfway between acceptor levels and the top of the valence band.

Effect of Temperature

With increasing temperature the doping becomes less important as all the acceptors have already received the electrons. As temperature increases, the thermal generation of electron–hole pairs become important. As a result of this the number of holes in valence band increases. The electron–hole pairs are intrinsic carriers, and their number increases with temperature. A temperature is finally attained when the numbers of electrons are nearly equal to the number of holes. The p-type semiconductor behaves like an intrinsic semiconductor. This causes the Fermi level of p-type semiconductor to approach the Fermi level of an intrinsic semiconductor as shown in Fig. 10.9.

Effect of Concentration

In p-type semiconductor, the density of holes in valence band must be equal to density of ionized acceptor levels n_a, that is $n_a = n_h$. From Eq. (10.15)

$$n_a = n_h = N_V \exp\left[-\frac{(E_F - E_V)}{k_B T}\right] \tag{10.38}$$

where

$$N_V = 2\left(\frac{2\pi m_h^* k_B T}{h^2}\right)^{\frac{3}{2}}$$

Taking log of Eq. (10.38)

$$-\frac{(E_F - E_V)}{k_B T} = \ln\frac{n_a}{N_V}$$

$$E_F = E_V + k_B T \ln\frac{N_V}{n_a} \tag{10.39}$$

From Eq. (10.39) it is seen that in p-type semiconductor, the Fermi level is slightly above the top of valence band. As the concentration of acceptor increases, the Fermi level decreases and approaches the top of the valence band as shown in Fig. 10.10.

10.8   Scattering

In a perfect crystal of a semiconductor, all the atoms are arranged in the designated position and are free from defects and imperfections. In real crystals there is deviation from ideal structure. The real crystal contains a number of defects and imperfections and further the atoms vibrate about their equilibrium position. These defects, imperfections and vibrations of the atoms cause the scattering of free charge carriers electrons and holes. These cause the conductivity of the semiconductor, or mobility of charge carriers to change. The charge carriers (electrons and holes) in a semiconductor are scattered by their interaction with.

(a) Ionized impurities
(b) Neutral impurities, dislocation vacancies, interstial etc.
(c) Alloy scattering
(d) Carrier–carrier scattering
(e) Surface and interface
(f) Lattice vibrations (phonons).

Ionized Impurities
Impurity atoms are added to intrinsic semiconductors to modify its electrical and optical properties. Let the semiconductor is lightly doped. Therefore the impurities are not very close together. The doped impurities (donor, acceptor) are ionized at room temperature. Ionized acceptor or donor provides the scattering potential. The potential produced by the charged acceptor or donor is screened by other free charge carriers. The effect of screening is to reduce the range of potential seen by the free charge carriers. The Coulomb interaction provides scattering and also alters the velocity of charged carrier. Let μ_I is the

mobility if only ionized impurity scattering existed, then mobility limited from ionized impurity scattering is

$$\mu_I = \frac{e\tau}{m^*} \propto \frac{T^{3/2}}{N_I}$$

where $N_I = N_d^+ + N_a^-$ is the total ionized impurity concentration.

The mobility changes with temperature. With increasing temperature, the thermal velocity of free charge carriers increases. Therefore, they spent less time near the ionized impurities. Thus free charge carriers spent less time in the vicinity of Coulomb force and therefore experiences smaller scattering effects resulting in larger value of μ_I.

Now consider a heavily doped semiconductor. In the degenerate limit, the Fermi level will be in the band. The ionized acceptors or donors are relatively close together. Because of large density of impurity, semiconductor will have large electron/hole density from ionized impurities. Thus there is probability that electrons/holes are scattered by an ionized impurities more frequently resulting in smaller value of μ_I.

Neutral Impurity Scattering

There are substitutional and dopant in the semiconductor which were not ionized in the semiconductor. The neutral impurities are quite similar to the scattering of charged carriers by hydrogen atom. The mobility associated with this type of scattering alone is given by

$$\mu_{\text{neutral}} = \frac{e\tau}{m^*} \propto N_{\text{imp}} T^{\frac{1}{2}}$$

where N_{imp} is density of neutral impurities. Unless N_{imp} is very high $\gg 10^{18}\text{cm}^{-3}$, there is very little effect of this scattering.

Dislocations are also scattering centres for charge carriers as a result of dilation they produce in the lattice. The probability for scattering is proportional to the number of dislocations line per cm^3 and proportional to temperature. Vacancies and interstitial also cause scattering.

Alloys Scattering

In alloys the atoms on anion sublattice is same while atoms on cation sublattice are randomly distributed for example in $(\text{InAs})_x(\text{GaAs})_{1-x}$, As is on anion sublattice while In and Ga are randomly distributed over cation lattice. Due to disorder, free charge carriers experience scattering as they move in the semiconductor lattice. The alloy scattering has no angular dependence of scattering rate and mobility μ_a due to alloy scattering is given by

$$\mu_a \propto T^{-\frac{1}{2}}$$

In obtaining this result it was assumed that alloys were cluster free and the smallest region in which disorder occurs was the unit cell.

Carrier–Carrier Scattering

Since electrons and holes are both charged particle, there is Coulombic scattering between them. Also there can be scattering between electrons themselves. The electron–electron and electron–hole scattering is quite important in materials where charge density exceeds 10^{18} cm^{-3}. The mobility due to carrier–carrier scattering varies as $T^{\frac{3}{2}}$.

Interface Scattering

Interface between two materials has varying degree of roughness. The interface roughness causes potential bumps in path of carriers, which causes carriers to scatter. The mobility is the inversely proportional to the nature of the surface field or the charge density in the channel.

Lattice Vibrations

The atoms in a semiconductor crystal vibrates about their mean equilibrium position by absorption of thermal energy. These vibrations are represented by phonons. In an ideal crystal the lattice provides a time independent periodic potential. The perfect periodic potential of the crystal allows charge carriers to move without scattering in the solid. However, thermal vibration causes a disruption of the periodic potential function resulting in scattering of free charge carriers by vibrating atoms of the crystal. This kind of scattering is known as phonon scattering or lattice scattering. In phonon scattering electron's energy and momentum are altered. In the scattering both acoustic and optical phonons participate depending on the temperature. At low temperature acoustic phonons dominate.

The vibration of atoms in lattice is related to thermal motion, therefore the rate at which scattering occurs depends on the temperature. The mobility μ_L due to scattering is found to be given by

$$\mu_L \propto T^{-\frac{3}{2}}$$

The mobility due to lattice scattering or phonon scattering increases as temperature decreases as expected because vibration decreases as temperature decreases. Thus probability of scattering event also decreases with increasing mobility.

In general, lattice scattering, ionic scattering and scattering by neutral impurities are all present in semiconductors. Let τ_L is the mean time between collisions due to lattice scattering then $\frac{dt}{\tau_L}$ is the probability of a lattice scattering event occurring in a differential time dt. Similarly, $\frac{dt}{\tau_I}$ is the probability of an ionized impurity scattering event occurring in differential time dt and $\frac{dt}{\tau_N}$ is the probability of neutral impurity scattering event occurring in the differential time dt. If these scattering processes are all independent, then the total probability of scattering event in the differential time dt is the sum of individual events.

Thus

$$\frac{dt}{\tau} = \frac{dt}{\tau_L} + \frac{dt}{\tau_I} + \frac{dt}{\tau_N}$$

$$\frac{1}{\tau} = \frac{1}{\tau_L} + \frac{1}{\tau_I} + \frac{1}{\tau_N}$$

where τ is mean time between scattering events. Since mobility is proportional to relaxation time therefore, the above equation can be written as

$$\frac{1}{\mu} = \frac{1}{\mu_L} + \frac{1}{\mu_I} + \frac{1}{\mu_N}$$

10.9 The p–n Junction

Single crystals of a semiconductor can be prepared in which one end is doped to make it p-type and the other end n-type. The change from p-type to n-type is taking place in a region whose thickness is of the order of 10^{-8} m. Such a unit is called p–n junction. It is important that this is made in one piece of semiconductor so that crystal lattice extends across the boundary. It is not sufficient just to have two pieces in contact.

The p-type material is made by doping with a concentration N_a of acceptors and n type by doping with a concentration N_d of donors. Thus there is a narrow region at the junction where the doping concentration varies from one extreme to the other and change can be considered discontinuous. As ionization potential of donor and acceptors are very small they are assumed to be completely ionized at room temperature.

There is an excess of hole concentration in the p-type and excess electron concentration in the n-type, when they are in the same crystal holes will diffuse to the right and electrons to the left each giving a positive current to the right. After crossing the junction they combine with each other leaving the immobile ions in the neighbourhood of the junction unneutralized (these are also called uncovered charges). The positive and negative charges of immobile ions produce an electric field across the junction. This field is directed from n side to p side and is termed as barrier field. This field opposes the diffusive motion of electrons and holes (majority carriers). Equilibrium is established at a barrier field and is just sufficient to stop further diffusion of majority carriers. Under this condition there is no movement of carriers across the junction. Since the vicinity of the junction is depleted of mobile charges it is called the depletion or space charge region. The thickness of this region is about 10^{-6} m on either side of the boundary. As depletion region is devoid of mobile charges it has much higher resistivity than that of the bulk material. Thus external voltage applied appears almost wholly across the barrier layer, there is little potential variation in the bulk material. Due to existence of a barrier, field across the junction, transfer of an electron from n side to p side require the expenditure of an amount of

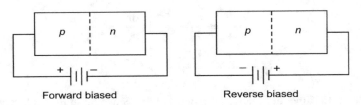

Figure 10.11 *p–n* junction biasing

energy which is termed as the barrier energy. This barrier energy depends on the width
of depletion layer. Same is the case for holes moving from *p* to *n* region.

The *p* material have electron as minority carriers and n type have holes as minority
carriers. When no external voltage is applied, that is, when *p–n* junction is unbiased the
barrier field acts in such a way that the electrons from p side and the holes from n side
can easily cross the junction. The movement of minority carriers, constitute a current
flow. Under equilibrium condition this current is exactly counter balanced by an equal
and opposite flow of majority carriers that gain of energy from the thermal sources is
sufficient to cross the barrier. The barrier potential for Ge is 0.3 V and for Si it is 0.7 V.

A Biased p–n Junction

When the positive terminal of a battery is connected to the *p*-type side and the negative
terminal to the *n*-type side of a *p–n* junction, the junction allows a large current to flow
through it. In this case the *p–n* junction is said to be forward biased (Fig. 10.11). When
the terminal of the battery is reversed, that is positive terminal is connected to *n*-type side
and negative terminal to the *p*-type side, the junction allows a very small current to flow
through it. Under this condition the *p–n* junction is called reverse biased (Fig. 10.11).

Forward Biased p–n Junction

In forward biased *p–n* junction, the forward biasing voltage exerts a force on the holes of
p-type side and on the electrons of the *n*-type side. This force causes holes and electrons to
move toward the junction. This reduced the width of the depletion region and height of
the potential barrier, that is, barrier energy. The magnitude of barrier energy is reduced
by *eV* (*V* is applied voltage). Due to the reduction in the barrier height the current flow
due to majority carriers. The minority carriers current flowing in the direction opposite to
that of majority carrier current is not affected by the forward bias.

Reversed Biased p–n Junction

In this case the voltage applied to the junction causes the holes in the *p*-type side and
electrons in the *n*-type side to move away from the junction. This increases the width of
the depletion region and height of the barrier. The amount of the increase in barrier energy

is given by eV. Due to the increase in barrier height a negligible number of majority carriers will be able to cross the junction and current would be practically zero. However, the majority carriers which travel down the potential barrier remain unaffected and give a small current.

p–n Junction Current

The total current flowing through a *p–n* junction by the application of a voltage V across the junction is given by

$$I = I_S\left[\exp\left(\frac{eV}{\eta k_B T}\right) - 1\right] \tag{10.40}$$

where I_s is reverse current, η is numerical constant depending on the material, it is one for Ge and 2 for Si. When V is positive, the junction is forward biased and when V is negative, the junction is reverse biased. At 300 K, I is given by

$$I = I_S\left[\exp\left(\frac{39V}{\eta}\right) - 1\right] \tag{10.41}$$

A typical plot of I versus V is shown in Fig. 10.12. Forward current is typically in the range of *mA* while reverse current is in the range of μA or less. *I–V* curve of *p–n* junction is nonlinear.

In the reverse biasing of semiconductor diodes, the reverse saturation current remains constant even at high voltage. However, in a diode known as Zener diode, at a particular reverse bias voltage, current increases abruptly (Fig. 10.13). This sudden increase of current is due to avalanche multiplication. In this an electron near the junction is sufficiently accelerated by the electric field so that on collision it ionizes atoms resulting in producing new electron–hole pair. These electrons, in turn, produce charge carriers by collision and thus large number of charge carriers so produced gives large current without increasing applied voltage. The other mechanism is termed as Zener tunneling or Zener

Fig. 10.12 *I–V* characteristic curve of *p–n* junction diode

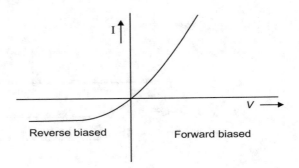

Fig. 10.13 *I–V* characteristic
of a Zener diode

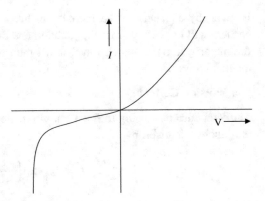

breakdown. In this mechanism of Zener tunneling, electrons from the valence band of the
p-type tunnel to the conduction band on the *n*-side.

10.10 Hall Effect

When a magnetic field is applied perpendicular to a conductor carrying current, a voltage
is developed across the conductor in the direction perpendicular to both, the current and
the magnetic field (Fig. 10.14).

Let the electric current be made to flow along the *x*-axis of a long metal strip and the
magnetic field **B** be at right angle to the strip in the *z*-direction. The force acting on the
charge carriers due to the electric field is

$$F_e = -e\,\boldsymbol{E}$$

The force due to the magnetic field

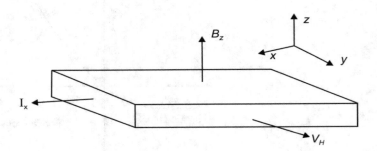

Fig. 10.14 Hall effect

$$F_B = -ev \times B$$

The total force, called Lorentz force, on the charge carrier is

$$F = -e[E + v \times B] \qquad (10.42)$$

The direction of magnetic force is given by right hand rule. The v and B lie in a plane, if we curl our fingers from v to B the direction of thumb gives the direction of force. The force due to magnetic field is acting downwards. As a result of this the electrons are accumulated at the bottom face ($-y$ direction) and causes a potential difference. This potential difference causes an electric force on electrons which opposes the Lorentz magnetic force. The accumulation process continues until the electron force called Hall force balances the Lorentz magnetic force, that is,

$$F_H = F_M \qquad (10.43)$$

If E_H is Hall electric field then

$$F_H = eE_H$$

and hence

$$evB = eE_H$$

$$E_H = vB$$

The current density

$$J_x = nev$$

n is number of electrons per unit volume and

$$\frac{E_H}{J_x} = \frac{vb}{nev} = \frac{B}{ne} \qquad (10.44)$$

$$E_H = \frac{BJ_x}{ne} \propto J_x B \qquad (10.45)$$

The coefficient of proportionality is called Hall coefficient

$$R_H = \frac{E_H}{J_x B} = -\frac{1}{ne} \qquad (10.46)$$

R_H is negative if the charge carriers are electrons and positive if the charge carriers are holes

$$R_H = \frac{E_H}{J_x B} = \frac{1}{pe} \tag{10.47}$$

If V_H is the Hall voltage developed across the specimen of thickness d then

$$V_H = d E_H = d R_H J_x B \tag{10.48}$$

If b is the width of the specimen than bd will be cross section and current density is given by

$$J_x = \frac{I_x}{bd}$$

$$V_H = d R_H \frac{I_x}{bd} B = \frac{R_H I_x B}{b}$$

$$R_H = \frac{V_H b}{I_x B} \tag{10.49}$$

R_H can be measured when V_H, I_x, B and b are known experimentally V_H will have opposite polarity for n- and p-type semiconductors. The unit of R_H is m²/coulomb.

For semiconductors, the calculations of average speeds lead to the conclusion that

$$R_H = -\frac{3\pi}{8ne} = -\frac{1.18}{n_e} \tag{10.50}$$

For n-type materials, the conductivity is given by

$$\sigma_n = n e \mu_n$$

$$\mu_n = \frac{\sigma_n}{ne} = -\frac{\sigma_n R_H}{1.18} \tag{10.51}$$

for p-type semiconductors

$$\mu_p = \frac{\sigma_p}{pe} = \frac{\sigma_p R_H}{1.18} \tag{10.52}$$

Mobility and Hall angle

Mobility is defined as

$$\mu = \frac{v_x}{E_x} \tag{10.53}$$

Therefore

$$v_x = \mu E_x \tag{10.54}$$

hence

$$v = \frac{E_H}{B}$$

$$\frac{E_H}{B} = \mu E$$

$$E_H = \mu E B \qquad (10.55)$$

Comparing it with Eq. (10.46)

$$R_H J_x B = \mu E B$$

$$\mu = \frac{R_H J_x}{E} = \sigma R_H \qquad (10.56)$$

where the conductivity σ is

$$\sigma = \frac{J_x}{E}$$

But

$$\mu = \frac{v}{E} = \frac{E_H}{BE} = \frac{E_H}{E} \times \frac{1}{B} = \varphi \frac{1}{B} \qquad (10.57)$$

where

$$\varphi = \frac{E_H}{E} \qquad (10.58)$$

φ is called Hall angle and

$$\varphi = \mu B \qquad (10.59)$$

The Hall effect can be used to determine the sign of charge carriers. Its value is positive for p type material and negative for n type materials. If a sample contain an equal number of electrons and holes then mobility of the charge carriers determine the sign of Hall coefficient. It is also used in determining the concentration of charge carriers. Since Hall voltage is proportional to the magnetic field density B for a given current through the sample, magnetic field can be measured.

Solved Examples

Example 1 In III-V compounds, will *Zn* act as a donor or acceptor?

Solution

Zn belongs to column II of periodic table. It will substitute for an element of group III (small difference in charge) and will act as an acceptor. Similarly, if the material is doped with element of group VI such as S, Se, then the group VI element will substitute for group V element (small difference in charge) and will act as donor.

Example 2 In III-V compounds, will Si act as a donor or acceptor?

Solution

Si belong to column IV of periodic table. If It will substitute for an element of group III (small difference in charge) it will act as a donor. If It will substitute for an element of group V (small difference in charge) it will act as an acceptor.

Example 3 Calculate the approximate binding energy for Si. For Si, $m^* = 0.2\,m$, $K = 11.7$.

Solution
We have

$$E_n = \left(\frac{m^*}{m}\right)\frac{13.6\ \text{eV}}{n^2 K^2}$$

Substituting the values

$$E_n = (0.2)\frac{13.6\ \text{eV}}{(11.7)^2} = 0.0198 eV$$

Example 4 For intrinsic GaAs, the electrical conductivity is $10^{-6}\ \Omega^{-1}\ m^{-1}$. The electron and hole mobilities are respectively 0.85 and 0.04 m^2/V-s. Obtain the intrinsic carrier concentration.

Solution

Since the material is intrinsic, $n_e = n_h = n_i$ and the conductivity is given by the relation

$$\sigma = n_i |e|(\mu_e + \mu_h)$$
$$n_i = \frac{\sigma}{|e|(\mu_e+\mu_h)} = \frac{10^{-6}\text{ohm}^{-1}\text{m}^{-1}}{(1.6\times10^{-19}\text{C})(0.85+0.04)\text{m}^2(V-s)^{-1}} = 7.0 \times 10^{12}\text{m}^{-3}$$

Example 5 The resistivity of a doped semiconductor is 8.9×10^3 Ω m. The Hall coefficient is 3.6×10^{-4} m ^{3}/C. Assume single carrier concentration, find the mobility μ and concentration n of charge carriers.

Solution

We have

$$\sigma = ne\mu$$

$$\mu = \frac{\sigma R_H}{1.18} = \frac{R_H}{\rho \times 1.18} = \frac{3.6 \times 10^{-4} \times 10^3}{8.9 \times 1.18} = 0.0342 \, \text{m}^2 \text{V}^{-1} \text{s}^{-1}$$

$$R_H = \frac{1.18}{ne}$$

$$n = \frac{1.18}{eR_H} = \frac{1.18 \times 10^{19} \times 10^4}{1.6 \times 3.6} = 2.048 \times 10^{22} / \text{m}^3$$

Example 6 In an n-type semiconductor, the Fermi level lies 0.4 eV below the conduction band. If the concentration of donor atoms is doubled, find the new position of the Fermi level. Assume $k_B T = 0.03$ eV.

Solution

For n-type semiconductor we have $n_e \approx n_d$. Therefore we can write

$$n_e = n_d = 2\left(\frac{2\pi m_e^* k_B T}{h^2}\right)^{\frac{3}{2}} \exp\left[\frac{E_F - E_C}{k_B T}\right]$$

For the first case, let $n_d = n_d$, $E_F = E_F$ and in the second case $n_d = n_d'$, $E_F = E_F'$.
Substituting these values of donor atoms concentrations Femi energy

$$\frac{n_d}{n_d'} = \frac{1}{2} = \frac{\exp\left[\frac{E_F - E_C}{k_B T}\right]}{\exp\left[\frac{E_F' - E_C}{k_B T}\right]} = \exp\left[\frac{-0.4\text{eV} + (E_C - E'_F)}{0.03\text{eV}}\right]$$

Taking log on both side
$$-0.03\text{eV} \times \ln 2 = -0.4\text{eV} + (E_C - E'_F)$$
$$-0.03\text{eV} \times 0.6931 + 0.4 \text{ eV} = 0.37921 \text{ eV} = (E_C - E'_F)$$
$$(E_C - E'_F) = 0.379 \text{ eV}.$$

Objective Type Questions

1. Which of the following materials is not a semiconductor?
   (a) Silicone (b) Germanium (c) Gallium arsenide (d) Gallium nitride
2. Which of the following statement is incorrect?

(a) doping pure semiconductor material with small amount of donor impurities produces *n*-type semiconductor are called majority charge carriers

(b) the dominant charge carriers within a doped semiconductor are called majority charge carriers

(c) conduction within pure semiconductor is termed intrinsic conduction

(d) at room temperature, pure semiconductor make excellent conductor

3. Which of the following statement is incorrect about semiconductors?

(a) when heated they conduct by having electron flow through the conduction band and negative holes flow through the valence band

(b) they act as an insulator at 0 K

(c) when heated they conduct electrically through electrons in conduction band and holes in the valence band

(d) resistance decreases with increased temperature

4. In *p*-type material, minority carriers would be

(a) holes (b) dopants (c) slower (d) electrons

5. Electron pair bonding occurs when atom

(a) lack electrons (b) share holes (c) lack holes (d) share electrons

6. What electrical characteristic of intrinsic semiconductor materials is controlled by the addition of impurities?

(a) conductivity (b) resistance (c) power (d) all of the above

7. A semiconductor is electrically neutral because it has

(a) no majority carriers

(b) no minority carriers

(c) no free carriers

(d) equal number of positive and negative carriers

8. An intrinsic semiconductor at 0 K

(a) has large number of holes

(b) has a few holes and small number of electrons

(c) acts as an insulator

(d) shows metallic characteristics

9. When the temperature of intrinsic semiconductor is increased?

(a) resistance of semiconductor is also increased

(b) conduction is decreased

(c) energy of atoms is increased

(d) holes are created

10. A donor type impurity

(a) is used to obtain *p*-type semiconductor

(b) is used to obtain *n*-type semiconductor

(c) must possess three valence electron

(d) cannot be used in Si crystal

11. If small amount of antimony is added to *Ge*

(a) the resistance increased

(b) *Ge* will become a *p*-type semiconductor

(c) *Sb* becomes an acceptor impurity

(d) there will be no more free electrons than holes in a semiconductor

12. In an *n*-type semiconductor, the concentration of minority carriers mainly depends upon
    (a) doping technique (b) number of donor atoms (c) temperature of material (d) quality
    of intrinsic *Ge* or *Si*

13. In intrinsic semiconductor, Fermi level lies

    (a) in the middle of conduction band and valence band

    (b) near conduction band

    (c) near valence band

    (d) none of these

14. The conduction band of a semiconductor material may be
    (a) completely filled (b) partially filled (c) empty (d) either *b* or *c*

15. The energy band which possess the free electron is called
    (a) valence band (b) conduction band (c) forbidden band (d) none of these

16. What is described with the Hall effect?

    (a) a Hall voltage is induced at right angles to a current and the magnetic field

    (b) a Hall voltage induced at right angles to a current and parallel to the magnetic field

    (c) an external magnetic field produces an electric field in a current carrying body

    (d) produces a measurable magnetic field in its surrounding

17. The Fermi level in an intrinsic semiconductor shifted to the bottom of the conduction
    band if
    (a) $m_e^* > m_h^*$ (b) $m_e^* < m_h^*$ (c) $m_e^* = m_h^*$ (d) none of these

18. The Fermi level in an intrinsic semiconductor shifted to the top of the valence band if
    (a) $m_e^* > m_h^*$ (b) $m_e^* < m_h^*$ (c) $m_e^* = m_h^*$ (d) none of these

19. In the intrinsic conduction range (high temperature range)

    (a) the transition temperature is greater for small forbidden gap

    (b) the transition temperature is small for large forbidden gap

    (c) the transition temperature is greater for greater forbidden gap and impurity
        concentration

    (d) the transition temperature is greater for small impurity concentration

20. Which is not true?

    (a) for extrinsic range the transition temperature is higher the smaller the impurity
        activation energy

    (b) for extrinsic range the transition temperature is higher the greater the impurity
        activation energy

    (c) for intrinsic range, the transition temperature is higher the greater the width of
        forbidden band and the impurity concentration

    (d) at low temperatures impurity charge carriers are excited in doped semiconductors

Problems

1. For an intrinsic semiconductor with a band gap of 0.7 eV, determine the position of Fermi level at 300 K if $m_h{}^* = 6m_e{}^*$. Also calculate the density of holes.
2. Obtain the value of forbidden gap energy at 300 K. It is given that conductivity of the material is 2.12 /Ω m, mobility of electrons and holes are 0.36 $m^2 V^{-1} s^{-1}$ and 0.17 $m^2 V^{-1} s^{-1}$, respectively.
3. In an intrinsic semiconductor the effective mass of the electron is 0.07 m and that of hole is 0.4 m. Evaluate the intrinsic concentration of charge carriers at 300 K. The band gap is 0.7 eV.
4. In an n-type semiconductor, the Fermi level lies 0.3 eV below the conduction band at 300 K. If the temperature is increased to 330 K, find the new position of the Fermi level.
5. If small quantities of boron (valence 3) were added to pure silicon (valence 4), would you expect the room temperature electrical conductivity of the resulting material to be greater than, less than, or the same as that of pure silicon? Explain why
6. A semiconducting material, 1.2 cm long, 0.5 cm wide and 0.1 cm thick was placed in a magnetic flux of 0.5 T, the direction of which is perpendicular to the largest face. A current of 20 mA flows lengthwise through the material. The voltage measured across the width is found to be 37 μV. Obtain Hall coefficient.

Answers
Objective Type Questions

1. (b)	2. (d)	3. (a)	4. (d)	5. (d)	6. (d)	7. (d)	8. (c)
9. (c)	10. (b)	11. (d)	12. (c)	13. (a)	14. (d)	15. (c)	
16. (a)	17. (b)	18. (a)	19. (c)	20. (b)			

Dielectric Properties of Solids

11

11.1 The Dielectric Constant

A dielectric is a non-conductor of electricity. The dielectric when inserted between oppositely charged conducting planes increases the amount of charge that the plate will held at a given potential difference, that is, it increases the capacity of the condenser. Suppose that the capacitance of a capacitor with a vacuum between its conductors is C_0. When the space between the capacitors is filled with a material of dielectric constant K (Fig. 11.1) capacitance C of the dielectric filled capacitor is

$$C = K C_0 \tag{11.1}$$

where dielectric constant K is a dimensionless quantity.

Consider a parallel-plate capacitor, carrying a fixed charge q and is disconnected from the battery as shown in Fig. 11.1. The capacitance is C_0; the potential difference V_0 across its plates is given by

$$V_0 = \frac{q}{C_0} \tag{11.2}$$

Suppose a material with dielectric constant K is inserted between the plates as shown in Fig. 11.1. As a result of this the capacitance increases. Because the capacitor is disconnected from the battery, the charge q on the plate is not changed by the presence of dielectric. The potential difference, however, decreases to

© The Author(s) 2022
V. K. Jain, *Solid State Physics*, https://doi.org/10.1007/978-3-030-96017-9_11

Fig. 11.1 a A charged
capacitor with a vacuum
between its plate and **b** the
same capacitor with a
dielectric material filling the
space between its plate

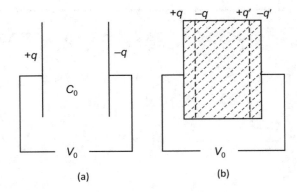

(a)   (b)

$$V = \frac{q}{KC_0} = \frac{V_0}{K} \tag{11.3}$$

The potential difference across a capacitor with a fixed charge decreases by a factor K because of the presence of a material with dielectric constant K. The electric field within the capacitor, when there is vacuum (air) between the plates, is

$$E_0 = \frac{V_0}{d} \tag{11.4}$$

d is separation between the plates of the capacitor. When the dielectric material is filled between the plates of the parallel-plate capacitor, the electric field is

$$E = \frac{V}{d} = \frac{V_0}{Kd} = \frac{E_0}{K} \tag{11.5}$$

Thus magnitude of electric field decreases with the introduction of dielectric. This can be understood on the microscopic level. Each molecule within the dielectric becomes polarized by the electric field E_0, which results from the charges on the plates of the capacitor. A molecule is polarized when its positively charged part is pushed in the direction of E_0 and the negatively charged part is pushed in the opposite direction. Thus there is a charge separation in each molecule of the dielectric material. There is a pile up of positive charges on the right face of the dielectric material and the negative charge on the left face of the dielectric. Since the dielectric material as a whole remains neutral, the positive charge on the right face of the dielectric material must be equal in magnitude to the negative charge on the left face of the dielectric material. These positive and negative charges set up an electric field E' which is in opposite direction to that of E_0. The resultant field in the dielectric E is the vector sum of E_0 and E'. It points in the same direction as E_0 but is smaller.

Terminology

Electric Dipole Moment
The electric dipole moment p of a charge distribution (assembly of charges) is defined as a vector from the centre of negative charge to the centre of positive charge. It has a magnitude qr, where q is magnitude of the positive or negative charge

$$p = qr \tag{11.6}$$

If each of the lattice point of a solid possesses a dipole moment, the total dipole moment of the sample would be

$$p = \sum_i q_i r_i \tag{11.7}$$

The electric field $E(r)$ due to a dipole at a point r is

$$E(r) = \frac{3(p \cdot r)r - r^2 p}{4\pi \epsilon_0 r^5}$$

Polarization
The polarization P in a solid is defined as the dipole moment per unit volume

$$P = \frac{p}{V} \tag{11.8}$$

It has the same unit as the surface charge density and its direction is same as that of electric field. In S.I. units, p is in coulomb/m^2.

Electric Field
The electric field E is defined as the force F per unit charge acting on a test charge. It has the same unit as the polarization

$$E = \frac{F}{q} \tag{11.9}$$

Depolarization Field
When a dielectric specimen is kept in an electric field E, the induced polarization field acts in such a way as to decrease the average field in the dielectric. In general, the induced polarized charges are non-uniform and therefore, the original field is modified differently in different portion of the specimen.

Polar Molecules

The molecules with a permanent electric dipole moment are called polar molecules. In such molecules the centre of charge of positive charges is at a different location than the centre of charge for the negative charges in the molecule. This separation of the centre of two types of charges means that the molecules act like a small electric dipole. Many molecules have permanent electric dipole moments due to some kind of asymmetry in their structure. The examples of some of the molecules are HCl, CO, HBr, ICl, H_2O, etc. The magnitude of electric dipole moments is of the order of 10^{-18} esu-cm. The unit 10^{-18} esu-cm is called the Debye unit for the electric dipole moments.

Non-polar Molecules

The molecules which have no permanent electric dipole moments are called non-polar molecules. In these molecules the centre of the positive charges and negative charges coincide. Symmetrical molecules such as O_2 and H_2 can have no dipole moment. The charge distribution can hardly be asymmetrical if the constituent atoms are identical. However, if the material containing non-polar molecules is placed in an electric field, the negatively charged electrons and their centre of charges are displaced slightly relative to the centre of the positive charge distribution. This results in an induced electric dipole moment in the molecules of the material.

11.2 Types of Polarization

Electronic Polarization

Since atoms consist of positive nuclei surrounded by negative electronic shells, an applied electric field would deform the electron shells around every nucleus and produce an electric dipole moment.

Consider an atom having a nucleus of charge Ze surrounded by electron cloud of charge −Ze. The electrons are assumed to be distributed uniformly inside a sphere of radius r. When an external electric field is applied, the centre of positive and the negative charge shifts by a distance d relative to each other. The force on electron cloud due to external field E is ZeE. This force is balanced by the force on the electron cloud due to Coulombic attraction between nuclear and electronic charges. Thus

$$\frac{(Ze)^2 d}{4\pi\epsilon_0 r^3} = ZeE$$

$$\alpha_e = Zed = 4\pi\epsilon_0 E r^3$$

Therefore, the polarizability α_e is proportional to r^3 and has dimensions of volume. Thus every atom in the material should have an electronic polarizability.

Ionic Polarization

Ionic polarization occurs only in materials that are ionic. An electric field displaces the positive cations in one direction and negative anions in the opposite direction from their equilibrium positions, giving a net electric dipole moment. The magnitude of the dipole moment for each pair is equal to the product of the relative displacement d and charge of each ion.

Orientational Polarization

Certain molecules have a permanent electric dipole moment. When these molecules condense into solid form, usually in weakly bound molecular crystal, they retain their electric dipole moment. The orientation of these dipoles is random in the absence of electric field. The applied electric field tends to orient these dipoles, giving rise to an effective polarization.

Within a solid which contains permanent electric dipoles all three of the above contributes to the polarizability, though to different extent. It is possible for one or more of these contributions to the total polarization to be either absent or negligible in magnitude relative to the others. For example, ionic polarization will not exist in covalently bonded materials in which no ions are present. Substances for which the polarization is entirely due to electronic displacement are necessarily elements such as diamond. For solid containing more than one type of atoms, for example, alkali halides, but no permanent dipoles, the contribution to polarization comes from electronic and ionic polarization. The substances composed of permanent electric dipole moment; the polarization is made up of electronic, ionic and dipolar contribution.

11.3 Gauss's Law in the Presence of a Dielectric

Gauss's law applies to any closed hypothetical surface (called a Gaussian surface) giving a connection between flux Φ for the surface and net charge q enclosed by the surface. It is

$$\epsilon_0 \oint \boldsymbol{E} \cdot \mathrm{d}\boldsymbol{S} = q \tag{11.10}$$

where ϵ_0 is permittivity constant.

Consider a parallel-plate capacitor formed of two parallel conducting plates of area A separated by a distance d. If we connect each plate to the terminal of the battery, a charge $+q$ will appear on one plate and a charge $-q$ on the other as shown in the Fig. 11.2. The electric field is

$$E_0 = \frac{\sigma}{\epsilon_0} = \frac{q}{\epsilon_0 A} \tag{11.11}$$

Fig. 11.2 Parallel-plate
condenser

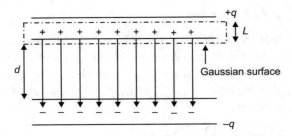

Fig. 11.3 Charges induced on
the surface of dielectric

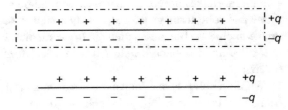

where σ is surface charge density. If the dielectric is present, positive and negative charges q' are induced at the surface of the dielectric as shown in Fig. 11.3.

Thus the net charge within the Gaussian surface is $q - q'$. According to Gauss law, the electric field is given by

$$\int \boldsymbol{E} \cdot \mathrm{d}\boldsymbol{S} = \frac{q - q'}{\epsilon_0}$$

$$\boldsymbol{E} \cdot A = \frac{q - q'}{\epsilon_0}$$

$$E = \frac{q}{\epsilon_0 A} - \frac{q'}{\epsilon_0 A} \tag{11.12}$$

Combining Eqs. (11.5) and (11.11)

$$E = \frac{E_0}{K} = \frac{q}{K\epsilon_0 A} \tag{11.13}$$

Substituting Eqs. (11.13) in Eq. (11.12)

$$\frac{q}{K\epsilon_0 A} = \frac{q}{\epsilon_0 A} - \frac{q'}{\epsilon_0 A}$$

$$\frac{q}{A} = E_0 \left(\frac{q}{K\epsilon_0 A} \right) + \frac{q'}{A} \tag{11.14}$$

$$q' = q\left(1 - \frac{1}{K}\right) \tag{11.15}$$

Thus the induced surface charge q' is always less in magnitude than the free charge q and equal to zero if no dielectric is present. The Gauss law for dielectric is then

$$\epsilon_0 \oint \boldsymbol{E} \cdot \mathrm{d}\boldsymbol{S} = q - q' \tag{11.16}$$

$q - q'$ is net charge within the Gaussian surface. Substituting Eq. (11.15) in (11.16)

$$\epsilon_0 \oint \boldsymbol{E} \cdot \mathrm{d}\boldsymbol{S} = q - q\left(1 - \frac{1}{K}\right) = \frac{q}{K}$$

$$\epsilon_0 \oint K \boldsymbol{E} \cdot \mathrm{d}\boldsymbol{S} = q \tag{11.17}$$

From Eq. (11.17) it is seen that flux integral contains a factor K and the charge contained within the Gaussian surface is free charge only. Induced charge has been taken into account by the introduction of K on the left hand side.

11.4 Three Electric Vectors

Consider a parallel-plate capacitor containing the dielectric. From Eq. (11.14)

$$\frac{q}{A} = \epsilon_0 \left(\frac{K}{K\epsilon_0 A}\right) + \frac{q'}{A}$$

The first term on the right hand side of the equation is the electric field strength E in the dielectric. The second term on the right hand side is the induced charge per unit area and is called polarization P or

$$P = \frac{q'}{A} \tag{11.18}$$

Multiplying the numerator and denominator on the right hand side of Eq. (11.18) by the thickness d of the dielectric

$$P = \frac{q'd}{Ad} = \frac{p}{V} \tag{11.19}$$

where $p = q'd$ is the induced electric dipole moment of the dielectric slab and $V = A\,d$ = volume of the dielectric slab. Equation (11.14) can be written as

$$\frac{q}{A} = \epsilon_0 E + P \tag{11.20}$$

where

$$E = \frac{q}{K \epsilon_0 A} \tag{11.21}$$

The quantity on the right hand side of the Eq. (11.20) is denoted by the symbol D and is called electric displacement, that is,

$$D = \epsilon_0 E + P \tag{11.22}$$

and

$$D = \frac{q}{A} \tag{11.23}$$

Since E and P are vectors, D must also be vector. Thus

$$D = \epsilon_0 E + P \quad \text{(in S.I. system)}$$

$$D = E + 4\pi P \quad \text{(in C.G.S. system)}$$

D is connected with free charge [Eq. (11.23)]; therefore, vector field of D is represented by lines of D, whereas the electric field E is connected with all charges that are actually present, whether free or polarization. The lines E reflect the presence of both kind of charges.

The vectors D and P can both be expressed in terms of E alone. We have

$$\frac{q}{A} = K \epsilon_0 \left(\frac{q}{K_0 A} \right) \tag{11.24}$$

From Eqs. (11.21), (11.23) and (11.24)

$$D = \frac{q}{A} = K \epsilon_0 E \tag{11.25}$$

The polarization from Eq. (11.14) is

$$P = \frac{q'}{A} = \frac{q}{A} \left(1 - \frac{1}{K} \right) \tag{11.26}$$

Substituting Eqs. (11.23) in (11.26)

$$P = D \left(1 - \frac{1}{K} \right) \tag{11.27}$$

$$P = D \left(\frac{K - 1}{K} \right) \tag{11.28}$$

From Eqs. (11.25) and (11.28)

$$P = \frac{K\epsilon_0}{K}[K - 1]E = \epsilon_0[K - 1]E \tag{11.29}$$

In vacuum $K = 1$ and $P = 0$ and Eq. (11.29) is

$$\frac{P}{\epsilon_0 E} = [K - 1] = \chi \tag{11.30}$$

where χ is electric susceptibility of the dielectric medium.

11.5 Concept of Local Molecular Fields

The electric field which is responsible for polarizing a molecule/atom of the dielectric is called the molecular field E_m. This is the electric field at a molecular position in the dielectric: it is produced by all external sources and by all polarized molecules in the dielectric with the exception of the molecule under consideration.

The local electric field at the molecule/atom is generally evaluated by a method suggested by Lorentz. The dielectric is imagined to be divided into two parts, and the contribution of each is consider separately. One part consists of a sphere whose size is so large that when considering the local field action on a molecule at the centre of the sphere, the effect of the molecules in the region outside the sphere may be evaluated regarding the region outside, the continuum. The approximation is satisfactory if the radius of the sphere is large compared with the intermolecular distances, so that sphere contains many molecules. Then the total field is

$$E_m = E_0 + E_1 + E_2 + E_3 \tag{11.31}$$

where.

E_0: external electric field, that is, field due to fixed charges external to dielectric.

E_1: field from a surface charge density on the outer surface of the dielectric.

E_2: field due to polarization charges lying on the surface of the sphere-on the inside of a spherical cavity cut out of dielectric.

E_3: field of molecules/atoms inside the cavity is due to dipoles lying within the sphere.
The E macroscopic is then

$$E_{\text{macro.}} = E_0 + E_1 = E$$

Let us suppose that the dielectric has been polarized by placing it in the uniform electric field between the plates of parallel-plate condenser (Fig. 11.4).

Fig. 11.4 Dielectric between
the plates of parallel-plate
capacitor

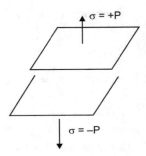

If the dimensions of the condenser plate are large compared with their separation, then
by Gauss law

$$E_0 = \frac{\sigma}{\epsilon_0} \tag{11.32}$$

where σ is surface charge density.

Now E_1 is the field arising from a surface charge density on the outer surface of the
dielectric that are in contact with the capacitor plates. The density of these bound charges
on the surface next to positive capacitor plate is $-P$ and $+P$ on the surface next to negative
plate. E_1 has the value

$$E_1 = \frac{P}{\epsilon_0} \tag{11.33}$$

To calculate the contribution of E_2, we note that as a consequence of the symmetry of
the problem, only field component parallel to E should be considered. If dA is the surface
area of sphere of radius r lying between θ and $\theta + d\theta$, θ is the direction with reference
to the direction of the applied field then from Fig. 11.5

$$dA = 2\pi(PQ) \cdot (QR) \tag{11.34}$$

From Fig. 11.5

Fig. 11.5 Calculation of E_2

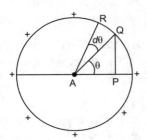

$$PQ = r \sin \theta \tag{11.35}$$

$$QR = rd\theta \tag{11.36}$$

Hence

$$dA = 2\pi r \sin \theta r d\theta = 2\pi r^2 \sin \theta d\theta \tag{11.37}$$

The charge dq on the surface dA is equal to the normal component of the polarization multiplied by the surface area. Therefore,

$$dq = P \cos \theta dA \tag{11.38}$$

The field due to this charge at A denoted by dE_2 in the direction $\theta = 0$ is then

$$dE_2 = \frac{dq \times 1 \times \cos \theta}{4\pi \int \epsilon_0} \tag{11.39}$$

From Eqs. (11.38) and (11.39)

$$dE_2 = \frac{2\pi r^2 \sin \theta P \cos \theta \cos \theta d\theta}{4\pi \epsilon_0 r^2} = \frac{2\pi r^2 \sin \theta P \cos^2 \theta}{4\pi \epsilon_0 r^2} d\theta \tag{11.40}$$

On integration

$$\int dE_2 = \frac{P}{2\epsilon_0} \int_0^\pi \cos^2 \theta \sin \theta d\theta \tag{11.41}$$

Let

$$\cos \theta = x, \; -\sin \theta d\theta = dx$$

Hence

$$E_2 = -\frac{P}{2\epsilon_0} \int_1^{-1} x^2 dx = -\frac{P}{2\epsilon_0} \left(\frac{x^3}{3}\right)\Bigg|_1^{-1} = \frac{P}{3\epsilon_0} \tag{11.42}$$

The contribution of field E_3 due to dipoles/charges inside the cavity is more difficult to calculate since it depends on how the molecules are arranged within the sphere. Lorentz showed that for a cubical array of molecules (as in simple cubic crystals) $E_3 = 0$. However, E_3 is not necessarily zero for all kinds of cubic crystals.

From Eqs. (11.32), (11.33) and (11.42), the molecular field is

$$E_m = \frac{\sigma}{\epsilon_0} - \frac{P}{\epsilon_0} + \frac{P}{3\epsilon_0} = \frac{D}{\epsilon_0} - \frac{P}{\epsilon_0} + \frac{P}{3\epsilon_0} \qquad (11.43)$$

Since $D = \sigma$, the free charge density on the condenser plate

$$E_m = \frac{D - P}{\epsilon_0} + \frac{P}{3\epsilon_0} \qquad (11.44)$$

$$E_m = E + \frac{P}{3\epsilon_0} \qquad (11.45)$$

where $E = \frac{D-P}{\epsilon_0}$ is the macroscopic electric field in the dielectric.

11.6 Clausius Mossotti Relation

The dipole moment of molecule per unit polarizing field is called molecular polarizability, thus dipole moment p is

$$p = \alpha E_m \qquad (11.46)$$

where α is molecular polarizability. If there are N atoms per unit volume, then polarization vector P

$$P = Np = N\alpha E_m \qquad (11.47)$$

From Eqs. (11.45) and (11.47)

$$P = N\alpha \left(E + \frac{P}{3\epsilon_0} \right) \qquad (11.48)$$

But

$$P = \epsilon_0 (K - 1)E \qquad (11.49)$$

Equating Eqs. (11.48) and (11.49)

$$\epsilon_0 (K - 1) = N\alpha \left[E + \frac{\epsilon_0 (K - 1)E}{3\epsilon_0} \right]$$

$$\frac{N\alpha}{3\epsilon_0} = \frac{K - 1}{K + 2} \qquad (11.50)$$

This is Clausius–Mossotti relation, which co-relates the dielectric constant with microscopic polarizability. Multiplying both sides of Eq. (11.50) by molar volume

$$\frac{N\alpha}{3\epsilon_0}\left(\frac{M_A}{\rho}\right) = \frac{K-1}{K+2}\left(\frac{M_A}{\rho}\right) \tag{11.51}$$

where M_A is molecular weight and

$$\frac{NM_A}{\rho} \quad \text{is Avogadro number NA .} \tag{11.52}$$

Thus

$$\frac{N_A\alpha}{3\epsilon_0} = \frac{K-1}{K+2}\left(\frac{M_A}{\rho}\right) = P_m \tag{11.53}$$

where P_m is molar polarization. The refractive index n of the dielectric medium is

$$n = \sqrt{K} \tag{11.54}$$

From Eqs. (11.53) and (11.54)

$$\alpha = \frac{3\epsilon_0}{N}\left(\frac{n^2-1}{n^2+2}\right) \tag{11.55}$$

The Clausius–Mossotti relation is a link between the macroscopic (K) and microscopic parameters (α) and is only accurate for small densities. We now can calculate the dielectric constants of all materials if we calculate α or alternatively by measuring dielectric constant, α can be calculated.

11.7 Orientational Polarization

Consider a gas containing a large number of identical molecules each with a permanent electric dipole moment $\textbf{p}$. In the absence of an external applied electric field, the dipole moments are randomly oriented resulting in zero electric dipole moment. When external electric field $\textbf{E}$ is applied it exerts a torque on each of the electric dipole and will tend to

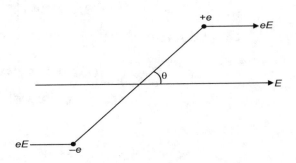

Fig. 11.6 Illustrating the torque exerted on a dipole by an external field

orient the dipole along the direction of $\boldsymbol{E}$ (Fig. 11.6). However, this alignment is opposed by the motion of the molecule.

The potential energy of the electric dipole $\boldsymbol{p}$ in the external field $\boldsymbol{E}$ is

$$E_d = -\boldsymbol{p} \cdot \boldsymbol{E} = -pE \cos \theta \tag{11.56}$$

where θ is the angle between $\boldsymbol{p}$ and $\boldsymbol{E}$. The dipole moment at right angle to $\boldsymbol{E}$ will not contribute to E_d. The probability that an electric dipole is inclined at an angle θ to the electric field direction in thermal equilibrium is given by Maxwell–Boltzmann distribution function

$$A \exp\left[-\frac{E_d}{k_B T}\right] = A \exp\left[\frac{pE \cos \theta}{k_B T}\right] \tag{11.57}$$

where A is a constant to be determined from the total number of dipoles.

The number dN of magnetic dipole moments which makes an angle between θ and $\theta + \mathrm{d}\theta$ with $\boldsymbol{E}$ is

$$\mathrm{d}N = 2\pi \sin \theta \mathrm{d}\theta A \exp\left[\frac{pE\cos\theta}{k_B T}\right] \tag{11.58}$$

The total number of dipoles is

$$N = \int \mathrm{d}N = 2\pi A \int_0^\pi \sin \theta \exp\left[\frac{pE\cos\theta}{k_B T}\right] \mathrm{d}\theta \tag{11.59}$$

The mean electric dipole moment in the direction of $\boldsymbol{E}$ is

$$p\langle \cos \theta \rangle = \int p \cos \theta \mathrm{d}n = 2\pi Ap \int_0^\pi \sin \theta \cos \theta \exp\left[\frac{pE\cos\theta}{k_B T}\right] \mathrm{d}\theta \tag{11.60}$$

From Eq. (11.59)

$$A = \frac{N}{2\pi \int_0^\pi \sin \theta \exp\left[\frac{pE\cos\theta}{k_B T}\right] \mathrm{d}\theta} \tag{11.61}$$

Substituting Eqs. (11.61) in (11.60)

$$p\langle \cos \theta \rangle = \frac{\int_0^\pi p \sin \theta \cos \theta \exp\left[\frac{pE\cos\theta}{k_B T}\right] \mathrm{d}\theta}{\int_0^\pi \sin \theta \exp\left[\frac{pE\cos\theta}{k_B T}\right] \mathrm{d}\theta} \tag{11.62}$$

Let

$$\frac{pE \cos \theta}{k_B T} = y, \quad -\frac{pE \sin \theta}{k_B T} \mathrm{d}\theta = \mathrm{d}y; \quad \frac{pE}{k_B T} = x \tag{11.63}$$

Substituting Eqs. (11.63) in (11.62)

$$\langle \cos\theta \rangle = \frac{N}{x}\frac{\int_{-x}^{x} y\exp[y]dy}{\int_{-x}^{x}\exp[y]dy} = \frac{N}{x}\frac{\int_{-x}^{x} yd\big[\exp(y)\big]}{\int_{-x}^{x} d\big[\exp(y)\big]}$$

$$\langle \cos\theta \rangle = \frac{N}{x}\left[\frac{y\exp(y)-\exp(y)}{\exp(y)}\right]\Bigg|_{-x}^{x} = \frac{N}{x}\left[\frac{x\exp(x)-\exp(x)+x\exp(-x)+\exp(-x)}{\exp(x)-\exp(-x)}\right]$$

$$\langle \cos\theta \rangle = N\left[\frac{\exp(x)+\exp(-x)}{\exp(x)-\exp(-x)}\right] - \frac{N}{x}\left[\frac{\exp(x)-\exp(-x)}{\exp(x)-\exp(-x)}\right]$$

$$\langle \cos\theta \rangle = N\left[\frac{\exp(x)+\exp(-x)}{\exp(x)-\exp(-x)} - \frac{1}{x}\right] = N\left[\coth x - \frac{1}{x}\right] = L(x) \tag{11.64}$$

where

$$L(x) = \coth x - \frac{1}{x} \tag{11.65}$$

$L(x)$ is called Langevin function. Figure 11.7 shows a plot of $L(x)$ versus x. For high electric field strength and low temperatures, the function approaches the saturation value unity. This corresponds to complete alignment of the electric dipoles in the direction of the electric field.

If the field strength is not too high and temperature is not too low, we can assume $x\ll 1$ or $pE \ll k_{\mathrm{B}}T$. Under these conditions

$$\exp(x) = 1 + x + \frac{x^2}{2!} + \frac{x^3}{3!} + \cdots \tag{11.66}$$

$$\exp(-x) = 1 - x + \frac{x^2}{2!} - \frac{x^3}{3!} + \cdots \tag{11.67}$$

Substituting Eqs. (11.66) and (11.67) in Eq. (11.65)

Fig. 11.7 Langevin function $L(a)$. For $a \ll 1$, the slope is $1/3$

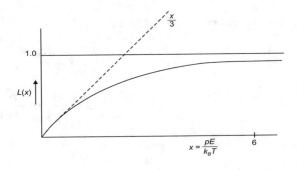

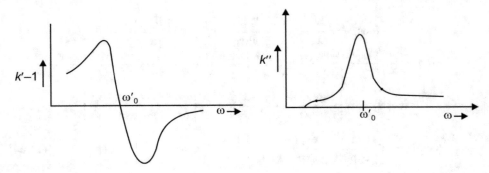

Fig. 11.8 Behaviour of $K'-1$ and K'' as function of frequency in the vicinity of the resonance frequency ω'_0

$$L(x) = \left[\frac{\exp(x) + \exp(-x)}{\exp(x) - \exp(-x)} - \frac{1}{x} \right]$$

$$= \left[\frac{\left(1 + x + \frac{x^2}{2!} + \frac{x^3}{3!} + \cdots\right) + \left(1 - x + \frac{x^2}{2!} - \frac{x^3}{3!} + \cdots\right)}{\left(1 + x + \frac{x^2}{2!} + \frac{x^3}{3!} + \cdots\right) - \left(1 - x + \frac{x^2}{2!} - \frac{x^3}{3!} + \cdots\right)} - \frac{1}{x} \right]$$

$$L(x) = \left[\frac{2\left(1 + \frac{x^2}{2!} + \cdots\right)}{2\left(x + \frac{x^3}{3!} + \cdots\right)} - \frac{1}{x} \right] = \frac{1}{x}\left(1 + \frac{x^2}{2!} + \cdots\right)\left(1 + \frac{x^2}{3!} + \cdots\right)^{-1} - \frac{1}{x}$$

$$L(x) = \frac{1}{x}\left(1 + \frac{x^2}{2!} + \cdots\right)\left(1 - \frac{x^2}{3!} + \cdots\right) - \frac{1}{x} = \frac{1}{x}\left(1 - \frac{x^2}{3!} + \frac{x^2}{2!} - \frac{x^5}{12} + \cdots\right) - \frac{1}{x} = \frac{x}{3} \quad (11.68)$$

and

$$\langle p \cos\theta \rangle = \frac{Npx}{3} = \frac{Np^2 E}{3k_{\mathrm{B}}T} \quad (11.69)$$

and orientational polarization P_o

$$P_o = \frac{Np^2 E}{3k_{\mathrm{B}}T}$$

$$\alpha_o = \frac{P_o}{NE} = \frac{p^2}{3k_{\mathrm{B}}T} \quad (11.70)$$

11.8 Classical Theory of Electronic Polarization

In the classical theory when an electron cloud is displaced from its original position around its nucleus, then a simple harmonic force exists which tend to restore the electron cloud to its original position. The restoring force determining the displacement is in first approximation proportional to the displacement itself. Beside restoring force there is also a collisions and a damping force which results from the fact that particle emits radiation as a consequence of its acceleration. We assume that this particle experienced an oscillating electric field whose wavelength is large compared with atomic dimensions so that field acting on an electron in a given atom is independent of its position with respect to the nucleus which is assumed stationary. The equation of motion is

$$m\frac{d^2x}{dt^2} + \gamma m\frac{dx}{dt} + \beta x = -eE_{\text{local}} \tag{11.71}$$

where γ is damping constant, β is force constant. Writing

$$\omega_0^2 = \frac{\beta}{m} \quad \text{and} \quad E_{\text{local}} = E_0 \exp(i\omega t) \tag{11.72}$$

In Eq. (11.71)

$$m\frac{d^2x}{dt^2} + \gamma m\frac{dx}{dt} + m\omega_0^2 x = -eE_0 \exp(i\omega t) \tag{11.73}$$

Let

$$x = x_0 \exp(i\omega t) \tag{11.74}$$

$$\frac{dx}{dt} = i\omega x_0 \exp(i\omega t) = i\omega x \tag{11.75}$$

$$\frac{d^2x}{dt^2} = -\omega^2 x \tag{11.76}$$

Substituting Eqs. (11.74)–(11.76) in Eq. (11.73)

$$-m\omega^2 x + i\gamma m\omega x + m\omega_0^2 x = -eE_0 \exp(i\omega t) \tag{11.77}$$

$$x_0 = -\frac{e}{m}\left[\frac{E_0}{(\omega_0^2 - \omega^2) + i\gamma\omega}\right] = -\frac{eE_0}{m}\left[\frac{(\omega_0^2 - \omega^2) - i\gamma\omega}{(\omega_0^2 - \omega^2)^2 + \gamma^2\omega^2}\right] \tag{11.78}$$

The dipole moment p is

$$p = -ex_0 = \frac{e^2 E_0}{m} \left[\frac{(\omega_0^2 - \omega^2) - i\gamma\omega}{(\omega_0^2 - \omega^2)^2 + \gamma^2\omega^2} \right] \tag{11.79}$$

The polarization due to N dipoles per unit volume is

$$p = \alpha E = -ex \tag{11.80}$$

We have for electronic polarization

$$\alpha_e = \frac{p}{E} = -\frac{ex_0}{E_0} = \frac{e^2}{m} \left[\frac{(\omega_0^2 - \omega^2) - i\gamma\omega}{(\omega_0^2 - \omega^2)^2 + \gamma^2\omega^2} \right] \tag{11.81}$$

From Clausius–Mossotti relation [Eq. (11.50)]

$$\frac{N\alpha}{3\epsilon_0} = \frac{K-1}{K+2}$$

$$K - 1 = \frac{N\alpha}{3\epsilon_0}(K+2)$$

$$K\left(1 - \frac{N\alpha}{3\epsilon_0}\right) = 1 + \frac{2N\alpha}{3\epsilon_0}$$

$$K = \frac{1 + \frac{2N\alpha}{3\epsilon_0}}{1 - \frac{N\alpha}{3\epsilon_0}} = 1 + \frac{\frac{N\alpha}{\epsilon_0}}{1 - \frac{N\alpha}{3\epsilon_0}} \tag{11.82}$$

Substituting the value of α_e from Eq. (11.81) for α in Eq. (11.82)

$$K = 1 + \frac{\frac{Ne^2}{m\epsilon_0}\left[\frac{1}{(\omega_0^2-\omega^2)+i\gamma\omega}\right]}{1 - \frac{Ne^2}{3m\epsilon_0}\left[\frac{1}{(\omega_0^2-\omega^2)+i\gamma\omega}\right]} = 1 + \frac{\frac{Ne^2}{m\epsilon_0}}{\left[(\omega_0^2-\omega^2)+i\gamma\omega\right] - \frac{Ne^2}{3m\epsilon_0}}$$

$$K = 1 + \frac{Ne^2}{m\epsilon_0\left[(\omega_0^2-\omega^2)+i\gamma\omega - \frac{Ne^2}{3m\epsilon_0}\right]} \tag{11.83}$$

Let

$$\omega_0^2 - \frac{Ne^2}{3m\epsilon_0} = \omega'^2 \tag{11.84}$$

Substituting Eqs. (11.84) in (11.83)

$$K = 1 + \frac{Ne^2}{m\epsilon_0\left[(\omega'^2-\omega^2)+i\gamma\omega\right]} = 1 + \frac{Ne^2}{m\epsilon_0}\left[\frac{(\omega'^2-\omega^2)-i\gamma\omega}{(\omega'^2-\omega^2)^2+\gamma^2\omega^2}\right] \tag{11.85}$$

The real part of K is

$$K' = 1 + \frac{Ne^2}{m\epsilon_0}\left[\frac{\gamma\omega}{\left(\omega'^2 - \omega^2\right)^2 + \gamma^2\omega^2}\right] \qquad (11.86)$$

The imaginary part of K is

$$K'' = \frac{Ne^2}{m\epsilon_0}\left[\frac{\gamma\omega}{\left(\omega'^2 - \omega^2\right)^2 + \gamma^2\omega^2}\right] \qquad (11.87)$$

The energy absorbed by the dielectric is proportional to K''. K'' contain damping factor γ, if there is no damping, there will be no absorption. Figure 11.8 shows $K'-1$ and K'' as a function of frequency.

11.9   Behaviour of Dielectrics in Alternating Electric Field

The dielectric polarization P may be considered as the sum of three contributions

$$P = P_e + P_a + P_d \qquad (11.88)$$

where e, a and d refer, respectively, to electronic, atomic and dipolar polarization. Consider a material, which have permanent dipole moment. When $E = 0$, the dipole is randomly oriented and continually changing in orientation as a result of collisions. When a field is switched on, the dipole tends to align up in the field. But it takes time; the time required to reach the equilibrium polarization is called the relaxation time τ. The relaxation time is defined as the time required to come to within $1/e$ of the equilibrium value. Since a finite amount of time τ is required for the dipoles to reach equilibrium, if the applied electric field changes in a time short compared to τ, the dipole system cannot follow. For example if $\tau = 5$ s. and an electric field were turned on momentarily for 0.01 s. the dipole would hardly respond to it. Consequently, the polarization due to permanent dipoles dies off at high frequencies. In particular if $E = E_0 \exp(i\omega t)$, then $P_d \to 0$ when $\omega \gg 1/\tau$.

The permanent dipole moments in the dielectric materials is thus unable to follow the electric field and the contribution to the polarizability tends to zero if the frequency of the applied electric field is in the radio frequency region (10^6–10^{11} Hz) or more. At higher frequency in the region of infrared (10^{11}–10^{14} Hz), the relatively heavy positive and negative ions cannot follow the electric field variation and contribution to polarizabilities from ionic polarization stops and only electronic polarization remains. At ultra-high frequencies, the electric field is changing its orientation so fast that even electrons fails to follow the field alternation and thus the electronic contribution to polarization tends to

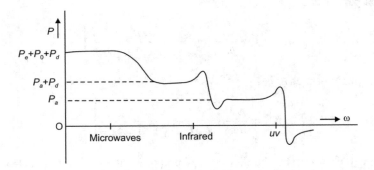

Fig. 11.9 Real part of the total polarization P as a function of frequency for a dipolar substance with a single atomic and electronic resonance frequency

zero. As a result the total polarization tends to zero at very high frequencies. Figure 11.9 shows the total polarization P as a function of frequency for a dipolar substance.

11.10 The Complex Dielectric Constant

When a dielectric is placed in an alternating electric field, the polarization P and displacement D vary periodically. In general P and D may lag behind in phase relative to E. Let

$$E = E_0 \cos \omega t \qquad (11.89)$$

and

$$D = D_0 \cos(\omega t - \delta) = D_0 \cos \omega t \cos \delta + D_0 \sin \omega t \sin \delta$$

$$D = D_1 \cos \omega t + D_2 \sin \omega t \qquad (11.90)$$

where δ is the phase angle and

$$D_1 = D_0 \cos \delta \qquad (11.91)$$

$$D_2 = D_0 \sin \delta \qquad (11.92)$$

For most dielectric, D_0 is proportional to E_0 but the ratio D_0/E_0 is generally frequency dependent. Introducing two frequency dependent dielectric constants

$$K'(\omega) = \frac{D_1}{E_0} = \frac{D_0}{E_0}\cos\delta \qquad (11.93)$$

$$K''(\omega) = \frac{D_2}{E_0} = \frac{D_0}{E_0}\sin\delta \qquad (11.94)$$

writing

$$K^* = K' - iK'' \qquad (11.95)$$

and expressing D and E as complex quantities

$$E = E_0\exp(i\omega t) \qquad (11.96)$$

$$D = D_0\exp[i(\omega t - \delta)] \qquad (11.97)$$

From Eq. (11.97)

$$D = D_0[\cos(\omega t - \delta) + i\sin(\omega t - \delta)]$$

$$D = D_0[\cos\omega t\cos\delta + \sin\omega t\sin\delta + i\sin\omega t\cos\delta - i\cos\omega t\sin\delta]$$

$$D = D_0[(\cos\omega t + i\sin\omega t)\cos\delta - i(\cos\omega t + i\sin\omega t)\sin\delta]$$

$$D = D_0\big[\exp(i\omega t)\cos\delta - i\exp(i\omega t)\sin\delta\big] = [D_0\cos\delta - iD_0\sin\delta]\exp(i\omega t)$$

$$D = \left[\frac{D_0}{E_0}\cos\delta - i\frac{D_0}{E_0}\sin\delta\right]E_0\exp(i\omega t)$$

$$D = \big[K' - iK''\big]E_0\exp(i\omega t) \qquad (11.98)$$

$$D = K^*E_0\exp(i\omega t) \qquad (11.99)$$

From Eqs. (11.93) and (11.94)

$$\tan\delta = \frac{\sin\delta}{\cos\delta} = \frac{K''(\omega)}{K'(\omega)} \qquad (11.100)$$

Since K' and K'' are frequency dependent, phase angle δ is also frequency dependent. Consider the current density in the capacitor

$$J = \frac{dD}{dt} = \frac{d}{dt}[D_1\cos\omega t + D_2\sin\omega t] = \omega[-D_1\sin\omega t + D_2\cos\omega t] \qquad (11.101)$$

The energy dissipated per second in the dielectric per unit volume is

$$W = \frac{1}{T} \int_0^T J E \, dt = \frac{\omega}{2\pi} \int_0^{2\pi/\omega} J E \, dt \tag{11.102}$$

Substituting Eqs. (11.89) and (11.101) into (11.102)

$$W = \frac{\omega^2}{2\pi} \int_0^{2\pi/\omega} [-D_1 \sin \omega t + D_2 \cos \omega t] E_0 \cos \omega t \, dt$$

$$W = \frac{\omega^2}{2\pi} E_0 \int_0^{2\pi/\omega} \left[-D_1 \sin \omega t \cos \omega t + D_2 \cos^2 \omega t \right] dt$$

$$W = \frac{\omega^2}{4\pi} E_0 \int_0^{2\pi/\omega} [-D_1 \sin 2\omega t + D_2(1 + \cos 2\omega t)] dt$$

$$W = \frac{\omega^2}{4\pi} E_0 \left. \frac{D_1 \cos 2\omega t}{2\omega} \right|_0^{2\pi/\omega} + \frac{\omega^2}{4\pi} E_0 \left. \frac{D_2 t}{2} \right|_0^{2\pi/\omega} + \frac{\omega^2}{4\pi} E_0 \left. \frac{D_2 \sin 2\omega t}{2\omega} \right|_0^{2\pi/\omega}$$

$$W = \frac{\omega^2}{4\pi} E_0 \left[\frac{D_1}{2\omega}(\cos 4\pi - \cos 0) + \frac{D_2 2\pi}{\omega} - 0 + \frac{D_2}{2\omega}(\sin 4\pi - \sin 0) \right]$$

$$W = \frac{\omega}{2} D_2 E_0 \tag{11.103}$$

Substituting Eq. (11.94) into Eq. (11.103)

$$W = \frac{E_0^2 \omega K''}{2} \tag{11.104}$$

The energy losses are thus proportional to $\sin \delta$. $\sin \delta$ is called the loss factor and δ is the loss angle. For small, δ, $\sin \delta = \tan \delta = \delta$ hence for small δ, $\tan \delta$ is also called loss factor.

The dielectric loss at low frequencies is mainly due to DC resistivity. But at high frequencies the dielectric loss is mostly due to dipole rotation or to ionic transitions from the lower energy states to higher energy states.

11.11 Ferroelectricity

The dielectrics show a linear relationship between polarization and applied electric field. In some materials this relationship exhibits hysteresis effects (Fig. 11.10), and these materials are called ferroelectrics. A ferroelectric is spontaneously polarized, that is, it is polarized in the absence of an external applied electric field. In general, the direction of spontaneous polarization is not the same throughout a macroscopic crystal. The crystal consists of a number of domains; within each domain the polarization has a specific direction. However, this direction varies from one domain to another. On the basis of the domain concept, the occurrence of the hysteresis in the P versus E relationship can be explained as follows:

We assume that the crystal has equal numbers of domains with oppositely directed polarization resulting in overall polarization equal to zero. When an electric field is applied parallel to a crystallographic direction, the domains whose polarization is more nearly parallel to the field direction have a lower energy so that they grow in size at the expense of the antiparallel domains and thus polarization increases (OC). When all domains are aligned in the direction of the applied electric field (AB), the polarization saturates and the crystal has become a single domain. A further increase in the polarization with increasing applied electric field results from normal polarization; the rotation of domain vector may also involved if the external field does not coincide with one of the possible direction of spontaneous polarization. The extrapolation of the linear part BC to zero external field gives the spontaneous polarization P_s. When the applied field for a crystal corresponding to point A is reduced, the polarization of the crystal decreases but for zero applied field their remains the remanent polarization P_r where P_r refers to the crystal as a whole. In order to remove the remanent polarization, the polarization of approximately half the crystal must be reversed and this occurs only when a field in the opposite direction is applied. The field required to make the polarization zero again is called the coercive field E_c. As the magnitude of the reverse field is increased further, the saturation value of polarization in the reverse direction is reached (EF). Reversing the field again then traces out the curve FEB, and so forth. If the coercive field is larger than the breakdown field of the crystal, no change in the direction of spontaneous polarization can be obtained, that is under those conditions solid cannot be treated as ferroelectric.

Fig. 11.10 Schematic representation of hysteresis in the polarization versus applied field relationship

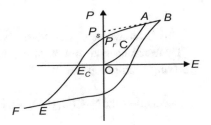

The ferroelectric properties of ferroelectric disappear above a critical temperature T_c called ferroelectric Curie temperature. Transition, from ferroelectric phase to non-ferroelectric phase, causes anomalies in other physical properties. Thus for a first-order transition, there will be a latent heat; for a second order transition, the specific heat will exhibit discontinuity.

The dielectric constant in a ferroelectric is not a constant but depends on the field strength at which it is measured. This is because of nonlinear relationship between P and E. The dielectric constant near the transition temperature is very large $\sim 10^4$ to 10^5. In order to understand the behavior of ferroelectrics, consider Clausius–Mossotti relation given by Eq. (11.50). Rearranging the relation, we have

$$K - 1 = \frac{N\alpha}{3\epsilon_0}(K + 2)$$

$$K = \frac{2N\alpha + 3\epsilon_0}{3\epsilon_0 - N\alpha} \tag{11.105}$$

The dielectric constant tends to infinity corresponding to a finite polarization in the absence of an applied electric field when $3\epsilon_0 = N\alpha$. From Eq. (11.105) it is seen that the dielectric constant is very sensitive to small deviation of $N\alpha$ from $3\epsilon_0$. Let the deviation is represented as

$$N\alpha = 3\epsilon_0(1 - \delta) \quad \text{for} \quad \delta \ll 1 \tag{11.106}$$

From Eqs. (11.105) and (11.106)

$$K = \frac{6\epsilon_0(1 - \delta) + 3\epsilon_0}{3\epsilon_0 - 3\epsilon_0(1 - \delta)} = \frac{3 - 2\delta}{\delta} \approx \frac{3}{\delta} \tag{11.107}$$

The deviation of δ is linearly dependent on the temperature near T_c, we assume that

$$\delta \cong \frac{3}{C}(T - T_c) \tag{11.108}$$

where C is constant of proportionality.

Substituting Eq. (11.108) in Eq. (11.107)

$$K = \frac{C}{T - T_c} \tag{11.109}$$

This is known as Curie–Weiss law for ferroelectrics.

There are several groups of crystals that are known to be ferroelectrics. The crystals typifying each group are.

(a) Rochelle salt, $NaK(C_4H_4O_6)\cdot 4H_2O$

(b) Potassium dihydrogen phosphate, KH_2PO_4
(c) Barium titanate, BaT_iO_3
(d) Guanidine compounds such as $C(NH_2)_3Al(SO_4)_2 \cdot 6H_2O$

Each of these groups contains a number of isomorphous or isostructural crystals all of which are ferroelectrics.

11.12 Piezoelectricty

An unusual property exhibited by a few materials is piezoelectricity. In these materials polarization is induced and an electric field is established across a specimen by the application of external forces. Reversing the sign of an external force (i.e. from tension to compression) reverses the direction of the field. The converge effect of piezoelectricity effect is that when such crystalline solids are subjected to an external field along specific orientations, and then these solids get strained. The piezoelectricity is possible in only those crystals which do not have a centre of inversion symmetry, because the centre of symmetry cancels out all possible polarization. In total there are 21 classes of the crystals which lack a centre of inversion symmetry; of these 21 classes, 20 are piezoelectric. Piezoelectric materials include titanates of barium and lead, tourmaline, lead zirconate, ammonium dihydrogen phosphate, quartz, etc.

11.13 Dispersion

According to Maxwell theory of electromagnetic field, the refractive index at a frequency is

$$v = \frac{c}{n} \qquad (11.110)$$

where v is the phase velocity of electromagnetic waves and is

$$v = \frac{c}{\sqrt{K\mu_r}} \qquad (11.111)$$

where $\mu_r = 1$ thus $n^2 = K$.

In frequency region where dielectric absorbs energy from electromagnetic field, the refractive index is frequency dependent. Such a phenomenon is called dispersion. For visible region, the refractive index increases with increase in frequency. This is known as normal dispersion. It is observed that refractive index decreases with increasing frequency at frequency near to characteristic frequencies of maximum absorption. This is called anomalous dispersion.

Solved Examples

Example 1 For helium gas, the value of dielectric constant at 273 K is 1.000074. Obtain the electric susceptibility at this temperature.

Solution
From Eq. (11.30)

$$[K - 1] = \chi$$

$$\chi = [1.000074 - 1] = 0.000074$$

Example 2 The dielectric constant of helium at 273 K and 1 atmosphere pressure is 1.000074. Find the dipole moment induced in each helium atom when the gas is an electric field of intensity 100 V/m. $\epsilon_0 = 8.854 \times \frac{10^{-12}c^2}{Nm^2}$

Solution
From Eq. (11.30)

$$K = \chi + 1 = \frac{P}{\epsilon_0 E} + 1$$

$$P = \epsilon_0(K - 1)E$$

If n is the number of helium atoms/m^3 and p is the dipole moment per atom, then $P = np$

$$p = \frac{P}{n} = \frac{\epsilon_0(K - 1)E}{n} = \frac{8.854 \times 10^{-12} \times 0.000074 \times 100}{n} = \frac{8.854 \times 10^{-16} \times 74}{n}$$

Since at NTP, 1 g atom of gas occupies 22.4 l, that is 22.4×10^{-3}m^3 and 6×10^{23} atoms (Avogadro number)

$$p = \frac{8.854 \times 74 \times 10^{-16} \times 22.4 \times 10^{-3}}{6 \times 10^{23}} = 24.46 \times 10^{-40}C - m$$

Objective Type Questions

1. The unit of dipole moment per unit volume is
   (a) Coulomb/meter (b) Coulomb/meter^2 (c) Coulomb/meter^3 (d) Coulomb
2. The dipole moment per unit volume of a solid is the sum of all individual dipole moments and is called
   (a) polarization of the solid (b) permittivity of the solid (d) none of these.

3. The polarization P in a solid dielectric is related to the electric field E and displacement vector D by the relation
    (a) $E = \epsilon_0 D + P$ (b) $D = E + \epsilon_0 P$ (c) $D = \epsilon_0 E + P$ (d) $D = \epsilon_0 (E + P)$

4. The losses in a dielectric subjected to *a.c.* field are determined by
    (a) real part of the complex dielectric constant
    (b) imaginary part of the complex dielectric constant
    (c) both real and imaginary parts of the complex dielectric constant
    (d) none of these

5. In a dielectric the power loss is proportional to
    (a) ω (b) ω^2 (c) $\frac{1}{\omega}$ (d) $\frac{1}{\omega^2}$

6. If τ is the relaxation time, and ω is the frequency of *a.c.* field; the contribution of dipolar polarization vanishes if
    (a) $\omega = \frac{1}{\tau}$ (b) $\omega \ll \frac{1}{\tau}$ (c) $\omega \gg \frac{1}{\tau}$ (d) none of these.

7. In infrared frequency region, the contribution towards polarization is due to
    (a) dipolar polarization only (b) ionic polarization only (c) dipolar and ionic polarization (d) electronic polarization

8. A single point charge q is imbedded in a dielectric of dielectric constant K. At a point inside the dielectric a distance r from the point charge, what is the magnitude of the electric field
    (a) $\frac{q}{4\pi\epsilon_0 r^2}$ (b) $\frac{Kq}{4\pi\epsilon_0 r^2}$ (c) $\frac{q}{4\pi\epsilon_0 K r^2}$ (d) none of these

9. A parallel-plate capacitor has charges q and $-q$ in its two plates. A dielectric slab with $K = 3$ is then inserted into the space between the plates. Rank the following electric field magnitudes in order from largest to smallest (i) the field before the slab is inserted (ii) the resultant field after the slab is inserted (iii) the field due to bound charges
    (a) i, ii, iii (b) i, iii, ii (c) ii, i, iii (d) iii, ii, i

10. Which molecule is not polar?
    (a) H_2O (b) CO_2 (c) ICl (d) CO

Problems

1. An isotropic material relative permittivity is placed normal to a uniform external electric field with an electric displacement vector 5×10^{-4} m^2. If the volume of the slab is 0.5 m^3 and magnitude of the polarization is 4×10^{-4} m^2, find the value of the total dipole moment of the slab.

2. Obtain an expression for the energy stored in a dielectric.

3. Consider a dielectric slab of thickness b and dielectric constant K placed between the plates of a parallel plate capacitor of plate area A and separation d. A potential difference V_0 is applied across the plates of parallel plate capacitor with no dielectric present. The battery is then disconnected and dielectric slab is introduced. Calculate (a) capacitance before the slab is introduced (b) free charge (c) electric field strength in the dielectric (d) D and P. It is given that $A = 100$ cm^2, $d = 1.0$ cm, $b = 0.50$ cm, $K = 7$, $V_0 = 100$ V.

4. Explain the behavior of a dielectric in an *a.c.* field and discuss the dielectric losses.
5. Explain the mechanisms contributing to dielectric polarization.

Answers
Objective Type Questions

   1. (b)   2. (a)   3. (c)   4. (b)   5. (a)   6. (c)   7. (d)   8. (c)
   9. (b)

Magnetic Properties of Matter

12

12.1 Angular Momenta and Magnetic Moment of Atoms

Every electron in an atom has two possible kinds of angular momenta, one due to orbital and other due to its spin motion. The magnitude of the orbital angular momentum vector for a single electron is given by

$$l = \hbar\sqrt{l(l+1)} \quad \text{where} \quad l = 0, 1, 2, 3, n-1$$

Similarly the magnitude of the spin-angular momentum vector for a single electron is

$$s = \hbar\sqrt{s(s+1)} \quad \text{where} \quad s = 1/2$$

For an electron which has both orbital and spin-angular momentum there is a quantum number j associated with the total (orbital plus spin) angular momentum. This is also a vector quantity whose magnitude is given by

$$j = \hbar\sqrt{j(j+1)} \tag{12.1}$$

where j can take the values

$$j = l+s, l+s-1, \ldots.|l-s| \tag{12.2}$$

The number of possible values of j is equal to the smaller of the two numbers $2s+1$ and $2l+1$. The angular momentum l and s interact magnetically. If there is no magnetic field, the total angular momentum j is conserved in magnitude and direction, and, l and s precess around the direction of their resultant j.

© The Author(s) 2022
V. K. Jain, *Solid State Physics*, https://doi.org/10.1007/978-3-030-96017-9_12

12.1.1 Orbital Magnetic Moment

Consider an electron moving with velocity v in circular Bohr orbit of radius r. A charge circulating in a loop constitutes a current of magnitude

$$i = \frac{e}{T} = \frac{ev}{2\pi r} \tag{12.3}$$

where $T = 2\pi \, r/v$ is the orbital period of the electron. From electromagnetic theory it is known that at large distance from the loop, the magnetic field due to the loop is the same as that of a magnetic dipole located at the centre of the loop. For a current i in a loop of area A, the magnitude of the orbital magnetic dipole moment μ of the equivalent dipole is

$$\mu = iA = \frac{evr}{2} \tag{12.4}$$

The direction of the magnetic dipole moment is perpendicular to the plane of the orbit. The angular momentum l of electron moving in a circular orbit is $r \times p$, where r is the position vector of the electron and p is its momentum. For circular motion r and p are perpendicular, hence

$$l = rp = mrv \tag{12.5}$$

From Eqs. (12.4) and (12.5)

$$\frac{\mu_l}{l} = \frac{e}{2m} \tag{12.6}$$

$$\frac{\mu_l}{l} = \frac{g_l \beta}{\hbar} \tag{12.7}$$

where $\beta = e \, \hbar /2 \, m = 0.927 \times 10^{-23}$ amp-m^2 $= 0.927 \times 10^{-20}$ erg/gauss is called the Bohr magneton and $g_l = 1$. g_l is called orbital g factor. The ratio of μ_l to l does not depend on the size of the orbit or on the orbital frequency. Since l is quantized and equal to $\hbar \, [l \, (l + 1)]^{1/2}$, hence

$$\mu_l = \beta g_l \sqrt{l(l + 1)} \tag{12.8}$$

Since electron charge is negative, Eq. (12.6) is written as

$$\mu_l = -\frac{el}{2m} = -\frac{g_l \beta l}{\hbar} \tag{12.9}$$

because of the negative charge of the electron, the direction of μ_l is opposite to that of l.

12.1.2 Spin Magnetic Moment

The classical picture of an electron spinning on its own axis indicates that there is a magnetic moment associated with this angular momentum also. The magnetic moment due to spin is given by

$$\mu_s = -\frac{g_s \beta s}{\hbar} = -g_s \beta \sqrt{s(s+1)} \tag{12.10}$$

where μ_s is the magnetic moment due to electron spin, g_s is simply referred to as the g value of the electron. The value of g_s was through early experiments, to be exactly 2 but its value was found to be 2.0023.

According to the classical theory, the ratio of magnetic and mechanical moment of an electron in an orbit is given by Eq. (12.9) and the ratio of magnetic and mechanical moment for the spinning electron is given by Eq. (12.10).

In case the atom has more than one valence electron, it is assumed that the interaction between the spin-angular momentum of the electrons on one hand, the interaction between their orbital motions on the other hand is large compared with the interaction between spin and orbital angular momenta of each electron, that is, LS coupling holds. The magnetic moment μ_L coupled with the total orbital angular momentum L of all electrons is then

$$\mu_L = -\frac{e}{2m} L \tag{12.11}$$

The magnetic moment μ_S coupled with the total spin-angular momentum S is then

$$\mu_S = -2 \frac{e}{2m} S \tag{12.12}$$

In the absence of a magnetic field, the vector L and S precess together around their resultant J. When a magnetic field H is applied, L and S couple with it and in the absence of coupling between L and S, the latter precesses independently around H. However, in weak magnetic field, energy corresponding to coupling of L and S with H is smaller than spin–orbit interaction energy. Therefore, under such condition H does not perturb the coupling between L and S. The L and S precess about their resultant J. Because of torque, J precesses around H at a rate small compared with the precession rate of L and S about J.

The total electronic magnetic moment $\mu = \mu_L + \mu_S$ of the atom is not oriented in the same direction as the total angular momentum $J = L + S$ because of different dependence of μ_L and μ_S on L and S, respectively Since L and S precess rapidly about J, μ_L and μ_S precess rapidly as well, causing μ to precess about $-J$ at the same rate. Therefore, the component of μ perpendicular to $-J$ averages to zero and the component parallel to $-J$ remains a constant in magnitude μ_J given by

$$\mu_J = -g \frac{e}{2m} J \tag{12.13}$$

where

$$g = 1 + \frac{J^2 + S^2 - L^2}{2J^2} = 1 + \frac{J(J+1) + S(S+1) - L(L+1)}{2J(J+1)} \tag{12.14}$$

g is called Lande's splitting factor

The magnetic interaction energy resulting from the interaction between the electronic magnetic moment of the atom and an external magnetic field H (directed along the z-axis) is

$$E = -\mu_J \cdot H \tag{12.15}$$

Substituting Eq. (12.14) in Eq. (12.15)

$$E = g\frac{e}{2m}J \cdot H = g\frac{e}{2m}JH \cos(J, H) \tag{12.16}$$

$J\cos (JH)$ is the projection of J on H and equal to J_z. The allowed values of J_z are $M\hbar$, where M is magnetic quantum number and can take values from $+J$ to $-J$, a total of $2J + 1$ values. Equation (12.16) takes the form

$$E = g\frac{e\hbar}{2m}MH = g\beta MH \tag{12.17}$$

where

$$M = J, J - 1, \ldots, -(J - 1), -J \tag{12.18}$$

12.2   DIA, PARA, FERRI and Ferromagnetic Materials

The externally applied field sometimes called the magnetic field strength is designated by **H**. The magnetic induction or magnetic flux density denoted by B represents the magnitude of the internal field strength within a substance that is subject to an **H** field. The units for **B** are tesla. The magnetic field strength and flux density are related according to

$$B = \mu H \tag{12.19}$$

where μ is called the permeability. The permeability has dimension of weber per ampere meter (Wb/A-m) or henries per meter (H/m). In vacuum

$$B_0 = \mu_0 H \tag{12.20}$$

where μ_0 is the permeability of vacuum and it has a value $4\pi \times 10^{-7}$ (1.257×10^{-6})H/m. When a material medium is placed in a magnetic field the medium is magnetized and a magnetic moment is induced. The magnetization vector **M** is magnetic dipole moment per unit volume. The magnetic induction inside the medium is

$$B = \mu_0 H + \mu_0 M \tag{12.21}$$

where $\mu_0 H$ is due to external field and $\mu_0 M$ is due to magnetization of the medium.

The magnetization is induced by the field, then we assume that **M** is proportional to **H**

$$M = \chi H \tag{12.22}$$

The proportionality constant χ is known as magnetic susceptibility of the medium. Here we have taken χ to be isotropic. But real crystals are anisotropic therefore χ is a tensor. Substituting Eq. (12.22) in Eq. (12.21)

$$B = \mu_0 H + \mu_0 \chi H = \mu_0(1 + \chi) H = \mu_0 \mu_r H = \mu H \tag{12.23}$$

where

$$\mu_r = 1 + \chi \quad \text{and} \quad \mu = \mu_0(1 + \chi) \tag{12.24}$$

Diamagnetism

Diamagnetism is very weak form of magnetism and is non-permanent and persists only when an external magnetic field is applied. It is caused by a change in the orbital motion of the electron when an external magnetic field is applied. The magnitude of the induced moment is very small. The direction of the induced magnetic moment is opposite to that of the applied field. The relative permeability is slightly less than unity. The magnetic susceptibility is negative, that is, the magnitude of **B** within the diamagnetic substance is less than that in a vacuum. When a diamagnetic substance is placed between the poles of a strong electromagnet it is attracted towards regions where field is weak. The susceptibility of solid substance is of the order of -10^{-5}. Diamagnetism is found in all substances. Since it is very weak, it can be observed only when other types of magnetism are totally absent. The diamagnetism is observed in ionic and covalent crystals. The other examples are organic solids like benzene, naphthalene, metals like copper, gold, silver and bismuth, atoms with rare gas configuration like helium, neon, argon, etc.

Paramagnetism

Paramagnetism occurs when substances possesses a permanent dipole moment by virtue of incomplete cancellation of electron spin and/or orbital magnetic moments. These magnetic dipole moments are randomly oriented in the absence of magnetic field resulting in no net magnetization. When the magnetic field is applied, the dipole moments try to align with the magnetic field and the induced magnetization tends to increase the magnetic field.

Paramagnetism is generally fairly weak because the alignment forces are relatively small compared with the forces from the thermal motion which try to disturb the alignment. The paramagnetism is sensitive to the temperature, the lower the temperature the stronger the effect. There is more alignment at low temperatures when the effect of thermal motion is less.

Paramagnetism occurs in atom and ions having odd number of electrons since they possess angular momentum and therefore must be paramagnetic. Molecules and molecular ions such as NO and NO_2, have odd number of electrons and are therefore paramagnetic. The molecules such as O_2, although having an even number of electrons, have a ground state with a partially filled molecular shell and is thus paramagnetic. Transition group ions with incomplete $3d$, $4d$, $5d$, $4f$ or $5f$ shell show paramagnetism. However, not all the valence states of these transition metal ions are paramagnetic. The most commonly observed paramagnetic ions are V^{4+}, VO^{2+}, Ti^{3+}, Cr^{3+}, Mn^{2+}, Fe^{3+}, Fe^{2+}, Co^{2+}, Ni^{2+}, Cu^{2+}, Pd^{2+}, Ru^{2+}, Os^{4+}, Gd^{3+}, Eu^{2+}, Mo^{5+}, Ir^{4+} etc. Paramagnetism is also shown by donors and acceptors in semiconductors such as arsenic donor impurities in germanium. Colour centres (e.g. V_k, F centres) are also paramagnetic.

Ferromagnetism

Ferromagnetic materials exhibit parallel alignment of moments resulting in large net magnetization even in the absence of a magnetic field (Fig. 12.1). Unlike paramagnetic materials, the atomic moments in these materials exhibit very strong interactions. These interactions are produced by electronic exchange forces and result in a parallel or antiparallel alignment of atomic moments. Exchange forces are very large, equivalent to a field on the order of 1000T (T). The exchange force is a quantum mechanical phenomenon due to the relative orientation of the spins of two electrons. The elements Fe, Ni and Co and many of their alloys exhibit ferromagnetism. This indicates that it is related to the partially filled $3d$ and $4f$ shells. Two distinct characteristics of ferromagnetic materials are

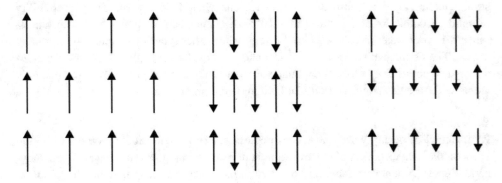

Fig. 12.1 Various types of magnetic ordering. Each arrow indicates a magnetic moment and its orientation **a** ferromagnetic, **b** antiferromagnetic and **c** ferrimagnetic

their (1) spontaneous magnetization and the existence of (2) magnetic ordering temperature known as Curie temperature. Above the Curie temperature the magnetic moments are oriented randomly resulting in a zero net magnetization. In this region the substance is paramagnetic. The spontaneous magnetization is the net magnetization that exists inside a uniformly magnetized microscopic volume in the absence of a field. The magnitude of this magnetization, at 0 K, is dependent on the spin magnetic moments of electrons. A related term is the saturation magnetization which is defined as the maximum induced magnetic moment that can be obtained in a magnetic field (H); beyond this field no further increase in magnetization occurs. Saturation magnetization is an intrinsic property, independent of particle size but dependent on temperature. There is a big difference between paramagnetic and ferromagnetic susceptibility. As compared to paramagnetic materials, the magnetization in ferromagnetic materials is saturated in moderate magnetic fields and at high (room-temperature) temperatures and susceptibility are ~10^5. Even though electronic exchange forces in ferromagnets are very large, thermal energy eventually overcomes the exchange and produces a randomizing effect. This occurs at a particular temperature called the Curie temperature (T_C). Below the Curie temperature, the ferromagnet is ordered and above it, disordered. The saturation magnetization goes to zero at the Curie temperature.

Antiferromagnetism

In antiferromagnetic materials coupling between adjacent atoms or ions causes an antiparallel alignment resulting in vanishing of magnetic moment over a finite volume (Fig. 12.1). For example, manganese oxide (MnO) shows antiferromagnetic behavior. MnO is an ionic compound, having both Mn^{2+} and O^{2-} ions. No net magnetic moment is associated with O^{2-} ions since there is a total cancellation of spin and orbital magnetic moments. The Mn^{2+} ions possess a net magnetic moment that is predominantly due to spin magnetic moment. The Mn^{2+} ions are arranged in the crystal structure such that moments of adjacent ions are antiparallel and thus possesses no net magnetic moment. The antiferromagnetism vanishes above a critical temperature known as Neel temperature.

Ferrimagnetism

In ferrimagnetism, the neighbouring magnetic moments tend to align themselves antiparallel to one another, but their magnitude is not the same, hence there is a net magnetic moment (Fig. 12.1). The examples of ferrimagnetic materials are garnets, ferrite, etc. The ferrimagnetic materials show spontaneous magnetization, remanence and other properties similar to ordinary ferromagnetic materials, but the spontaneous moment does not correspond to the value expected for full parallel alignment of the dipoles.

12.3   Classical Langevin Theory of Diamagnetism

Consider an electron of mass m, charge e moving with a velocity v about the nucleus in a circular orbit of radius r. The frequency of revolution is

$$\omega = \frac{\text{velocity}}{\text{circumference of the orbit}} = \frac{v}{2\pi r} \tag{12.25}$$

The current i is

$$i = e\frac{v}{2\pi r} \tag{12.26}$$

The magnetic moment of the orbiting electron is

$$\mu = i \times \text{Area of the orbit} = i \times \pi r^2 \tag{12.27}$$

Substituting the value of i from Eq. (12.26) in Eq. (12.27)

$$\mu = e\frac{v}{2\pi r} \times \pi r^2 = \frac{1}{2}evr \tag{12.28}$$

Let the magnetic field of induction B is applied perpendicular to the plane of the orbit. Then flux φ in the circuit is

$$\varphi = \pi r^2 B \tag{12.29}$$

If the magnetic flux varies with time, the induced e.m.f. will be produced in the circuit given by

$$V = -\frac{d\varphi}{dt} = -\frac{d}{dt}(\pi r^2 B) = -\pi r^2\frac{dB}{dt} \tag{12.30}$$

The induced e.m.f. is equivalent to an electrostatic field E given by

$$2\pi r E = -\pi r^2\frac{dB}{dt}$$

$$E = -\frac{1}{2}r\frac{dB}{dt} \tag{12.31}$$

The force on the electron due to this electrostatic field is

$$F = eE \tag{12.32}$$

The effect of this force is to cause a change in the velocity by dv and is given by

$$m\frac{dv}{dt} = eE = -\frac{1}{2}er\frac{dB}{dt}$$

$$dv = -\frac{er}{2m}dB \tag{12.33}$$

From Eq. (12.28)

$$v = \frac{2\mu}{cr} \tag{12.34}$$

Differentiating Eq. (12.34)

$$dv = \frac{2}{cr}d\mu = -\frac{er}{2m}dB$$

$$d\mu = \frac{e^2r^2}{4m}dB \tag{12.35}$$

From the definition of magnetic susceptibility

$$\chi = \frac{dM}{dH} = -\frac{\mu_0 e^2 r^2}{4m} \tag{12.36}$$

In an atom of atomic number Z, there are Z electrons. Therefore orbit of each electron should be considered. The magnetic susceptibility of the atom will be Z times the value obtained for a single electron. Since all the electrons may not be rotating in the circular orbit of radius r, to consider the effect, it will be necessary to replace r^2 by an average value $\langle R^2 \rangle$. For orbit in the xy plane

$$\langle r^2 \rangle = \langle x^2 \rangle + \langle y^2 \rangle \tag{12.37}$$

$$\langle R^2 \rangle = \langle x^2 \rangle + \langle y^2 \rangle + \langle z^2 \rangle \tag{12.38}$$

For spherical symmetry

$$\langle x^2 \rangle = \langle y^2 \rangle = \langle z^2 \rangle \tag{12.39}$$

Therefore

$$\langle x^2 \rangle = \langle y^2 \rangle = \langle z^2 \rangle = \frac{1}{3}\langle R^2 \rangle \tag{12.40}$$

And from Eqs. (12.37) and (12.40)

$$\langle r^2 \rangle = \frac{1}{3}\langle R^2 \rangle + \frac{1}{3}\langle R^2 \rangle = \frac{2}{3}\langle R^2 \rangle \tag{12.41}$$

The magnetic susceptibility is thus from Eqs. (12.36) and (12.41)

$$\chi = -Z\frac{\mu_0 e^2}{4m}\frac{2}{3}\langle R^2\rangle \tag{12.42}$$

For N atoms per unit volume

$$\chi = -ZN\frac{\mu_0 e^2}{6m}\langle R^2\rangle \tag{12.43}$$

From Eq. (12.43) it is seen that diamagnetism is independent of temperature and field strength. For larger atoms the diamagnetism is large as diamagnetism is proportional to atomic number Z. The same formula can be obtained from the quantum mechanical approach in which the average value of $\langle R\rangle$ is determined by the wave function and since the susceptibility is proportional to $\langle R^2\rangle$, it follows that the outer electrons make the largest contribution.

12.4   Classical Langevin Theory of Paramagnetism

Paramagnetism occurs in those substances where the individual atoms, ions or molecules possess a permanent magnetic dipole moment $\boldsymbol{\mu}$. When the magnetic dipole moment is placed in an external magnetic field $\boldsymbol{H}$, the magnetic potential energy is

$$E = -\boldsymbol{\mu} \cdot \boldsymbol{H} = -\mu H \cos\theta \tag{12.44}$$

where θ is the angle between $\boldsymbol{\mu}$ and $\boldsymbol{H}$.

The probability that a magnetic dipole is inclined at an angle θ to the magnetic field direction in thermal equilibrium is given by Maxwell–Boltzmann distribution function

$$A\exp\left[-\frac{E}{k_B T}\right] = A\exp\left[\frac{\mu H \cos\theta}{k_B T}\right] \tag{12.45}$$

where A is a constant to be determined from the total number of dipoles.

The number dn of magnetic dipole moments which makes an angle between θ and θ + dθ with $\boldsymbol{H}$ is

$$dn = 2\pi \sin\theta d\theta A\exp\left[\frac{\mu H\cos\theta}{k_B T}\right] \tag{12.46}$$

The total number of dipoles is

$$n = \int dn = 2\pi A \int_0^\pi \sin\theta \exp\left[\frac{\mu H\cos\theta}{k_B T}\right] d\theta \tag{12.47}$$

The mean magnetic moment in the direction of H is

$$\langle \mu \rangle = \int \mu \cos\theta \, dn = 2\pi A\mu \int_0^\pi \sin\theta \cos\theta \exp\left[\frac{\mu H \cos\theta}{k_B T}\right] d\theta \tag{12.48}$$

From Eq. (12.47)

$$A = \frac{n}{2\pi \int_0^\pi \sin\theta \exp\left[\frac{\mu H \cos\theta}{k_B T}\right] d\theta} \tag{12.49}$$

Substituting Eq. (12.49) in Eq. (12.48)

$$< \mu > = n\mu \frac{\int_0^\pi \sin\theta \cos\theta \exp\left[\frac{\mu H \cos\theta}{k_B T}\right] d\theta}{\int_0^\pi \sin\theta \exp\left[\frac{\mu H \cos\theta}{k_B T}\right] d\theta} \tag{12.50}$$

Let

$$\frac{\mu H \cos\theta}{k_B T} = y, \quad -\frac{\mu H \sin\theta}{k_B T} d\theta = dy; \quad \frac{\mu H}{k_B T} = x \tag{12.51}$$

Substituting Eq. (12.51) in Eq. (12.50)

$$< \mu > = \frac{n\mu}{x} \frac{\int_{-x}^x y \exp[y] dy}{\int_{-x}^x \exp[y] dy} = \frac{n\mu}{x} \frac{\int_{-x}^x y \, d[\exp(y)]}{\int_{-x}^x d[\exp(y)]}$$

$$< \mu > = \frac{n\mu}{x} \left[\frac{y \exp(y) - \exp(y)}{\exp(y)}\right]\Bigg|_{-x}^x = \frac{n\mu}{x} \left[\frac{x \exp(x) - \exp(x) + x \exp(-x) + \exp(-x)}{\exp(x) - \exp(-x)}\right]$$

$$< \mu > = n\mu \left[\frac{\exp(x) + \exp(-x)}{\exp(x) - \exp(-x)}\right] - \frac{n\mu}{x} \left[\frac{\exp(x) - \exp(-x)}{\exp(x) - \exp(-x)}\right]$$

$$< \mu > = n\mu \left[\frac{\exp(x) + \exp(-x)}{\exp(x) - \exp(-x)} - \frac{1}{x}\right] = n\mu \left[\coth x - \frac{1}{x}\right] = n\mu L(x) \tag{12.52}$$

where

$$L(x) = \coth x - \frac{1}{x} \tag{12.53}$$

$L(x)$ is called Langevin function. Figure 12.2 shows a plot of $L(x)$ versus x. For high magnetic field strengths and low temperatures, the function approaches the saturation value unity. This corresponds to complete alignment of the magnetic dipoles in the direction of the magnetic field.

If the field strength is not too high and temperature is not too low, we can assume $x \ll 1$ or $\mu H \ll k_B T$. Under these conditions

$$\exp(x) = 1 + x + \frac{x^2}{2!} + \frac{x^3}{3!} + \cdots . \tag{12.54}$$

Fig. 12.2 Langevin function

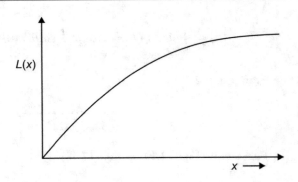

$$\exp(-x) = 1 - x + \frac{x^2}{2!} - \frac{x^3}{3!} + \cdots \qquad (12.55)$$

Substituting Eqs. (12.54) and (12.55) in Eq. (12.53)

$$L(x) = \left[\frac{\exp(x) + \exp(-x)}{\exp(x) - \exp(-x)} - \frac{1}{x} \right]$$

$$= \left[\frac{\left(1 + x + \frac{x^2}{2!} + \frac{x^3}{3!} + \cdots\right) + \left(1 - x + \frac{x^2}{2!} - \frac{x^3}{3!} + \cdots\right)}{\left(1 + x + \frac{x^2}{2!} + \frac{x^3}{3!} + \cdots\right) - \left(1 - x + \frac{x^2}{2!} - \frac{x^3}{3!} + \cdots\right)} - \frac{1}{x} \right]$$

$$L(x) = \left[\frac{2\left(1 + \frac{x^2}{2!} + \cdots\right)}{2\left(x + \frac{x^3}{3!} + \cdots\right)} - \frac{1}{x} \right] = \frac{1}{x}\left(1 + \frac{x^2}{2!} + \cdots\right)\left(1 + \frac{x^2}{3!} + \cdots\right)^{-1} - \frac{1}{x}$$

$$L(x) = \frac{1}{x}\left(1 + \frac{x^2}{2!} + \cdots\right)\left(1 - \frac{x^2}{3!} + \cdots\right) - \frac{1}{x}$$

$$= \frac{1}{x}\left(1 - \frac{x^2}{3!} + \frac{x^2}{2!} - \frac{x^5}{12} + \cdots\right) - \frac{1}{x} = \frac{x}{3} \qquad (12.56)$$

And

$$M = \frac{n\mu x}{3} = \frac{n\mu^2 H}{3k_B T} \qquad (12.57)$$

and

$$\chi = \frac{M}{H} = \frac{n\mu^2}{3k_B T} = \frac{C}{T} \qquad (12.58)$$

where C is known as Curie constant given by

$$C = \frac{n\mu^2}{3k_B} \qquad (12.59)$$

Equation (12.58) is known as Curie's law. This predicts that susceptibility is independent of applied magnetic field.

12.5 Quantum Theory of Paramagnetism

According to quantum theory, the magnetic dipole moment of a given atom or ion is not freely rotating, but can take only discrete orientations with respect to the applied magnetic field as discussed in Sect. 12.1. Suppose there are n atoms per unit volume, and total angular momentum of each atom is J (obtained from combination of total orbital angular momentum L and spin-angular momentum S under LS coupling). The possible components of the magnetic moment according to Eq. (12.18) are

$$Mg\beta \quad \text{where} \quad M = J, J-1, \ldots, -(J-1), -J$$

The magnetic potential energy is given by Eq. (12.44). The magnetization is therefore $M = N \times$ statistical average of the magnetic moment component per atom along H

$$M = N \frac{\sum_{M=-J}^{M=J} Mg\beta \exp\left[\frac{Mg\beta H}{k_B T}\right]}{\sum_{M=-J}^{M=J} \exp\left[\frac{Mg\beta H}{k_B T}\right]} \tag{12.60}$$

Since M can take only discrete values, summation is used in place of integral. Let us consider the case

$$\frac{Mg\beta H}{k_B T} \ll 1 \tag{12.61}$$

That is, at normal field strength (~ 1 T) and temperature (~ 300 K). Under these conditions

$$\exp\left[\frac{Mg\beta H}{k_B T}\right] = 1 + \frac{Mg\beta H}{k_B T} + \cdots \tag{12.62}$$

Substituting Eq. (12.62) in Eq. (12.60)

$$M = Ng\beta \frac{\sum_{M=-J}^{M=J} M\left(1 + \frac{Mg\beta H}{k_B T} +\right)}{\sum_{M=-J}^{M=J} \left(1 + \frac{Mg\beta H}{k_B T} +\right)} \tag{12.63}$$

Since

$$\sum_{M=-J}^{M=J} M^2 = \frac{J(J+1)(2J+1)}{3} \tag{12.64}$$

$$\sum_{M=-J}^{M=J} 1 = (2J + 1) \tag{12.65}$$

$$\sum_{M=-J}^{M=J} M = 0 \tag{12.66}$$

Substituting Eqs. (12.64–12.66) in Eq. (12.63)

$$M = Ng\beta \left[\frac{\frac{g\beta H J(J+1)(2J+1)}{3k_B T}}{(2J+1)} \right] = \frac{Ng^2\beta^2 H J(J+1)}{3k_B T} \tag{12.67}$$

The paramagnetic susceptibility is therefore

$$\chi = \frac{M}{H} = \frac{Ng^2\beta^2 J(J+1)}{3k_B T} \tag{12.68}$$

The result is identical with Eq. (12.58) because the total magnetic moment associated with J is given by

$$\mu_J^2 = g^2\beta^2 J(J+1) \tag{12.69}$$

The relation (12.68) is written as

$$\chi = \frac{N\beta^2 p_{\text{eff}}^2}{3k_B T} \tag{12.70}$$

where

$$p_{\text{eff}} = g\sqrt{J(J+1)} \tag{12.71}$$

The susceptibility is given by Curie law

$$\chi = \frac{C}{T} \tag{12.72}$$

where

$$C = \frac{Ng^2\beta^2 J(J+1)}{3k_B} \tag{12.73}$$

Let us now consider the case when condition (12.61) is not satisfied, that is, at high magnetic fields and at low temperatures. Suppose

$$\frac{g\beta H}{k_B T} = x \tag{12.74}$$

Substituting Eq. (12.74) in Eq. (12.60)

$$M = Ng\beta \frac{\sum_{M=-J}^{M=J} M \exp[Mx]}{\sum_{M=-J}^{M=J} \exp[Mx]} = Ng\beta \frac{d}{dx}\left[\ln\left(\sum_{M=-J}^{M=J} \exp(Mx)\right)\right]$$

$$M = Ng\beta \frac{d}{dx}\left[\ln(\exp(Jx) + \exp(J-1)x + \cdots + \exp(-Jx))\right]$$

$$M = Ng\beta \frac{d}{dx}\left[\ln\left(\frac{\exp(Jx)(1 - \exp(-(2J+1)x))}{1 - \exp(-x)}\right)\right] \tag{12.75}$$

Since $\sum_{M=-J}^{M=J} \exp(Mx)$ is a geometrical progression with $2J + 1$ terms, common ratio $\exp(-x)$ and first term $\exp(Jx)$.

Multiplying numerator and denominator of Eq. (12.75) by $\exp(x/2)$

$$M = Ng\beta \frac{d}{dx}\left[\ln\left(\frac{(\exp(J + \frac{1}{2})x - \exp(-(J + \frac{1}{2})x))}{\exp(\frac{x}{2}) - \exp(-\frac{x}{2})}\right)\right]$$

$$M = Ng\beta \frac{d}{dx}\left[\ln\left(\frac{\sinh(J + \frac{1}{2})x}{\sinh \frac{x}{2}}\right)\right]$$

$$M = Ng\beta \frac{d}{dx}\left[\ln \sinh\left(J + \frac{1}{2}\right)x - \ln \sinh \frac{x}{2}\right]$$

$$M = Ng\beta \left[\frac{2J+1}{2} \coth\left(J + \frac{1}{2}\right)x - \frac{1}{2}\coth \frac{x}{2}\right]$$

$$M = Ng\beta J\left[\frac{2J+1}{2J} \coth\left(\frac{2J+1}{2J}\right)a - \frac{1}{2J}\coth \frac{a}{2J}\right] = Ng\beta J B(a) \tag{12.76}$$

where $x = a/J$ and

$$B(a) = \left[\frac{2J+1}{2J} \coth\left(\frac{2J+1}{2J}\right)a - \frac{1}{2J}\coth \frac{a}{2J}\right] \tag{12.77}$$

$B(a)$ is known as Brillouin function. In the limit $J \to \infty$, $B(a) \to L(x)$ as there are an infinite number of J values, corresponding to an infinite number of moment orientations. Thus summation over a large number of terms can be replaced by integration. $B(a)$ is zero for $a = 0$ and tends towards unity for large a. However the shape of the curve in between depends on the value of J for the atoms involved.

For normal magnetic field strength and at room temperature, a is small and we have

$$\coth\left(\frac{2J+1}{2J}\right)a \to \frac{2J}{(2J+1)a} + \frac{(2J+1)}{2J}\frac{x}{3} \tag{12.78}$$

$$\coth \frac{a}{2J} \to \frac{2J}{a} + \frac{a}{2J}\cdot\frac{1}{3} \tag{12.79}$$

Substituting Eqs. (12.78) and (12.79) in Eq. (12.77)

$$B(a) = \left[\frac{1}{a} + \frac{(2J+1)^2}{4J^2}\frac{a}{3} - \frac{1}{a} - \left(\frac{1}{2J}\right)^2\frac{a}{3} \right] = \frac{J(J+1)a}{3J^2} \qquad (12.80)$$

Let us consider the paramagnetism of rare earth ions and transition ions.

Rare earth ions

Rare earth ions in a crystal are found to obey Curie law with an effective Bohr magneton given by Eq. (12.71). In the rare earth ions, the $4f$ shell is partially filled. The outer $5p$ shell is completely filled while $5d$ and $6s$ shells are still further out are stripped of their electrons to form ionic crystal. Thus only $4f$ is the incomplete shell and due to this magnetic behaviour occurs. The $4f$ shell lies deep within the ion and is screened by the outer $5p$ and $5d$ shell and are not appreciably affected by the crystalline electric field produced by the other ions in the crystal. Thus the rare earth ions behave like free ions. In the calculations of p_{eff}. It is assumed that only the ground state of angular momentum J is occupied.

Iron group ions

The magnetic properties of iron group elements are due to electrons in the incomplete $3d$ shell. For example, the electronic configuration of *Fe* is $3d^5 4s^2$. The electronic configuration of Fe^{3+} is $3d^5$ with the ground state $^6S_{5/2}$. The electrons in the outermost shell $3d$ interact strongly with the crystalline electric field produced by neighbouring ions. The orbital motion is locked into the crystalline electric field and cannot orient itself in an external magnetic field. The electron spin has no direct interaction with the crystalline electric field and thus orient itself freely in the external magnetic field. The orbital motion is thus quenched leaving only the spin moment to contribute to the magnetism. Therefore the effective Bohr magneton is given by $g\sqrt{S(S+1)}$ when J is replaced by S in Eq. (12.71) as orbital angular momentum is quenched.

12.6 Ferromagnetism

In developing the theory of paramagnetism, it was assumed that interactions between different paramagnetic ions were not considered. Such interactions are of two type (A) magnetic dipole–dipole interaction arising from the magnetic field due to one dipole acting on the another (b) exchange interaction between the electrons in different paramagnetic ions of the same nature or between the electrons of different atoms in chemical bonding. The exchange interaction outweighs the magnetic dipole–dipole interaction in ordinary substances. The former interaction decreases more rapidly than the latter as the atomic separation increased. For larger atomic separations (~0.6 nm) both the interactions are small and they can have an appreciable effect on the magnetic properties well below $1\ K$.

The mechanism of exchange forces was proposed by Heisenberg in 1928. The exchange forces are electrostatic in origin and leads to coupling between electronic spins. The interaction energy is

$$E = -2\mathcal{I} s_i \cdot s_j \tag{12.81}$$

The quantity $\mathcal{I}$ is known as exchange energy. The coupling between the spins depends on the cosine of the angle between two spin vectors. For two separate atoms with total spin vectors S_i and S_j we have the interaction energy

$$E = -2\mathcal{I} \sum_i \sum_j s_i \cdot s_j = -2\mathcal{I} \sum_i s_i \sum_j s_j = -2\mathcal{I} S_i \sum_j s_j = -2\mathcal{I} S_i \cdot S_j \tag{12.82}$$

which depends only on the relative orientation of the two total spin vector S_i and S_j. For closed shell for electrons $S = 0$, therefore exchange interaction vanishes.

Consider an ion in which J is a good quantum number. The projection of S on to J is also a constant of motion. On the other hand the components of S perpendicular to J are precessing rapidly and therefore its contribution to the scalar product averaged out to zero. Now we have

$$g J = L + 2S \tag{12.83}$$

where is given by Eq. (12.14) and

$$J = L + S \tag{12.84}$$

From Eqs. (12.83) and (12.84)

$$S = (g - 1)J \tag{12.85}$$

Substituting Eq. (12.85) in Eq. (12.82)

$$E = -2\mathcal{I} S_i \cdot S_j = -2\mathcal{I}(g - 1)^2 J_i \cdot J_j = -2\mathcal{I}' J_i \cdot J_j \tag{12.86}$$

In a solid, any magnetic ion is surrounded by other magnetic ions with each of the magnetic ion will have an exchange interaction. The total interaction is therefore

$$E_i = -2 J_i \sum_j J_{ij} \cdot' J_j \tag{12.87}$$

The magnetic dipole moment of each ion is proportional to angular momentum and is given by

$$\mu = g\beta J \tag{12.88}$$

If all ions have the same g, then exchange energy is

$$E_i = -2\left(\frac{\mu_i}{g\beta}\right) \sum \mathcal{I}'_{ij}\left(\frac{\mu_j}{g\beta}\right) \tag{12.89}$$

In the presence of magnetic field, each ion of a paramagnetic or ferromagnetic substance will have an average magnetic dipole moment along the direction of magnetization while the fluctuating components in other direction will be averaged out as at any instant the contributions from different neighbours will be as often positive as negative. Thus replacing the vector sum over the neighbouring dipole moments by a sum over the average moment per neighbours μ_j and further assuming that all ions are identical, the only important interaction is with z equivalent neighbours (having the same exchange interaction energy $\mathcal{I}'$) we have

$$E_i = -2\left(\frac{\mu}{g\beta}\right) \sum \mathcal{I}'\left(\frac{\mu_j}{g\beta}\right) = -2\left(\frac{\mu}{g\beta}\right)\mathcal{I}'\left(\frac{\mu}{g\beta}\right) = -\frac{2z\mathcal{I}'}{Ng^2\beta^2}\boldsymbol{\mu}\cdot\boldsymbol{M} = -\boldsymbol{\mu}\cdot\boldsymbol{H}_{int}$$
$$\tag{12.90}$$

where $\boldsymbol{M} = N\boldsymbol{\mu}$, N is number of ions per unit volume and

$$\boldsymbol{H}_{int} = \frac{2z\mathcal{I}'}{Ng^2\beta^2}\boldsymbol{M} = \lambda\boldsymbol{M} \tag{12.91}$$

The effect of exchange forces is thus represented by effective internal field $\boldsymbol{H}_{int}$ which is proportional to intensity of magnetization.

The total magnetic field acting on an ion is thus

$$\boldsymbol{H}_{eff} = \boldsymbol{H} + \lambda\boldsymbol{M} \tag{12.92}$$

where $\boldsymbol{H}$ is the external magnetic field. For small magnetization compared with the saturation value and assuming Curie law (12.58) still holds we have by replacing $\boldsymbol{H}$ by $\boldsymbol{H}_{eff}$ in Eq. (12.58)

$$M = \frac{C}{T}H_{eff} = \frac{C}{T}(H + \lambda M)$$

$$M\left(1 - \frac{C\lambda}{T}\right) = \frac{C}{T}H$$

$$\chi = \frac{M}{H} = \frac{C}{T - C\lambda} = \frac{C}{T - \theta} \tag{12.93}$$

where θ is Weiss constant given by

$$\theta = \lambda C \tag{12.94}$$

The relation (12.93) is known as Curie–Weiss law, representing the behavior of paramagnetic substances for $T > \theta$. For $T = \theta$, the susceptibility is infinite. Since χ and maximum value of M is finite (which is obtained when all the magnetic dipoles are aligned parallel to one another) and thus χ is infinite if H is zero or the substance is magnetized even in the absence of external magnetic field. The spontaneous magnetization due to internal field is a characteristic of ferromagnetism and the temperature θ is boundary between paramagnetic behaviour at $T > \theta$ and ferromagnetic behaviour when $T < \theta$.

The temperature below which spontaneous magnetism appears is known as the Curie point. The constant C is given by Eq. (12.73). The Weiss constant is

$$\theta = \lambda C = \frac{2z\mathcal{I}'}{Ng^2\beta^2} \times \frac{Ng^2\beta^2 J(J+1)}{3k_B} = \frac{2z\mathcal{I}'J(J+1)}{3k_B} \tag{12.95}$$

The sign of θ is same as that of $\mathcal{I}'$. Thus a positive value of $\mathcal{I}'$ is required to give a vanishing denominator in Eq. (12.93) and a state in which the electron spins are parallel to each other. The exchange coupling gives a lower energy for any pair of electrons when their spins are parallel, provided $\mathcal{I}'$ is positive. If $\mathcal{I}'$ is negative, the state of lower energy corresponds to antiparallel spins, θ is also negative and the denominator of Eq. (12.93) does not vanish. Nevertheless a basic arrangement of antiparallel spins occurs. This phenomenon is called antiferromagnetism.

12.7 Weiss Theory of Spontaneous Magnetization

The internal field in a ferromagnetic substance is very large and magnetism approaches its saturation value even at room temperature. The assumption that the magnetization is small and proportional to the effective is used in deriving

$$\chi = \frac{C}{T - \theta}$$

for susceptibility above $T > \theta$ cannot be used below the Curie point.

We will retain the concept of internal field in calculating the magnetization by using the Brillouin function. From Eq. (12.76)

$$M = Ng\beta J B(a) = M_s B(a)$$

$$M_s = Ng\beta J \tag{12.96}$$

$$\frac{M}{M_s} = B(a) \tag{12.97}$$

where M_s is saturation magnetization per unit volume and

$$a = \frac{g\beta J H_{\text{eff}}}{k_B T} = \frac{M_s H_{\text{eff}}}{N k_B T} = \frac{M_s(H + \lambda M)}{N k_B T} \tag{12.98}$$

$$M_s H + \lambda M M_s = a N k_B T \tag{12.99}$$

Dividing by λM_s^2

$$\frac{M}{M_s} = \frac{a N k_B T}{\lambda M_s^2} - \frac{H}{\lambda M_s} \tag{12.100}$$

The value of the magnetization under any given condition of H_{eff} and T may be found by eliminating the parameter a between the Eqs. (12.97) and (12.100). However, this cannot be done analytically, but the general behavior of the magnetism can be found from a graphical solution. For $H = 0$, Eq. (12.100) gives

$$\frac{M}{M_s} = \frac{a N k_B T}{\lambda M_s^2} \tag{12.101}$$

A plot of M/M_s versus a (Fig. 12.3) gives a straight line which passes through the origin and intersects the curve given by Eq. (12.97) at this point. Thus one possible value of magnetization is always zero. If T is high, the slope of

$$\frac{M_0}{M_s} = \frac{a N k_B T}{\lambda M_s^2} \tag{12.102}$$

is so great that this is only point of intersection and the substance must therefore be unmagnetized in zero magnetic field. This corresponds to the paramagnetic behavior for $T > \theta$. As T is lowered the slope of the line given by Eq. (12.101) decreases until at a certain temperature T_c it is tangential to the curve of given by Eq. (12.97) at the origin. For small a

Fig. 12.3 Graphical solution of the Eqs. (12.97) and (12.100) for spontaneous magnetization

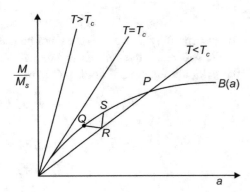

$$B(a) = \frac{a(J+1)}{3J} \tag{12.102}$$

On equating this value to given by Eq. (12.101)

$$\frac{aNk_BT_C}{\lambda M_s^2} = \frac{a(J+1)}{3J}$$

$$T_C = \frac{\lambda M_s^2(J+1)}{3JNk_B} \tag{12.103}$$

Substituting the value of M_s from Eq. (12.96)

$$T_C = \frac{\lambda(Ng\beta J)^2(J+1)}{3JNk_B} = \frac{\lambda Ng^2\beta^2 J(J+1)}{3k_B} = C\lambda = \theta \tag{12.104}$$

where θ is the Weiss constant.

At still lower temperatures, the slope of the line is less than the initial slope of the B (12.97) and there will be two points of intersection and two possible value of magnetization, one zero and the other finite. If we imagine the magnetization at any instant to correspond to the point Q in B (12.97) then the internal field produced by the magnetization corresponds to the point R and this field produces the greater magnetization correspond to point S on B (12.97). Thus the magnetization will increase until the point P is reached where the two curves intersect. Above P, the curves cross and further increase in the magnetization would produce an internal field insufficient to sustain the increase magnetization. It thus appears that the state of spontaneous magnetization corresponding to the point P is stable while unmagnetized state is unstable.

The value of spontaneous magnetization is determined by the intersection of B (12.97) of the line corresponding to the Eq. (12.101) and the slope of the line depends on the temperature, it is obvious the whole of the curve B (12.97) will be traced out as we lower the temperature from T_C to 0 K. From Eq. (12.96)

$$\frac{\lambda M_s^2}{Nk_B} = \frac{3\theta J}{J+1} \tag{12.105}$$

and thus

$$\frac{M_0}{M_s} = a\left(\frac{T}{\theta}\right)\left(\frac{J+1}{3J}\right) \tag{12.106}$$

Eliminating a between Eq. (12.106) and Eq. (12.97)

$$\frac{M_0}{M_s} = f\left(\frac{T}{\theta}\right) \tag{12.107}$$

Fig. 12.4 A plot of M_0/M_s versus T/θ

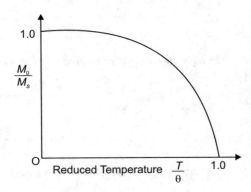

where $f\left(\frac{T}{\theta}\right)$ is same for all substances with the same value of J. This function is plotted in Fig. 12.4 for the special case of $J = \frac{1}{2}$.

12.8 Ferromagnetic Domains

Weiss theory is able to explain the spontaneous magnetization in a ferromagnetic substance. It is not able to explain the fact that unmagnetized state is stable as is well known that a piece of iron can be demagnetized on dropping it. Further, in a single crystal the magnetization can be restored by applying an external field of less than 10^{-4} T, although the internal field is about 10^3 T. Some explanation for hysteresis is also needed. To overcome these difficulties Weiss introduced the concept of domains of magnetization within the solid. According to Weiss, any ferromagnetic material is composed of small volume regions in which there is a mutual alignment in the same direction of all magnetic dipole moments. Such a region is called domain, and each one is magnetized to its saturation magnetization. Adjacent domains are separated by domain boundaries or walls, across which the direction of magnetization gradually changes. This boundary or wall is known as Bloch wall. Each domain contains about 10^{17}–10^{21} atoms and a piece of unmagnetized domain contains many domains all spontaneously magnetized, but the direction of magnetization of different domains is oriented at random. These domains are aligned in the presence of external magnetic field.

In the presence of an external magnetic field, the domains that have net magnetic moments parallel to the direction of the field have their energy reduced, whereas those domains that do not have their energy increased. The crystal's energy can be lowered if all the domains align themselves parallel to the applied magnetic field. This can be obtained in the following ways:

(i) The direction of magnetization of an entire domain changes at once

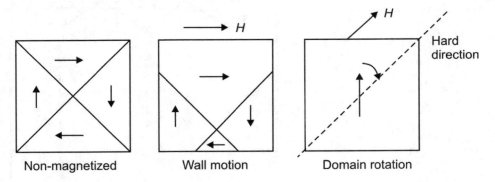

Fig. 12.5 Domain structure

(ii) A domain that is favourably oriented grows in size at the expense of the less favourable oriented domains (Fig. 12.5)

(iii) In strong magnetic applied fields the magnetization rotates towards the direction of the field (Fig. 12.5)

The magnetization of single crystals along certain direction is invariables accompanied by change in their physical dimensions. It can expand or contract along the direction of magnetization depending on the crystal. For iron it expands while for *Ni* it contracts. This phenomenon is called magnetostriction.

The thickness of Bloch wall is not infinitely small, but it has a finite value, that is, the spin orientation changes gradually in the transition region (Fig. 12.6). In this manner the spin reversal is accomplished over a number of steps and hence the spin orientation between two neighbouring moments is rather small. This leads to reduction in exchange energy associated with the wall.

It is observed that in a ferromagnet, there are crystallographic direction of easy magnetization in which saturation is reached spontaneously and direction of hard magnetization in which an external field is needed to reach the saturation of magnetization. For Ni, the direction of easy magnetization is [111] and hard magnetization is [100] while for

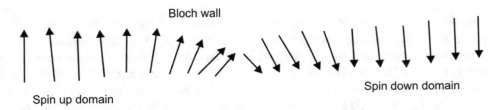

Fig. 12.6 Bloch wall

Fig. 12.7 Closure domains

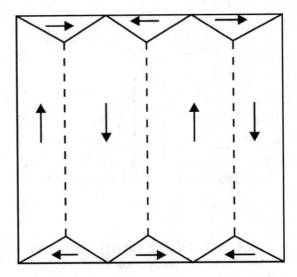

iron reverse is true. This phenomena is known as magnetic anisotropy. The difference in energy between the easy and hard direction is called the magnetic anisotropy energy. The effect of this energy on the Bloch wall is to reduce its thickness, because the thicker the wall, the more dipoles point in the hard direction. The exchange energy favours a thick wall, and anisotropic energy favours a thin wall.

Close examination of the domain structure reveals the presence of small transverse domains near the end of the sample (Fig. 12.7). These are called closure domains, as they have effect of closing the magnetic loop between two adjacent domains, resulting in a further decrease in magnetostatic energy. The experimental evidence for the existence of domains is provided by 'Bitter powder patterns' which are obtained by putting a drop of a colloidal suspension of ferromagnetic particles on the carefully prepared surface of the specimen. There are strong local magnetic fields near the domain boundaries. The particles are collected there, and the domain may be observed under a microscope.

12.9   *B-H* Curve

Flux density B and the field density H are not proportional for ferromagnets and ferrimagnets. If the material is initially unmagnetized, then B varies as a function of H. Ferromagnetic substances are characterized by Curie point at which its property changes abruptly. Above Curie temperature the susceptibility is independent of field strength and follows approximately a Curie–Weiss law (Eq. 12.93). Below the Curie temperature the behaviour is quite different. Very large values of magnetization are produced by quite small fields, and magnetization varies quite nonlinearly with the field strength. Figure 12.8

Fig. 12.8 *B* versus *H* hysteresis curve

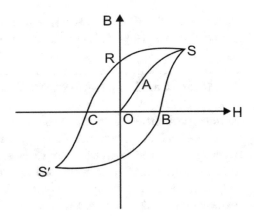

shows a characteristic plot of magnetic induction *B* as a function of field H in a sample of iron.

If iron is initially unmagnetized, and a field of slowly increasing magnitude is applied, *B* follows the path OAS known as magnetization curve. In a field of a few hundred ampere/meter the value of *B* is practically constant at about 1.5 Wb/m². If the magnetic field H is now reduced, the induction *B* does not return along the magnetization curve but follows the path SR, and even at $H = 0$, corresponding to the point *R* in the Fig. 12.8, *B* is still near the saturation value. The value of *B* at this point is known as the residual induction, and the retention of magnetization in zero field is known as remanence.

On applying a reverse field the value of *B* falls and finally becomes zero (point *C*); the value of the field reverse induction is set up which quickly reaches the saturation value. The reverse field H at which B = 0 is called coercivity. Finally if the reverse field is gradually removed and a positive field applied the induction traces out the curve in the direction S'BS. This curve RCS'BSR is called the hysteresis curve. It shows that the change in the magnetic induction (**B**) is always behind the change in the applied magnetic field.

The *B-H* curve can be understood on the basis of domain theory of Weiss. In an unmagnetized polycrystalline specimen domains are oriented at random, so that there is no resultant magnetic moment in any direction. When a field is applied, domains where magnetization is parallel or at a small angle with the field grow at the expenses of those where the magnetization is antiparallel or nearly so, so that the boundary between domains is displaced. Initially (*OA* in the *B-H* curve) the magnetization of the substance as a whole proceeds by small (reversible) boundary displacement, but the steeper part *AB* of the magnetization curve is due to larger (irreversible) displacements. Above the knee of the curve, magnetization proceeds by rotation of the direction of magnetization of whole domains; such a process is rather difficult and the increase in the magnetization is relatively slow. When the applied field is reduced, there is little change in the domain structure so that the magnetization remains quite high until reverse fields are applied, thus giving rise to the hysteresis described above. Area of *B-H* curve is proportional to work that should be used to remagnetize a ferromagnet of unit volume. In the course of remagnetization, the work is completely transformed into heat.

12.10 Soft and Hard Magnetic Materials

Ferromagnetic materials are classified as either soft or hard on the basis of their hysteresis characteristic. Soft magnetic materials are used in devices that are subjected to alternating magnetic fields and in which energy losses must be low for example transformer core. Therefore, the relative area within the hysteresis loop must be small; it is characteristically thin and narrow. Consequently, a soft magnetic material must have a high initial permeabilities and a low coercivity so that a material possessing these properties may reach its saturation magnetization with a relatively low applied magnetic field and still has low hysteresis energy losses. A low value of coercivity corresponds to the energy movement of domain walls as the magnetic field changes magnitude and/or direction. The energy losses may also result from the electric currents that are induced in a magnetic material by a magnetic field that varies in magnitude and direction with time; these are called eddy current. Therefore, it is desirable to minimize these energy losses in soft magnetic materials by increasing the electrical resistivity. In ferromagnetic materials this is accomplished by forming solid solution alloys iron-silicon and iron nickel alloys. The hysteresis characteristic of soft magnetic materials may be enhanced for some application by an appropriate heat treatment in the presence of a magnetic field. Using such a technique, a square hysteresis loop may be produced which is desirable in some magnetic applications. Soft magnetic materials are also used in generators, motors dynamos and switching circuits. Examples of soft magnetic materials are supermalloy (an alloy of Ni, Fe, Mo, Mn), 45 permalloy (55Fe, 45Ni).

Hard magnetic materials are used in permanent magnets which must have a high resistance to demagnetization. In terms of B-H curve, a hard magnetic material has a (a) high remanence, (b) high coercivity, (c) high saturation flux density, (d) low initial permeability and (e) high hysteresis energy losses, that is, high value of the area of B-H curve. To obtain hard material domain wall movement should be difficult. Examples of hard magnetic materials are Magnico (iron based material containing 24% Co, 14% Ni, 8% Al and 3% Cu; residual magnetic induction is 1.35T and coercivity is ~5x10^5 A/m) high carbon steels, low alloy tungsten, chromium steels, etc.

Solved Examples

Example 1 Why $KCr(SO_4)_2.12H_2O$ is paramagnetic?

Solution

In the solid state the paramagnetism occurs in salts of the transition group ions and the magnetic moment is associated with the metallic ion itself. Thus in $KCr(SO_4)_2.12H_2O$

only the Cr^{3+} ion have permanent magnetic moment, the other ions ($K+$, $SO_4{}^{2-}$) and water molecules giving only a diamagnetic contribution to the susceptibility.

Example 2 Obtain p_{eff}^2. for $(NH_4)_2Ni(SO_4)_2.6H_2O$. The ground state of Ni^{2+} is $^3F_4$. The experimental value is 9.7.

Solution

From Eq. (12.71)

$$p_{eff} = g\sqrt{J(J+1)}$$

From the ground state of the ion

$$2S + 1 = 3, \quad \text{hence} \quad S = 1; \, J = 4; \quad \text{for} \quad F \quad \text{state} \quad L = 3,$$

Substituting these values in Eq. (12.14)

$$g = 1 + \frac{J(J+1) + S(S+1) - L(L+1)}{2J(J+1)} = 1 + \frac{4(4+1) + 1(1+1) - 3(3+1)}{2 \times 4(4+1)}$$

$$= 1 + \frac{1}{4} = \frac{5}{4}$$

and

$$p_{eff}^2 = g^2 J(J+1) = \left(\frac{5}{4}\right)^2 \times 4(4+1) = 31.3$$

However this value of p_{eff}^2 is quite different from those of the experimental value. *Ni* belongs to $3d$ group and in the crystal its orbital angular momentum is quenched by the crystalline electric field and only spin moment contributes to paramagnetism. Therefore, we should use spin contribution only. For such cases

$$p_{eff}^2 = 4S(S+1) = 4 \times 1(1+1) = 8$$

which is close to the experimental value.

Example 3 Obtain p_{eff} for Nd^{3+} ion whose ground state is $^4I_{9/2}$ and experimentally determined value of p_{eff} is 3.5.

Solution

From Eq. (12.71)

$$p_{\text{eff}} = g\sqrt{J(J+1)}$$

From the ground state of the ion

$$2S + 1 = 3 = 4, \quad \text{hence} \quad S = 3/2; \ J = 9/2; \quad \text{for } I \text{ state } L = 6$$

Substituting these values in Eq. (12.14)

$$g = 1 + \frac{J(J+1) + S(S+1) - L(L+1)}{2J(J+1)}$$

$$= 1 + \frac{\frac{9}{2}(\frac{9}{2}+1) + \frac{3}{2}(\frac{3}{2}+1) - 6(6+1)}{2 \times \frac{9}{2}(\frac{9}{2}+1)} = 1 + \frac{99 + 15 - 168}{198} = 0.7272$$

From Eq. (12.71)

$$p_{\text{eff}} = 0.7272\sqrt{6(6+1)} = 3.617$$

which is close to experimental value.

Example 4 Derive an expression for paramagnetic susceptibility of conduction electrons (Pauli Paramagnetism).

Solution

The paramagnetism of conduction electrons can be explained on the basis of free electron theory. The electron gas obeys Fermi Dirac distribution. When such a gas is placed in a magnetic field, the energy of the electron with spin along it is

$$E = \frac{p^2}{2m} - \mu H$$

The energy of electron with spin antiparallel to magnetic field is then

$$E = \frac{p^2}{2m} + \mu H$$

where μ is the magnetic moment of the electron. At $T = 0$ K, all levels up to Fermi level will be filled up. The kinetic energy of the electrons with spin parallel to magnetic field will be

$$E_F + \mu H$$

and the kinetic energy of the electrons with spin antiparallel to the magnetic field will be

$$E_F - \mu H$$

The number N_1 of electrons with spin parallel to the magnetic field is

$$N_1 = \frac{V}{4\pi^2 \hbar^3}(2m)^{\frac{3}{2}} \int_0^{E_F + \mu H} E^{\frac{1}{2}} dE$$

$$N_1 = \frac{V}{4\pi^2 \hbar^3}(2m)^{\frac{3}{2}} \frac{E^{\frac{3}{2}}}{\frac{3}{2}}\Big|_0^{E_F + \mu H} = \frac{V}{6\pi^2 \hbar^3}(2m)^{\frac{3}{2}}(E_F + \mu H)^{\frac{3}{2}}$$

Similarly the N_2 of electrons with spin antiparallel to the magnetic field is

$$N_2 = \frac{V}{6\pi^2 \hbar^3}(2m)^{\frac{3}{2}}(E_F - \mu H)^{\frac{3}{2}}$$

The net magnetic moment is

$$M = \mu(N_1 - N_2) = \frac{\mu V}{6\pi^2 \hbar^3}(2m)^{\frac{3}{2}}\left[(E_F + \mu H)^{\frac{3}{2}} - (E_F - \mu H)^{\frac{3}{2}}\right]$$

$$M = \frac{\mu V}{6\pi^2 \hbar^3}(2m)^{\frac{3}{2}} E_F^{\frac{3}{2}}\left[\left(1 + \frac{\mu H}{E_F}\right)^{\frac{3}{2}} - \left(1 - \frac{\mu H}{E_F}\right)^{\frac{3}{2}}\right]$$

For small value of H

$$M = \frac{\mu V}{6\pi^2 \hbar^3}(2m)^{\frac{3}{2}} E_F^{\frac{3}{2}}\left[1 + \frac{3}{2}\frac{\mu H}{E_F} - 1 + \frac{3}{2}\frac{\mu H}{E_F}\right] = \frac{\mu^2 V}{2\pi^2 \hbar^3}(2m)^{\frac{3}{2}} E_F^{\frac{1}{2}} H$$

The susceptibility is therefore

$$\chi = \frac{M}{VH} = \frac{\mu^2}{2\pi^2 \hbar^3}(2m)^{\frac{3}{2}} E_F^{\frac{1}{2}}$$

From the relation of Fermi energy

$$E_F = \frac{\hbar^2}{2m}\left(\frac{3\pi^2 N}{V}\right)^{\frac{2}{3}}$$

We have

$$2m = \frac{\hbar^2}{E_F}\left(\frac{3\pi^2 N}{V}\right)^{\frac{2}{3}}$$

substituting this value of m in the expression for susceptibility

$$\chi = \frac{\mu^2}{2\pi^2} \frac{1}{E_F} \left(\frac{3\pi^2 N}{V} \right) = \frac{3}{2} \frac{\mu^2 n}{E_F}$$

where $n = N/V$.

The susceptibility is independent of temperature and depends on the density of electron gas.

Objective Type Questions

1. Basic source of magnetism

    (a) Charged particles alone (b) Movement of charged particles (c) Magnetic dipoles (d) Magnetic domains
2. Units for magnetic flux density
    (a) Wb/m^2 (b) $Wb/A.m$ (c) A/m (d) $Tesla/m$
3. Magnetic permeability has units as
    (a) Wb/m^2 (b) $Wb/A.m$ (c) A/m (d) $Tesla/m$
4. Magnetic permeability has units as
    (a) $Tesla$ (b) $Henry$ (c) $Tesla/m$ (d) $Henry/m$
5. Magnetic field strength's units are
    (a) Wb/m^2 (b) $Wb/A.m$ (c) A/m (d) $Tesla/m$
6. Example for diamagnetic materials
    (a) superconductors (b) alkali metals (c) transition metals (d) Ferrites
7. Example for paramagnetic materials
    (a) superconductors (b) alkali metals (c) transition metals (d) Ferrites
8. Example for ferromagnetic materials
    (a) super conductors (b) alkali metals (c) transition metals (d) Ferrites
9. Example for antiferromagnetic materials
    (a) salts of transition elements (b) rare earth elements (c) transition metals (d) Ferrites
10. Example for ferrimagnetic materials
    (a) salts of transition elements (b) rare earth elements (c) transition metals (d) Ferrites
11. Magnetic susceptibility paramagnetic materials is
    (a) $+10^{-5}$ (b) -10^{-5} (c) 10^5 (d) 10^{-5} to 10^{-2}
12. Magnetic susceptibility diamagnetic materials is
    (a) $+10^{-5}$ (b) -10^{-5} (c) 10^5 (d) 10^{-5} to 10^{-2}
13. Magnetic susceptibility ferromagnetic materials is
    (a) $+10^{-5}$ (b) -10^{-5} (c) 10^5 (d) 10^{-5} to 10^{-2}
14. Typical size of magnetic domains
    (a) 1–10 mm (b) 0.1–1 mm (c) 0.05 mm (d) 0.001 mm
15. Typical thickness of Bloch walls _____ (nm).
    (a) 0.1–1 nm (b) 1–10 nm (c) 10–50 nm (d) 100 nm

16. Example for soft magnets
    (a) 45 Permalloy (b) CrO_2 (c) Fe–Pd (d) Alnico
17. Example for hard magnet
    (a) 45 Permalloy (b) CrO_2 (c) Fe–Pd (d) Alnico
18. Example for magnetic material used in data storage devices
    (a) 45 Permalloy (b) CrO_2 (c) Cunife (d) Alnico
19. If the magnetic moments of neighbouring atoms spontaneously align, we may not observe
    (a) ferromagnetic behaviour.
    (b) antiferromagnetic behaviour.
    (c) ferrimagnetic behaviour.
    (d) paramagnetic behaviour.
20. Magnetic domains are
    (a) surface'-near regions of a bulk ferromagnet where the magnetization direction changes somewhat because of the surface.
    (b) regions in a bulk ferromagnet with a net magnetization different from neighbouring domains.
    (c) something induced in a ferromagnetic material as soon as a magnetic field is switched on.
    (d) a fine structure of the internal magnetization required because of entropy concerns.
21. Without magnetic domains
    (a) the entropy of the material would be too low.
    (b) the enthalpy of the material would be too low.
    (c) the enthalpy of the material would be too large.
22. If a ferromagnetic material is exposed to an *increasing* magnetic field, the following effects will be observed:
    (a) all magnetization vectors in the domains turn increasingly in field direction.
    (b) first, the magnetization vectors in 'fitting' domains (=relatively small misalignment) turn in external field direction; then the ones with larger misalignment.
    (c) 'fitting' domains grow, 'unfitting' domains shrink while first keeping their magnetization direction; finally the magnetization turns in field direction.
    (d) none of these
23. You are given a material which produces no initial magnetic field when in free space. When it is placed in a region of uniform magnetic field, the material produces an additional internal magnetic field parallel to the original field. However, this induced magnetic field disappears when the external field is removed. What type of magnetism does this material exhibit?
    (a) diamagnetism (b) paramagnetism (c) ferromagnetism (d) none of these
24. Once again, you are given an unknown material that initially generates no magnetic field. When this material is placed in a magnetic field, it produces a strong internal

magnetic field, parallel to the external magnetic field. This field is found to remain even after the external magnetic field is removed. Your material is which of the following?

(a) diamagnetic

(b) paramagnetic

(c) ferromagnetic

25. What type of magnetism is characteristic of most materials?

(a) ferromagnetism

(b) paramagnetism

(c) diamagnetism

(d) no magnetism

26. Which is not paramagnetic?

(a) $KFe(SO_4)_2.12H_2O$

(b) $(NH_4)_2Zn(SO_4)_2.6H_2O$

(c) $(NH_4)_2Mn(SO_4)_2.6H_2O$

(d) $KCr(SO_4)_2.12H_2O$

Problems

1. Briefly describe the phenomenon of magnetic hysteresis and why it occurs for ferromagnetic materials.

2. Cite the difference between hard and soft magnetic materials.

3. Do all atoms have a net magnetic moment? Why or why not?

4. Distinguish between diamagnetism, paramagnetism and ferromagnetism. Derive an expression for diamagnetic susceptibility on the basis of classical theory.

5. Distinguish between ferromagnetic, ferromagnetic and antiferromagnetic materials.

6. Discuss quantum theory of paramagnetism and explain how it removes the shortcoming of the Langevin's theory.

7. Obtain p_{eff}^2 for $Cr(SO_4)_2.12H_2O$. The ground state of Cr^{3+} is $^4F_{3/2}$. The experimental value is 14.8.

8. Obtain p_{eff} for Dy^{3+} ion whose ground state is $^6H_{15/2}$ and experimentally determined value of p_{eff} is 10.6.

Answers
Objective Type Questions

1. (b)	2. (a)	3. (b)	4. (d)	5. (c)	6. (a)	7. (b)	8. (c)
9. (a)	10. (d)	11. (d)	12. (b)	13. (c)	14. (c)	15. (d)	16. (a)
17. (d)	18. (b)	19. (d)	20. (b)	21. (c)	22. (c)	23. (b)	24. (b)
25. (c)	26. (b)						

Magnetic Resonance

13

13.1 Electron Spin Resonance

Electron spin resonance (*ESR*) and electron paramagnetic resonance (*EPR*) are synonymous terms. This is an important method for obtaining information about paramagnetic substances. Essentially it forms a branch of high-resolution spectroscopy using frequencies in the microwave region ($v \sim 10^9 - 10^{11}$ Hz). *ESR* has been defined as the form of spectroscopy concerned with microwave-induced transitions between magnetic energy levels of electron having a net angular momentum. *ESR* differs from simple microwave spectroscopy in being concerned with paramagnetic materials.

Since *ESR* requires the presence of the unpaired electrons in the sample being studied; its range of applications is restricted to paramagnetic substances and to substances that can be converted to a paramagnetic form with sufficient stability for a spectrum to be observed. Paramagnetism occurs in atom and ions having odd number of electrons since they possess angular momentum and therefore must be paramagnetic. Molecules and molecular ions such as NO and NO_2 have odd number of electrons and are therefore paramagnetic. The molecules, such as O_2 although having an even number of electrons, have a ground state with a partially filled molecular shell and are thus paramagnetic. Transition group atoms or ions with incomplete 3*d*, 4*d*, 5*d*, 4*f* or 5*f* shell show *ESR*. However, not all the valence states of these transition metal ions are paramagnetic. Donors and acceptors in semiconductors such as arsenic donor impurities in Germanium. Colour centres (e.g. V_k, *F* centres). Organic and inorganic radicals. Activators and coactivators in phosphors, such as self-activated ZnS, radiation damage centres and conduction electrons.

© The Author(s) 2022
V. K. Jain, *Solid State Physics*, https://doi.org/10.1007/978-3-030-96017-9_13

13.1.1 Resonance Condition

For a free electron, the only magnetic moment is that which is associated with the spin. Classically the energy of interaction between the magnetic moment μ_e and magnetic field **B** is

$$E = -\mu_e \cdot \mathbf{B} \tag{13.1}$$

The magnetic moment is a vector, which is collinear with spin-angular momentum vector **S**. The operators for the vectors are related by

$$\mu_e = -\gamma \mathbf{S} \tag{13.2}$$

where γ is the magnetogyric ratio, or the ratio between magnetic and mechanical moments. The value of γ is given by g_e $(e/2m)$, g_e is positive dimensionless factor (2); however, quantum electrodynamical calculations show it to be 2.0023. μ_e is antiparallel to **S** because of the negative charge of the electron.

Equation (13.2) can be written as

$$\mu_e = -g_e \beta \mathbf{S} \tag{13.3}$$

To obtain the quantum mechanical Hamiltonian, we replace μ_e by the appropriate operator in Eq. (13.1)

$$H = g_e \beta \mathbf{S} \cdot \mathbf{B} \tag{13.4}$$

If the magnetic field is assumed to be in z-direction, $B_x = B_y = 0$ and $B_z = B$ and scaler product becomes

$$H = g_e \beta S_z B \tag{13.5}$$

where β is Bohr magneton given by e $\hbar/2$ m The eigenvalues of Eq. (13.5) are just the multiples of the eigenvalues of S_z and are given by

$$E = g_e \beta B M \tag{13.6}$$

where $M = 1/2$ and $-1/2$.

Substituting the values of M in Eq. (13.6)

$$E = \pm \frac{1}{2} g_e \beta B \tag{13.7}$$

The negative sign corresponds to magnetic moment aligned parallel to the magnetic field and hence spin antiparallel to it. The difference in energy between the two states is given by

Fig. 13.1 Energy level of an
electron in a magnetic field.
Application of microwave
radiation of energy $h\nu$ causes
resonance at a field given by
$h\nu/g_e\beta$

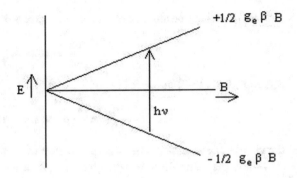

$$\Delta E = h\nu = E_2 - E_1 = g_e\beta B \tag{13.8}$$

The separation between two levels increases linearly with the magnetic field
(Fig. 13.1). Transition between the two levels can be induced by a microwave magnetic
field B_1 at right angle to B. The transitions, which give rise to ESR spectra, are magnetic
dipole in origin and parity does not change. The selection rule for the magnetic quantum
number M is $\Delta M = \pm 1$. This is in contrast to the intrinsically far more strongly allowed
electric dipole transitions observed in electronic spectra where the parity of the states
must differ.

Substituting the values of constants in Eq. (13.8) and $g_e = 2$, we have

$$B(mT) = 35.724\nu(GHz) \tag{13.9}$$

The conventional ESR spectrometer operates at different frequency bands, for example
at X-band ($\nu = 9$ GHz, $\lambda = 3.0$ cm), S-band ($\nu = 3$ GHz, $\lambda = 9.0$ cm), K-band ($\nu =
24$ GHz, $\lambda = 1.2$ cm), Q-band ($\nu = 35$ GHz, $\lambda = 0.8$ cm), E-band ($\nu = 70$ GHz, $\lambda =
0.4$ cm).

Equation (13.8) shows that magnetic field B brings forth the appearance of spin sublevel and determines the energy difference between them. The microwave frequency
energy quantum $h\nu$ causes transitions from $M = -1/2$ state to $M = +1/2$ state, attended
by absorption of energy, and produces an absorption signal. The g value defines the change
in the position of the absorption line in the spectrum under given $h\nu$ and B depending on
the features particular to the state of paramagnetic electron in the study of the specimen,
thus presenting the characteristic of the substances in condition of resonance.

13.1.2 Description of ESR by Precession

When a magnetic dipole is placed in a magnetic field the torque $\mathbf{N}$ acting on it is

$$\mathbf{N} = \mu_e \times \mathbf{B} \tag{13.10}$$

Rate of change of angular momentum is torque, if $\mathbf{S}$ is the angular momentum

$$\frac{d\mathbf{S}}{dt} = \boldsymbol{\mu}_e \times \mathbf{B} \tag{13.11}$$

From Eqs. (13.2) and (13.11)

$$\frac{d\boldsymbol{\mu}_e}{dt} = -\gamma\boldsymbol{\mu}_e \times \mathbf{B} = \gamma\mathbf{B} \times \boldsymbol{\mu}_e \tag{13.12}$$

the change in $\boldsymbol{\mu}_e$ at any time is perpendicular to both $\boldsymbol{\mu}_e$ and $\mathbf{B}$.

Taking magnetic field in z-direction, i.e. $\mathbf{B} = B\mathbf{k}$ and $\boldsymbol{\mu}_e = (\mu_x\mathbf{i} + \mu_y\mathbf{j} + \mu_z\mathbf{k})$, we have

$$\frac{d\boldsymbol{\mu}_e}{dt} = \gamma(B\mathbf{k}) \times (\mu_x\mathbf{i} + \mu_y\mathbf{j} + \mu_z\mathbf{k}) = \gamma(\mu_xB\mathbf{j} - \mu_yB\mathbf{i}) \tag{13.13}$$

since Eq. (13.13) relates two vectors, the components on each side must be identical. Using Eq. (13.13), we have

$$\frac{d\mu_x}{dt} = -\gamma\mu_yB$$

$$\frac{d\mu_y}{dt} = \gamma\mu_xB$$

$$\frac{d\mu_z}{dt} = 0 \tag{13.14}$$

These equations are satisfied by

$$\mu_z = \text{constant}$$

$$\mu_x = \cos\omega_L t$$

$$\mu_y = \sin\omega_L t \tag{13.15}$$

where

$$\omega_L = \gamma B \tag{13.16}$$

This $\boldsymbol{\mu}$ precess about $\mathbf{B}$ with a constant frequency ω_L making a fixed angle with the direction of static magnetic field B (Fig. 13.2). The frequency ω_L is known as Larmor frequency.

Let us consider a second coordinate system (x', y', z') rotating about z-axis at an angular velocity ω (Fig. 13.3). The connection between the laboratory and rotating frames is given by

$$\left(\frac{d\mathbf{S}}{dt}\right)_{\text{lab}} = \left(\frac{d'\mathbf{S}}{dt}\right)_{\text{rot}} + \boldsymbol{\omega} \times \mathbf{S} \tag{13.17}$$

Fig. 13.2 Precession of magnetic moment μ around a constant magnetic field B

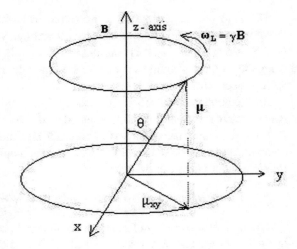

Fig. 13.3   Fixed laboratory coordinates x, y, z and rotating coordinates x', y', z' for the angular velocity ω along z-axis. The primed and unprimed axes being the same at $t = 0$

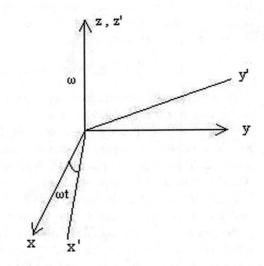

$$\left(\frac{d'\mathbf{S}}{dt}\right)_{rot} = \left(\frac{d\mathbf{S}}{dt}\right)_{lab} + \mathbf{S} \times \boldsymbol{\omega} = \gamma \boldsymbol{\mu} \times (\mathbf{B} - \frac{\boldsymbol{\omega}}{\gamma}) \tag{13.18}$$

$$= \gamma \mathbf{B}' \times \boldsymbol{\mu} \tag{13.19}$$

with

$$\mathbf{B}' = \mathbf{B} - \frac{\boldsymbol{\omega}}{\gamma} \tag{13.20}$$

The time rate of change of μ measured in the rotating system can be calculated-; therefore, as if the rotating system was a stationary one having an effective magnetic field $\mathbf{B'}$ along z-axis. If $\mathbf{B'}$ is constant and $\omega = \gamma \, \mathbf{B}$, the effective field $\mathbf{B'}$ vanishes. The magnetic moment μ would see no torque in the rotating frame and would remain constant with respect to it.

ESR experiments do not use a constant B but rather employ time-dependent components, which are usually small, compared to the constant component. Let total magnetic field has a z-component which is of constant magnitude B, and it also has an oscillating component along x-axis and with a peak to peak amplitude of $4B_1$ such that $B_1 \ll B$. The time-dependent field can be written as

$$\mathbf{B}_1 = \mathbf{i} 2B_1 \cos \omega t = B_1 (\mathbf{i} \cos \omega t + \mathbf{j} \sin \omega t) + B_1 (\mathbf{i} \cos \omega t - \mathbf{j} \sin \omega t) \qquad (13.21)$$

The first term is a field of constant amplitude B_1 rotating about z-axis at frequency ω in the same sense as electronic precession. The second term is a similar field but with the opposite sense of the electronic precession. The component rotating opposite to the sense of the electronic precession does not cause resonance. Neglecting this component the total field $\mathbf{B_e}$ can be written as

$$\mathbf{B}_e = \mathbf{i} B_1 \cos \omega t + \mathbf{j} B_1 \sin \omega t + \mathbf{k} B \qquad (13.22)$$

B_1 is constant in a coordinate system that rotates about the z-direction. In this rotating frame, it is shown by Eq. (13.19) that static field can be written as $\mathbf{B'} = \mathbf{B} - \omega/\gamma$. B_1 is along x' -axis of rotating coordinate system. B' is along z- or z'-axis. When $\mathbf{B} = \omega/\gamma$ the condition for magnetic resonance occurs and $B_e = B_1$. For away from magnetic resonance B' is either nearly parallel to B or antiparallel to B as $B_1 \ll B$. Thus, the motion of magnetic moment is then of a precession about B_1 with angular velocity γB_1, and every half cycle of this motion, it changes from being parallel to antiparallel and back again. In the rotating system, this motion takes place in the plane normal to B. Since $B_1 \ll B$, the precession about B_1 occurs at a much lower velocity than that at which B_1 rotates in the laboratory system. Thus, in the latter system, the motion of the angular momentum and magnetic moment consists of a rapid motion about B at an angle to B which varies from 0 to π and back again. Thus when $\mathbf{B} = \omega/\gamma$ the magnetic moment assume to be initially parallel to B can be completely reversed by an application of rotating field B_1, and it will occur no matter how small is the value of B_1.

13.1.3 Relaxation Mechanisms

Under thermal equilibrium conditions, the populations of the two levels with $M = -1/2$ and $M = 1/2$ are determined by the Boltzmann distribution

$$\frac{N_1}{N_2} = \exp\left(\frac{h\nu}{k_B T}\right) \tag{13.23}$$

where $h\nu$ is the energy difference between the two levels. N_1 and N_2 are the population of $M = -1/2$ and $M = 1/2$ levels, respectively. However, $N_1 - N_2$ is very small. The phenomenon of *ESR* absorption depends on this difference. The a priori probability of transition from lower level to upper level equals that from transition from upper level to lower level. Because of $N_1 > N_2$ results in an excess of upward transitions and a net absorption of energy from microwave field. However, this leads to an increase of N_2, which will continue until $N_1 = N_2$ and net absorption of energy will tend to zero. That this situation does not occur it means that there must therefore be other mechanisms by means of which energy absorbed and stored in the upper level can be dissipated in such a manner as to allow return to lower level to maintain the population difference. Such mechanisms are called relaxation processes.

(a) Spin–Lattice Relaxation

In the absence of magnetic field, the electron spins are randomly oriented in space and their resultant moment is zero. When a static magnetic field **B** is applied, the spins become aligned parallel or antiparallel to **B**, with slightly less spins having the parallel alignment to give the population difference. As a result of this, magnetization M_z is observed. However, a certain time interval is needed to obtain equilibrium value, which is same as that needed for the spins to become again randomly oriented when the magnetic field is suddenly switched off. This reorientation time is measured by T_1. Thus, the spin–lattice relaxation time T_1 measures the characteristic time for recovery of the magnetization of the paramagnetic system along the static field direction after the equilibrium has been disturbed.

The interpretation of the spin–lattice relaxation time is based on the interaction between the spins and lattice vibrations (phonons). The spin magnetic moment is not influenced directly by vibrations of the lattice. The coupling of the lattice vibrations with magnetic spin states occurs indirectly through residual spin–orbit coupling. The crystalline electric field in ionic compounds or chemical bonds in molecular free radical raise the degeneracy of the orbital states, usually leaving an orbital singlet as ground state. The orbital singlet behaves like an atomic S state. However, orbital magnetic field is not completely quenched because of a second-order admixture of the singlet orbital ground state with higher orbital states. Therefore, there is slight orbital magnetic field acting on the spin moment. The orbital moments are strongly coupled to the lattice through strong crystalline electric fields. If the lattice is vibrating, interatomic distances vary and therefore crystalline field also varies. As a result of this, the residual orbital magnetic field acting on the electron spin is effectively modulated by all vibrational modes (we are concerned with the phonons having frequency 3–30 GHz, the region in which ESR work is usually carried out). The probability that a magnetic spin transition will be induced by such modulated at resonance frequency is proportional to the square of the residual orbital field component that are transverse to **B**. In general, one expect large value of T_1 with decrease of temperature because of the freezing of lattice vibrations. Three processes are proposed for spin–lattice relaxation.

(i) *Direct Process*

It involves phonons of the same frequency as the *ESR* resonance quantum $h\nu$. Only a small fraction of the normal distribution of the thermal energy is concentrated in vibrational frequencies as low as the *ESR* frequencies. Therefore, phonons at the resonance frequencies are normally scarce. The process is prominent at low temperatures.

(ii) *Raman Process*

This is a two-phonon process. In this process, lattice phonons are scattered by the spins, with the ESR frequency add to or subtracted from the frequency of scattered phonons. For this method to be effective, there must be lattice modes having difference frequencies equal to the ESR frequency. Such vibrations, however, can be in the higher frequency region, more densely populated at room temperature. In this process, many phonons pair can participate since the only requirement is that their frequency difference is equal to ESR frequency. Effectiveness of this process decreases as the temperature is decreased since the vibration at higher frequencies would be frozen out at low temperatures.

(iii) *Orbach Process*

This involves absorption of a phonon by direct process to excite the spin system from the upper level to a much higher level at an energy δ above the ground doublet, then dropping back into lower Zeeman level of the ground state by emitting another phonon of slightly different energy (Fig. 13.4). Thus, the paramagnetic ion is indirectly transferred from upper Zeeman level to the lower Zeeman level of the ground doublet. It is more restricted than the Raman process because two specific phonons are involved.

At liquid *He* temperature only the direct process is usually significant, whereas Raman and Orbach processes become dominant at higher temperatures.

(b) **Spin–Spin Relaxation**

It contains all the mechanisms whereby the spins can exchange energy among themselves, rather than giving it back to the lattice, or molecular system as a whole.

Fig. 13.4 Energy level diagram to illustrate the Orbach process. Level b and *a* relax by way of two direct transitions involving a third level c

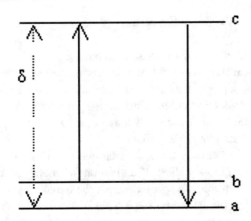

(i) *Dipole Interaction*

It arises from the influence of the magnetic field of one of the paramagnetic ion on the dipole moments of neighbouring ions. The actual local field at any given site will depend on the arrangement of the neighbours and the direction of their dipole moments. If the external magnetic field acts on the paramagnetic compound, the local field at each ion must be added vectorially to it. If the local field is small compared with the external field (which might be ~tesla), only the component of the former parallel to the latter is important. The size of this component varies from site to site, giving a random displacement to the resonance frequency of each ion. If the paramagnetic ions are identical, so that they process at the same frequency in the external magnetic field, there is an additional resonance interaction. The processing components of one magnetic dipole set up an oscillatory field at another dipole which is just at right frequency to cause magnetic resonance transitions and vice versa. The mutual interaction produces resonance transitions that are equivalent to the exchange of quanta between neighbouring ions. Thus, the quanta are exchanged among neighbouring ions by mutual spin flip and spins are in thermal equilibrium amongst themselves. If this equilibrium is disturbed, it is re-established exponentially with time constant T_2 which is called spin–spin relaxation time. $T_1 \sim 10^{-6}$ s and $T_2 \sim 10^{-10}$ s. T_2 is independent of temperature.

(ii) *Exchange Coupling*

This is important only in undiluted crystals where the paramagnetic sites are so close together that the orbital of the unpaired electrons overlap. Here the spins interact electrostatically through a short-range interaction, which is called exchange interaction. This results in a change in width of absorption lines in the ESR spectrum.

(iii) **Cross-Relaxation**

Let us consider a sample having two spin systems with different resonance frequencies. If the tails of the two absorption curves corresponding to the spin systems overlap, a flip-flop spin exchange can take place in the region of the overlap. This process is known as cross-relaxation. Since spin–spin relaxation times are often much shorter than spin–lattice relaxation time, this may be more effective process for dissipating the energy than direct transfer to the lattice.

Let us assume, for example $T_1{}^X$ (for species X) is very long and that $T_1{}^Y$ (for species Y) is very short. If a quantum of energy absorbed by a spin of type X is given to a spin Y through a spin exchange, the quantum will have a high probability of being transferred to the lattice by the Y spin before it is transferred back to the X spin through a reverse exchange (because of short $T_1{}^Y$). The spin–lattice relaxation time of the X spin system is thus effectively reduced.

13.1.4 Bloch Equations

Consider a magnetic material in the absence of an external magnetic field. The magnetic dipoles are then oriented at random, and there is no resultant magnetization. Suppose now that suddenly a magnetic field B is applied. A certain period will elapse before the magnetization M has reached its thermal equilibrium value M_0. The built-up of the magnetization may be described by relaxation time T

$$\frac{dM}{dt} = \frac{M_0 - M}{T} \tag{13.24}$$

determine the rate of growth of M. The rate of change of z-component of magnetization for a collection of electron spins from Eqs. (13.12) and (13.24)

$$\frac{dM_z}{dt} = \gamma (\mathbf{B} \times \mathbf{M})_z + \frac{M_0 - M_z}{T_1} \tag{13.25}$$

M_z is the z-component of magnetization at time t; T_1 is the longitudinal or spin–lattice relaxation time. The x- and y-components of M both decay to zero in the static field and we have

$$\frac{dM_x}{dt} = \gamma (\mathbf{B} \times \mathbf{M})_x - \frac{M_x}{T_2} \tag{13.26}$$

$$\frac{dM_y}{dt} = \gamma (\mathbf{B} \times \mathbf{M})_y - \frac{M_y}{T_2} \tag{13.27}$$

T_2 is time characterizing the loss of phase coherence of the spins; the transverse or spin–spin relaxation time. Equations (13.25)–(13.27) are the Bloch equations.

In an *ESR* or *NMR* c experiment, the total applied magnetic field B_e consists of a steady component B in the z-direction, and a rotating component in the xy plane as given by Eq. (13.22). Substituting Eq. (13.22) in Eqs. (13.25)–(13.27)

$$\frac{dM_z}{dt} = \gamma(B_x M_y - B_y M_x) + \frac{M_0 - M_z}{T_1} = \gamma(B_1 M_y \cos \omega t - B_1 \sin \omega t M_x) + \frac{M_0 - M_z}{T_1}$$
$$(13.28)$$

$$\frac{dM_x}{dt} = \gamma(B_y M_z - M_y B_z) - \frac{M_x}{T_2} = \gamma(B_1 \sin \omega t M_z - M_y B) - \frac{M_x}{T_2} \qquad (13.29)$$

$$\frac{dM_y}{dt} = \gamma(-B_x M_z + B_z M_x) - \frac{M_y}{T_2} = \gamma(-B_1 \cos \omega t M_z + B M_x) - \frac{M_y}{T_2} \qquad (13.30)$$

To solve these equations, it is convenient to transform to a coordinate system rotating about the z-axis at a frequency ω, that is,

$$u = M_x \cos \omega t + M_y \sin \omega t \qquad (13.31)$$

$$v = M_x \sin \omega t - M_y \cos \omega t \qquad (13.32)$$

u represents in-phase and v represents $\pi/2$ out of phase (lag) with B_1, and they are called the in-phase and out of phase components of the magnetization. Differentiating Eq. (13.31)

$$\frac{du}{dt} = \frac{dM_x}{dt} \cos \omega t - \omega M_x \sin \omega t + \frac{dM_y}{dt} \sin \omega t + \omega M_y \cos \omega t \qquad (13.33)$$

From Eqs. (13.29), (13.30) and (13.33)

$$\frac{du}{dt} = \gamma(-M_y B + M_z B_1 \sin \omega t) \cos \omega t - \frac{M_x}{T_2} \cos \omega t - \omega M_x \sin \omega t$$
$$+ \gamma(-M_z B_1 \cos \omega t + M_x B) \sin \omega t - \frac{M_y}{T_2} \sin \omega t + \omega M_y \cos \omega t$$

$$\frac{du}{dt} = -\gamma M_y B \cos \omega t + \gamma M_z B_1 \sin \omega t \cos \omega t - \frac{M_x}{T_2} \cos \omega t$$
$$- \omega M_x \sin \omega t - \gamma M_z B_1 \cos \omega t \sin \omega t$$
$$+ \gamma M_x B \sin \omega t - \frac{M_y}{T_2} \sin \omega t + \omega M_y \cos \omega t$$

$$\frac{du}{dt} = \gamma B(-M_y \cos \omega t + M_x \sin \omega t) + \omega(M_y \cos \omega t - M_x \sin \omega t)$$
$$- \frac{1}{T_2}(M_x \cos \omega t + M_y \sin \omega t)$$

$$\frac{du}{dt} = \gamma B v - v \omega - \frac{u}{T_2} \qquad (13.34)$$

Similarly from Eqs. (13.29)–(13.32)

$$\frac{dv}{dt} = -\gamma\,B\,u + \gamma\,M_z B_1 + u\,\omega - \frac{v}{T_2} \tag{13.35}$$

From Eqs. (13.28) and (13.32)

$$\frac{dM_z}{dt} = -\gamma B_1 v + \frac{M_0 - M_z}{T_1} \tag{13.36}$$

In the steady state, all the time derivatives of Eqs. (13.34)–(13.36) vanishes, that is, from Eq. (13.34)

$$(\gamma\,B - \omega)v = \frac{u}{T_2}$$
$$u = (\gamma\,B - \omega)v T_2 \tag{13.37}$$

From Eq. (13.35)

$$(\gamma\,B - \omega)u = \gamma\,M_z B_1 - \frac{v}{T_2} \tag{13.38}$$

From Eqs. (13.37) and (13.38)

$$(\gamma\,B - \omega)^2 v\,T_2 = \gamma\,M_z B_1 - \frac{v}{T_2}$$
$$[1 + (\gamma\,B - \omega)^2 T_2^2]v = \gamma\,M_z B_1 T_2 \tag{13.39}$$

From Eq. (13.36)

$$\gamma\,B_1 v = \frac{M_0 - M_z}{T_1}$$
$$M_z = M_0 - \gamma\,B_1 v\,T_1 \tag{13.40}$$

From Eqs. (13.39) and (13.40)

$$[1 + (\gamma\,B - \omega)^2 T_2^2]v = \gamma(M_0 - \gamma\,B_1 v\,T_1)B_1 T_2 = \gamma\,M_0 B_1 T_2 - \gamma^2 B_1^2 v\,T_1 T_2$$
$$v[1 + \gamma^2 B_1^2 T_1 T_2 + (\gamma\,B - \omega)^2 T_2^2] = \gamma\,B_1 M_0 T_2$$
$$v = \frac{\gamma\,B_1 M_0\,T_2}{1 + \gamma^2 B_1^2 T_1 T_2 + (\gamma\,B - \omega)^2 T_2^2} \tag{13.41}$$
$$v = \frac{\gamma\,B_1 M_0 (1/T_2)}{(1/T_2)^2 + \gamma^2 B_1^2 (T_1/T_2) + (\gamma\,B - \omega)^2}$$
$$v = \frac{\gamma\,B_1 M_0 (1/T_2)}{(1/T_2)^2 + \gamma^2 B_1^2 (T_1/T_2) + (\omega_L - \omega)^2}$$

where $\gamma\,B = \omega_L$. From Eqs. (13.37) and (13.41)

$$u = \frac{\gamma\,B_1 M_0 (\omega_L - \omega)}{(1/T_2)^2 + \gamma^2 B_1^2 (T_1/T_2) + (\omega_L - \omega)^2} \tag{13.42}$$

From Eqs. (13.40) and (13.41)

$$M_z = M_0 - \frac{\gamma^2 B_1^2 M_0 (T_1/T_2)}{(1/T_2)^2 + \gamma^2 B_1^2 (T_1/T_2) + (\omega_L - \omega)^2}$$

$$M_z = \frac{M_0[(1/T_2)^2 + (\omega_L - \omega)^2]}{(1/T_2)^2 + \gamma^2 B_1^2 (T_1/T_2) + (\omega_L - \omega)^2} \tag{13.43}$$

Multiplying Eq. (13.31) by $\cos \omega t$ and Eq. (13.32) by $\sin \omega t$ and adding

$$M_x = u \cos \omega t + v \sin \omega t \tag{13.44}$$

On substituting Eqs. (13.41) and (13.42) in Eq. (13.44)

$$M_x = \frac{\gamma B_1 M_0 (\omega_L - \omega) \cos \omega t}{(1/T_2)^2 + \gamma^2 B_1^2 (T_1/T_2) + (\omega_L - \omega)^2} + \frac{\gamma B_1 M_0 (1/T_2) \sin \omega t}{(1/T_2)^2 + \gamma^2 B_1^2 (T_1/T_2) + (\omega_L - \omega)^2}$$

$$M_x = \frac{\gamma B_1 M_0[(\omega_L - \omega) \cos \omega t + (1/T_2) \sin \omega t]}{(1/T_2)^2 + \gamma^2 B_1^2 (T_1/T_2) + (\omega_L - \omega)^2} \tag{13.45}$$

Multiplying Eq. (13.31) by $\sin \omega t$ and Eq. (13.32) by $\cos \omega t$ and subtracting the two equations

$$M_x = \frac{\gamma B_1 M_0[(\omega_L - \omega) \sin \omega t - (1/T_2) \cos \omega t]}{(1/T_2)^2 + \gamma^2 B_1^2 (T_1/T_2) + (\omega_L - \omega)^2} \tag{13.46}$$

Equations (13.45) and (13.46) show that the magnetization has a constant component in the direction of the applied magnetic field and a rotating component in the xy plane.

It is convenient to use linearly polarized radiation in place of circularly polarized. For a linearly polarized field, we can write

$$B_x = 2B_1 \cos \omega t, \quad B_y = 0 \tag{13.47}$$

The linearly polarized field can be decomposed into two counter-rotating components, one corresponding to the circularly polarized light with frequency ω and another of the same magnitude but going at a frequency $-\omega$. Only the component at frequency $+\omega$ has the right sense to induce transition. Thus, component at $-\omega$ can be neglected. The effect of this field can be described in terms of complex susceptibility

$$\chi = \chi' - i\chi'' \tag{13.48}$$

χ' and χ'' are often referred to as Bloch susceptibility. Let B_x of Eq. (13.47) is the real part of a complex field B_e, that is

$$B_e = 2B_1 \exp(i\omega t) = 2B_1(\cos \omega t + i \sin \omega t) \tag{13.49}$$

Then M_x can be taken as the real part of a complex magnetization M_e

$$M_e = \chi B_e \tag{13.50}$$

From Eqs. (13.48) and (13.49)

$$M_e = 2B_1\chi' \cos \omega t + 2B_1\chi'' \sin \omega t \tag{13.51}$$

On comparing Eqs. (13.48) and (13.50)

$$\chi' = \frac{1}{2}\left[\frac{\gamma\, M_0(\omega_L - \omega)}{(1/T_2)^2 + \gamma^2 B_1^2(T_1/T_2) + (\omega_L - \omega)^2}\right] \tag{13.52}$$

$$\chi'' = \frac{1}{2}\left[\frac{\gamma M_0(1/T_2)}{(1/T_2)^2 + \gamma^2 B_1^2(T_1/T_2) + (\omega_L - \omega)^2}\right] \tag{13.53}$$

13.1.5 Paramagnetic Ion in a Crystalline Field

When the paramagnetic ion is incorporated into a solid, there are interaction between it and its surrounding. The ions (ligands) surrounding the paramagnetic ion produce a strong electrostatic field (crystalline field) at the ion site. This electrostatic field reflects the symmetry of the site of the paramagnetic ion. The electrons localized on the ions and moving in this field experiences a Stark splitting in their orbital levels, and their degeneracies are partially lifted depending on the symmetry of the site. In the case of local cubic symmetry groups 1, 2, 3 or fourfold degenerate levels are found, but in case of lower symmetry, levels very often are either singlet or degenerate in pairs only. An important theorem concerning the residual degeneracy is due to Kramer's: In a system containing an odd number of electrons, at least twofold degeneracy must remain in the absence of a magnetic field.

13.1.6 Effective Spin

The results of ESR measurements are summarized in a so-called spin Hamiltonian. The formalism is adopted to the conditions of paramagnetic ions inserted in crystals. The ground state of these ions is degenerate or is split not more than some cm^{-1}. The next higher state is several hundred or thousand cm^{-1} higher. In *ESR*, transitions between the lowest levels are studied. The ground state is described by a spin quantum number S which sometimes differs from the spin quantum number of levels in the ground state multiplet which is put equal to $2S + 1$. S is termed as effective spin or fictitious spin.

13.1.7 Features of ESR Spectra

(i) The g-Factor

The parameter, which may be associated with a single absorption line, is its actual resonance position. If microwave frequency is held constant, then Eq. (13.8) shows that only two parameters that can vary are the values of the externally applied magnetic field and the value of the g-factor. Hence, a measurement of the value of B at which resonance occurs determine the value of the g-factor associated with particular unpaired electron in the atom or molecule in which it is residing.

An electron with no orbital angular momentum has a g value equal to that of the free electron spin, i.e. equal to 2.0023. If on the other hand, the electron is moving in an atomic orbital associated with a single atom, it may possess considerably orbital angular momentum and this will shift the g value away from that appropriate to the free spin. The reason for this shift in g value is that relationship between magnetic moment and angular momentum is different for spin motion and for orbital motion.

In the case of an electron associated with a free atom, which has no external magnetic or electric field acting on it, the resultant g value can be derived directly in terms of the quantum numbers defining the total spin and orbital magnetic moments. The g value is in fact then identical with Lande's splitting factor. If the unpaired electron is associated with an atom contained within a solid crystalline lattice, then internal crystalline electric field, which arises from the interaction between atom and its surrounding, will also be acting on it. These electric fields act on the orbital states of the atom and can vary radically alter their energies, and, as a result, the simple Lande's factor can no longer be applied. Here the position of the resonance for a fixed frequency is determined by the competitive effects of the environment, which electric field tend to 'quench' (the electronic orbital motion interacts strongly with the crystalline fields and becomes decoupled from the spin, a process called 'quenching') the orbital angular momentum. The more complete the quenching, the closer the g-factor approaches the free electron value. For example, $g = 2.0036$ in the free radical α, α'-diphenyl-β-picryl hydrazyl ($DPPH$) which is very close to the free electron value of 2.0023. It equals 1.98 in many chromium compounds, and it sometimes exceeds 6 for Co^{2+}. Thus, the amount of quenching varies with the spin system.

In the general case, therefore the quantitative field frequency relation is not known a priori and we write the generalized resonance condition $h\nu = g\,\beta_e B$, where g is the g-factor and is determined from the ESR spectrum. The above relation can be regarded as giving the separation between eigenvalues of a spin Hamiltonian $H = \beta_e\,\mathbf{S.g.B}$, so that g is a measure of the effective magnetic moment associated with an angular momentum $\mathbf{S}$ such that there are $(2S + 1)$ energy levels in an applied magnetic field. Since the magnitude of the orbital part of the magnetic moment depends on the crystal field, its magnitude is usually different for different directions of applied magnetic field $\mathbf{B}$ and shows an angular variation, which follows the symmetry of the crystal field. The g value may then be anisotropic by an amount, which depends on the magnitude of the orbital contribution to the moment, and on the asymmetry in the crystalline field. The two general procedures for determining g are (i) the direct method where frequency and magnetic field are measured independently at the resonance absorption peak and g is calculated from Eq. (13.8), (ii) the comparison method where an unknown specimen is measured along with a system of known g value (e.g. *DPPH*).

(ii) **Fine Structure**

The spectra of oriented systems of total spin greater than one-half often show ($2S$) features, indicating unequal separation between the $2S + 1$ Zeeman levels. This structure is called fine structure. The fine structure reflects a splitting of the $2S + 1$ levels even in the absence of a magnetic field, a phenomenon called zero-field splitting. If trivalent chromium ions existed in free state and were not subjected to the electrostatic fields inside the crystal or molecule, then all possible orientations of $S = 3/2$ total spin vector would have same energy, and hence, all would be degenerate on the energy level diagram, as shown in Fig. 13.5. The application of an external magnetic field would then cause the spins to take up different quantized orientations with respect to this field. The possible orientations vary from those with component of $M = +3/2$ in the direction of the field through those with orientation corresponding to $M = +1/2, -1/2$ to the other extreme of $-3/2$. The energies of these different orientations will diverge with the application of external magnetic field, by amounts proportional to the resolved spin quantum number. Transition from one state to another takes place according to selection rule $\Delta M = \pm1$. It is evident that for a given frequency all the possible transitions will take place at the same value of the externally applied magnetic field, and hence, a single absorption line will be produced from the overlapping transitions (Fig. 13.5).

Fig. 13.5 Energy levels for Cr^{3+} ion ($S = 3/2$). At a given frequency, all transitions take place at the same value of B

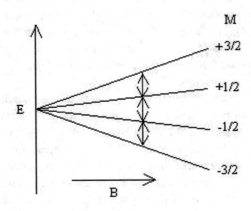

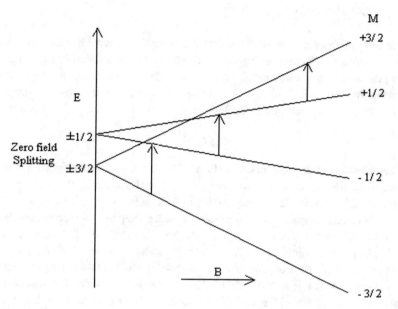

Fig. 13.6 Energy levels of Cr^{3+} in a crystal. At a given frequency, the three allowed transitions take place at different values of B

If the Cr^{3+} ion is present in a crystal, it will be acted on by the internal electrostatic field and the effect of the latter on the energy levels of the ion is profound. First of all, they split the ground state into a number of components. The number of such components and the extent of the splitting depends on the symmetry of crystalline electric field and on its strength but electric field itself cannot remove the twofold degeneracy associated with the states of the same resolved quantum numbers. The action of the internal electric field will therefore be as indicated on the left hand side of (Fig. 13.6), and zero-field electronic splitting will be produced between $\pm 1/2$ and $\pm 3/2$ levels as indicated. Application of external magnetic field will now remove the twofold degeneracy of these spin levels, and these will diverge in the way as shown in Fig. 13.6. At high enough values of B, they will be arranged in an energy order $+3/2$ to $-3/2$ as indicated. The allowed transitions between them no longer occur at the same value of B and three separate transitions will now be obtained. The separation between these is a direct reflection of the separation produced in zero magnetic field by the electrostatic internal fields and can therefore be used as a measure of this quantity. The intensity of fine structure lines is proportional to $[S(S + 1) - M(M - 1)]$ where S is the spin of the ion, and M is greater value of M for level between which the resonance transition is taking place. For determination of g value, to a first approximation the average of fine structure lines position gives the value of B to be used in Eq. (13.8).

(iii) **Hyperfine Structure**

If the nucleus of the paramagnetic ion has a magnetic moment, it will interact with the electronic moment, resulting in hyperfine structure in the *ESR* spectrum. The nuclear moment produces a magnetic field B_N at the magnetic electron. Thus, an unpaired electron experiences the magnetic field, which is applied externally as well as the magnetic field due to the nuclear magnetic moment. The resonance condition (Eq. 13.8) is modified to

$$h\nu = g\beta_e|\mathbf{B} + \mathbf{B_N}| \qquad (13.54)$$

The nuclear moment is space quantized in the direction of the internal magnetic field (which is due largely to the electrons) and can take up $2I + 1$ positions, $\mathbf{I}$ being the nuclear spin. There are therefore $(2I + 1)$ possible values of $\mathbf{B_N}$ and the same number of total magnetic field which satisfy resonance condition [Eq. (13.8)] when the frequency ν is constant. The nuclear moments are smaller than the electron moments by a factor of 10^3. Therefore, according to Eq. (13.23) in a given magnetic field the nuclear spin states are more nearly equally populated than are the electron spin states. Therefore, at room temperature there will be an equal number of nuclei in all of the $(2I + 1)$ orientations. Hence, there will be an equal number of unpaired electrons experiencing the $(2I + 1)$ different values of the total magnetic field resulting in hyperfine lines of equal intensity in each electronic transition. To a first approximation, the separation between hyperfine lines is equal.

Therefore, to a first approximation hyperfine coupling constant can be determined by taking the average of the separations between hyperfine lines. Further spacing between the components is independent of field. The centre of a hyperfine group defines the position of fine structure line. Thus, one can distinguish a hyperfine structure from fine structure where the lines have unequal intensities (Fig. 13.7). The selection rule for hyperfine transition is $\Delta m = 0$, i.e. in electronic transition, the orientation of nuclear spins remain unchanged.

The magnitude of the hyperfine coupling depends on whether the interacting electron is in an *s* or a *p* or *d* orbital. Since *s* orbital has high electron density at the nucleus, the

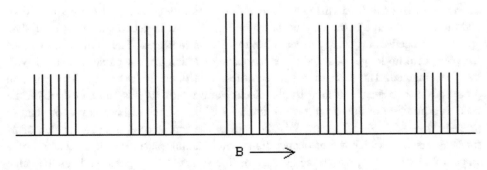

Fig. 13.7 A stick diagram showing the ESR spectrum of Mn^{2+} in a single crystal. $S = I = 5/2$

hyperfine coupling constant will be large. Because of spherical symmetry of s orbital, A will be independent of direction. This interaction is called the isotropic hyperfine coupling (A_{iso}) or the Fermi contact interaction. The p and d orbitals, where there is no electron density at the nucleus, the electron is to be found at some distance away from the nucleus. The interaction between it and then nucleus will be of two magnetic dipoles, and consequently, it will be small and depend on the direction of the orbital relative to the applied magnetic field as well as their separation. This interaction is called anisotropic hyperfine coupling. In liquids, anisotropic coupling is averaged to zero and only isotropic coupling A_{iso} is observed.

The isotropic interaction is determined by the expression

$$A_{iso} = \frac{8\pi}{3} g_N \beta_N |\psi(0)|^2 \text{gauss} \tag{13.55}$$

$|\psi(0)|^2$ is the electron density at the nucleus. A_{iso} will be more intensive, the greater the magnetic moment of the nucleus and the higher the density of the s electrons on the nucleus. A_{iso} will be a direct measure of the s character of the unpaired electron since only s orbitals have nonzero values of the wave function at the nucleus. The anisotropic hyperfine coupling is given by

$$A_{anis} = \frac{g_e \beta \, g_N \beta_N}{2r^3} < 3\cos^2\alpha - 1 > < 3\cos^2\theta - 1 > \tag{13.56}$$

where r represents the unpaired electro–nucleus distance, α is the angle between r and one of the principal axes of the anisotropic tensor A; θ is the angle between this same axis and the magnetic field. The sign $<>$ denotes the mean value.

Nuclei with $I \geq 1$ have electric quadrupole moment Q because their charge distribution deviate from spherical symmetry. The quadrupole moment can interact with electric field gradient in an atom, molecule or crystal. The electric field gradient is set up by the anisotropic distribution of electric charges on the paramagnetic ion and its immediate neighbours. The effect of quadrupole interaction on the resonance spectrum is usually complicated because it is accompanied by a large magnetic hyperfine structure interaction. Therefore, there is a competition between the electric field gradient and the magnetic field from the unpaired electrons, in fixing the orientations of the nucleus. The result is that, for a general orientation of B, the intensities and spacing of the usual $2I + 1$ hyperfine lines corresponding to $\Delta m = 0$ are modified by Q and there are also additional lines corresponding to the normally forbidden transitions with $\Delta m = \pm 1, \pm 2$, etc.

The nuclear magnetic moment β_N interacts directly with the applied magnetic field B with energy $\sim \beta_N B$.

(iv) Ligand Hyperfine Structure (Superhyperfine Structure)

The unpaired electrons can also interact with the nuclei of the near neighbouring atoms, if their nuclei are magnetic resulting in the observation of a structure called superhyperfine or ligand hyperfine structure. Under high field approximation, the isotropic spectrum may be obtained to first order by splitting the unperturbed electronic levels into $2I_i + 1$ for nucleus i, then each of these levels is further split into $2Ij + 1$ levels by nucleus j and so on. Thus, the total number of energy levels for a given value of M is $(2I_i + 1)(2I_j + 1)(2I_k + 1)$ ... or can be written as

$$\prod_i (2I_i + 1) \tag{13.57}$$

Since transitions between hyperfine levels occur with no change of nuclear spin quantum number, i.e. $\Delta m = 0$, this is also the maximum number of observable ESR lines. Since each interacting nuclei of spin I causes $(2I + 1)$ number of hyperfine energy levels for a given electronic level. For n nuclei of spin I, the maximum number of levels is

$$N = (2I + 1)^n \tag{13.58}$$

when the different nucleus have equivalent hyperfine coupling constant, the number of hyperfine components will be

$$N = 1 + 2\sum_i I_i = 1 + 2T \tag{13.59}$$

where $T = \sum I_i$. If n equally coupling nuclei have the same spin I, the number of hyperfine components are

$$N = 2nI + 1 \tag{13.60}$$

When unpaired electron has more than one coupling nucleus the relative intensities of the ESR hyperfine components are not necessarily equal because of the possibility that the components arising from the different nuclei may fall at the same frequency. If an unpaired electron is coupled to several equivalent protons, the relative intensities of such a multiplet can be calculated from probability theory. For $I = 1/2$, each nucleus must point up ($m = 1/2$) or down ($m = -1/2$) in the static magnetic field B. The outside components of the multiplet are given by the combination of the n nuclei pointing either all up or all down. The outermost components are assigned unit intensity. The component next to these will have all of the nuclei except one pointing in the same direction. Thus $k = n - 1$ must point in the same direction. This can happen n-1 time. For the next component, all of the nuclei except two will be pointing out in the same direction and this can happen $n - 2$ time and so on. Generally, the number of combination of n thing taken k at a time is

$$^nC_k = \frac{n!}{(n - k)!} \tag{13.61}$$

13.2 Nuclear Magnetic Resonance

Nuclear magnetic resonance (*NMR*) has been defined as the form of spectroscopy concerned with radio frequency-induced transitions between magnetic energy levels of the nucleus having a nuclear angular momentum. This is the same type of magnetic resonance phenomenon that occurs in *ESR* except that it is the magnetic dipole moment of the nucleus that is involved. As a result of mass difference, the nuclear magnetic moment is smaller than that of electron. The resonance is correspondingly lower in frequency and falls in the radio frequency region. Another important difference from *ESR* is that there is no longer a restriction to molecules with unpaired electrons. Any molecule with a magnetic nucleus will give an *NMR* spectrum. The *NMR* and *ESR* are similar in the sense that while *NMR* deals with nuclear ground state; *ESR* deals with electronic ground state.

NMR requires the presence of nuclear magnetic moment associated with nonzero nuclear spin. The occurrence of nonzero nuclear spin is common in the periodic table, and thus, *NMR* can be observed in isotopes of most elements. The nuclei can be divided into the following categories:

(a) All the nuclei with $I = 0$ with magnetic moment and quadrupole moment are zero.
(b) All the nuclei with $I = 1/2$ have magnetic moment but no quadrupole moment.
(c) All the nuclei with $I \geq 1$ posses both magnetic and quadrupole moments.

The nuclei of category (a) do not show *NMR* signal. They include nuclei with an even Z and even mass number A. The nuclei in this category include C^{12}, O^{16}, O^{18}, Si^{28}, S^{32} and Ca^{40}. The nuclei of categories (b) and (c) show *NMR* signal. The possibility of studying the *NMR* spectra of nuclei having spin I different from zero is conditioned by the magnetic moment and natural abundance.

The splitting of the ground level of the nucleus in the presence of magnetic field is small because of low values of nuclear magnetic moment. Hence, the population of these sublevels are very close to each other. For example, at normal temperature population of lower sublevel of proton with $m = 1/2$ is only 1.0000066 times greater than that of higher sublevel with $m = -1/2$. Because of this difference, the *NMR* transitions with absorption occur. The intensity of such transitions is low, and therefore, the specimen to be studied should contain large number of nuclei or is of considerable bulk ($\sim 0.5~cm^3$).

13.2.1 The Principle of the Phenomenon

Consider a nucleus having a magnetic moment μ_N. Classically, the energy of interaction between the nuclear magnetic moment and magnetic field **B** is

$$E = -\mu_N \cdot \mathbf{B} \tag{13.62}$$

The magnetic moment is a vector, which is collinear with angular momentum $\mathbf{J}$. The operator for the vectors is related by

$$\mu_N = \gamma \mathbf{J} \tag{13.63}$$

where $\gamma = g_N \frac{e}{2M_p}$ is called gyromagnetic ratio, g_N is nuclear g-factor, and M_p is mass of the proton. γ varies with the state it can have positive or negative value and is characteristic of the nuclei.

We define a dimensionless angular momentum operator $\mathbf{I}$ by

$$\mathbf{J} = (h/2\pi)\mathbf{I} \tag{13.64}$$

I^2 has eigenvalues $I (I + 1)$ where I is an integer or half integer. The component I_z has eigenvalue m where m can take values $I, I - 1, I - 2, ..., -I$. From Eqs. (13.63) and (13.64)

$$\mu_N = \gamma (h/2\pi)\mathbf{I} \tag{13.65}$$

The quantum mechanical Hamiltonian from Eqs. (13.62) and (13.65)

$$H = -\gamma (h/2\pi)\mathbf{I} \cdot \mathbf{B} \tag{13.66}$$

If the magnetic field is assumed to be in the z-direction, i.e. $B_x = B_y = 0$ and $B_z = B$, the Hamiltonian

$$H = -\gamma (h/2\pi)I_z B \tag{13.67}$$

The eigenvalues of Hamiltonian are just the eigenvalues of I_z, that is

$$E_m = -\gamma (h/2\pi)Bm \tag{13.68}$$

For $I = 3/2$, $m = 3/2, 1/2, -1/2, -3/2$, the energies from Eq. (13.68) are

$$E_{3/2} = -\frac{3\gamma h}{4\pi} B$$

$$E_{1/2} = -\frac{\gamma h}{4\pi} B$$

$$E_{-1/2} = \frac{\gamma h}{4\pi} B$$

$$E_{-3/2} = \frac{3\gamma h}{4\pi} B \tag{13.69}$$

The energy levels for $I = 3/2$ in a constant magnetic field are shown in Fig. 13.8. The separation between levels of Eq. (13.68) is

Fig. 13.8 Energy levels of
Eq. (13.69)

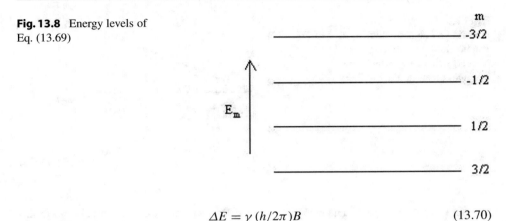

$$\Delta E = \gamma\,(h/2\pi)B \qquad\qquad (13.70)$$

The separation between the levels increases linearly with B. Since energy is related to frequency, hence Eq. (13.70) is

$$\Delta E = h\nu = \gamma(h/2\pi)B$$

$$\nu = \frac{\gamma B}{2\pi} \qquad\qquad (13.71)$$

$$\omega_L = \gamma B \qquad\qquad (13.72)$$

ω_L is called Larmor frequency. For H^1 in normal magnetic field (2.35–18.6 T), the frequency is in the range 100–800 MHz. Putting the value of γ in Eq. (13.71)

$$\nu = g_N \frac{eB}{4\pi\,M_p}$$

$$h\nu = g_N \frac{eh}{4\pi\,M_p} B = g_N\beta_N B \qquad\qquad (13.73)$$

where $\beta_N = eh/4\pi M_p$ is called nuclear magneton. Equations (13.71), (13.72) or (13.73) are called resonance condition. The resonance for NMR may be achieved by varying either B or by varying driving frequency. Figure 13.9 shows the energy levels for a proton ($I = 1/2$).
From Eq. (13.68)

$$E_{\pm 1/2} = \pm(-1)\frac{h}{4\pi}\gamma B_0 = \pm(-1)\frac{1}{2}g_N\beta_N B_0 \qquad\qquad (13.74)$$

The separation between the two levels increases linearly with the magnetic field and transition between them can be induced by magnetic field component B_1 of radio frequency, which is at right angle to B. The transitions, which give rise to *NMR* spectra, are

Fig. 13.9 Energy levels of proton in a magnetic field. Application of radio frequency of energy $h\nu$ causes resonance at a field given by Eq. (13.73)

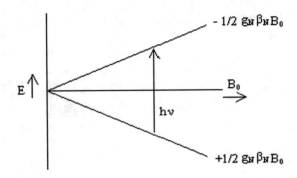

magnetic dipole in origin. The selection rule for the magnetic quantum number m is $\Delta m = \pm 1$.

The phenomenon of *NMR* can also be understood on the basis of precession as explained for ESR in Sect. 13.1.2, but with the difference that electronic magnetic moment is replaced by nuclear magnetic dipole moment. Similarly, quantum mechanical description of NMR phenomenon is on the line of ESR as described in Sect. 13.1.3 with nuclear moments replacing electronic moments.

Two basic branches exist in *NMR* spectroscopy

(a) High-resolution *NMR* is usually used to study molecules with spin I nuclei dissolved in low viscosity solvents.
(b) Broadline *NMR* is observed in solids.

The most obvious difference between the *NMR* spectra of liquids and solids lies in the linewidth of the observed resonance. In solids, the width of the lines lies within the range 1–100 kHz, whereas in liquids it is in the range 1–100 Hz. The difference is due to rapid molecular motion in liquids or solutions greatly reducing the magnetic coupling of the nuclear dipole. However, this coupling is preserved in the solids. The *NMR* spectra of liquids are very sharp, and as a result, the very small but important electron-coupled spin–spin interaction (Sect. 13.2.4 and chemical shifts Sect. 13.2.3) can be observed. On the other hand, the anisotropic interactions are lost. The broadline *NMR* studies these interactions. The interaction between the spin of the two nuclei is the dipole–dipole coupling between their magnetic moments. The effect of dipolar field upon a nucleus will depend upon where this nucleus is located within the dipolar field. This shift in *NMR* frequency due to dipolar field is

$$\Delta\omega_{\text{dip.}} = \text{constant}\ \frac{1 - 3\cos^2\theta}{r^3} \tag{13.75}$$

where θ is the angle between the line connecting the interacting nuclei and external magnetic field and r is the internuclear distance. The constant of proportionality is the magnitude of the magnetic moment of the interacting nuclei. Thus, anisotropic shift in *NMR* resonance frequency, which is sensitive to internuclear distance, provides a measure of nuclear geometries. The main application of broad line *NMR* is to the determination of internuclear distance and other crystal parameters.

13.2.2 Relaxation Mechanisms

Spin–lattice relaxation

Each nuclear spin in the system is not entirely isolated from rest of the assembly of the molecules commonly referred to as the lattice. Each nucleus sees a number of nearby nuclei both in the same molecule and the other molecules. The neighbouring nuclei are in motion relative to the observed nucleus. The electronic and nuclear motion of the molecule produces fluctuating magnetic fields. The observed nuclear magnetic moment, which is precessing about the applied magnetic field, also experiences the fluctuating magnetic fields of its neighbours. Since the motion of each molecule and its neighbours are random, therefore there will be a broad range of frequencies describing them. The fluctuating magnetic fields will have components of right frequency in a direction perpendicular to the static field. This oscillating field causes transition from upper level to lower level. The lattice absorbs the energy exchanged in this process and the population difference between the levels is maintained.

The relaxation time in *NMR* is defined as the time required for the excited state population to fall $1/e$ of its original population. Spin–lattice relaxation time is also known as longitudinal relaxation time and is denoted by the symbol T_1. $T_1 \approx 10^{-2}$–10^4 s for solids and from 10^{-2} to 10^2 s for liquids.

Spin–spin relaxation

In this mechanism, spins can exchange energy among themselves, rather than giving it back to the lattice. The spin exchange process is brought about when a precessing nucleus generates a localized magnetic field at the site of the neighbouring nucleus with a different nuclear spin orientation. The magnetic field produced by the precessing nucleus can be resolved into a static component along the direction of static field and into an oscillating component. This oscillating component has a frequency corresponding to resonance frequency as the nuclear dipole was precessing with Larmor frequency about the static field. This will induce a transition in the neighbouring nucleus. If this spin exchange or flip-flop occurs, the nuclei will exchange magnetic energy states with no overall change in the energy of the system but with a shortening of the lifetime of each. This will have a pronounced effect on the uncertainties in the energy level separation. These spin–spin relaxation times are known as transverse relaxation times and are denoted by the symbol T_2. For solids $T_2 \approx 10^{-4}$ s and for liquids $T_2 \approx T_1$.

From uncertainty relation

$$\Delta E \Delta \tau \sim (h/2\pi) \tag{13.76}$$

$$\Delta \nu \sim \frac{1}{2\pi \Delta \tau} \tag{13.77}$$

where $\Delta \tau$ is relaxation time. $\Delta \tau \approx 10^{-4}$ s, $\Delta \nu \approx 0.11 \times 10^4$ Hz. When T_2 is more important relaxation time, it can be calculated from the shape of the absorption line. The relaxation time T_2 is

$$\frac{1}{T_2} = \pi \Delta \nu_{1/2} Hz \tag{13.78}$$

where $\Delta \nu_{1/2}$ is the half-width of the line at half the absorption peak height.

13.2.3 Chemical Shift

The principal effect underlying the usefulness of *NMR* technique is chemical shift. This refers to the fact that the field at the nucleus is not external field B but one which is modified by the chemical environment. The origin of the chemical shift is that a nucleus in a molecule is surrounded by an electric charge cloud that is a reflection of a chemical bonding about the nucleus. The shape of this charge cloud is usually not spherical but is of complicated form. External magnetic field induces motion in the charge cloud such that an additional magnetic field called local field is developed. This field is proportional to B and algebraically added to it. Thus, the molecular electron cloud effectively shields the nucleus from the external field. This shielding interaction causes the magnetic field seen by the nucleus in the molecule to be different from B. The resonance condition given by Eq. (13.72) takes the form

$$\nu = \frac{\gamma B_{\text{eff}}}{2\pi} = \frac{\gamma}{2\pi}(B - B_{\text{local}}) = \frac{\gamma}{2\pi}B(1 - \sigma_i) \tag{13.79}$$

B_{eff} is the algebraic sum of B and local field. σ_i is the shielding constant for nuclei i. The σ is less than 10^{-5} for proton and less than about 10^{-3} for most other nuclei. Since the magnitude of shielding depends on the orientation of the molecule relative to B, σ is a second rank tensor. However, for gases and liquids orientational dependence is averaged out and σ may be treated as a scalar. It is convenient to regard σ as the sum of positive diamagnetic contribution σ_d and a negative paramagnetic contribution σ_p and hence write $\sigma = \sigma_d + \sigma_p$. σ is positive if diamagnetic contribution dominates and is negative if the paramagnetic contribution dominates.

The effect of screening is to decrease the spacing of the nuclear magnetic energy levels of H^1 as the effective field seen by the nucleus is decreased. At constant frequency of radio frequency field, the value of B will have to be increased to achieve the resonance condition given by Eq. (13.73). Thus, if the resonance absorption positions are expressed on a scale with B increasing from left to right, the absorption peaks for more shielded nucleus will be on the right hand side of the spectrum. This is because more screened nucleus requires higher field to satisfy resonance condition.

The frequency for resonance for a nucleus in a reference compound in analogy with Eq. (13.79) is

$$v_r = \frac{\gamma B}{2\pi}(1 - \sigma_r) \tag{13.80}$$

Difference of frequency for the resonating nucleus in a sample and in the reference compound (having the same nucleus) is obtained from Eqs. (13.79) and (13.80)

$$v_i - v_r = \frac{\gamma B}{2\pi}(\sigma_r - \sigma_i) \tag{13.81}$$

From Eqs. (13.80) and (13.81)

$$\frac{v_i - v_r}{v_r} = \frac{\sigma_r - \sigma_i}{1 - \sigma_r} \tag{13.82}$$

Since $1 \gg \sigma_r$,

$$\frac{v_i - v_r}{v_r} = \sigma_r - \sigma_i = \delta \tag{13.83}$$

δ is called chemical shift. It is the frequency difference in Hz divided by the operating frequency in MHz. The value of chemical shift is expressed in parts per million (ppm). The use of dimensionless scale unit has the great advantage as the chemical shift values so expressed are independent of the value of B (or v) of any particular NMR spectrometer. However, the spacing of the chemical shift position between two given nuclei depends on B (or v).

The chemical shift scale for a given nucleus is established by choosing a substance as a standard and defining its chemical shift as zero. For proton, C^{13} and Si^{29}, tetramethylsilane $(CH_3)_4Si$ (TMS) is used as a reference material because (i) it is miscible with most of the organic solvents, (ii) it is inert and (iii) highly volatile (~300 K) and therefore can easily be removed after the measurement have been made. In TMS, protons are highly shielded and give a narrow resonance line. For water-soluble substances, DSS (2,2-dimethyl-2 silapentane-5-sulphonate) may be used as a reference. The proton in the methyl group of DSS gives a strong resonance. The compounds C_6F_6, $CFCl_3$ and trifluoroacetic acids are used as a reference for F^{19} while Cl^-(aq.) and H_2O are used as reference for Cl^{35} and O^{17}, respectively.

In the H^1 NMR spectrum, the less screened region is on the left hand side of the spectrum with TMS on the right. If B is kept constant, then the effective magnetic field

Table 13.1 Typical proton chemical shifts

Compound	δ
TMS	0
CH_4	0.2
CH_3 in C_2H_5OH	1.17
$(CH_3)_2CO$	2.1
$(CH_3)_2O$	3.2
CH_2 in C_2H_5OH	3.59
C_2H_4	5.5
C_6H_6	7.2
$CHCl_3$	7.2
CH_3OCHO	8.03
CH_3CHO	9.716

seen by the nucleus goes on increasing as one moves away from TMS signal position on the left. Thus to observe resonance, the frequency according to resonance condition should also increase. Thus, frequency increases to the left of TMS. On the right of TMS signal, the nucleus is more screened, and hence, effective field seen by the nucleus is also less. Therefore, for this low magnetic field, the resonance frequency is also low. The frequency therefore decreases on the right of TMS signal. Alternatively, if the frequency is kept constant and B is varied, then highly screened TMS will show signal at low magnetic field. Thus, magnetic field increases from left to right. The left hand side region of TMS is termed as high frequency and low magnetic field region while right hand side of TMS is low frequency, high field region. In H^1 NMR, all those protons on high field of TMS have negative chemical shift value. This scale is called δ scale. An alternate system, which is generally used for defining the position of resonance relative to the reference, is tau (τ) scale. On this scale, the reference is assigned the arbitrary position of 10 and the values of other resonance are given by $\tau = 10 - \delta$.

The proton resonance shifts are restricted to a very small range (Table 13.1). The H^1 chemical shifts in organic compounds can be correlated with the electronegativity of neighbouring groups, type of carbon bonding and hydrogen bonding. Protons attached to or near electronegative groups or atoms such as O, OH, halogens, CO_2H, NH_3^+ and NO_2 experience a lower density of shielding electrons and are resonant at lower value of B_o. Protons removed from such groups appear at higher value of B. For example, in low-resolution spectrum, that is, showing only chemical shifts, of ethyl alcohol furnishes an illustration of these effects. Since oxygen is more electronegative than C, the electron density on H in O–H bond will be less than on in the C–H bond. The OH proton of CH_3CH_2OH is least shielded as it is attached to the more electronegative oxygen atom. Out of CH_2 and CH_3, proton of CH_2 is expected to be less shielded as it is directly attached to the oxygen atom. The observed lines are shown on a

Fig. 13.10 A stick diagram showing the low-resolution H^1 NMR spectrum of ethyl alcohol

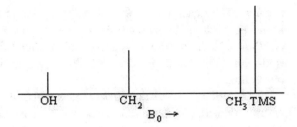

OH CH_2 CH_3 TMS

$B_0 \rightarrow$

stick diagram in Fig. 13.10. The number of protons in OH, CH_2 and CH_3 is 1, 2 and 3, respectively. Therefore, the intensity of lines should also be in the same ratio, which is observed experimentally.

13.2.4 Spin–Spin Coupling

The interaction between the spins of the neighbouring nuclei in a molecule may cause the splitting of the lines in the *NMR* spectrum. This interaction occurs via the bonding electrons. The coupling between the nuclei is a two-stage process, in which the first nucleus A creates a magnetic field affecting its electron shell, which in turn influence the electron shell around the other nucleus B. This changes the magnetic field at the nucleus B and hence changes the nuclear frequency. The effect is mutual. The magnitude of the effect for a particular pair of nuclei depends on the following factors:

(i) The coupling of two nuclei via polarization of the intervening electron spins depends upon the electron density at each nucleus. As s electrons have appreciable density at the nuclei, therefore spin–spin interaction depends on s character in the bonds. (ii) Coupling of two nuclei A and B occur via the orbital motion of the valence electrons. The nuclear magnetic moment of a nucleus A will induce currents in its electron cloud, and the resulting magnetic field will be felt by the other nucleus B and vice versa. This coupling may be unimportant for proton- proton interaction. (iii) Coupling between equivalent nuclei or group of equivalent nuclei is not directly observable. A group of nuclei is chemically equivalent if nuclei in the group are related by a symmetry operation of the molecule and have same chemical shift. The nuclei experiencing different environment or having different chemical shifts are called non-equivalent. (iv) The magnetic moment of the nuclei and is directly proportional to $\gamma_A \gamma_B$ where γ_A and γ_B are the magnetogyric ratios of the interacting nuclei A and B. (v) The coupling generally decreases rapidly as the number of interacting bonds increases and is not usually observed over more than four or five bonds.

The magnitude of the coupling interaction is measured in Hz. It is called coupling constant and is denoted by the symbol *J*. The coupling constant can be either positive or negative. The magnitude of the multiplet separation resulting from spin–spin interaction is independent of the strength of the applied magnetic field. Although the absorption positions of a group (in Hz) changes with applied magnetic field strength, the spacing of spin–spin coupled multiplet does not. The spacing is determined by the spin–spin coupling interaction.

For describing systems of nuclear spins within the molecule, the following notation is usually adopted. Non-equivalent nuclei of the same species which have chemical shift difference comparable to the mutual coupling are denoted by A, B, C, ..., etc. The another group of nuclei separated from this group (A, B, C, ...) by large chemical shifts are denoted by X, Y, Z, ..., etc. The nuclei of second group have chemical shift difference among themselves comparable to mutual coupling. Equivalent nuclei are denoted by the same symbol, and the number of equivalent nuclei as subscript, e.g. PF_3, would be referred to as AX_3.

Let us consider the following example. There are two nuclei A and B, three bonds away from one another in a molecule. Assume that nuclear spin of B is 1/2. In the external magnetic field, spin of B is either parallel or antiparallel to the magnetic field. The A nuclei in the sample can each experience one of the two different perturbations. The magnetic field at the nucleus A will be either greater or smaller than the external magnetic field by a constant amount due to the influence of nucleus B. A similar effect would be observed by B due to A. The NMR spectrum for nuclei A and B reflects the splitting observed in the energy level diagram. The absorption line of A is split into two components centred on δ_A and absorption line of B is split into two components centred on δ_B. The separation between two split absorption lines is called the J coupling constant or the spin–spin splitting constant and is a measure of the magnetic interaction between two nuclei.

Knight Shift

Knight discovered that the resonance frequency in metals is higher than the nuclei of the same isotope in chemical compounds in the same magnetic field. This effect is due to the local field produced at the position of the nuclei by the paramagnetism of the conduction electrons.

13.3 Mössbauer Effect

Isomer Nuclear Transitions

A nucleus has discrete energy levels. When a transition from upper level to lower energy level takes place, gamma rays are emitted. The time during which the nucleus remains in a state determines its mean lifetime for that state. Two nuclei with equal charge and mass number but in different excited states are called isomer nuclei. The lifetime of the excited state is ~10^{-6} to 10^{-8} s. The gamma emission of Fe^{57} nucleus with energy of 14.4 keV occurs as a result of isomer transition from the excited state to ground state of a Fe^{57} nucleus. Since in the case of a Fe^{57} an isomer with energy of 14.4 keV has a mean lifetime τ of only about 1.4×10^{-7} s, in practice cobalt isotope Co^{57} with a half-life of 270 days is taken as a source of gamma ray irradiation; then through electron capture Co^{57} transforms into an excited Fe^{57} isomer. In the case of tin, its isomer with a long lifetime (250 days) is used.

Nuclear Gamma Resonance Fluorescence

Consider a nucleus in its excited state whose energy is E. Nucleus emits gamma rays as a result of transition from excited state to the ground state. The energy of this gamma ray photon is $E\gamma = h\nu$. If this gamma ray photon is directed on another identical nucleus in its ground state, the photon may be absorbed, resulting in lifting of the nucleus from ground state to its excited state. The process, which is possible only because the energy of photon is exactly equal to the energy of the excited state of the second nucleus, is the case of resonance absorption. However, the resonance absorption does not take place in the nucleus, because when the nucleus emits gamma rays, the nucleus recoils and the energy of the emitted gamma rays is reduced by the amount of the recoil energy. The energy of the emitted gamma rays is

$$E_e = E - E_R$$

E_e energy of the emitted photon
$E_R=$ recoil energy of the emitter.

Similarly, the absorbing nucleus recoils forward as it absorbs the photon, acquiring some translational kinetic energy, and consequently, if the absorption is to take place, the photons energy must be slightly greater than E, that is,

$$E_a = E + E_R$$

where E_a is the energy of the absorbing nucleus.

Figure 13.11 shows the position of E_e and E_a relative to the hypothetical recoil-free situation. Since $E_e < E_a$, the emitted photon does not appear to have enough energy to excite the second nucleus. Therefore, resonance absorption is not usually observed in nuclear case.

Let us consider an isolated atom in the gas phase and the energy difference between the ground state E_g and excited state E_e is given by

Fig. 13.11 Position of E_e and E_a relative to recoil-free situation

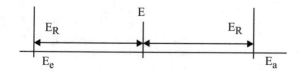

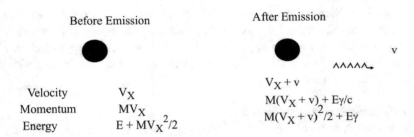

Fig. 13.12 Energy and momentum are conserved in the gamma emission process

$$E = E_e - E_g \tag{13.84}$$

Let the emitting atom of mass M is moving with a velocity V_x in the x-direction (Fig. 13.12). The linear momentum of the atom before emission of gamma rays is MV_x. After emission of gamma ray, assumed in x-direction, the linear momentum of the system (gamma ray plus de-excited nucleus) must still equal to MV_x, that is,

$$M V_x = \frac{h\nu}{c} + M(V_x + v) \tag{13.85}$$

where $(V_x + v)$ is the velocity of the atom after emission of the gamma ray, v is vector, and therefore, it can be negative. From Eq. (13.85)

$$v = -\frac{h\nu}{Mc} \tag{13.86}$$

recoil energy E_R is therefore by using Eq. (13.86) is

$$E_R = \frac{M v^2}{2} = \frac{E^2}{2Mc^2} \tag{13.87}$$

before emission of gamma ray, the total energy above the ground state nucleus at rest is $(E + MV_x^2/2)$. After emitting the gamma ray of energy $E\gamma$, the nucleus will have a new velocity $(V_x + v)$ due to recoil. The total energy of the system is $E\gamma + M(V_x + v)^2/2$. From conservation of energy

$$E + \frac{1}{2}M V_x^2 = E_\gamma + \frac{1}{2}M(V_x + v)^2 \tag{13.88}$$

$$\delta E = E - E_\gamma = \frac{1}{2}M v^2 + M v V_x \tag{13.89}$$

or

$$E = E_R + E_D \tag{13.90}$$

where δE is difference between energy of the nuclear transition E and energy of the emitted gamma ray photon $E\gamma$. This difference depends (i) on the recoil kinetic energy which is independent of the velocity V_x (ii) on the term $E_D = MvV_x$ which is proportional to the atom velocity V_x and is the Doppler effect energy.

The mean kinetic energy per translational degree of freedom of a free atom in a gas with random thermal motion is given by

$$< E_k > = M < V_x^2 > \approx \frac{1}{2} k_B T \qquad (13.91)$$

$<V_x{}^2>$ is the mean square velocity of the atom, k_B is the Boltzmann constant and T is the absolute temperature. From Eq. (13.91), we have

$$(< V_x^2 >)^{1/2} = \left(\frac{2 < E_k >}{M} \right)^{1/2} \qquad (13.92)$$

using $E_D = MvV_x$ and Eq. (13.91) we have mean broadening

$$< E_D > = M v (< V_x^2 >)^{1/2} = 2 (E_R < E_k >)^{1/2} \qquad (13.93)$$

The gamma ray distribution (Eq. (13.90)) is displaced by E_R and broadened by twice the geometric mean of the recoil energy and the average thermal energy (Eq. (13.93)). The distribution, which is Gaussian, is shown in Fig. 11.4. Using Eqs. (13.86) in (13.93), we have

$$< E_D > = E_\gamma \left(\frac{2 < E_k >}{Mc^2} \right)^{1/2} \qquad (13.94)$$

for gamma rays of energy 10^4 eV, and a mass $M = 100$ amu, it is found that $E_R = 5.4 \times 10^{-4}$ eV and $<E_D> \sim 5 \times 10^{-3}$ eV at 300 K. Thus, the resonance overlap for free atom resonance is small.

Mössbauer Effect

In 1958, Mössbauer used an Ir^{191} gamma ray source of 129 keV, for which the Doppler broadening at room temperature is about twice the value of E_R, and therefore, the lines in Fig. 11.2 overlapped a little and resonance fluorescence could be observed. He expected that on cooling, the emitter and absorber, the absorption should decrease because of the decrease in Doppler width and thus in the overlap. However, he observed an increase in absorption. The absorption was as much as would be expected if there had been no recoil at all. The qualitative explanation of this fact is that at sufficiently low temperature an atom in a solid cannot recoil individually. The crystal as a whole absorbs the recoil momentum. The effective mass in Eq. (13.87) is therefore the mass of the crystal, which is so much larger than that of atom that the recoil energy is completely negligible. From Eq. (13.94), $<E_D>$ will also be negligible.

We know that the nucleus is not bound rigidly in crystal, but is free to vibrate. The above or it can be transferred to the lattice by increasing recoil energy of a single nucleus can be taken by either by the whole crystal as discussed the vibrational energy of the crystal. The vibrational energy levels of the crystal are quantized. Therefore, it can be excited only if the recoil energy corresponds closely with the allowed values, thus ensuring that the whole crystal recoils, leading to negligible recoil energy. Thus, the necessary condition for the Mössbauer effect to occur is that the nucleus emitting the gamma ray photon should be in an atom, which has established vibrational integrity with the solid matrix. The vibrational energy of the lattice as a whole can change by discrete amounts $0, \pm \hbar \omega, \pm 2 \hbar \omega, \ldots$ If $E_R < \hbar \omega$, no transfer of energy will take place as either zero or $\hbar w$ units of vibrational energy but nothing intermediate can be transferred. When an average is taken over many emission processes, the energy transfer per event is exactly the free atom recoil energy. Let f be fraction of events which takes place without transferring energy to the lattice ($E_R < \hbar \omega$), then a fraction $(1 - f)$ will transfer one photon energy $\hbar \omega$, neglecting two, three, etc., quantum transitions to a first approximation and therefore

$$E_R = (1 - f)\hbar \omega \tag{13.95}$$

or

$$f = 1 - \frac{E_g}{\hbar \omega} \tag{13.96}$$

Only those events give rise to the Mössbauer effect. f is often called Mössbauer coefficient. The probability that recoilless emission or absorption occurs is then given by general expression for the elastic scattering process

$$f = \exp[-4\pi^2 < s^2 > /\lambda^2] \tag{13.97}$$

where λ is the wavelength of the gamma rays, $<s^2>$ is mean square thermal displacement of the appropriate atom in the direction of photon emission. Equation (13.97) can be written as

$$f = \exp[-E_\gamma^2 < s^2 > /\hbar^2 c^2] \tag{13.98}$$

$E\gamma$ is gamma photon energy. $<s^2>$ is calculated on the basis of Einstein or Debye model. When $<s^2>$ is calculated on the basis of Debye model, f is called Debye–Waller factor. The value of f becomes large for smaller values of $E\gamma$ and the smaller $<s^2>$. $<s^2>$ depends on the firmness of binding and on temperature. The reduction of force constant causes reduction of vibrational frequency of the lattice and hence increases in the probability of the recoil. The displacement of the nucleus must be smaller than wavelength. This is why Mossbauer effect is not observed in gases and non-viscous liquids. The smaller $<s^2>$ implies lower temperature and larger Debye or Einstein temperature. Because f decreases as E increases and there is an upper limit to $E\gamma \sim 150$ keV but in practice most Mossbauer nuclei have gamma ray energies < 100 keV. On the other hand gamma photon with $E\gamma < 10$ keV are strongly absorbed by matter, so this energy forms an approximate lower limit.

The absorption process can be modulated by rigidly moving either the emitter or the absorber. If the emitter moves towards the absorber with a velocity v, the frequency ν of emitted gamma ray undergoes a Doppler shift

$$\nu = \nu_0 (1 + \frac{v}{c}) \qquad (13.99)$$

where ν_0 is the frequency of radiation from a stationary emitter. If the emitter and absorber are tuned in the beginning, the motion of the emitter (or absorber) causes detuning and reduces absorption. On the other hand, if the absorber and emitter are detuned at the beginning, the motion of emitter (or absorber) can be so adjusted as to bring in the desired tuning. If E_e and E_a are the energies of the emitter and absorber such that $E_e = h\nu_0$, $E_a = h\nu$, then using Eq. (13.99) we have

$$\frac{E_a - E_e}{E_a} = \frac{h(\nu - \nu_0)}{h\nu} = \frac{v}{c} \qquad (13.100)$$

The above expression gives the velocity of emitter (or absorber) required to establish tuning.

Mössbauer spectra are measured by using spectrometer consisting of an emission source, a specimen (absorber), a detector for the resonant gamma rays transmitted through the absorber and a system capable of moving the emitter (or absorber). Emitter, e.g. Fe^{57}, is prepared by embedding the radioactive isotope, Co^{57}, in metallic iron, stainless steel and other host materials. A Mössbauer spectrum of a sample reveals resonance fluorescence of the same nucleus (the same isotope), which is incorporated in the emitter, and emits gamma rays. Therefore, by means of Fe^{57} emission it is possible to investigate only the iron-making part of the absorber. Detection is done by gamma ray counter. The motion of the emitter relative to the absorber (required to equalize by means of Doppler effect) and the energy of gamma ray quanta emitted by the emitter and absorbed by the specimen are measured. The number of counts in the detector is measured as a function of the velocity. The minimum of the transmission of gamma ray quanta corresponds to the maximum of the absorption by the absorber (specimen). This corresponds to the Doppler velocity at which the resonance absorption occurs. The Mössbauer spectrum is therefore a velocity spectrum and represents the number of gamma rays quanta recorder by the counter at different Doppler velocities of emitter relative to the absorber.

For nuclei to be used for Mössbauer spectroscopy depend upon the following:

(i) The utilized radioactive isotope (parent nucleus) must have fairly long half-life. (ii) isotope (parent nucleus) must have fairly long half-life. For a given element, a source of gamma ray emission must exist—a parent nucleus whose decay gives rise to the appearance of isomer nuclei. (iii) The Mössbauer emission energy lies within the range of a few keV up to lower hundred of *keV*. (iv) The lifetime of the isomer excited level should be in the range $\sim 10^{-6}$–10^{-3} s. The lifetime determines the width Γ of lines and ratio $\Gamma/E \sim 10^{-10}$–10^{-14}. Longer lifetime reduces Γ/E. Therefore, Doppler velocity becomes too low and measurable only with difficulty. Conversely, with rising Γ/E, the resonance selectivity diminishes.

13.3.1 Isomer (Chemical) Shift

The Coulombic interaction between the nuclear charge and the electron charge alters the energy separation between the ground state and the excited state of the nucleus. Therefore, it causes a slight shift in the position of the observed resonance line. This shift will be different in different chemical compounds (Fig. 13.13). For this reason, it is generally known as the isomer shift or chemical shift or isomer chemical shift.

The isomer shift is calculated by assuming the nucleus to be a uniformly charged sphere of radius R and the electronic charge density is uniform over nuclear dimensions. At a distance r from the centre of the atomic nucleus, the electrostatic potential is given by (for $r > R$)

$$V = \frac{Ze}{4\pi\,\varepsilon_0 r} \tag{13.101}$$

and if $r < R$

$$V' = \frac{Ze[3 - (r^2/2R^2)]}{(4\pi\,\varepsilon_0)2R} \tag{13.102}$$

Equation (13.91) is for the hypothetical point nucleus. The s electrons ($l = 0$) also have a finite density within the atomic nucleus. The energy state of the nucleus is affected by the charge originating from the electrons in the sphere of radius R. The energy difference

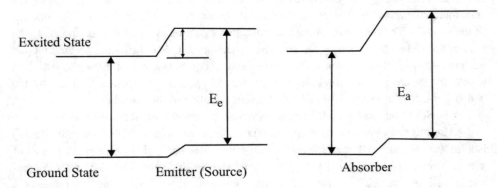

Fig. 13.13 Isomer shift

E is given by

$$\delta E = \int\limits_0^\infty \rho(V' - V)4\pi)^2 dr$$

$$\delta E = \int\limits_0^R \frac{4\pi\,\rho\,Z\,e}{(4\pi\,\varepsilon_0)R}\left[\frac{3}{2} - \frac{r^2}{2R^2} - \frac{R}{r}\right]r^2 dr \tag{13.103}$$

or

$$\delta E = -\frac{2\pi\,Z\,\rho\,e\,R^2}{5(4\,\pi\,\varepsilon_0)} = \frac{2\pi\,Z\,e^2}{5(4\,\pi\,\varepsilon_0)}|\psi_s(0)|^2 R^2 \tag{13.104}$$

where $\rho = -e|\psi_s(0)|^2$ is the charge density originating from the s electrons in the atomic nucleus.

The atomic nuclei in excited states usually have a different radius from those in the ground state; they may be either smaller or larger. The radius of excited state of I^{129} is greater than the radius of ground state of I^{129}. On the other hand, the radius of the excited state of Fe^{57} is smaller than the radius of ground state of Fe^{57}. The energy difference δE will be different in excited and ground states. For excited state, Eq. (13.104) is

$$\delta E_e = \frac{2\pi\,Ze^2}{5(4\,\pi\,\varepsilon_0)}|\psi_s(0)|^2 R_e^2 \tag{13.105}$$

and for the ground state, Eq. (13.104) is

$$\delta E_g = \frac{2\pi\,Ze^2}{5(4\pi\,\varepsilon_0)}|\psi_s(0)|^2 R_g^2 \tag{13.106}$$

On subtracting Eqs. (13.106) and (13.105), we obtain

$$\delta E_e - \delta E_g = \frac{2\pi\,Ze^2}{5(4\pi\,\varepsilon_0)}|\psi_s(0)|^2 (R_e^2 - R_g^2) \tag{13.107}$$

by subtraction, the interaction energy resulting from the charges at a distance $r > R$ from the centre of the atomic nucleus is eliminated, as these interaction energies are approximately the same in the excited and ground states; accordingly, the integration in Eq. (13.103) is between the limit 0 and R. The magnitude of $|\psi_s(0)|^2$ depends somewhat on electron configuration, chemical environment and other factors vary from compound to compound.

An experiment usually determines the shift in gamma ray energies between absorber and the emitter, and the shift δ is given by

$$\delta = E_{ab} - E_{em} = \frac{2\pi Z e^2}{5(4\pi \varepsilon_0)}\left[|\psi_s(0)|^2_{ab} - |\psi_s(0)|^2_{em}\right](R_e^2 - R_g^2) \tag{13.108}$$

Now

$$R_e^2 - R_g^2 = (R_e - R_g)(R_e + R_g) \tag{13.109}$$

using $\Delta R = (R_e - R_g)$ and $R = (R_e + R_g)/2$ in Eq. (13.109)

$$R_e^2 - R_g^2 = \frac{2\Delta R R^2}{R} \tag{13.110}$$

Substituting Eq. (13.110) in Eq. (13.108), we have

$$\delta = E_{ab} - E_{em} = \frac{2\pi\,Ze^2}{5(4\pi\,\varepsilon_0)}\left[|\psi_s(0)|^2_{ab} - |\psi_s(0)|^2_{em}\right]2R^2\frac{\Delta R}{R}$$
$$\delta = k\frac{\Delta R}{R}\left\{|\psi_s(0)|^2_{ab} - |\psi_s(0)|^2_{em}\right\} \tag{13.111}$$

where $k = [4\pi\,R^2 Z e^2/(4\pi\,\varepsilon_o)5]$. The expression for the isomer shift has two factors (a) $\Delta R/R$ includes only nuclear parameters and is assumed to be constant for nuclei in different emitters and absorbers. (b) Second factor $\left[\psi_s(0)^2_{ab} - \psi_s(0)^2_{em}\right]$ is an atomic parameter linked with the difference in the distribution of the s electron density in sample relative to the emitter. If $\Delta R/R$ is positive, and a positive isomer shift implies an increase in s electron density at the nucleus in going from emitter to absorber. If it is negative, the same (positive) shift signifies a decrease in s electron density. Electrons in $1s$, $2s$, $3s$ .... shell contribute to $|\psi_s(0)|^2$ but in decreasing amounts as the principal quantum number n increases. However, the inner shells are not markedly affected by chemical bonding. Therefore, the principal influence on the isomer shift will be by the outermost occupied s orbitals. Shielding by other electrons effectively increases the s-radial function and decreases the s density at the nucleus.

The isomer shift is thus a measure of the s-electron density at the nucleus of absorber as compared with their density at the emitter nucleus. For the Fe^{57} nucleus, $\Delta R/R$ is negative, and therefore, a positive isomer shift corresponds to a decreased electron density on the nucleus of the absorber. The s-electron density, however, becomes subject to the effect produced by the $3d$ electrons of the iron. The effect manifests itself by shielding the s-electrons on the nucleus. Therefore, an increase in the number of d electrons leads to a reduction of the s-electron density at the nucleus, and consequently, to a greater isomer shift. For example, in comparison with metallic iron (configuration $3d^6 4s^2$ or $3d^7 4s^1$), the density of s-electron diminishes in compounds of Fe^{2+} (electron configuration $3d^6$) or Fe^{3+} (electronic configuration $3d^5$) and a positive isomer shift is observed. Isomer shift for Fe^{2+} is 1.3–1.4 mm/s and for Fe^{3+} it is 0.4–0.5 mm/s. In Fe^{2+} compounds, the number of $3d$ electrons is greater than in Fe^{3+} compounds, and therefore, the s-electrons densities at the nucleus for Fe^{2+} are smaller and isomer shifts are generally larger.

13.3.2 Quadrupole Splitting

Any nucleus with a spin $I \geq 1$ has a non-spherical charge distribution and hence a quadrupole moment. In Mössbauer spectra, the quadrupole splitting becomes manifest in the presence of two conditions: (i) When the Mössbauer nucleus displays the quadrupole moment in the ground or excited (isomer) states (ii) The presence of electric field gradient (EFG) at the site of nucleus position.

The asymmetric of the electric charge about the nucleus gives rise to EFG. The EFG acting on the nucleus is given by

$$Q = q_v(1 - R) + q_L(1 - \gamma_\infty) \tag{13.112}$$

The aspherical distribution of lattice charges will contribute q_L, which is enhanced at the nucleus through polarization of inner electron shell by a Sternheimer factor $(1 - \gamma_\infty)$. The unfilled valence electron shells in transition elements produce a contribution q_v which is shielded from the nucleus by a factor $(1 - R)$.

The interaction between the nuclear electric quadrupole moment Q and EFG is expressed by the Hamiltonian

$$\mathbf{H} = \mathbf{Q} \cdot \nabla \mathbf{E}$$

and the energy is given by

$$E_Q = \frac{e^2 q Q [3m_I^2 - I(I + 1)]}{4I(2I - 1)} \text{eV} \tag{13.113}$$

In the ground state, the Fe^{57} nucleus has a spin $I = 1/2$ and therefore has no quadrupole moment. In the excited state, Fe^{57} has a nuclear spin $I = 3/2$ and therefore has a quadrupole moment. The isomer level (excited level with $I = 3/2$) splits into two sublevels with $m_I = \pm 3/2, \pm 1/2$. Reversal of the sign of m_I will not change the nuclear charge distribution. Therefore, EFG will not completely lift the fourfold degeneracy of the $I = 3/2$ state. The quartet will split into two doublets while $I = 1/2$ state will remain degenerate. For $I = 3/2$, the energy of the sublevels $m_I = \pm 3/2, \pm 1/2$ is

$$E_Q(\pm 3/2) = \frac{e^2 q Q}{4} \text{eV}$$

$$E_Q(\pm 1/2) = -\frac{e^2 q Q}{4} \text{eV}$$

and

$$\Delta = E_Q(\pm 3/2) - E_Q(\pm 1/2) = \frac{e^2 q Q}{2} \text{eV} \tag{13.114}$$

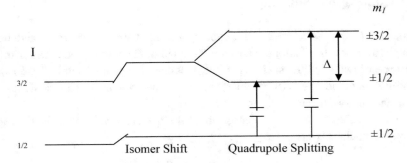

Fig. 13.14 Quadrupole splitting in Fe^{57}

The splitting of the level is shown in Fig. 13.14. The transitions between these energy levels take place according to selection rule $\Delta m_I = 0, \pm 1$ and resulting in two lines. The separation between the transitions $m_I = \pm 1/2 \leftrightarrow \pm 1/2$ (σ transition) and $m_I = \pm 1/2 \leftrightarrow \pm 3/2$ (π transitions) is given by Eq. (13.114). The relative probability of the transitions $m_I = \pm 1/2 \leftrightarrow \pm 1/2$ and $m_I = \pm 3/2 \leftrightarrow \pm 3/2$ is identical, but shows an angular dependence. In monocrystals, for transitions $m_I = \pm 3/2 \leftrightarrow \pm 1/2$ the angular dependence is of the form $3/2(1 + \cos^2\theta)$, and for $m_I = 1/2 \leftrightarrow \frac{1}{2}$, it is of the form $(1 + 3/2 \sin^2\theta)$. In polycrystalline sample, the intensities are equal.

13.3.3 Magnetic Hyperfine Structure

The energy levels of a nucleus, possessing a magnetic moment, are perturbed by the magnetic field. In the presence of magnetic field, the nuclear spin magnetic moment experiences a dipolar interaction with magnetic field, i.e. Zeeman effect. The magnetic field at the nucleus is local one; i.e. it does not extend over the whole of the crystal but arises due to the interaction of the nucleus itself with its own electrons. The internal magnetic field at the nucleus is

$$B = B_{contact} + B_{orbital} + B_{dipole}$$

$B_{contact}$ arises due to the interaction of the nucleus with s-electrons. $B_{orbital}$ is due to orbital magnetic moment of the atom, and $B_{dipolar}$ is dipolar field due to spin of these electrons. The field is almost completely linked with the interaction of the nucleus with s-electrons, this being known as Fermi contact term. The s-electrons are subject to the action of unpaired d electrons, which polarize the s-electrons even of the occupied shells. The greatest number of unpaired d electrons, e.g. five d electrons in Fe^{3+}, causes a strong polarization of the s shell and large magnetic field $\sim$40–60 T. Decrease in the number of unpaired d electrons (e.g. four unpaired d electrons in Fe^{2+}) leads to decrease of the magnetic field to ~20–40 T.

The nuclear magnetic moment $\boldsymbol{\mu_n}$ interact with the field and Hamiltonian is given by

$$H = -\boldsymbol{\mu}_s \cdot \mathbf{B} \tag{13.115}$$

and

$$\boldsymbol{\mu}_n = \beta_N g_I \mathbf{I} \tag{13.116}$$

where β_N is nuclear magneton.

On substituting Eq. (13.116) in Eq. (13.115)

$$H = -g_I \beta_N \mathbf{I} \cdot \mathbf{B} \tag{13.117}$$

and corresponding energy is

$$E = -g_I \beta_N m_I B \tag{13.118}$$

where $m_I = I, I - 1, ..., -I$. Thus, magnetic field splits the nuclear level with spin I into $2I + 1$ components. For Fe^{57}, $g_I = 0.1806$ and -0.1033 for ground state and the excited state, respectively. The transitions between these sublevels give rise to six lines according to the selection rules $\Delta m_I = 0, \pm 1$. The line positions are related to the splitting of the energy levels, but the line intensity is related to the angle between Mossbauer gamma rays and nuclear spin moment. The outer middle and inner lines are related by 3: $4\sin^2\theta/(1 + \cos^2\theta)$: 1. The outer and inner lines are always in the same proportion but the middle lines can vary in relative intensity between 0 and 4. In polycrystalline samples, the value averages to 2.

The splitting of the nuclear levels in the magnetic field is quite similar to the splitting of the levels in nuclear magnetic resonance (*NMR*). However, there is difference between the NMR and Mössbauer effect as stated below (i) The magnetic field in *NMR* is superimposed outer field generated by the magnet ~1 T while in Mössbauer effect the intensity of the magnetic field ~20–60 T, (ii) In *NMR*, the ground state alone that comes under consideration while in Mössbauer effect one considers both ground and excited states, (iii) *NMR* results from transitions between neighbouring magnetic sublevels during absorption of a radio frequency quantum ~MHz (10^{-7}–10^{-8} eV) while in Mössbauer effect gamma transitions from magnetic sublevels of the ground level to magnetic sublevels of the excited level occur. In *NMR*, the selection rules are $\Delta m_I = \pm 1$, and on the other hand, selection rules are $\Delta m_I = 0, \pm 1$ for Mössbauer effect.

Solved Examples

Example 1 Calculate the ESR resonance frequency of an unpaired electron ($g = 2$) in a magnetic field of 0.3375 T.

Solution
From Eq. (13.8)

$$v = \frac{g\,\beta\,B}{h} = \frac{2 \times 9.27 \times 10^{-24} \times 0.3375}{6.626 \times 10^{-34}} = 9.446\,\text{GHz}$$

Example 2 Calculate the g value if the methyl radical shows *ESR* at 0.337 T in an *ESR* spectrometer operating at 9.45 GHz.

Solution
From Eq. (13.8)

$$g = \frac{h\nu}{\beta\,B} = \frac{6.626 \times 10^{-34} \times 9.45 \times 10^{9}}{9.27 \times 10^{-24} \times 0.337} = 2.0044$$

Example 3 The ESR spectrum of an unknown sample shows a single line at $g = 2.344$ at a frequency. The *DPPH* with $g = 2.0036$ shows a line at 0.3375 T at the same frequency.
   Determine the magnetic field value of *ESR* line of unknown sample.

Solution
The resonance condition (Eq. 13.8) for unknown sample becomes

$$h\nu = 2.344\beta \times B(T)$$

For DPPH, the resonance condition is

$$h\nu = 2.0036 \times \beta \times 0.3375\,T$$

Equating the above two equations

$$B(T) = \frac{2.0036 \times 0.3375\,T}{2.0185} = 0.335T$$

Example 4 The separation of two lines (splitting) in a free radical *ESR* spectrum is given by 2.672 mT and $g = 2.0050$. Express the splitting in *MHz* and in cm^{-1}.

Solution
Using Eqs. (13.8) and (13.9), we have

$$B(mT) = \frac{71.448\, v\,(\text{GHz})}{g}$$

Substituting the values of g and B in the above, we have

$$\frac{71.448\, v\,(\text{GHz})}{2.005} = 2.672 mT$$

$$v(\text{GHz}) = \frac{2.005 \times 2.672 mT}{71.448} = 75 \times 10^{-3}\,\text{GHz}$$

To convert frequency into wavenumber, divide the frequency by the velocity of light, i.e. splitting (in cm^{-1}) is

$$\frac{v}{c} = \frac{75 \times 10^6\,\text{Hz}}{2.9979 \times 10^{10}\,\text{cm s}^{-1}} \approx 25.0175 \times 10^{-4}\,\text{cm}^{-1}$$

Example 5 Compute the population difference of two states of an electron spin in a magnetic field of 0.4 T at 300 K.

Solution
From Eqs. (13.8) and (13.23)

$$\frac{N_{-1/2}}{N_{1/2}} = \exp(hv/k_B T) = \exp(g\,\beta\,B/k_B T)$$

For $g = 2$, the above equation is

$$\frac{N_{-1/2}}{N_{1/2}} = \exp\left(\frac{2 \times 9.2741 \times 10^{-24}\,JT^{-1} \times 0.4T}{1.3807 \times 10^{-23}\,JK^{-1} \times 300\,\text{K}}\right) \approx 1.00179279$$

Since

$$N_{-1/2} + N_{1/2} = 1$$
$$N_{-1/2} - 1.00179279\,N_{1/2} = 0$$

Solving these two equations

$$N_{-1/2} = 0.500896$$
$$N_{1/2} = 0.499104$$

and

$$N_{-1/2} - N_{1/2} = 1.792 \times 10^{-3}$$

Example 6 The ESR spectrum of Gd^{3+} in $Nd_2Mg_3(NO_3)_{12}.24H_2O$ single crystals shows seven fine structure lines corresponding to allowed selection rule $\Delta M = \pm 1$. Determine the effective spin of Gd^{3+} in this single crystal.

Solution
According to $\Delta M = \pm 1$ transitions can take place between adjacent levels. Therefore, there must be eight levels between which seven fine structure lines corresponding to allowed transitions are observed. Effective spin can be determined by equating the number of levels to $2S + 1$. Thus, $2S + 1 = 8$ or effective spin $S = 7/2$.

Example 7 Determine the relative intensity of the five fine structure transitions of Mn^{2+} in a single crystal. The effective spin is 5/2.

Solution
The intensity of a fine structure transition is proportional to $S(S + 1) - M(M - 1)$. For $S = 5/2$, the values of M are 5/2, 3/2, 1/2, −1/2, −3/3, −5/2. The intensity of $5/2 \rightarrow 3/2$ transition is $[S(S + 1) - M(M - 1)] = 35/4 - 15/4 = 5$.
    Similarly,

> The intensity of $3/2 \rightarrow 1/2$ transition is 8
> The intensity of $1/2 \rightarrow -1/2$ transition is 9
> The intensity of $-1/2 \rightarrow -3/2$ transition is 8
> The intensity of $-3/2 \rightarrow -5/2$ transition is 5

Therefore, the relative intensities of five fine structure transitions are 5:8:9:8:5.

Example 8 The ESR spectrum of VO^{2+} in $(NH_4)_2SeO_4$ show eight lines of equal intensity at 276.2 mT, 296.8 mT, 317.7 mT, 339.6 mT, 359.6 mT, 380.4 mT, 401.8 mT and 422.8 mT. Determine the effective spin and nuclear spin of VO^{2+}. Also determine the value of hyperfine coupling constant and g value of VO^{2+} in this sample given that *DPPH* line (with $g = 2.0036$) occurs at 335.0 mT.

Solution
Since all the lines of the spectrum are of equal intensity, therefore it is hyperfine structure and number of lines are equal to $2I + 1$. Hence, $I = 7/2$. Further, there is only one set of eight hyperfine lines which are centred at an electronic transition which is between two levels and hence $S = 1/2$. To a first approximation, the hyperfine splitting is determined by taking the average of the separation between the lines which is $(20.6 + 20.9 + 21.9 + 20 + 20.8 + 21.4 + 21)/7 = 20.94$ mT. The position where the hyperfine structure is centred is obtained to a first approximation by taking the average of positions of all the eight lines

which is 349.36 mT. Hence, g value can be obtained from the expression

$$h\nu = g_{DPPH}\,\beta\,B_{DPPH} = g_{sample}\,\beta\,B_{sample}$$
$$g_{sample} = \frac{g_{DPPH}\,B_{DPPH}}{B_{sample}} = \frac{2.0036 \times 335\,\text{mT}}{349.36\,\text{mT}} = 1.9212$$

Example 9 The ESR spectrum of a radical with two equivalent nuclei of a particular kind is split into five lines of intensity ratio 1:2:3:2:1. What is the spin of the nuclei?

Solution
According to Eq. (13.60), the total number of lines for equivalent nuclei is $2nI + 1$. Therefore, equating this to five gives $2nI = 4$. Since there are two nuclei, hence $4I = 4$ or $I = 1$. The spin of the nuclei is therefore 1.

Example 10 Calculate the magnetic field needed to satisfy the resonance condition for Hg^{199} nucleus in a 17.83 MHz radio frequency field ($g_{Hg} = 5.585$, $\beta_N = 5.05082 \times 10^{-27}$ JT^{-1}, $h = 6.6262 \times 10^{-34}$ Js).

Solution
From Eq. (13.73)

$$B = \frac{h\nu}{g_{Hg}\,\beta_N} = \frac{6.6262 \times 10^{-34}\,Js \times 17.83 \times 10^{6}\,\text{Hz}}{5.585 \times 5.05082 \times 10^{-27}\,JT^{-1}} \approx 0.4188T$$

Example 11 Calculate the frequency needed to satisfy the resonance for Na^{23} in a magnetic field of 2.348 T ($g_{Na} = 1.478$).

Solution
From Eq. (13.73)

$$\nu = \frac{g_{Na}\,\beta_N\,B}{h} = \frac{1.478 \times 5.05082 \times 10^{-27}\,JT^{-1} \times 2.348T}{6.6262 \times 10^{-34}\,Js} \approx 26.45\,\text{MHz}$$

Example 12 What is the Larmor frequency for F^{19} at 10 T ($\gamma_F = 2.5162 \times 10^{8}$ s^{-1} T^{-1})?

Solution
From Eq. (13.71)

$$\nu = \frac{\gamma_{proton}\,B}{2\pi} = \frac{2.5162 \times 10^{8}\,s^{-1}\,T^{-1} \times 10T}{2 \times 3.14} = 400.69 \times 10^{6}\,\text{Hz} = 400.69\,\text{MHz}$$

Example 13 A line in the H^{1} NMR spectrum of a molecule had a width of 0.3 Hz. What is the effective relaxation time?

Solution

According to Eq. (13.78)

$$T_2 = \frac{1}{\pi \, \Delta\nu_{1/2}} = \frac{1}{3.14 \times 0.3 Hz} \approx 1.06 \, ms.$$

Example 14 The chemical shift of the CH_3 proton in acetaldehyde is 2.2 and that of CHO proton is 9.8. What is the difference in the local magnetic field between the two regions of molecule, when the external magnetic field is 1.5 T.

Solution

From Eq. (13.83)

$$\delta = \frac{\nu_{sample} - \nu_{ref}}{\nu_{ref}}$$

$$2.2 = \frac{\nu_{CH_3} - \nu_{ref}}{\nu_{ref}}$$

$$9.8 = \frac{\nu_{CHO} - \nu_{ref}}{\nu_{ref}}$$

On subtracting

$$9.8 - 2.2 = 7.6 = \frac{\nu_{CHO} - \nu_{CH_3}}{\nu_{ref.}}$$

From Eq. (13.73)

$$\nu = \frac{g_H \beta_N B}{h} = \frac{5.585 \times 5.05082 \times 10^{-27} JT^{-1} \times 1.5T}{6.6262 \times 10^{-34} Js} = 63.857 \, \text{MHz}$$

$$\nu_{CHO} - \nu_{CH_3} = 63.857 \times 7.6 = 485.31 \, \text{Hz}$$

From resonance condition (13.73)

$$h\nu_{CH_3} = g_H \beta_N B_{CH_3}$$

$$h\nu_{CHO} = g_H \beta_N B_{CHO}$$

and

$$h(\nu_{CHO} - \nu_{CH_3}) = g_H \beta_N (B_{CHO} - B_{CH_3}) = g_H \beta_N \left[(B_{loc})_{CH_3} - (B_{loc})_{CHO} \right]$$

$$(B_{loc})_{CH_3} - (B_{loc})_{CHO} = \frac{h(\nu_{CHO} - \nu_{CH_3})}{g_H \beta_N} = \frac{6.6262 \times 10^{-34} Js \times 485.31 \, \text{Hz}}{5.585 \times 5.051 \times 10^{-27} JT^{-1}} \approx 11.4 \, \mu T$$

Example 15 At what frequency shift from TMS would a group of nuclei with $\delta = 9.8$ resonate in a spectrometer operating at 100 MHz?

Solution

From Eq. (13.83)

$$\delta = \frac{(\nu_i - \nu_r)\text{Hz}}{\nu(\text{MHz})}$$
$$\nu_i - \nu_r = 9.8 \times 100\,\text{Hz} = 980\,\text{Hz}$$

Example 16 A proton spectrometer operating at 200 MHz was used to measure the frequency separation of the resonance of benzene and TMS, which was found to be 1460 Hz, the benzene being to high frequency. What is the chemical shift of benzene on the δ scale? What would be the frequency separation and chemical shift be if the sample is measured with a spectrometer operating at 60 MHz?

Solution

From Eq. (13.83)

$$\delta = \frac{1460\,\text{Hz}}{200\,\text{MHz}} = 7.3\,\text{ppm}$$

Since the chemical shift is independent of spectrometer frequency, therefore the chemical shift would also be 7.3 ppm at 60 MHz. The frequency separation between benzene signal and TMS signal would be from Eq. (10.22)

$$\nu_{\text{benzene}} - \nu_{\text{TMS}} = 7.3\,\text{ppm} \times 60 = 438\,\text{Hz}$$

Example 17 What is the Doppler shift of the gamma ray frequency to an outside observer when a free Mössbauer nucleus of mass 1.67×10^{-25} kg emits a gamma ray of 0.1 nm wavelengths?

Solution

From momentum conservation $M v = \frac{h}{\lambda}$, the recoil velocity is

$$v = \frac{h}{M\lambda} = \frac{6.626 \times 10^{-34}\,js}{1.67 \times 10^{-25}\,Kg \times 0.1 \times 10^{-9}m} = 39.67 m/s$$

$$\text{Doppler shift} \quad \Delta v = \frac{v}{\lambda} = \frac{39.67 m/s}{0.1 \times 10^{-9}m} = 3.967 \times 10^{11}\,Hz$$

Example 18 The lifetime of Fe^{57} (excited) state is 1.5×10^{-7} s. The excited state is 14.4 keV above the ground state. Determine the linewidth Γ.

Solution

We have $\Gamma\tau \sim \hbar$

$$\Gamma \sim \frac{\hbar}{\tau} = \frac{6.626 \times 10^{-34} Js}{2 \times 3.14 \times 1.5 \times 10^{-7} s} = 0.703 \times 10^{-27} J$$

$$= \frac{0.703 \times 10^{-27}}{1.602 \times 10^{-19}} eV = 4.4 \times 10^{-9} eV$$

Example 19 For a Fe^{57} nucleus with emission energy of $E = 14,400$ eV, find the recoil energy.

Solution

We have

$$E_R = \frac{(h\nu)^2}{2Mc^2}$$

Considering the mass of Fe^{57} to be 57 amu and $h\nu = 14,400$ eV we have

$$E_R = \frac{(14,400 \text{ eV})^2}{2 \times 57 \times 931 \times 10^6 \text{ eV}} = 1.953 \times 10^{-5} \text{ eV}$$
$$(1 \text{ amu} = 931 \times 10^6 \text{ eV})$$

Example 20 A particular Mössbauer nucleus has spins 7/2 and 5/2 in its excited and ground states, respectively. Into how many lines will the gamma ray spectrum split if (a) the nucleus is under the influence of an internal electric field gradient, but no magnetic field is applied. (b) there is no electric field gradient at the nucleus but an external magnetic field is applied (c) both an internal electric field gradient and an external magnetic field are present.

Solution

(a) Under the action of internal electric field gradient, the excited state is having $I = 7/2$ split into four levels with $m_I = \pm 7/2, \pm 5/2, \pm 3/2$ and $\pm 1/2$. On the other hand, ground state will split into three components with $m_I = \pm 5/2, \pm 3/2, \pm 1/2$. To obtain the number of allowed transitions, the following procedure is adopted. The values of $\pm m_I$ for both initial and final states are written down in two rows with equal values of m_I directly below or above each other. For the $I = 7/2 \leftrightarrow 5/2$ transitions, they are as follows

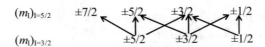

In this array, the vertical arrows indicate $\Delta m_I = 0$ transitions and the diagonal arrows indicate $\Delta m_I = \pm 1$ transitions. The total number of arrows indicates the number of allowed transitions. For $I = 7/2 \leftrightarrow 5/2$, the number of quadrupole transitions is eight in numbers.

(b) In the presence of magnetic field, the degeneracy of the energy level of ground state and excited state is completely lifted. Excited state splits into eight components with $m_I = 7/2, 5/2, 3/2, 1/2, -1/2, -3/2, -5/2, -7/2$ and ground state split into six components with $m_I = 5/2, 3/2, 1/2, -1/2, -3/2, -5/2$. To obtain the number of allowed transitions, the procedure given in (a) is adopted. The total number of allowed transitions for $I = 7/2 \leftrightarrow 5/2$ is the number of magnetic hyperfine transitions which are 18 in numbers.

(c) Effect of electric field is only to change the spacing between the levels, and thus, again 18 lines will be observed.

Example 21 In a Mössbauer experiment, gamma rays emitted in transitions from 0.129 meV first excited state to the ground state of Ir^{191} (a) consider the recoil of the nucleus when it emits the gamma rays, and determine the downward shift in the energy of the gamma ray that results from the energy taken by the nuclear recoil (b) then compare this energy shift to the width of the first excited state of Ir^{191}, which has a measured lifetime of 1.4×10^{-10} s.

Solution

(a) Recoil energy is

$$E_R = \frac{(h\nu)^2}{2Mc^2}$$

Since the sum of gamma ray energy E and nuclear recoil energy must be equal to the energy available in gamma decay, i.e. 0.129 meV, energy of the first excited state of the decaying nucleus. Therefore, E is less than the energy of the first excited state by an amount E_R. This is the downward shift ΔE in the energy of the gamma ray due to nuclear recoil, i.e.

$$\Delta E = - \text{ recoil energy } E_R = -\frac{E^2}{2Mc^2}$$

because M is so large, ΔE is very small compared to E; we may evaluate it approximately by setting $E = 0.129$ meV. Using the relation 931 MeV $= 1$ amu, we have

$$\Delta E = -\frac{(1.29 \times 10^6 eV)^2}{2 \times 191 \times 931 \times 10^6 eV} = -4.7 \times 10^{-2} eV$$

(b) If the lifetime τ of the first excited state of Ir^{191} is 1.4×10^{-10} s, the width Γ from

$\Gamma\tau \sim \hbar$ is

$$\Gamma = h/\tau = \frac{6.6 \times 10^{-16} eV - s}{2 \times 3.14 \times 1.4 \times 10^{-10} s} = 7.5 \times 10^{-7} eV$$

clearly, the gamma rays emitted by the decay from the first excited state of Ir^{191} emitter nucleus cannot excite a Ir^{191} absorber nucleus from the ground state to its first excited state. The nuclear recoil shift of the gamma ray is larger by a factor of 10^4 than the width of the state it is supposed to excite. Therefore, the gamma ray is thrown completely out of resonance, and the resonant absorption is destroyed.

Objective Type Questions

1. The *ESR* spectrum of Fe^{3+} in a single crystal show five fine structure lines corresponding to allowed selection rules $\Delta M = \pm 1$. The effective spin of Fe^{3+} in this single crystal is
   (a) 5 (b) 10 (c) 2 (d) 5/2
2. The ESR spectrum of a radical with two equivalent nuclei of a particular kind is split into five lines of intensity ratio 1:2:3:2:1. The spin of the nuclei is
   (a) 5 (b) 1 (c) 3/2 (d) 2
3. The ESR spectrum of VO^{2+} in a single crystal shows eight lines of equal intensity. (a) What is effective spin and nuclear spin of VO^{2+}
   (a) 1/2, 7/2 (b) 7/2, 1 (c) 3/27/2, 7/2 (d) 7/2, 1/2
4. The ESR is not observed for
   (a) H atom (b) He (c) Gd^{3+} (d) O_2
5. The frequency associated with K-band is
   (a) 3 GHz (b) 9 GHz (c) 24 GHz (d) 35 GHz
6. The allowed selection rules for fine structure transition in ESR spectra are
   (a) $\Delta M = \pm 1$ (b) $\Delta m = \pm 1$ (c) $\Delta M = 0$ (d) $\Delta m = 0$
7. Which of the following nuclei will have a magnetic moment?
   (a) $_1D^2$ (b) $_8O^{16}$ (c) $_6C^{12}$ (d) $_{16}S^{32}$
8. A NMR transition is shifted from reference in 400 MHz spectrometer by 529 Hz. The chemical shift is
   (a) 5.29 (b) 1.32 (c) 7.56 (d) 1.76
9. Which is not true?
   (a) Mössbauer effect is not observed in gases and non-viscous liquids.
   (b) two nuclei with equal charge and mass number but in different excited states are called isomer nuclei
   (c) in Mössbauer effect both ground and excited states are considered
   (d) in Mössbauer effect, the gamma rays energies lie between 200 and 500 keV
10. Which is false?
    (a) isomer shift of Fe^{2+} is 1.3–1.4 cm/s
    (b) the principle influence on isomer shift is due to outermost occupied s orbitals
    (c) the ground state of Fe^{57} nuclei is 1/2

(d) the excited state of Fe^{57} nuclei has a nonzero quadrupole moment
11. The allowed selection rules for Mössbauer spectra are
    (a) $\Delta M = \pm 1$ (b) $\dot{\Delta}m_I = \pm 1$ (c) $\Delta m_I = 0, \pm 1$ (d) $\Delta m_I = 0$
12. The allowed selection rules for NMR spectra are
    (a) $\Delta M = \pm 1$ (b) $\Delta m_I = \pm 1$ (c) $\Delta m_I = 0, \pm 1$ (d) $\Delta m_I = 0$
13. Which is false?
    (a) the Mössbauer spectrum is a velocity spectrum
    (b) in Mössbauer spectroscopy parent nucleus must have fairly long half-life
    (c) the energy and momentum are conserved in the gamma emission process
    (d) gamma transitions occur between magnetic sublevels of the ground level of the nucleus
14. Which is false?
    (a) the principal influence on isomer shift is due to outermost occupied s orbitals
    (b) the ground state spin of Fe^{57} nuclei is 3/2
    (c) excited state of Fe^{57} nuclei has a nonzero quadrupole moment
    (d) isomer shift of Fe^{2+} is 1.3–1.4 cm/s
15. At what frequency shift from TMS would a group of nuclei with chemical shift 5 resonant in a spectrometer operating at 100 MHz?
    (a) 500 MHz (b) 500 Hz (c) 20 MHz (d) 20 Hz
16. Which is true for the chemical shift of CH_3Cl, CH_3Br, CH_3F and CH_3I?
    (a) $CH_3F > CH_3Cl > CH_3Br > CH_3I$
    (b) $CH_3Cl > CH_3Br > CH_3I > CH_3F$
    (c) $CH_3Br > CH_3I > CH_3F > CH_3Cl$
    (d) $CH_3I > CH_3F > CH_3Cl > CH_3Br$
17. A line in the H^1 NMR spectrum of a molecule had a width of 10 Hz. The effective relaxation time is
    (a) 0.1 ms (b) 10 ms (c) 1.6 ms (d) 3.2 ms

Problems

1. I^{127} has a nuclear spin of 5/2. Calculate the energies of the nuclear spin states in a magnetic field of 5 T ($g_I = 1.118$).
2. A radical containing two equivalent proton with hyperfine constant 2.0 mT and 2.6 mT gives a spectrum centred on 332.5 mT. At what fields do the hyperfine lines occur and what are their relative intensity.
3. Calculate the frequency needed to satisfy the resonance for F^{19} in a magnetic field of 2.3487 T ($g_F = 5.255$).
4. The ESR spectrum of a radical with a single magnetic nucleus is split into four lines of equal intensity. What is the spin of the nuclei?
5. S^{32} has a nuclear spin of 3/2. Calculate the energies of the nuclear spin states in a magnetic field of 7.50 T ($g_S = 0.4289$).

6. Predict the form of the ESR spectrum of radical containing three equivalent N^{14} nuclei ($I = 1$).

7. The ESR spectrum of Cu^{2+} in $Rb_2C_2O_4 \cdot H_2O$ shows four lines of equal intensity at 268.9 mT, 282.3 mT, 295.5 mT and 307.2 mT. Determine the effective spin and nuclear spin of Cu^{2+}. Also determine the value of hyperfine coupling constant and g value of Cu^{2+} in this sample given that DPPH line (with $g = 2.0036$) occurs at 337.5 mT.

8. Calculate the energy difference between the lowest and highest nuclear spin state of Cl^{35} nucleus in a magnetic field of 3 T ($I = 3/2$, $g_{Cl} = 0.4555$).

9. What is the value of γ for F^{19} ($I = 1/2$, $g_F = 5.255$)?

10. In a magnetic field of 2T what fraction of the protons have their spins line up with the field at 298 K.

11. Which of the two chemically different types of proton in CH_3COOH resonate at lower magnetic field or higher frequency?

12. A Zn^{67} nucleus has spins 5/2 and 1/2 in its excited and ground states, respectively. Into how many lines will the gamma ray spectrum split if (a) the nucleus is under the influence of an internal electric field gradient, but no magnetic field is applied. (b) there is no electric field gradient at the nucleus but an external magnetic field is applied (c) both an internal electric field gradient and an external magnetic field are present. For a Zn^{67} nucleus, the lifetime of the isomer state is 9400 ns. Determine the natural linewidth and recoil velocity.

Answers

Objective Type Questions

1. (d)	2. (b)	3. (a)	4. (b)	5. (c)	6. (a)	7. (a)	8. (b)
9. (c)	10. (a)	11. (c)	12. (b)	13. (d)	14. (b)	15. (b)	16. (a)
17. (d)							

Superconductivity

14

14.1   Introduction

The electrical resistivity of a conductor is controlled by two mechanisms, namely electron scattering from impurities and defects in crystal lattice, and interaction between electrons and lattice vibration. The former process dominates at low temperatures and the latter dominates at higher temperatures. The conductivity of a metal is therefore increased on cooling due to decreased electron–phonon interaction.

After the discovery of liquefied helium, allowing materials to be cooled to within 4 K of absolute zero, it was discovered by Dutch physicist Heike Kamerlongh Onnes in 1911; that when pure mercury was cooled to 4.15 K, its resistance suddenly dropped to zero (Fig. 14.1). He also found that the addition of impurities to his mercury sample failed to change this state of affair. The resistivity still dropped to an apparently zero value rather than tending to a finite value that would be expected by conventional theory. Onnes referred to this new state of mercury, in which the electrical resistance vanishes as the superconducting state. The temperature Tc at which the transition to the superconducting state occurs is often referred to as the transition temperature or the critical temperature.

In 1913, it was discovered that lead became superconducting at 7.2 K. It was then 17 years until niobium (Nb) was to superconduct at higher temperature 9.2 K. The highest known temperature T at which a material went superconducting increased slowly as scientists found new material with higher value of T, but it was in 1986 that a Ba–La–Cu–O system was found to be superconducting at 35 K by far the highest then found. Soon materials were found that could superconduct above 77 K, the temperature of liquid nitrogen which is safer and much less expensive than liquid helium as a refrigerant. Although high temperature superconductors are more useful above 77 K, the term technically refers to those materials that superconduct above 30–40 K. In 1994, $HgBa_2Ca_2Cu_3O_{8+x}$ showed superconducting behaviour at 164 K under 30 GP of pressure.

© The Author(s) 2022
V. K. Jain, *Solid State Physics*, https://doi.org/10.1007/978-3-030-96017-9_14

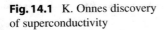

Fig. 14.1 K. Onnes discovery
of superconductivity

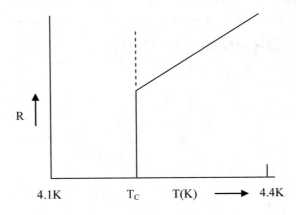

$$4.1K \qquad\qquad T_C \qquad T(K) \longrightarrow 4.4K$$

14.2   Superconducting System

Superconductivity occurs in a wide variety of materials including simple elements, various
metallic alloys, some heavily doped semiconductors, certain ceramic compounds, etc.
Superconducting systems can be grouped as

(i)   Elemental superconductors: Superconductivity is displayed by non-transition metal
      (e.g. Be, Al, Pb, Sn, etc.) transition metals (e.g. Nb, Mo, Zn) semiconductor elements
      (Si at 8.31 K at a pressure of 165 K bar, Ge, Se and Te), semimetal (Bi). Supercon-
      ductivity does not occur in noble metals like gold and silver nor in ferromagnetic
      metals (Fe, Co, Ni).

(ii) Alloys: A large number of alloys exhibit superconductivity, for example Nb–Ti and
      $Nb_3Al_{0.75}Ge_{0.25}$.

(iii) Binary Compounds: Some binary compounds also show superconductivity. These are
      the so-called A-15 compounds. The examples are V_3Ga, V_3Si, Nb_3Sn.

(iv) Heavy Fermion Superconductor: The main feature of this kind of superconductors is
      low Tc and large effective mass of electron (several hundred times the mass of free
      electron). Examples are $CeCu_2$, $SiUBe_{13}$.

(v)   Organic Superconductors: These form an unusual class of conducting materials. They
      are insulators in normal states. The examples are tetramethyl tetraselenafulvalence
      (TMTSF) tetrathiafulvalene.

(vi) High Temperature Superconductors: These oxide superconductors are defective
      perovskite-like cuprate materials. These materials have high T_c (>90 K). Examples
      are $La_{2-x}Ba_xCuO_4$, $YBa_2Cu_3O_{7-x}$, etc.

14.3 Elementary Properties of Superconductors

Most of the physical properties of superconductor vary from material to material, such as the heat capacity and the critical temperature at which superconductivity is destroyed. On the other hand, there is a class of property that is independent of the underlying material. For example, all superconductors have exactly zero resistivity to low applied current when there is no magnetic field present.

14.3.1 Critical Magnetic Field

Onnes discovered that application of magnetic field destroyed superconductivity. Consider a material in its superconductivity state. It is observed that when a sufficiently strong external magnetic field is applied to the material, then the normal resistivity of the sample is restored. It is found that there is a critical value of the magnetic field such that, if the applied magnetic field is below the critical value, the superconducting state is maintained. On the other hand, if the applied magnetic field is above the critical value, the normal state of the material is restored.

The critical magnetic field B_c is a function of temperature T of the material and is approximately given by

$$B_c = B_0 \left[1 - \left(\frac{T}{T_0} \right)^2 \right]$$ (14.1)

where B_0 is constant and depends on the material. T_c is the critical temperature of the material. For mercury, the critical field varies roughly as shown in Fig. 14.2. The critical field becomes zero at the critical temperature (4.2 K for Hg). This is because above this temperature the normal resistivity of Hg is restored.

Fig. 14.2 Critical magnetic field as a function of temperature. S is superconducting and N is normal state

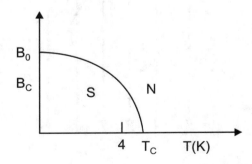

14.3.2 Meissner Effect

Meissner and Ochsenfeld discovered in 1933 that when a superconducting material is cooled below the critical temperature in a magnetic field, the magnetic flux is expelled from the interior of the superconductor (Fig. 14.3). In other words, the magnetic field penetrates for only a short distance called penetration depth after which it decays rapidly to zero. The penetration depth is of the order of few hundred nanometres. The limit of external magnetic field strength at which a superconductor can expel the magnetic field is known as the critical field strength.

Meissner performed the experiment with a solid sphere. If the sphere is first cooled and a field is applied, afterwards the lines of forces do not penetrate into the material. The magnetic field inside the sphere remains zero, and this means that there must be surface current flowing which compensates for the external field. However, if the surface is cooled in the applied field, it is observed that the final result is the same as before, that is, surface current start, which reduces the field inside the sphere.

The Meissner effect can be understood by considering the cause of destruction of magnetic field in a bulk superconductor as the existence of the electric current flowing on the surface of the material. The current produces a magnetic field which will exactly cancel the applied magnetic field. This means that inside the superconductor

$$\boldsymbol{B}_{in} = \mu_0(\boldsymbol{H} + \boldsymbol{M}) = 0$$
$$\text{or} \quad \boldsymbol{M} = -\boldsymbol{H} \tag{14.2}$$

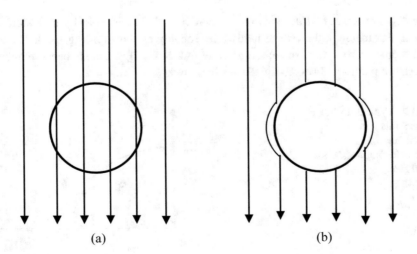

(a)   (b)

Fig. 14.3 Meissner effect **a** normal $T > T_c$ or $B > B_c$, **b** superconducting $T < T_c$ or $B < B_c$

This gives a negative susceptibility for the superconductor. Thus, superconductor behaves like a perfect diamagnet, and all the flux penetrating the superconductor is abruptly expelled in the transition from normal to superconducting state.

The Meissner effect is sometimes confused with the perfect diamagnetism, one would expect in a perfect electric conductor. According to Lenz's law, when a changing magnetic field is applied to a conductor, it will induce an electric current in the conductor that creates an opposing magnetic field. In a perfect conductor, an arbitrary large current can be induced and the resulting magnetic field exactly cancels the applied magnetic field. The Meissner effect is distinct from perfect diamagnetism because a superconductor expels all magnetic fields, not just those that are changing. Suppose we have a material in its normal state, containing a constant internal magnetic field. When the material is cooled below the critical temperature, one would observe the abrupt expulsion of the internal magnetic field which is not expected on the basis of Lenz's law.

14.3.3 Isotope Effect

In the study of superconductors of different isotopic composition, it was discovered that the transition temperature of a superconductor metal is different for different isotopes (Fig. 14.4). It has been found experimentally that the superconducting transition temperature varies as

$$T_c \propto M^{-\alpha} \tag{14.3}$$

The exponent a is 0.5 or slightly less for most non-transition metal, M is the isotope mass. Since Debye temperature θ_D is proportional to the velocity of sound which depends $M^{-1/2}$, therefore the transition temperature can be related to Debye temperature as $T_c \propto \theta_D$ (Fig. 14.4).

Fig. 14.4 Superconducting transition temperature Tc as a function of isotopic mass for Hg. $A = Hg^{203.4}$, $B = Hg^{202}$, $C = Hg^{200.7}$, $D = Hg^{200.6}$, $E = Hg^{199.7}$, $F = Hg^{198}$

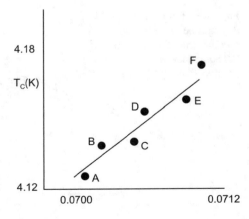

The dependence of critical temperature for superconductivity upon isotopic mass is direct evidence for interaction between the electrons and the lattice. This supported the BCS theory of lattice coupling of electron pairs. Isotope effect disapproves (i) superconductivity is due to the interaction of free electrons (ii) superconductivity is due to electron in a rigid potential field corresponding to infinitely heavy atoms.

14.3.4 Specific Heat

The specific heat C_V in a normal conductor at low temperature consists of two contribution C_V^e and C_V^L corresponding to electrons in conduction band and from lattice, respectively. Thus,

$$C_V = C_V^e + C_V^L = \gamma T + \beta T^3 \qquad (14.4)$$

C_V^e is linearly proportional to T while lattice specific heat is proportional to T^3. Figure 14.5 shows variation of C_V^e versus temperature.

The heat capacity of a superconductor material is quite different in the normal and superconducting state. In the normal state (produced at temperatures below the transition temperature by applying a magnetic field greater than the critical field), the heat capacity is determined primarily by the normal electrons (with a small contribution from the thermal vibration of the crystal lattice) and is nearly proportional to the temperature.

In zero magnetic field, there appears a discontinuity in the specific heat at the transition temperature. At temperature just below the transition temperature, the heat capacity is larger than in the normal state. It decreases more rapidly with decreasing temperature, however, and at temperature well below the transition temperature varies as $\exp(-\Delta/k_B T)$ where Δ is constant. Such an exponential temperature dependence is a hallmark of a system with a gap in the spectrum of the allowed energy states.

Fig. 14.5 Temperature dependence of electronic specific heat of a conductor in the normal (N) and superconducting (S) state

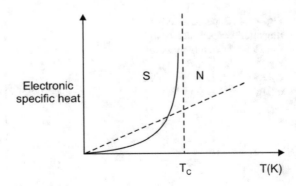

14.3.5 Thermal Conductivity

A large electric conductivity is ordinarily accompanied by a large thermal conductivity, as in the case of copper. However, the thermal conductivity of a pure superconductor is less in the superconducting state than in the normal state and at very low temperatures approaches zero. Crudely speaking, the explanation for the association of infinite electrical conductivity with vanishing thermal conductivity is that of a transport of heat requires the transport of disorder (entropy). The superconducting state is one of perfect order (zero entropy) and so there is no disorder to transport and therefore no thermal conductivity. Further, the thermoelectric effects disappear in the superconducting state.

14.4   London Equations and Penetration Depth

F. London and H. London in 1935 showed that the magnetic flux lines are not completely expelled from a superconductor but remain confined in a thin surface layer. The flux change so produced induces a flow of supercurrent in a closed circuit within the thin layer.

According to London theory, electrons in a superconductor may be considered a mixture of two groups: superconducting electrons and normal electrons. Superconducting electrons experience no scattering and have zero entropy and a long coherence length or spatial extension. The number density of superconducting electrons n_s decreases with increasing temperature and eventually becomes zero at $T = T_c$. At $T = 0$ K, it is the other way round; that is, n_s is equal to the total n_o, density of conduction electrons. A flow of superconducting electrons meets no resistance. Such a resistance cannot generate a constant electric field in a superconductor because if it did it would cause the superconducting electrons to accelerate infinitely. Therefore, under stationary conditions, that is with an electric field, the normal electrons are at rest.

The equation of motion of superconducting electrons accelerated by electric field E is

$$m\frac{\mathrm{d}v_s}{\mathrm{d}t} = -eE \tag{14.5}$$

where v_s is the mean velocity of superconducting electrons. The current density j_s of superconducting electron is

$$j_s = -en_s v_s \tag{14.6}$$

Differentiating Eq. (14.6) and using Eq. (14.5)

$$\frac{\mathrm{d}j_s}{\mathrm{d}t} = -en_s\frac{\mathrm{d}v_s}{\mathrm{d}t} = \frac{e^2 n_s}{m}E \tag{14.7}$$

This is known as first London equation. Taking curl of Eq. (14.7)

$$\frac{d}{dt}\left(\text{curl } \boldsymbol{j}_s\right) = \frac{e^2 n_s}{m}\text{curl } \boldsymbol{E}$$

$$\text{curl } \boldsymbol{E} = \frac{m}{n_s e^2}\frac{d}{dt}\left(\text{curl } \boldsymbol{j}_s\right) \tag{14.8}$$

Using Maxwell equation

$$\text{curl } \boldsymbol{E} = -\frac{\partial \boldsymbol{B}}{\partial t} \tag{14.9}$$

in Eq. (14.8)

$$\frac{\partial}{\partial t}\left(\frac{m}{n_s e^2}\text{curl } \boldsymbol{j}_s + \boldsymbol{B}\right) = 0 \tag{14.10}$$

Since the time derivative vanishes, the quantity in the bracket of Eq. (14.10) is constant. The static magnetic field and the static current density are related to each other through Eq. (14.8). Any time dependence of $\boldsymbol{B}$ and $\boldsymbol{j}_s$ could be trivial solution of Eq. (14.8). This describes the behaviour of an ideal conductor but not the perfect diamagnetism (i.e. expulsion of the magnetic field) associated with the Meissner effect. On integrating Eq. (14.10) and putting constant of integration equal to zero, we have

$$\text{curl } \boldsymbol{j}_s = -\frac{e^2 n_s}{m}\boldsymbol{B} \tag{14.11}$$

Combining the fourth Maxwell equation (for time independent field)

$$\text{curl } \boldsymbol{B} = \mu_0 \boldsymbol{j}_s \tag{14.12}$$

Taking curl of Eq. (14.12)

$$\text{curl curl } \boldsymbol{B} = \mu_0 \text{ curl } \boldsymbol{j}_s \tag{14.13}$$

Using the identity

$$\text{curl cule } \boldsymbol{A} = \nabla(\nabla \cdot \boldsymbol{A}) - \nabla^2 \boldsymbol{A} \tag{14.14}$$

We have from Eq. (14.13)

$$\nabla(\nabla \cdot \boldsymbol{B}) - \nabla^2 \boldsymbol{B} = \mu_0 \text{ curl } \boldsymbol{j}_s \tag{14.15}$$

Now $\nabla \cdot \boldsymbol{B} = 0$. Hence, Eq. (14.15) is

$$\nabla^2 \boldsymbol{B} = -\mu_0 \text{ curl } \boldsymbol{j}_s \tag{14.16}$$

From Eqs. (14.11) and (14.16)

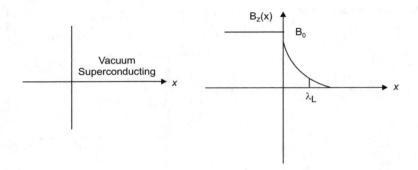

Fig. 14.6 Exponential attenuation of a magnetic field in a superconducting half plane

$$\nabla^2 \boldsymbol{B} = \frac{n_s e^2 \mu_0 \boldsymbol{B}}{m} \tag{14.17}$$

Now consider a simple geometry, namely the boundary between a superconducting half space and vacuum (Fig. 14.6). For a magnetic field parallel to the surface, Eq. (14.17) is

$$\frac{d^2 B_z}{dx^2} - \frac{1}{\lambda_L^2} B_z = 0 \tag{14.18}$$

$$\text{where } \lambda_L = \left(\frac{m}{\mu_0 n_s e^2} \right)^{\frac{1}{2}} \tag{14.19}$$

is called London penetration depth. The solution of Eq. (14.18) is

$$B_z(x) = B(0) \exp\left(-\frac{x}{\lambda_L} \right) \tag{14.20}$$

This predicts that the magnetic field in a superconductor decays exponentially from whatever value it possesses at the surface. The penetration depth ranges from 30 to 500 nm depending on the material. The penetration depth also depends on the temperature and becomes quite large as temperature approaches the transition temperature. The observation can be written in the form (Fig. 14.7)

$$\frac{\lambda_L(T)}{\lambda_L(0)} = \left[1 - \left(\frac{T}{T_c} \right)^4 \right]^{-\frac{1}{2}} \tag{14.21}$$

Putting the value of λ_L from Eq. (14.19) in Eq. (14.21)

$$\left(\frac{n_0}{n_s} \right)^{\frac{1}{2}} = \frac{1}{\left[1 - \left(\frac{T}{T_c} \right)^4 \right]^{\frac{1}{2}}}$$

$$n_s = n_0 \left[1 - \left(\frac{T}{T_c} \right)^4 \right] \tag{14.22}$$

Fig. 14.7 Variation of λ_L with T

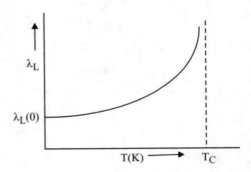

Thus, density of superconducting electrons decreases with increase of temperature and is zero at $T = T_c$.

14.5   Classification of Superconductors (Type I and Type II)

Superconductors can be divided into two classes: type I (soft) and type II (hard) super-conductors. The characteristic property, on which the classification is based takes into account the way in which the transition from the superconducting state to normal state proceeds when the applied field is greater than the critical magnetic field (Fig. 14.8).

In type I superconductors, the superconductivity is almost destroyed when the strength of the applied magnetic field rises above the critical value H_c. These are often called soft because of their tendency to give away the low magnetic field. The examples of type I superconductors are In, Al, Hg, Pb, etc.

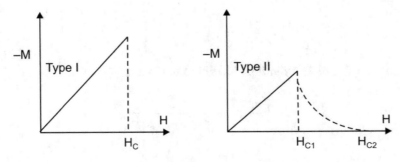

Fig. 14.8 Magnetization curve for type I and type II superconductors

In type II superconductors, raising the applied field past a critical value H_{C_1} leads to a mixed state in which an increasing amount of magnetic flux penetrates the material but there remains no resistance to the flow of electric current as long as the current is not too large. At a second critical field strength, H_{C_2} superconductivity is destroyed. The examples of type II superconductors are YBCO ($YBa_2Cu_3O_7$), thallium-barium-calcium copper oxide, V, Nb, Ta, Sn, Ge, etc.

In general type I, superconductivity is exhibited by most of the elements that exhibit superconductivity. Most alloys that exhibit superconductivity show type II behaviour.

14.6 BCS Theory

The microscopic theory of superconductivity was developed mainly by Bardeen, Cooper and Schrieffer in 1957. The abbreviation for it is the BCS theory. The BCS theory has successfully described the measured properties of type I superconductors. The key concept of the BCS theory is the pairing of electrons close to Fermi level into Cooper pair through interaction with the crystal lattice. Let us understand the formation of Cooper pair.

When an electron having wave vector k_F travels through a lattice, it exerts an electrostatic attraction on the positively charged ions in the lattice. This causes the ions to be distorted away from the equilibrium positions (Fig. 14.9) in the lattice, towards the path of the electron. This results in a region of net positive charge in the lattice. The ions in the lattice are much more massive than the electrons and as such they respond to any change in position more slowly than the electrons. Thus, the distortion of the lattice remains, for a while when the initial electron is long gone from that part of the lattice while the region of positive charge remains, a second electron with opposite spin and wave vector $-k_F$, can feel an electrostatic attraction towards the region. In this way, the motion of the second electron is correlated with that of the first electron. There are a lot of such electron pairs called Cooper pairs in superconductor. The angular momentum of such a pair is zero. The effective diameter of Cooper pair is 10^{-7}–10^{-6} m. There are about 10^6 centre of masses of Cooper pairs inside the effective volume of one such pair.

The Cooper pairs can behave very differently from single electrons which are fermions and must obey Pauli exclusion principle. The Cooper pair acts more like bosons which can condense into the same energy level. The formation of a Cooper pair results in the reduction of energy of two electrons by an amount equal to the binding energy of the electrons in the pair E_b. This means that a conduction electron which is in a normal metal had a maximum energy E_F at $T = 0$ K, in the superconducting state has an energy $E_b/2$ less (the energy per pair being E_b less), since this is the energy that must be spent to break up the pair and move the electrons to the normal state. Therefore, in the one electron spectrum there must be a gap of energy $2E_b$ between the upper level of a coupled electron and lower level corresponding to the normal state. This energy gap is of the order of 10^{-3} eV. This energy gap prevents the kind of collision interaction which leads to the

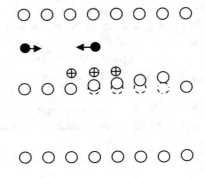

Lattice of superconducting material

A passing electron attracts the lattice causing a slight ripple towards its path

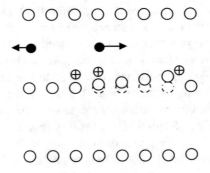

Another electron passing in the opposite direction is attracted to that displacement

Fig. 14.9 Lattice of superconducting material. **a** A passing electron attracts the lattice causing a slight ripple towards its path. **b** Another electron passing in the opposite direction is attracted to that displacement

ordinary resistivity. In a superconductor, the current is made of Cooper pairs rather than individual electrons. For temperatures such that the thermal energy is less than the band gap the material exhibits zero resistivity.

14.7   Variation of Superconducting Energy Gap with Temperature

The origin of band gap in superconductor is due to Fermi gas while in semiconductors or insulators is due to the lattice. At 0 K, when all the Fermi electrons are in pairs the

Fig. 14.10 Variation of band
gap with temperature

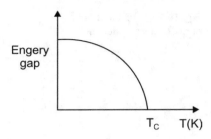

width of the energy band gap is maximum. The rise in temperature is accompanied by
the generation of phonons capable in the act of scattering of transmitting energy to the
electrons sufficient to break up the Cooper pair. At low temperatures, the concentration
of such phonons is not large and breaking of a pair is rare event. The breaking of some
Cooper pairs cannot lead to the disappearance of the gap for the remaining pairs but make
it somewhat narrower. With further rise in temperature the phonon concentration grows
very rapidly, their mean energy growing as well. This results in an increase in the rate of
breaking of Cooper pairs and thus a decrease in the energy gap width for the remaining
Cooper pairs. At some temperature T_c, the gap disappears completely (Fig. 14.10) and
the material returns to the normal state. Thus, the critical temperature for the transition of
a metal to superconductive state should be large, the greater the energy gap at 0 K. The
BCS theory gives the following approximate dependence of energy gap on T

$$\text{Energy gap} \cong 3.2\, k_B\, T_C \left(1 - \frac{T}{T_C} \right)^{\frac{1}{2}} \tag{14.23}$$

14.8 Tunnelling and the Josephson Effect

Consider two pieces of metals, in the superconducting and the normal states, respectively,
separated by a very thin insulating film of thickness of about 5 nm (Fig. 14.11). The film
acts as a potential barrier for the flow of electrons across the junction. Since the film
is thin, therefore, it does not completely stop the flow of electrons and hence current.
According to quantum mechanics, there is some probability that electrons are able to
tunnel a thin potential barrier.

If a small voltage V is applied across the junction such that it is directed to the left,
the energy band of the left side is raised by an amount eV. However, the electrons are not
able to flow to the right because the states lying horizontally across the insulating film
are already occupied. For further increase in voltage, the energy band of superconductor
is raised by $\Delta_0/2$ than the corresponding horizontal states on the right are empty and the
electrons (current) proceeds to flow. The voltage at which the current begins to flow is

Fig. 14.11 Tunnelling in a
normal–superconductor
junction

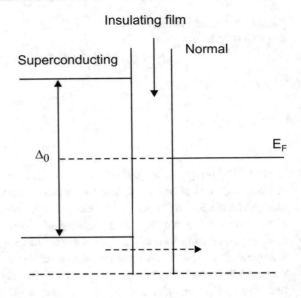

$$\frac{\Delta_0}{2} = eV \qquad (14.24)$$

this is known as normal or single electron tunnelling.

Josephson in 1962 postulated that Cooper pairs could tunnel through a thin insulating barrier (of thickness ~1 nm) between two superconducting electrodes without breaking up the pair. The Cooper pair on each side of the junction can be represented by a wave function similar to a free particle wave function. The Cooper pair in the two superconductors develops a phase correlation and there is a tunnelling of these through the barrier and a supercurrent flows across the junction in the absence of a voltage. The effect of the insulating film is merely to introduce a phase difference θ between the two parts of wave function on opposite side of the junction. The current I across the junction is given by

$$I = I_c \sin\theta \qquad (14.25)$$

where I_c is critical current; that is, maximum supercurrent that the junction can sustain, θ is quantum mechanical phase difference across the junction and $I_c \sim 100\ \mu A^{-1}$ -mA and depends on the material, its temperature and the dimension of the barrier.

The phase difference changes with time when a *DC* voltage is applied to a Josephson junction. According to quantum mechanics, the phase of the wave function is

$$\Delta\theta = \frac{Et}{\hbar} \qquad (14.26)$$

E is total energy of the system. The additional phase difference experienced by the Cooper pair as it tunnel across the junction is obtained by putting

$$E = (2)eV \tag{14.27}$$

The factor 2 is introduced as the system involves a pair of electrons. Equation (14.26) is then

$$\Delta\theta = \frac{2eVt}{\hbar} \tag{14.28}$$

and the current is

$$I = I_c \sin(\theta + \Delta\theta) = I_c \sin\left(\theta + \frac{2eVt}{\hbar}\right) \tag{14.29}$$

which represent an alternating current. This indicates that a static potential lead to an *AC* current having the frequency

$$\omega = \frac{2eV}{\hbar}$$
$$\nu = \frac{2eV}{h} \tag{14.30}$$

occurs at the junction, since the relation between ν and V involves fundamental constants and frequency can be measured very accurately. Therefore, the Josephson junction has become the standard voltage measurement. Numerically

$$\nu = 483.6\, V \text{ GHz} \tag{14.31}$$

Since V is usually of the order of several millivolt, the Josephson frequency falls in the microwave range. The standard voltage is now defined as the voltage required to produce a frequency of 483.5979 MHz.

The application of the Josephson effect is in determination of fundamental constant as can be seen from Eq. (14.30) that the frequency include the ratio $2e/h$. Another application of the effect is superconducting quantum interference device (SQUID). It consists of two superconductors separated by thin insulating layers to form two parallel Josephson junction. This device may be used to determine very small magnetic field $\sim 10^{-13}$ T.

14.9 High Temperature Superconductors

Superconductivity is observed in various kinds of materials. It is a low-temperature phenomenon and material has to be cooled below liquid He temperature. The liquid He is expensive and difficult to handle. There was a search for high temperature superconductors. High temperature superconductor's materials exhibit transition temperature well above few degrees Kelvin. In 1986, Bednorz and Müller discovered that La–Ba–Cu–O

Table 14.1 High temperature superconductors with transition temperature

High temperature superconductor	Transition temperature (K)
Nb_3Sn	18.0
Nb_3Ge	23.2
$Bi_2Sr_2Ca_2Cu_3O_{10}$	107
$Tl_2Ba_2Ca_2Cu_3O_{10}$	125
$HgBa_2Ca_2Cu_3O_8$	135
MgB_2	39
$Sr_{0.5}Sm_{0.5}FeAsF$	56
H_3S	~203 K at pressure of 155 GPa
LaH_{10}	250 K at pressure of 170 GPa

system shows superconductivity up to transition temperature which is 35 K. With this discovery, there is surge of efforts to obtain the materials having high transition temperature. As a result, many systems were discovered showing transition temperature even above liquid N_2 temperature (77 K). Some of the materials with their transition temperature are given in Table 14.1.

The main focus of the high temperature superconductivity work was on oxides, though work has also been done on other materials. The high temperature superconductors can be broadly grouped into the following categories

1. **Cuprates**: These are materials made up of oxygen and copper layers. Cuprates can be broadly classified as
   (a) **HgHTS**: The general formula is $Hg_mBa_2Ca_{n-1}Cu_nO_{2n+m+2}$ with $m = 1, 2$ and $n = 1, 2, 3, \ldots$. These are also termed as HBCCO. T_C is found as high as 135 K depending on index n. It is seen that T_C increases as index n increases up to $n = 3$, and then it usually decreases as n is further increased. With the application of pressure is observed that T_C can be increased to ~ 153 K.
   (b) **Tl-HTS**: The general formula is $Tl_mBa_2Ca_{n-1}Cu_nO_{2n+m+2}$ with $m = 1, 2$; $n = 1, 2, 3, \ldots$. They are also termed as TBCCO. Table 14.2 gives various TBCCO with transition temperature. The numbering system represents number of atoms of Tl, Ba, Ca and Cu. T_C is found as high as 133 K. T_C first increases up to $n = 3$ and then decreases as n further increases.
2. **BiHTS**: The general formula is $Bi_mSr_2Ca_{n-1}Cu_nO_{2n+m+2}$ with $m = 1, 2$; $n = 1, 2, 3, \ldots$. They are also termed as BSCCO. The numbering system represents number of atoms of Bi, Sr, Ca and Cu. The Bi-2201, Bi-2212 and Bi-2223 differ from each other in the number of CuO_2 planes. There is one CuO_2 plane in Bi-2201, two in Bi-2212 and three in Bi-2223. The coordination of Cu is different in these three superconductors. In Bi-2201, copper is octahedrally coordinated with oxygen. In Bi-2212, copper is

Table 14.2 TBCCO with transition temperatures

Kind of superconductor	Chemical formula	Transition temperature (K)
Tl-1212	$TlBa_2\,CaCu_2\,O_7$	82
Tl-1223	$TlBa_2\,Ca_2\,Cu_3\,O_9$	133
Tl-1234	$TlBa_2\,Ca_3\,Cu_4\,O_{11}$	127
Tl-2201	$Tl_2\,Ba_2\,CuO_6$	90
Tl-2212	$Tl_2\,Ba_2\,CaCu_2\,O_8$	110
Tl-2223	$Tl_2\,Ba_2\,Ca_2\,Cu_3\,O_{10}$	128
Tl-2234	$Tl_2\,Ba_2\,Ca_3\,Cu_4\,O_{12}$	119

coordinated with five oxygen while in Bi-2223, one copper is coordinated by four oxygen atoms in square planar configuration and another copper atom is coordinated with five oxygen in pyramidal configuration. There is a general trend that transition temperature increases with the number of CuO_2 planes till $n = 3$ and then start decreasing. T_C is found as high as 110 K depending on index n.

3. **123-HTS**: The general formula is $REBa_2Cu_3O_{7-\delta}$, where Re are rare earth elements. They are also termed as RBCO. T_C is found as high as 96 K depending on index n.

4. **Electron doped superconductors**: There are electron doped superconductors with transition temperatures as high as 28 K. The general formula is $(Ln_{1-x}M_x)_2CuO_4$ where Ln : La, Nd, Pr, Sm, Eu and M : Ce, Th. T_C is found as high as 28 K.

A common characteristic of these cuprates or oxide high temperature superconductors is that they have a layered structure consisting of CuO_2 layers separated by the insulating layers. Superconducting occurs in CuO_2 plane and insulating layers supply charge carriers to CuO_2 plane. Figure 14.12 shows the structure of $YBa_2Cu_3O_7$. The unit cell of $YBa_2Cu_3O_7$ is three pseudo-cubic elementary perovskite unit cell. The Y and Ba are stacked in the sequence Ba-Y-Ba along the c-axis. The Cu is at the corners of the unit cell. The Cu has two different coordinations with respect to oxygen. There are four possible crystallographic sites for oxygen. The role of Y plane is to serve as a spacer between two CuO_2 planes.

Properties of High Temperature Oxide Superconductors

1. Cuprates are brittle in nature.
2. The properties of the normal state of these materials are highly anisotropic.
3. Hall coefficient is positive indicating that the charge carriers are holes. However, in $(Ln_{1-x}M_x)_2CuO_{4-\delta}$ the charge carriers are electrons.
4. The isotopic effect is almost absent in these materials.
5. Cuprates have their structure derived from the ideal perovskite structure.

Fig. 14.12 Structure of
$YBa_2Cu_3O_7$

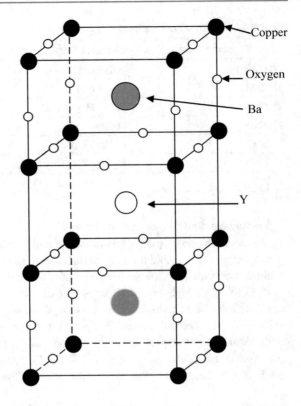

6. Copper exists in the mixed valence state involving a partial oxidation of Cu^{2+} into Cu^{3+}.
7. The structure of cuprate superconductors has one or more CuO_2 planes. Superconductivity is taking place between these planes. Each copper atom in the plane is strongly bonded to four oxygen atoms at a distance of 0.19 nm.
8. The presence of CuO_2 layers determines most of the properties of cuprate.
9. When Y is replaced by rare earth ions in $YBa_2Cu_3O_7$, transition temperature virtually remains the same.
10. They are all type II superconductors.
11. The coherence length (the distance over which density of superconductor carriers rise from zero to its value) and penetration depth are highly anisotropic. Further, the magnitude of coherence length is relatively small in comparison with normal–superconductors.
12. They show metallic behaviour which is highly anisotropic.
13. They show conductivity mainly in CuO_2 plane and conductivity is much smaller in a direction perpendicular to the CuO_2 plane.

2. MgB_2

Magnesium boride, MgB_2, has a superconductivity temperature of 39 K. It falls some-where between low temperature and high temperature superconductors. At ordinary temperature, it behaves like a metal. The structure contains hexagonal honeycombed planes of boron atoms separated by planes of magnesium atoms. The unit cell contains one formula unit. In the structure, boron-boron bonds are much shorter than interlayer distances. Transition temperature increases by 1 K when ^{11}B is replaced with ^{12}B. The transition temperature decreases with the application of pressure. The rate of decrease of transition temperature with pressure is 2 K/GPa. It is two band gap superconductor and classified as Type II superconductor. It is brittle in nature.

In addition to oxide high temperature superconductors and magnesium boride, the discovery of superconductivity in fullerenes, iron-based superconductors and hydride superconductors triggered a lot of interest in recent years. Iron-based superconductor also called ferropnictides (IBSCS) is iron-containing compounds. They have layers of iron and a pnictide, such as As or P. In the system Fe is tetrahedrally coordinated by As(P) atoms. Eremets and co-workers have shown that under very high pressure (100 GPa) H_3S acts as a high temperature superconductor with transition temperature ~ 203 K. The pressure needed to achieve superconductivity is beyond limit were applications are realized.

The mechanism of high temperature superconductivity is not completely understood. BCS theory has certain limitation in explaining the superconductivity in high temperature superconductors.

Solved Examples

Example 1 The penetration depth of mercury at 3.5 K is about 75.0 nm. Estimate the penetration depth at 0 K. Tc for Hg is 4.12 K.

Solution

Given $\lambda_L(3.5 \text{ K}) = 75.0$ nm, $T_c = 4.12$ K, $T = 3.5$ K. From Eq. (14.21)

$$\lambda_L(0\ K) = \lambda_L(T\ K)\left[1 - \left(\frac{T}{T_c}\right)^4\right]^{\frac{1}{2}}$$

$$\lambda_L(0\ K) = 75.0\,\text{nm}\left[1 - \left(\frac{3.5}{4.12}\right)^4\right]^{\frac{1}{2}} = 51.9\,\text{nm}$$

Example 2 The critical temperature T_c for Hg with isotopic mass 199.5 is 4.185 K. Calculate the critical temperature when its isotopic mass changes to 203.4.

Solution

The critical temperature varies with isotopic mass as

$$T_c M^{\frac{1}{2}} = \text{constant}$$

Therefore,

$$T_{c_2} = T_{c_1} \frac{M_1^{\frac{1}{2}}}{M_2^{\frac{1}{2}}} = \frac{4.185 \text{ K} \times (199.5)^{\frac{1}{2}}}{(203.4)^{\frac{1}{2}}} = 4.14 \text{ K}$$

Example 3 Estimate the magnetic field strength necessary to destroy superconductivity in a sample of lead at 4.2 K. The critical magnetic field at 0 K is 0.80 T and transition temperature $T_c = 7.2$ K.

Solution

From Eq. (14.1)

$$B_c = B_0 \left[1 - \left(\frac{T}{T_0} \right)^2 \right] = 0.80 \left[1 - \left(\frac{4.2}{7.2} \right)^2 \right] \text{tesla} = 0.053 \text{ tesla}$$

Objective Type Questions

1. The critical temperature for Hg is
   (a) 4.15 K (b) 20 K (c) 70 K (d) 12.4 K
2. The critical magnetic field is given by
   (a) $B_c = B_0 \left[1 - \left(\frac{T_c}{T} \right)^2 \right]$ (b) $B_c = B_0 \left[1 + \left(\frac{T_c}{T} \right)^2 \right]$ (c) $B_c = B_0 \left[1 - \left(\frac{T}{T_c} \right)^2 \right]$ (d) $B_c = B_0 \left[1 + \left(\frac{T}{T_c} \right)^2 \right]$
3. The London penetration depth λ_L is given by
   (a) $\frac{m}{\mu n_s e^2}$ (b) $\left(\frac{m}{\mu n_s e^2} \right)^{\frac{1}{2}}$ (c) $\left(\frac{m}{\mu n_s e^2} \right)^{\frac{3}{2}}$ (d) $\left(\frac{m}{\mu n_s e^2} \right)^2$
4. The density of superconducting electrons
   (a) Increases with increase of temperature
   (b) Decreases with increase of temperature
   (c) Decreases with decrease of temperature
   (d) Independent of temperature
5. The band gap of superconductor is of the order of
   (a) MeV (b) keV (c) 1 eV (d) 10^{-3} eV

6. Cooper pair follows
   (a) Bose–Einstein statistics (b) Fermi–Dirac statistics (c) Maxwell–Boltzmann statistics (d) none of these
7. The number of centre of masses of Cooper pair inside the effective volume of one Cooper pair is about
   (a) 1 (b) 100 (c) 10^4 (d) 10^6
8. The effective diameter of copper pair is about
   (a) 1 m (b) 10^{-3}m (c) 10^{-6}m (d) 10^{-10}m
9. Which is not true?
   (a) superconducting electrons experiences no scattering (b) superconducting electrons have zero entropy (c) superconducting electrons have long spatial extension (d) the number of superconducting electrons increases with increase of temperature
10. Which has highest number of CuO_2 planes?
    (a) Bi-2201 (b) Bi-2212 (c) Bi-2223 (d) Bi-1212
11. Which of these has highest transition temperature?
    (a) Bi-2234 (b) LaH_{10} (c) H_3S (d) MgB_2
12. In which of the following, charge carriers are electrons
    (a) $(Ln_{1-x}M_x)_2CuO_{4-\delta}$ (b) Bi-2212 (c) Tl-1212 (d) Hg-1201
13. Cuprate have their structure derived from
    (a) Alkali halides (b) Perovskites (c) Ammonium Selenate (d) Cupric chloride
14. Electrical conductivity in cuprate is prominently
    (a) In the plane of CuO_2
    (b) Perpendicular to the plane of CuO_2
    (c) Isotropic
    (d) At $45°$ to the plane of CuO_2
15. In cuprate, each copper atom is strongly bonded to four oxygen at a distance of
    (a) 1.9 nm (b) 0.19 Å (c) 0.19 nm (d) 0.019 nm
16. Which is not true for cuprate superconductors?
    (a) They are brittle (b) Copper exists in mixed valence state (c) CuO_2 layer determine the superconducting properties (d) They are type I superconductors
17. Which is not correct for cuprates?
    (a) Coherence length is anisotropic
    (b) Penetration depth is anisotropic
    (c) Electrical conductivity is isotropic
    (d) Have one or more CuO_2 planes.

Problems

14.1 Outline some experimental facts about superconductivity.
14.2 Describe Meissner effect. Distinguish between type I and type II superconductors.

14.3 Derive the London equations and discuss how its solution explain Meissner effect and flux penetration.

14.4 Describe and discuss BCS theory of superconductivity.

14.5 What are high T_c superconductors? Give some examples.

14.6 Explain Josephson effect.

14.7 Obtain the value of Josephson frequency if the applied DC voltage to the junction is 10 mV.

14.8 Describe the properties of high temperature oxide superconductors.

14.9 Explain the structure of $YBa_2Cu_3O_7$.

Answers
Objective Type Questions

      1. (a)   2. (c)   3. (b)   4. (b)   5. (d)   6. (a)   7. (d)   8. (c)

      9. (d)   10. (c)   11. (b)   12. (a)   13. (b)   14. (a)   15. (c)   16. (d)

      17. (c)

Nanomaterials
15

15.1   Nanomaterials

Nanomaterials are cornerstones of nanoscience and nanotechnology. A nanometre, 10^{-9} m is about ten times the size of the smallest atom, such as hydrogen while a micron is barely larger than the wavelength of the visible light, thus invisible to the human eye. The range of scale from millimetre to nanometre is one million. Nanoscience is about the phenomena that occur in systems with nanometre dimensions. Some of the unique aspects of nanosystems arise solely from the tiny size of the system. The diameter of the hydrogen atom is about one-tenth of a nanometre. About 1000 hydrogen atom could be packed in a cube of size of 1 nm $\times$ 1 nm $\times$ 1 nm.

Nanostructural materials derive their special properties having one or more dimension (made small compared to a length scale critical to the physics of the process. For example, for electron transport, the length scale would be Fermi length ~1 Å and scattering length $\sim 10 - 100$ nm. For optical interaction, it would be half wavelength of the light in medium, $\sim 100 - 300$ nm, for magnetic interaction this length depend on the range (~0.1–10.0 nm) of exchange interaction and magnetic dipole interaction (micron), and for thermal properties, it is mean free path of phonons (a few hundred nm at 300 K to very large at low T).

Let us classify low dimensional structure. Consider quantum confinement effect. We obtain nonzero lowest energy and quantization of the allowed energy levels when one consider an electron in a box as discussed in Sect. 8.2. The energy levels for one dimensional box according to Eq. (8.33) are

$$E_n = \frac{n^2 \pi^2 \hbar^2}{2 m L^2} \quad (n = 1, 2, 3, \ldots \ldots)$$

The electron is confined to a region of space and occupies only discrete energy levels. The lowest state has energy

© The Author(s) 2022
V. K. Jain, *Solid State Physics*, https://doi.org/10.1007/978-3-030-96017-9_15

$$E_1 = \frac{\pi^2 \hbar^2}{2 m L^2} > 0$$

In addition to increasing the minimum energy, confinement also causes its excited state energies to become quantized proportional to n^2. The number of directions remaining free of confinement is used for classification of elementary low dimensional structure within three groups. These are quantum films, quantum wires and quantum dots.

Quantum Films
These are two dimensional structure in which quantum confinement acts only in one direction, which is the z-direction corresponding to the film thickness. Charge carriers are free to move in the xy direction of the film giving their total energy to be sum of the confinement induced and kinetic component

$$E_n = \frac{n^2 \pi^2 \hbar^2}{2 m L^2} + \frac{\hbar^2 k_x^2}{2m} + \frac{\hbar^2 k_y^2}{2m}$$

Quantum Wire
These are one dimension structure. Charge carriers are free to move only along the wire. Thus only the kinetic component along the confinement energy contributes to the total energy.

Quantum Dot
There is zero dimension structure in which carriers are confined in all three directions. Their energy states are quantized. The quantum dots are usually formed by a definite number of atoms. They are typically represented by atomic clusters or nanocrystallites. In a quantum dot the number of atoms is small, and hence its electron states are few sharply separated in energy. Quantum dot ranges from 2 to 10 nm in diameter (about the width of 50 atoms). Because of their small size, quantum dots display unique optical and electrical properties that are different in character than those of the corresponding material. The most apparent of these is the emission of photons under excitation which are visible to the human eye as light. The wavelength of these photon emissions depends not on the material from which the quantum dot is made, but its size. The smaller the dot, the closer it is to the blue end of the spectrum, and the larger the dot the closer to the red end. The different sized dots emit different colours. Dots can even be tuned beyond visible light, into the infrared or into ultraviolet. Quantum dot appear physically as a powder or in a solution. The examples of quantum dots are chalcogenides, CdSe, ZnS, etc.

Quantum dots have applications in various branches of science and technology. The quantum dots are used for biological labelling, also used as lasing element, as sensitizers in photovoltaic applications, as active layers in high emitting diodes, as memory elements and single electron transistors.

Carbon Nanotubes

Carbon nanotubes were discovered by Iijima in 1991 who observed them in high resolution electron microscope images of the by-products of combustion of carbonaceous materials. Carbon nanotube is self-organized nanostructure of carbon atoms with completed bonds. The diameter is ranging from ~1 nm up to several nanometres but with lengths that can approach millimetre or more. There are two kinds of carbon nanotubes. A single wall carbon nanotube which can be considered as a single sheet of graphite called grapheme that has been rolled up into a tube. The resulting nanotube can have metallic or semiconducting properties depending on the direction in which the sheet was rolled up. The fundamental band gap in semiconducting nanotubes ranges from 0.4 to 0.7 eV being dependent on small variation in the diameter and the bonding angle. In general the band gap decreases with an increase in the tube diameter. There is quantum confinement in the radial direction of the tube provided by the monolayer thickness of its wall. Electron can travel along the tube for long distance without being back scattered. The metallic carbon nanotubes have been found to behave like quantum dots. Multiwall nanotubes consist of several concentrically arranged single wall nanotubes. These are typically 10–40 nm in diameter. Thus the diameter in nm and length in mm, strong covalent structure, and large range structure, free of defects made these tubes very useful. Inter tube coupling within a multiwall nanotubes has a relatively small effect on its electronic band structure. As a consequence semiconducting and metallic tubes retain their character if they are a part of a multiwall nanotube. Nantubes (a) can behave like metals or semiconductors (b) can conduct electric current better than diamond and (d) rank among the strongest materials.

Carbon nanotubes are found naturally in combustion products of graphite electrodes and are also readily synthesized by a process of chemical vapour deposition using iron or cobalt seed particles as catalyst. The electronic properties of carbon nanotubes follow from the electronic properties of its parents; a graphite sheet. Given a particular wrapping vector with components n_1 and n_2, the diameter of the tube is given by

$$d = \left(n_1^2 + n_2^2 + n_1 n_2\right)^{\frac{1}{2}} 0.0783 \, \text{nm}$$

The folding of the sheet controls the electronic properties of nanotubes. Each carbon atom contributes three electrons to s-type bonding orbitals and a fourth to a p_z state. These p_z state hybridize to form π and π^* valence and conduction band that are separable by small energy gap semiconductor. However, there are certain symmetry directions in the crystal; the wave function in the valence band is required by the symmetry to be the same as the wave function of the conduction band. Thus for just these directions, the material behave like a metal. This limited conductivity gives grapheme properties intermediate between metal and semiconductor. It is a semimetal.

When the graphine sheet is folded into a tube, the wave vector component along the long axis of the tube can take any value. The component of wave vector perpendicular to the long axis is a quantized specifying

$$\pi D k_\perp = 2\pi n$$

D is diameter of the tube. $k_\perp$ is component of wave vector perpendicular to the length of the tube. The conductivity of the tube is determined by whether one of the allowed values of k is intersected with the **k** points at which the conduction band and the valence band meets. This carbon nanotube can be either metal or semiconductor.

Objective Type Questions

1. A zero dimensional object (a nanometre sized object) is called
   (a) quantum well (b) quantum wire (c) quantum dot (d) quantum film
2. Which is not true about carbon nanotubes (CNT)?
   (a) CNT behaves like metal or semiconductor
   (b) CNT can conduct electric current better than copper
   (c) CNT can transmit heat better than diamond
   (d) CNT rank among the weakest materials
3. The fundamental band gap in semiconducting nanotube ranges
   (a) 4–7 eV (b) 2–4 eV (c) 0.4–0.7 eV (d) 1–2 eV
4. The diameter of multiwall CNT is
   (a) ~80–100 nm (b) >100 nm (c) ~50–80 nm (d) ~10–40 nm
5. Which is false?
   (a) the band gap of semiconducting nanotubes increases with increase of the tube diameter
   (b) in nanotubes quantum confinement in the radial direction is provided by monolayer thickness of wall of nanotube
   (c) electrons cannot travel a long distance in nanotube without back scattering
   (d) semiconducting quantum dots are finding wide spread use as highly efficient fluorescent label.

Answers
Objective Type Questions

1. (c)   2. (d)   3. (c)   4. (d)   5. (c)

Optical Properties
16

16.1 Classification of Optical Processes

When a light beam is incident on an optical medium, the following process as shown in Fig. 16.1 takes place

 (i) Reflection
 (ii) Propagation
(iii) Transmission

As the light is incident on the optical medium, a part of it get reflected from the surface, while rest enters the medium and propagate through it. During propagation refraction, absorption, luminescence and scattering takes place. The amount of light transmitted depends on reflection and absorption by the medium. Refraction causes the light waves to propagate with a small velocity than in free space. Absorption occurs during the propagation if the frequency of the light is resonant with the transition frequencies of the atoms in the medium. Thus the beam will be attenuated as it passes through the medium. The propagating light in the medium can promote the atoms to excited states. The excited atoms de-excite to lower states with spontaneous emission of radiation resulting in phenomenon known as luminescence. The light is emitted in a random direction and has a lower frequency than the incoming beam. In the medium the light may change its direction and frequency after interacting with the medium. This phenomenon is known as scattering. The total number of photons is unchanged, but the number going in the forward direction decreases because light is being redirected in other directions.

The reflection at the surface is described by the coefficient of reflection or reflectivity. This is denoted by the symbol R and is defined as the ratio of reflected power to the power incident on the surface. The coefficient of transmission or transmittivity T is defined as

© The Author(s) 2022
V. K. Jain, *Solid State Physics*, https://doi.org/10.1007/978-3-030-96017-9_16

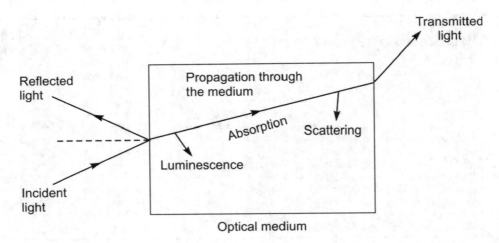

Fig. 16.1 Reflection, propagation and transmission of light beam incident on an optical medium

the ratio of transmitted power to the incident power. If there is only reflection (no other process like absorption or scattering) then by conservation of energy we have

$$T + R = 1 \tag{16.1}$$

If absorption is also taking place, then Eq. (16.1) is modified to

$$T + R + A = 1$$

where A is absorption

The propagation of the beam through a transparent medium is described by the refractive index n, defined as the ratio of velocity of light c in free space to velocity of light v in the medium

$$n = \frac{c}{v} \tag{16.2}$$

The material may also absorb light. This is denoted by absorption coefficient α and is defined as the fraction of the power absorbed in a unit length of the material. Consider a beam propagating in the z direction and its intensity (optical power per unit area) at a point z is $I(z)$. The intensity of light decreases as it passes through the material. The decrease of intensity when length increases by dz is given by

$$dI = -\alpha dz I(z) \tag{16.3}$$

The negative sign indicates a decrease of intensity with increasing value of the length of material. From Eq. (16.3)

$$\frac{dI}{I} = -\alpha dz$$

On integration

$$\int_{I_0}^{I(z)} \frac{dI}{I} = -\alpha \int_0^z dz$$

$$\ln\frac{I(z)}{I_0} = -\alpha z$$

$$I(z) = I_0 \exp(-\alpha z) \tag{16.4}$$

where $I(z) = I_0$ at $z = 0$. Since the absorption coefficient is a strong function of frequency, therefore, material may absorb a colour but not another. Equation (16.4) is known as Beer's law.

The beam of light which is incident on the front surface is partially reflected from the face and the rest of light enter the medium and propagate through it. The light entering the medium is then given by $(1 - R_1)$ where R_1 is reflectivity of surface 1. During its propagation through the material a part of it is absorbed by the medium as given by Eq. (16.4). Thus when the light strike the back face of the material, its intensity reduces to

$$(1 - R_1)\exp(-\alpha l)$$

where l is the length of material. On striking the back surface, part of it is reflected back and the rest is transmitted. If R_2 is the reflectivity of the back surface, then the light transmitted out of the material will be

$$T = (1 - R_2) \times (1 - R_1)exp(-\alpha l) \tag{16.5}$$

If the front and back surface have equal reflectivity, that is, $R_1 = R_2 = R$, Eq. (16.5) takes the form

$$T = (1 - R)^2 \exp(-\alpha l) \tag{16.6}$$

The absorption of an optical material is also specified by optical density or absorbance given by

$$\text{Optical density} = -\log_{10}\left(\frac{I_l}{I_0}\right) \tag{16.7}$$

From Eq. (16.4)

$$\frac{I_l}{I_0} = \exp(-\alpha l)$$

Taking log

$$\ln\frac{I_l}{I_0} = -\alpha l$$

$$-\log_{10}\left(\frac{I_l}{I_0}\right) = 0.434\alpha l \tag{16.8}$$

16.2 Refraction and Absorption

In the absence of absorption, the refractive index of the material is given by

$$n^2 = K \tag{16.9}$$

In the presence of absorption, the refractive index is a complex quantity and is given by

$$n^* = n + i\kappa \tag{16.10}$$

The real part of n^*, that is n, is the same as the normal refractive index. The imaginary part is called the extinction coefficient and is related with the absorption coefficient.

Consider the propagation of the wave in the z-direction. The spatial and time dependence of the electric field E is given by

$$E(z, t) = E_0 \exp[i(kz - \omega t)] \tag{16.11}$$

where k is wave vector, ω is the angular frequency, E_0 is the amplitude of E at $z = 0$. In a non-absorbing medium of refractive index n, the wavelength of the light is reduced by a factor of n compared to free space wavelength λ. Therefore,

$$k = \frac{2\pi}{(\lambda/n)} = \frac{2n\pi}{\lambda} = \frac{2n\pi}{(c/v)} = \frac{n(2\pi v)}{c} = \frac{n\omega}{c} \tag{16.12}$$

Generalizing Eq. (16.12) to the case of an absorbing medium (introducing complex refractive index)

$$k = \frac{2\pi}{(\lambda/n^*)} = \frac{2n^*\pi}{\lambda} = \frac{2n^*\pi}{(c/v)} = \frac{n^*(2\pi v)}{c} = \frac{n^*\omega}{c} \tag{16.13}$$

From Eqs. (16.11) and (16.13)

$$E(z, t) = E_0 \exp\left[i\omega\left(\frac{n^*z}{c} - t\right)\right] = E_0 \exp\left\{i\omega\left[\frac{(n + i\kappa)z}{c} - t\right]\right\}$$

$$E(z, t) = E_0 \exp\left(-\frac{\kappa \omega z}{c}\right) \exp\left[i\omega\left(\frac{nz}{c} - t\right)\right] \tag{16.14}$$

The factor $\exp\left(-\frac{\kappa \omega z}{c}\right)$ indicates that a nonzero κ attenuates the electric wave by a factor of e in a distance $\frac{c}{\omega \kappa}$.

The optical intensity of a light wave is proportional to the square of electric field, that is,

$$I \propto EE^* = |E|^2 \tag{16.15}$$

From Eqs. (16.14) and (16.15)

$$I \propto E_0^2 \exp\left(-\frac{2\kappa \omega z}{c}\right) \tag{16.16}$$

The intensity variation according to Beer's law is

$$I = I_0 \exp(-\alpha z) \tag{16.17}$$

A comparison of Eqs. (16.16) and (16.17) gives

$$\alpha = \frac{2\kappa \omega}{c} = \frac{4\pi \kappa}{\lambda} \tag{16.18}$$

where λ is free space wavelength. Equation (16.18) shows that absorption coefficient is directly proportional to the imaginary part of the refractive index.

The refractive index of a medium is related to relative dielectric constant K as

$$n = \sqrt{K} \tag{16.19}$$

If refractive index is complex, then K must also be complex. We define

$$K^* = K_1 + iK_2 \tag{16.20}$$

Thus in analogy with Eq. (16.19)

$$n^* = \sqrt{K^*}$$

$$n^{*2} = K^* = K_1 + iK_2 \tag{16.21}$$

From Eqs. (16.10) and (16.21)

$$K_1 = n^2 - \kappa^2 \tag{16.22}$$

$$K_2 = 2n\kappa \tag{16.23}$$

Adding the squares of Eqs. (16.22) and (16.23)

$$\left(n^2 - \kappa^2\right)^2 + 4n^2\kappa^2 = K_1^2 + K_2^2$$

$$\left(n^2 + \kappa^2\right)^2 = K_1^2 + K_2^2$$

$$n^2 + \kappa^2 = \left(K_1^2 + K_2^2\right)^{\frac{1}{2}} \tag{16.24}$$

From Eqs. (16.22) and (16.24)

$$2n^2 = \left[K_1 + \left(K_1^2 + K_2^2\right)^{\frac{1}{2}}\right]$$

$$n = \frac{1}{\sqrt{2}}\left[K_1 + \left(K_1^2 + K_2^2\right)^{\frac{1}{2}}\right]^{\frac{1}{2}} \tag{16.25}$$

Subtracting Eqs. (16.22) and (16.24)

$$2\kappa^2 = \left[-K_1 + \left(K_1^2 + K_2^2\right)^{\frac{1}{2}}\right]$$

$$\kappa = \frac{1}{\sqrt{2}}\left[-K_1 + \left(K_1^2 + K_2^2\right)^{\frac{1}{2}}\right]^{\frac{1}{2}} \tag{16.26}$$

For weakly absorbing medium, κ is very small and Eqs. (16.25) and (16.26) can be written as

$$n = \sqrt{K_1} \tag{16.27}$$

$$\kappa = \frac{K_2}{2n} \tag{16.28}$$

In a transparent medium such as glass in the visible region of the spectrum, the absorption coefficient is very small and κ and K_2 are negligible and hence both n^* and K^* can be taken as real numbers.

16.3   Reflection

Consider reflection of electric and magnetic waves at the interface between two materials having refractive indices n_1 and n_2 and absorption indices κ_1 and κ_2, respectively, as shown in Fig. 16.2. The Incident and reflected waves are in material 1 whereas transmitted wave is in material2. At the interface the continuity condition that electric and magnetic

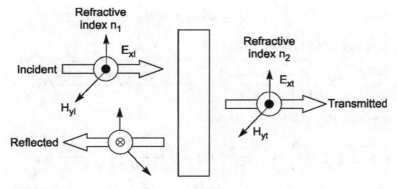

Fig. 16.2 Transmission and reflection of light at an interface between two mediums. All rays are normal to the interface. The symbol • for the magnetic fields of the incident and transmitted rays indicates that the field direction is out of page, while the symbol × for the reflected wave indicate that the field pointing into page

tangential components must be conserved. The incident wave is

$$E_{Ix} = E_{0I}\exp\left[i\left(\frac{\omega n_1^* z}{c} - \omega t\right)\right] \tag{16.29}$$

$$H_{Iz} = \frac{n_1^*}{\mu_r}\left(\frac{\epsilon_0}{\mu_0}\right)^{\frac{1}{2}} E_{0I}\exp\left[i\left(\frac{\omega n_1^* z}{c} - \omega t\right)\right] \tag{16.30}$$

The reflected wave will be

$$E_{yR} = -E_{0R}\exp\left[i\left(-\frac{\omega n_1^* z}{c} - \omega t\right)\right] \tag{16.31}$$

$$H_{yR} = \frac{n_1^*}{\mu_r}\left(\frac{\epsilon_0}{\mu_0}\right)^{\frac{1}{2}} E_{0R}\exp\left[i\left(-\frac{\omega n_1^* z}{c} - \omega t\right)\right] \tag{16.32}$$

The transmitted waves are

$$E_{yT} = E_{0T}\exp\left[i\left(\frac{\omega n_2^* z}{c} - \omega t\right)\right] \tag{16.33}$$

$$H_{yT} = \frac{n_2^*}{\mu_r}\left(\frac{\epsilon_0}{\mu_0}\right)^{\frac{1}{2}} E_{0T}\exp\left[i\left(\frac{\omega n_2^* z}{c} - \omega t\right)\right] \tag{16.34}$$

If E_x and H_y are conserved at the interface $z = 0$, then from Eqs. (16.29)-(16.34)

$$E_{0I} - E_{0R} = E_{0T} \tag{16.35}$$

$$n_1^* E_{0I} + n_1^* E_{0R} = n_2^* E_{0T} \tag{16.36}$$

On adding Eqs. (16.35) and (16.36)

$$E_{0I} = \frac{E_{0T}}{2}\left(1 + \frac{n_2^*}{n_1^*}\right) = \frac{(n_1^* + n_2^*)}{2n_1^*}E_{0T} = \frac{(n_2 + n_1) + i(\kappa_2 + \kappa_1)}{2(n_1 + i\kappa_1)}E_{0T} \tag{16.37}$$

On subtracting Eqs. (16.35) and (16.36)

$$E_{0R} = \frac{E_{0T}}{2}\left(\frac{n_2^*}{n_1^*} - 1\right) = \frac{(n_2^* - n_1^*)}{2n_1^*}E_{0T} = \frac{(n_2 - n_1) + i(\kappa_2 - \kappa_1)}{2(n_1 + i\kappa_1)}E_{0T} \tag{16.38}$$

The reflection coefficient from Eqs. (16.37) and (16.38)

$$R = \frac{(E_{0R})^2}{(E_{0I})^2} = \frac{[(n_2 - n_1) + i(\kappa_2 - \kappa_1)]^2}{[(n_2 + n_1) + i(\kappa_2 + \kappa_1)]^2} = \frac{(n_2 - n_1)^2 + (\kappa_2 - \kappa_1)^2}{(n_2 + n_1)^2 + (\kappa_2 + \kappa_1)^2} \tag{16.39}$$

If the interface is between vacuum and material of refractive index n and absorption index κ then

$$\kappa_1 = 0, n_1 = 1; n_2 = n, \kappa_2 = \kappa$$

$$R = \frac{(n-1)^2 + \kappa^2}{(n+1)^2 + \kappa^2} \tag{16.40}$$

For $\to 0$, Eq. (16.40) takes the form

$$R = \frac{(n-1)^2}{(n+1)^2} \tag{16.41}$$

Thus for a dielectric material without losses ($\kappa \to 0$), the reflection is solely determine by the refractive index.

The absorption coefficient and the reflection coefficients measurements are enough to determine the values of n and κ. The n and κ for reflection coefficient can be determined from the reflection coefficient and Kramers–Kronig relation between n and κ.

16.4 Kramers–Kronig Relationship

The refractive index and the absorption coefficients are not independent parameters but are related to each other. They are obtained from real and imaginary parts of the complex refractive index n^*. Kramers and Kronig by invoking the law of casuality (that an effect may not preceeds its cause) and using complex number analysis obtain the general relationship between real and imaginary parts of the complex refractive index. The relations are

$$n(\omega) = 1 + \frac{1}{\pi} P \int_{-\infty}^{\infty} \frac{\kappa(\omega')}{\omega' - \omega} d\omega' \qquad (16.42)$$

$$\kappa(\omega) = -\frac{1}{\pi} P \int_{-\infty}^{\infty} \frac{n(\omega') - 1}{\omega' - \omega} d\omega' \qquad (16.43)$$

where P indicates that principal part of the integral is taken. Kramers–Kronig relations can be used to calculate n from κ and vice versa.

16.5   Interband Transitions

In this process a photon is absorbed and an excited state is formed and a hole is left behind. Consider a material having a band gap E_g. A photon of energy E_g can excite an electron from valance band to conduction band leaving behind a hole in the valance band as shown in Fig. 16.3. The following features are characteristic of these kind of transitions:

(i)   The interband transition has a threshold energy at the energy gap.
(ii) The transitions are either direct or indirect. In both kind of transition the energy and momentum is conserved
(iii) The coupling between valence and conduction band.
(iv) Because of Pauli exclusion principle, an interband transition occurs from occupied state below the Fermi level to an unoccupied state above the Fermi level.

Fig. 16.3 Schematic diagram of an allowed interband transition

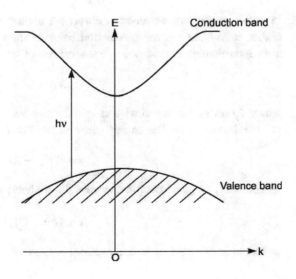

(v) Photons of a particular energy are more effective in producing an interband transition if the energy separation between the two bands is nearly constant over many k (wave vector) values. In such a case there are many initial and final states which can be coupled by the same photon energy. If photon have a small energy band width then it will be effective over many k values provided $E_C(k) - E_V(k)$ does not vary rapidly with k. Thus the interband transition to be most significant for k values near band extrema.

(vi) In semiconductors the interband transitions usually occur at frequencies above which free carrier concentrations are important.

Interband transitions for various class of materials have the following differences:

1. Insulators: In insulators the band gap is quite large; therefore at room temperature, no carriers are thermally excited across the band gap. Therefore, the interband transitions become important at relatively high photon energies (above the violet). Thus insulators frequently are optical transparent.

2. Semiconductors: In semiconductor the band gap is small; therefore at room temperature the carriers are thermally excited across the band gap. In semiconductors the interband transitions occur in the infrared and visible.

3. Metals: In the case of metals the interband transition occurs at frequency where free carrier effects are still important.

16.6 Direct Transitions

In direct transitions, an electron absorbs a photon and makes a vertical transition to a higher empty state in the conduction band as shown in Fig. 16.4. The conservation of momentum which correspond to conservation of wave vector is given by

$$k_f - k_i = \Delta k = 0 \tag{16.44}$$

where k_f and k_i are the final and initial wave vector in the conduction $E_C(k)$ and the valance $E_V(k)$ band. The energy conservation gives

$$h\nu = E_C - E_V \tag{16.45}$$

For small range of photon energies, the probability ρ for direct transition is given by

$$\rho \propto (h\nu - E_g)^{\frac{1}{2}} \tag{16.46}$$

Example of direct transition material is GaAs.

Fig. 16.4 Direct interband
transition

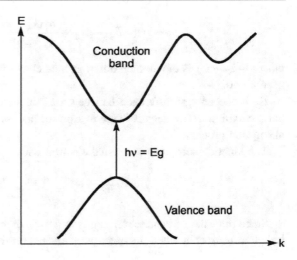

16.7   **Indirect Transitions**

In making indirect transitions, the semiconductor can either emit or absorb a phonon of
energy $h\nu_q$. From energy conservation

$$h\nu = E_f - E_i \pm h\nu_q \qquad (16.47)$$

where E_f and E_i are respectively, the energies of the final and initial states and $\pm$ sign
refer to phonon emission (+ sign) and absorption (− sign). Figure 16.5 shows an indirect
transition.

The conservation of energy can be written as

Fig. 16.5 Indirect interband
transition

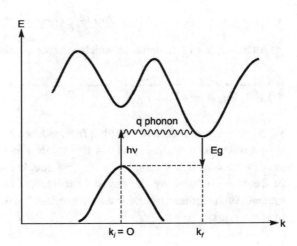

$$hv = E_g - hv_q + \frac{\hbar^2(k_e - k_C)^2}{2m_e^*} + \frac{\hbar^2 k_h^2}{2m_h^*} \tag{16.48}$$

where $\hbar(k_e - k_C)$ is the momentum of excited electron, $\hbar k_h$ is the momentum for the holes near $k = 0$.

E_g is the energy difference between the conduction band minimum and the valence band maximum. The negative sign in front of phonon energy hv_q corresponds to phonon absorption process.

The kinetic energy of the excited electron with crystal momentum $\hbar k_e$ is

$$E_e - E_C = \frac{\hbar^2(k_e - k_C)^2}{2m_e^*} \tag{16.49}$$

Since the valence band extremum is at $k = 0$, then $\hbar k$ is the crystal momentum for the hole that is created when the electron is excited corresponding to kinetic energy of hole

$$E_h = \frac{\hbar^2 k_h^2}{2m_h^*} \tag{16.50}$$

Thus conservation of energy

$$hv = E_g - hv_q + E_e - E_C + E_h \tag{16.51}$$

Conservation of momentum gives

$$q = k_e - k_h \tag{16.52}$$

where q is the wave vector of the absorbed phonon. The frequency dependence of indirect band transition probability involving phonon absorption is given by

$$\rho(hv) \propto \frac{\pi}{8}\left(hv - E_g + hv_q\right)^2 \tag{16.53}$$

An example of a material in which indirect transition is taking place is Si.

16.8   Excitons

When an electron makes a transition from valence band to conduction band and becomes a free electron, it leaves a hole in the valence band. However, it is also possible that excited electron is not completely free but coupled with the hole which it leaves behind in the valence band, by the attractive electrostatic Coulomb force and forms a bound system. Such a bound system is called exciton. It exists in semiconductors, insulators and in some liquids.

The exciton is similar to an excited hydrogen atom—in both cases there is an electron moving about a unit positive charge and the energy spectrum is discrete one as shown in Fig. 16.6. The excitation levels are near the conduction band. The binding energy of the free exciton is typically a few meV. Since excitons are electrically neutral, they do not contribute to additional charge carriers. The excitons wander through the lattice for some time. They collide with phonons, impurity centres and other lattice imperfections resulting in either recombination or decomposition. In the case of recombination, the ground state is restored and exciting energy is given to the lattice or emitted in the form of light. In the case of decomposition, free charges, electrons and holes are produced, and they contribute to the photoconductivity.

Since the exciton is bound electron–hole pair, the interaction between them is given by

$$V(r) = -\frac{e^2}{(4\pi\epsilon_0)Kr} \tag{16.54}$$

where K is dielectric constant of the material in which the exciton is formed. Let the effective mass of electron is m_e^* and that of hole is m_h^*. The charge on the electron is $-e$ and on hole is $+e$. The Hamiltonian for the exciton is

$$H = -\frac{\hbar^2}{2m_e^*}\nabla_e^2 - \frac{\hbar^2}{2m_h^*}\nabla_h^2 - \frac{e^2}{(4\pi\epsilon_0)Kr} \tag{16.55}$$

where the first and second term on the right hand side of Eq. (16.55) is kinetic energy of the electron and hole, respectively. The third term is the potential energy of the exciton. The Hamiltonian given by Eq. (16.55) is quite similar to that of hydrogen atom. Thus the energy eigenvalues of exciton can be written as

$$E_{ex} = E_g + \frac{\hbar^2 k^2}{2(m_e^* + m_h^*)} - \frac{\mu e^4}{2(4\pi\epsilon_0)^2 n^2 \hbar^2 K^2} \tag{16.56}$$

where μ is the reduced mass given by

Fig. 16.6 Exciton states in a semiconductor

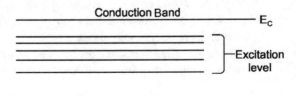

$$\frac{1}{\mu} = \frac{1}{m_e^*} + \frac{1}{m_h^*} \tag{16.57}$$

and n is principal quantum number and can take values 1, 2, 3, ...

In Eq. (16.56), E_g represent the minimum energy needed to create an unbound electron–hole pair from the ground state. The second term denotes the kinetic energy of translation of the exciton. Exciton may be formed by both direct and indirect transitions. For direct transitions $k \approx 0$ and the binding energy of exciton ($n = 1$) is by using Eq. (4.113)

$$E = -\frac{\mu e^4}{2(4\pi \epsilon_0)^2 \hbar^2 K^2} = -\frac{m\mu e^4}{2m(4\pi \epsilon_0)^2 \hbar^2 K^2} = -\frac{\mu}{mK^2}(13.6eV) \tag{16.58}$$

where 13.6 eV is the ionization or binding energy of the hydrogen atom. The radius of the nth orbit is given by

$$a_n = K\frac{m}{\mu}n^2 a_0 \text{where } a_o = 0.0529 \, \text{nm} \tag{16.59}$$

The model given above is applicable to weakly bound exciton known as Wannier–Mott exciton:

The exciton can be classified into two groups as follow:

1. Wannier–Mott Exciton:
    (i)   These are weakly bound (free) exciton.
    (ii) The binding energy is ~ 10 meV.
    (iii) These are common in inorganic semiconductors like GaAs, CdS, CuO_2.
    (iv) These are not localized.
    (v)   It moves in a medium of effective dielectric constant K.
2. Frenkel Exciton or Tightly Bound Exciton:
    (i)   These exciton are strongly (tightly) bound.
    (ii) The binding energy $\sim 0.1 - 1 \, \text{eV}$.
    (iii) Typically found in insulators and molecular crystals, for example, in rare gas crystals, alkali halides, aromatic molecular crystals.
    (iv) Model of bound electron–hole pair in dielectric medium breaks down when a_n is the order of interatomic distance.
    (v)   These are often localized on just one atomic/molecular site of the crystal but whole exciton may transfer throughout the crystal.

16.9   Colour Centres

Colour centre is a point defect which produces optical absorption bands in an otherwise transparent crystal. Colour centres are found in many inorganic crystals and in glasses. They are very common in naturally occurring minerals.

The colour centres for example F centres can be produced by irradiating the crystal with X-rays or γ-rays. The ionizing radiation ejects an anion, for example, Cl^-, in NaCl lattice, from its site and thus creating an anion vacancy as shown in Fig. 16.7. This anion vacancy which is surrounded by six positive ions in alkali halide lattice can trap an electron and thus maintaining the charge neutrality of the crystal. The trapped electron has a series of energy levels. It can absorb light and make transition from the ground state to excited states and thus giving rise colour to the crystal. The colour centres may be destroyed by heating or by bleaching with light that is by exposure to light corresponding to the spectra region in which the centres absorb the light.

The F centre was also produced by heating an alkali halide crystal in the vapour of same alkali metal or another alkali metal and then quenching to room temperature. It shows an absorption band in the visible and ultraviolet whereas the pure crystals were transparent. The absorption band is called F-band and the crystal has a coloured appearance; when NaCl is heated in a vapour of Na or K, it turns a yellow–brown colour. KCl heated in K vapour turns violet, LiF heated in vapour of Li look pink. F centre band is independent of added metal. The same band is formed when stoichiometric crystals are irradiated with X-rays or γ-rays. The colour centres can also be produced by electrolysis. The density of the crystal decreases when excess metal is introduced.

Let us now consider the model of F centres produced by heating alkali halides in vapours of alkali metal. The crystal, for example, NaCl when heated in the Na vapour absorbed metal atom at its surface for example at point A as shown in Fig. 16.8, the atom may then split up into sodium ion and an electron

$$Na \rightarrow Na^+ + e^-$$

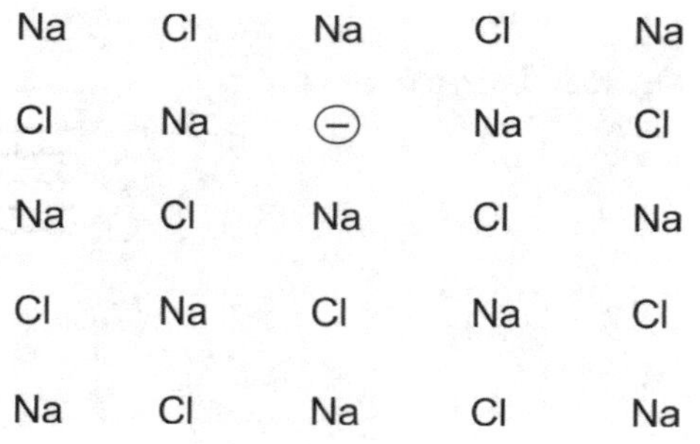

Fig. 16.7 Structure of F centre in NaCl

Na	Cl	Na	Cl	Na
Cl	Na	⊖	Na	Cl
Na	Cl	Na	Cl	Na
Cl	Na	Cl	Na	Cl
Na	Cl	Na	Cl	Na

Fig. 16.8 F centre model

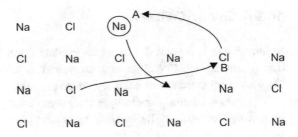

The sodium ion stays as a surface ion in the usual position on the face of a crystal of NaCl. The negative ion (Cl^-) from the lattice such as B may then jump into the position next to A forms the beginning of a new layer on the surface of the crystal. The electron and the vacancy at B caused by the motion of negative ion, diffuse into the crystal and the electron gets trapped in a region where there is a anion vacancy. The trapped electron is surrounded by six positive ions. Here it behaves as a particle in the three dimensional box and giving rise to quantized energy levels as shown in Fig. 16.9. Difference between energy levels of the box is proportional to $\frac{1}{L^2}$ where L is the length of the box. The absorption takes place between energy levels of the box. For larger L, the absorption energy is smaller. The absorption peak of F centers shifts towards longerwavelength (red shift) with increasing lattice parameter. Pressure causes the box to shrink and absorption blue shift. Reduced temperature has similar effect. These energy levels lie below the conduction band.

The allowed transitions between the energy levels cause the light absorption which coloured the crystal. The intensities of such absorptions are very large. Thus even low concentrations tend to give whole crystal a faint hue of the respective colour.

The technique of electron spin resonance (ESR) which is very sensitive probe for unpaired electrons is used to study the structure of F centre. ESR spectra of F centre in NaH show 19 lines of different intensities. The ESR spectra are explained on the basis of trapped electron coupled with six equivalent nearest neighbour Na^{23} ions (having nuclear spin $I = 3/2$). The number of lines according to Eq. (13.60), expected is

Fig. 16.9 Energy levels of F centre

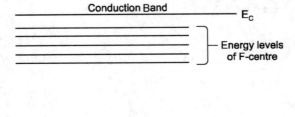

$$2nI + 1 = 2 \times 6 \times \frac{3}{2} + 1 = 19 \text{ is observed.}$$

The most common electron and hole colour centres observed in alkali halides are as follow:

1. F centres due to electron trapped at anion (that is halogen vacancy).
2. M centres due to two adjacent F centres (Fig. 16.10).
3. R_2 centres due to group of two neative ion vacancies with two associated electrons (Fig. 16.11).
4. V_k centres due to hole trapped by a pair of anions.
5. V_I Centres Due to a Hole Trapped at a Positive Ion Vacancy as Shown in Fig. 16.12.
6. H centres due to singly ionized halogen molecules at anion vacancy as shown in Fig. 16.13.

Fig. 16.10 M centre

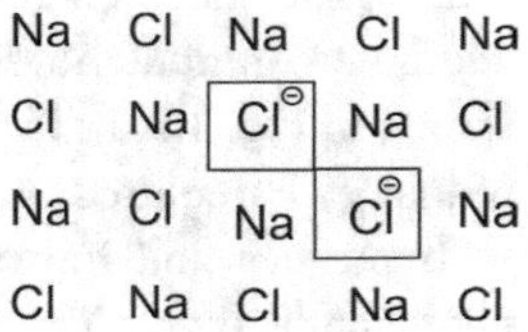

Fig. 16.11 R_2 centre

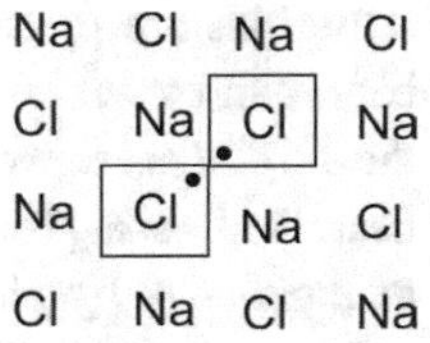

Fig. 16.12 V_1 centre

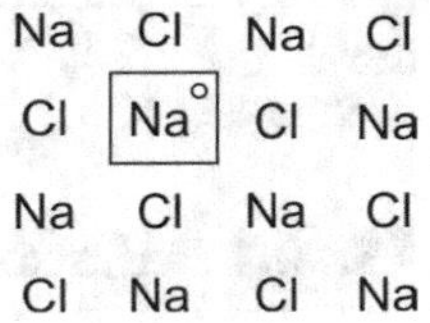

Fig. 16.13 H centre

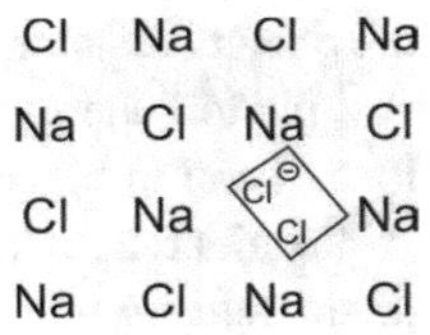

Fig. 16.14 Schematic diagram
of the experimental
arrangement for measuring
photocurrent

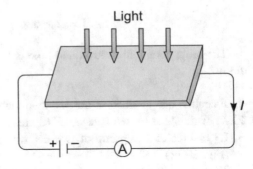

16.10 Photoconductivity

Photoconductivity is observed when light is incident on an insulator or semiconductor,
and the photon energy is sufficiently high to excite an electron from an occupied valence
state to an unoccupied conduction state. In such interband transition both the electron
and hole contribute to electrical conductivity if a voltage is applied across the sample as
shown in Fig. 16.14. In pure alkali halides, the energy band gap is $\sim 8\,\text{eV}$ and this much
of energy is required to produce an electron in the conduction band.

Impurities and imperfections in the insulating material also contribute towards the
photoconductivity. The presence of impurities causes discrete energy levels below the
conduction bands (donors) and above the valance band (acceptors). If acceptor and donor
impurities are present, then the photons of frequency less than the corresponding to the
band gap can cause the transition. Imperfections also introduce discrete energy levels in
the forbidden energy gap, which are often called traps. Thus photoconductivity may be
described as a process whereby electrons (or holes) are freed from one or other type
of band state by photon, pass some time in conduction (or valence) band in which they
act as current carriers and are finally recaptured by traps. When the incident radiation
is removed, photoconductivity decays because hole and electrons recombine with each
other.

16.10.1 Variation of Photoconductivity with Illumination

The variation of photoconductivity with photon energy is called the spectral response of
the photoconductor. Spectral response curve typically show a fairly well defined maximum
at a photon energy close to that is minimum energy required to excite an electron from
lower band to higher lying conduction band where it is free to contribute to conductivity.

Consider light of appropriate frequency is falling on a photoconductor. As a result
of this electron–holes pairs are formed in the material. Recombination occurs by direct
combination of electrons and holes. It is assumed that the mobility of holes is negligible.

The rate of change of electron concentration n is given by

$$\frac{dn}{dt} = L - Anp = L - An^2 \text{ (since } n = p) \tag{16.60}$$

where L is the number of photons absorbed per unit volume of the material per unit time. $Anp = An^2$ represent the loss of electron because of recombination, A is proportionality constant.

In equilibrium

$$\frac{dn}{dt} = 0 \tag{16.61}$$

$$L = Anp = An^2$$

For $n = p = n_0$

$$L = An_0^2 \tag{16.62}$$

The electron concentration is thus

$$n_0^2 = \frac{L}{A}$$

$$n_0 = \sqrt{\frac{L}{A}} \tag{16.63}$$

Using Eq. (16.63) the photoconductivity σ is given by

$$\sigma = n_0 e \mu = \mu e \sqrt{\frac{L}{A}} \tag{16.64}$$

where μ is electron mobility. For a given voltage the photoconductivity vary with light level L as $\sqrt{L}$. The actual exponential observe is 0.5–1.0. The graph between photocurrent and light irradiance is shown in Fig. 16.15. The plot shows that the photocurrent is almost

Fig. 16.15 Photoconductive response versus illumination level for a CdS crystal. Current $\propto (illumination)^n$ ($n = 0.92$ for low level of irradiance and $n = 0.58$ for high level of illumination

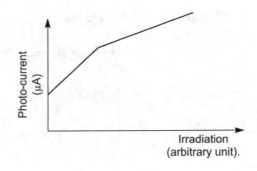

Photo-current (μA)

Irradiation (arbitrary unit).

linear function of the illumination intensity for low intensities but is nonlinear at high illumination levels. The nonzero value of the current in the absence of illumination is called dark current which is referred to as background current.

When the light is switched off suddenly ($L = 0$) then from Eq. (16.60)

$$\frac{dn}{dt} = -An^2$$

$$\frac{dn}{n^2} = -Adt \tag{16.65}$$

On integration

$$\frac{1}{n} = At + C \tag{16.66}$$

Let $n = n_0$ at $t = 0$ then $C = \frac{1}{n_0}$ and Eq. (16.66) is

$$\frac{1}{n} = At + \frac{1}{n_0}$$

$$\frac{1}{n} = \frac{n_0 At + 1}{n_0}$$

$$n = \frac{n_0}{n_0 At + 1} \tag{16.67}$$

The time t_0 in which electron concentration drops to $\frac{n_0}{2}$ is

$$\frac{n_0}{2} = \frac{n_0}{n_0 At_0 + 1}$$

$$n_0 At_0 + 1 = 2$$

$$t_0 = \frac{1}{An_0} \tag{16.68}$$

From Eqs. (16.63) and (16.68)

$$t_0 = \frac{1}{A(L/A)^{\frac{1}{2}}} = \frac{1}{\sqrt{LA}} = \frac{n_0}{n_0\sqrt{LA}} = \frac{n_0}{\sqrt{\frac{L}{A}}\sqrt{LA}} = \frac{n_0}{L} \tag{16.69}$$

From Eqs. (16.63) and (16.64)

$$\sigma = \mu e\sqrt{\frac{L}{A}} = n_0\mu e$$

$$n_0 = \frac{\sigma}{\mu e} \qquad (16.70)$$

From Eqs. (16.69) and (16.70)

$$t_0 = \frac{\sigma}{\mu e L} \qquad (16.71)$$

t_0 is called the response time. Thus response time is directly proportional to photo-conductivity at a given light level. Sensitive photoconductor should have long response time.

16.10.2 Sensitivity or Gain Factor (G)

It is defined as the ratio of the number of carriers crossing the specimen to the number of photons absorbed in the specimen. Let the thickness of the specimen is d and cross sectional area is unity, then

$$G = \frac{\text{Particle flux}}{Ld} = \frac{(J/e)}{Ld} \qquad (16.72)$$

But

$$J = \sigma E \text{and E} = \frac{V}{d} \qquad (16.73)$$

From Eqs. (16.63), (16.72) and (16.73)

$$G = \frac{(\sigma E/e)}{Ld} = \frac{\sigma V}{eLd^2} = \frac{n_0 e \mu V}{eLd^2} = \frac{n_0 \mu V}{Ld^2} = \sqrt{\frac{L}{A}} \frac{\mu V}{Ld^2} = \frac{\mu V}{d^2 \sqrt{LA}} \qquad (16.74)$$

Gain factor G is also defined as

$$G = \frac{T_e}{T_d} \qquad (16.75)$$

where T_e is the life time of an electron before recombination. From Eq. (16.69)

$$T_e = \frac{n_0}{L} = \frac{1}{L}\sqrt{\frac{L}{A}} = \frac{1}{\sqrt{LA}} \qquad (16.76)$$

and T_d is the transit time of an electron between the electrodes. If the distance between the electrodes is d, the velocity of electron is v

$$T_d = \frac{d}{v} \qquad (16.77)$$

and

$$v = \mu E = \mu \frac{V}{d} \tag{16.78}$$

From Eqs. (16.75)–(16.77)

$$G = \frac{T_e}{T_d} = \frac{\mu V}{d^2 \sqrt{LA}} \tag{16.79}$$

The expression for G is quite general and is not limited to a specific model. This model overestimates the value of gain while the experimental values are lower. This is as much as 10^8.

16.10.3 Effect of Traps

A trap is an energy level in the forbidden energy gap of the crystal capable of capturing either an electron or hole as shown in Fig. 16.16. The captured carrier may be reemitted at a subsequent time and may move to another trap as shown in Fig. 16.16. There are two types of traps:

1. The traps help electrons and holes to recombine and thereby try to restore equilibrium. These are known as recombination centre. In it thermal freeing of the captured carrier is less than the probability of recombination. These affect lifetime and photosensitivity.
2. The second kind of traps do not contribute directly in an important way to recombination but affects principally by freedom of motion of charge carriers of one sign or the other. Here probability of thermal freeing of trapped carriers is greater than probability of recombination. It affects the speed of response.

Fig. 16.16 Influence of traps in electron–hole recombination

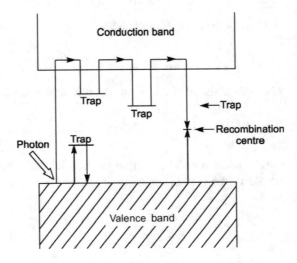

Consider a crystal having N electron traps levels per unit volume. If temperature is low, then it can assume that production of thermal charge carriers is negligible. Thus the rate of change of number of electrons n is

$$\frac{dn}{dt} = L - An(n+N) \qquad (16.80)$$

where n is electron concentration in the conduction band, and N is the trap concentration. It is assumed that proportionality constant A is same for electrons and holes. In the steady state

$$\frac{dn}{dt} = 0$$

$$L = An_0(n_0 + N)$$

$$\frac{L}{A} = n_0(n_0 + N) \qquad (16.81)$$

Usually in a crystal $\approx 10^{14} \text{cm}^{-3}$. Now consider the case in which $n_0 \ll 10^{14} \text{cm}^{-3}$. Thus for the case $n_0 \ll N$, we have from Eq. (16.81)

$$n_0 = \frac{L}{AN} \qquad (16.82)$$

The photoconductivity is now directly proportional to illumination level L. Consider the case $n_0 \gg N$, from Eq. (16.81)

$$n_0^2 = \frac{L}{A}$$

$$n_0 = \sqrt{\frac{L}{A}} \qquad (16.83)$$

which is the case if traps are missing. When the light is switched off, then from Eq. (16.80)

$$\frac{dn}{dt} = -An(n+N)$$

$$\frac{dn}{n(n+N)} = -A dt$$

$$\frac{1}{N}\left[\frac{1}{n} - \frac{1}{n+N}\right]dn = -A dt \qquad (16.84)$$

On integration

$$-\int \frac{dn}{n} + \int \frac{dn}{n+N} = NA \int dt$$

$$-\ln n + \ln(n+N) = NAt + C$$

$$\ln \frac{n+N}{n} = NAt + C \tag{16.85}$$

At $t = 0$, $n = n_0$ and from Eq. (16.85)

$$\ln \frac{n_0 + N}{n_0} = C \tag{16.86}$$

From Eqs. (16.85) and (16.86)

$$\ln \frac{n+N}{n} - \ln \frac{n_0 + N}{n_0} = NAt \tag{16.87}$$

For $n_0 \ll N$, Eq. (16.87) takes the form

$$-\ln \frac{N}{n} + \ln \frac{N}{n_0} = -NAt$$

$$\ln \frac{n}{n_0} = -NAt$$

$$n = n_0 \exp(-NAt) \tag{16.88}$$

The time t_0 for the signal or photocurrent to fall $(1/e)$ of its initial value is

$$NAt_0 = 1$$

$$t_0 = \frac{1}{NA} \tag{16.89}$$

A comparison of Eqs. (16.88) and (16.89) indicates that the presence of traps reduces the conductivity and also the response time.

16.11 Luminescence

When a material absorbs energy, a portion of it is reemitted in the form of photons having energy corresponding to the visible or near visible region. This phenomenon is called luminescence. Figure 16.17 shows luminescence process in an atom. The atom jumps to an excited state by absorption of a photon, and then relax to an intermediate state before re-emitting a photon by spontaneous emission. The photon emitted has a

Fig. 16.17 Luminescence
process in an atom

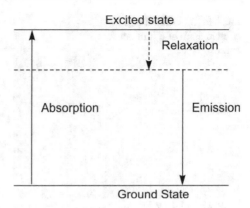

smaller energy than the absorbed photon. This term does not include the emission of black body radiation. Luminescent materials are usually referred to as phosphors. The phosphors exhibit luminescence because of the presence of (i) small amount of impurities called activator, for example, Tl in alkali halides, Ag in ZnS, Mn^{2+} in Zn_2SiO_4, anthracene activated with naphtacene or (ii) small stoichiometric excess of one of the constituents of the material. This is known as self-activation, for example, ZnS, CdS, ZnO etc.

Luminescence process is classified on the source of excitation energy. The most important luminescence process and their source of excitation are:

(i)   Photoluminescence: Its source of excitation is optical radiation.
(ii) Electroluminescence: Its source of excitation is electric fields or currents.
(iii) Cathodoluminescence: Its source of excitation is electron beam (or cathode rays).
(iv) Radioluminescence: Its source of excitation is energetic particles or high energy radiation.

The luminescent processes can also be classified on the basis of time that the light is emitted relative to initial excitation. If the emission is fast ($\leq 10^{-8}$s) is typical lifetime for an atomic excited state), then the process is fluorescence. If the emission process is slow and can last for minutes or hours, then the process is called phosphorescent and the materials showing it are called phosphors.

Consider a crystalline material in which a small amount of activator is incorporated. The energy levels of the activator in the crystalline material are localized in the forbidden gap. These localized levels can be grouped into two categories (a) the energy levels belonging to activators (b) the energy levels belonging the crystalline material which are under perturbing influence of the activators. These may be associated with the atoms of the crystalline material in the immediate vicinity of the activators and may also be associated with lattice defects such as vacancies.

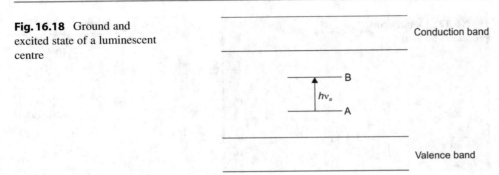

Fig. 16.18 Ground and excited state of a luminescent centre

Figure 16.18 shows the energy band picture of crystalline material which is incorporated with small amount of activator. A and B are two energy levels in the forbidden gap. A is the ground level and B is excited level. In the ground state, the level A is occupied and the excited state B is vacant. In the excited state B is occupied and A is vacant. The excitation from A to B can take place in the following ways:

(i)   A photon of appropriate frequency is absorbed by the electron in the ground state and being lifted to the excited state.
(ii) An exciton may be present in the crystalline material. It diffuses about in the material and may encounter a centre such as AB. As a result it may give off its energy thereby exciting the electron from the ground level A.
(iii) The electron from ground level A can also be excited by the motion of free electrons and holes. When a photon strikes a crystalline material it may produce holes and electrons. If the centre is in its ground state then level A can capture a hole from valence band and level B may trap an electron from the conduction band as shown in Fig. 16.19.

The electron returns from the excited state B to the ground state A by emission of photon with a frequency equal to absorption frequency provided Franck–Condon principle is

Fig. 16.19 Excitation is achieved by capture of a hole at A and of an electron at B

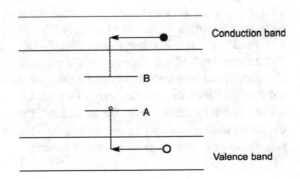

Fig. 16.20 Energy of the
ground state A and of an
excited state B as a function of
a configuration coordinate q

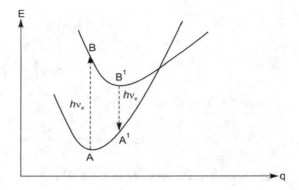

satisfied. The energy levels A and B are not represented by horizontal lines but by energy curves. The shape of these curves is different for different electronic states as shown in Fig. 16.20. The minimum of potential curve lies at different values of q for excited and ground state. Each value of q corresponds to a particular configuration of the nuclei in the vicinity of luminescence centre. The transition from A to B is very fast and nuclei remain stationary. After the absorption, the nuclei do not occupy the proper equilibrium position of the excited state B, and therefore system moves slowly to the minimum of B curve with emission of photons. The emission take place vertically and $v_e < v_a$. The excited state might also return to the ground state by means of a radiation less transition.

Decay Mechanism
Let $n(t)$ be the number of electrons in the excited state B and the probability that an electron from excited state B returns to ground state A with the emission of a photon is $\frac{1}{\tau}$ per second. If the luminescent centre is well screened from its surrounding, the average lifetime τ of excited state B is independent of temperature and other excited centres. The number of photons emitted per second or intensity of luminescence $I(t)$ is given by

$$I(t) = -\frac{dn}{dt} = \frac{n}{\tau} \tag{16.90}$$

Or

$$\frac{dn}{n} = -\frac{dt}{\tau}$$

On integration

$$\ln n = -\frac{t}{\tau} + C$$

At $t = 0, n = n_0$

$$\ln n_0 = 0 + C$$

$$\ln \frac{n}{n_0} = -\frac{t}{\tau}$$

$$n = n_0 \exp\left(-\frac{t}{\tau}\right) \tag{16.91}$$

From Eqs. (16.90) and (16.91)

$$I(t) = \frac{n_0}{\tau} \exp\left(-\frac{t}{\tau}\right) = I_0 \exp\left(-\frac{t}{\tau}\right) \tag{16.92}$$

where $I_0 = \frac{n_0}{\tau}$ is the intensity at $t = 0$. For dipole transition $\tau \approx 10^{-8}$s.

Temperature dependent exponential decay

Suppose in an excitation a transition from A to B take place. The electron from B may return to A by emission of photon or may fall into a metastable state M as shown in Fig. 16.21.

The energy difference between B and M is E. It is assumed that the transition from M to A is not allowed. When the excitation source is switched off, a certain number of electrons, n_0, remains in the metastable state. These electrons can make transitions to ground level A via excited level B. The probability per unit time for an electron in M to be excited to A will be given by

$$\frac{1}{\tau} = \frac{1}{\tau_0} \exp\left(-\frac{E}{k_B T}\right) \tag{16.93}$$

where $\frac{1}{\tau_0}$ represent the frequency. When electron reached excited level B, then probability of transition from B to A with emission of a photon is much larger than falling back into M. In such a case the intensity of the luminescence is determined by the rate at which a transition from M to B takes place. Therefore, Eqs. (16.90) and (16.92) are still applicable,

Fig. 16.21 Schematic representation of a centre with metastable state M

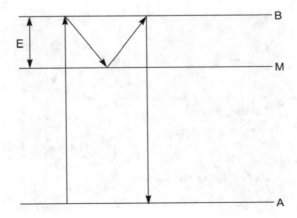

but since τ depends on T in accordance with Eq. (16.93) we have from Eqs. (16.92) and (16.93)

$$I(t) = \frac{n_0}{\tau}\exp\left(-\frac{t}{\tau}\right) = \frac{n_0}{\tau_0}\exp\left(-\frac{E}{k_B T}\right)\exp\left[\left(-\frac{t}{\tau_0}\right)\exp\left(-\frac{E}{k_B T}\right)\right] \qquad (16.94)$$

If the temperature is low, the intensity will be low. At high temperature the electrons in M levels may be boiled off at a high rate. By introducing the metastable states, the exponential decay of intensity and also exponential decay of τ with temperature can be explained.

Power-law decay
There are several decay laws. Let us consider simple power decay law. Suppose by excitation an electron from the luminescence centre makes transition to the conduction band. It is assumed when free electron and empty centre recombine, a photon is emitted. If there are n free electrons and n empty centres, the intensity would be given by

$$I(t) = -\frac{dn}{dt} = \alpha n^2 \qquad (16.95)$$

$$\frac{dn}{n^2} = -\alpha dt$$

On integration

$$\frac{1}{n} = \alpha t + C$$

At $t = 0; n = n_0$

$$\frac{1}{n_0} = 0 + C$$

Substituting the value of C

$$\frac{1}{n} = \alpha t + \frac{1}{n_0}$$

$$\frac{n_0}{n} = n_0 \alpha t + 1$$

$$n = \frac{n_0}{n_0 \alpha t + 1} \qquad (16.96)$$

From Eqs. (16.95) and (16.96)

$$I(t) = \frac{n_0}{(n_0 \alpha t + 1)^2} n_0 \alpha = \frac{n_0^2 \alpha}{(n_0 \alpha t + 1)^2} \qquad (16.97)$$

For larger values of t, the intensity decays as t^{-2}.

The important groups of luminescent crystalline solids are:

(i) Compounds which luminescence in the 'pure' state. Such compounds contain one ion or ion group per unit cell with an incompletely filled shell of electrons which is well screened from its surroundings. Examples are probably the manganous halides, samarium and gadolinium sulphate, molybdates, and platinocyanides.
(ii) The alkali halides activated with thallium or other heavy metals.
(iii) ZnS and CdS activated with Cu, Ag, Au, Mn, or with an excess of one of their constituents (self-activation).
(iv) The silicate phosphors, such as zinc orthosilicate (willernite, Zn_2SiO_4) activated with divalent manganese, which is used as oscilloscope screens.
(v) Oxide phosphors, such as self-activated ZnO and Al_2O_3 activated with transition metals.
(vi) Organic crystals, such as anthracene activated with naphtacene. These materials are often used as scintillation counters.

Solved Examples

Example 1 A complex refractive index of Ge at 400 nm is given by $4.141 + i2.215$. Calculate the absorption coefficient and reflectivity.

Solution
The absorption coefficient is given by Eq. (16.18)

$$\alpha = \frac{4\pi\kappa}{\lambda} = \frac{4 \times 3.14 \times 2.215}{400 \times 10^{-9}} \text{m}^{-1} = 6.96 \times 10^7 \text{m}^{-1}$$

The reflectivity is given by Eq. (16.40)

$$R = \frac{(n-1)^2 + \kappa^2}{(n+1)^2 + \kappa^2} = \frac{(4.141-1)^2 + (2.215)^2}{(4.141+1)^2 + (2.215)^2} = 47.1\%$$

Example 2 A salt absorb strongly at 60 μm. The complex dielectric constant at this wavelength is

$$K^* = -16.8 + i91.4$$

Calculate absorption coefficient and reflectivity.

Solution
From Eqs. (16.25) and (16.26)

$$n = \frac{1}{\sqrt{2}}\left[K_1 + \left(K_1^2 + K_2^2\right)^{\frac{1}{2}}\right]^{\frac{1}{2}}$$

$$\kappa = \frac{1}{\sqrt{2}}\left[-K_1 + \left(K_1^2 + K_2^2\right)^{\frac{1}{2}}\right]^{\frac{1}{2}}$$

Putting $K_1 = -16.8$ and $K_2 = 91.4$ in the above equations

$$n = \frac{1}{\sqrt{2}}\left[-16.8 + \left((-16.8)^2 + (91.4)^2\right)^{\frac{1}{2}}\right]^{\frac{1}{2}} = 6.17$$

$$\kappa = \frac{1}{\sqrt{2}}\left[16.8 + \left((-16.8)^2 + (91.4)^2\right)^{\frac{1}{2}}\right]^{\frac{1}{2}} = 7.41$$

From Eq. (16.18) the absorption coefficient is

$$\alpha = \frac{4\pi\kappa}{\lambda} = \frac{4 \times 3.14 \times 7.41}{60 \times 10^{-6}}\,\mathrm{m}^{-1} = 1.55 \times 10^6 \mathrm{m}^{-1}$$

From Eq. (16.40) the reflectivity is

$$R = \frac{(n-1)^2 + \kappa^2}{(n+1)^2 + \kappa^2} = \frac{(6.17-1)^2 + (7.41)^2}{(6.17+1)^2 + (7.41)^2} = 76.8\%$$

Example 3 The reflectivity of Si at 633 nm is 35% and the absorption coefficient is $3.8 \times 10^5 \mathrm{m}^{-1}$. Calculate the transmission coefficient of a sample with thickness of 10 μm.

Solution
We have $R = 0.35$ and $\alpha l = 3.8 \times 10^5 \times 10 \times 10^{-6} = 3.8$
From Eq. (16.6), the transmission coefficient is

$$T = (1 - R)^2 \exp(-\alpha l)\,T = (1 - 0.35)^2 \exp(-3.8) = 0.0095$$

Example 4 The fraction of non-reflected light that is transmitted through a 200 mm thickness of glass is o.98. Calculate the absorption coefficient of this material.

Solution
From Eq. (16.4)

$$\frac{I(z)}{I_0} = \exp(-\alpha z)$$

$$\ln\frac{I(z)}{I_0} = -\alpha z$$

$$\alpha = -\frac{1}{z}\ln\frac{I(z)}{I_0} = -\frac{1}{200}\ln(0.98) = 1.01 \times 10^{-4}\text{mm}^{-1}$$

Example 5 Determine the binding energy of exciton in GaAs (for $n = 1$). It is given that the dielectric constant of the material is 13, effective mass of electron is 0.067 m_e and the effective mass of hole is 0.45 m_e. Also determine the radius of $n = 1$ orbit.

Solution
From Eq. (16.58)

$$E = -\frac{\mu}{mK^2}(13.6eV)$$

From Eq. (16.57)

$$\frac{1}{\mu} = \frac{1}{m_e^*} + \frac{1}{m_h^*} = \frac{1}{0.067m_e} + \frac{1}{0.45m_e} = \frac{0.45 + 0.067}{0.45 \times 0.067m_e} = \frac{0.517}{0.03015m_e} = \frac{17.147}{m_e}$$

$$E = -\frac{\mu}{m_eK^2}(13.6\,\text{eV}) = \frac{m_e}{17.147m_e \times (13)^2}(13.6\,\text{eV}) = 4.69\,\text{meV}$$

Radius from Eq. (16.59) is

$$a_1 = K\frac{m_e}{\mu}a_0 = 13 \times \frac{17.147}{m_e} \times m_e \times 0.0529\text{nm} = 11.79\text{nm}$$

Objective Type Questions

1. Luminescence is because of
   (a) Photons emitted while excited electrons drop down
   (b) Knocking out of electrons by photons
   (c) Photons stimulated by photons
   (d) All.
2. Fluorescence occurs within
   (a) 10^{-5} s (b) 10^{-5} ms (c) 10^{-5} μs (d) 10^{-5} ns
3. Electroluminescence occurs in
   (a) Electrical conductors (b) Electrical insulators (c) p–n junction (d) All
4. Sum of these is unity
   (a) Reflection (b) Reflection plus refraction (c) Reflection plus absorption plus transmission (d) Any
5. Metal can
   (a) Reflect (b) Refract (c) Transmit (d) Any
6. Metals are
   (a) Transparent (b) Opaque (c) Translucent (d) None

7. Beers law relates

    (a) Light reflection (b) Light refraction (c) Light transmission (d) Light absorption

Problems

1. Obtain Beer's law.
2. Derive an expression for the reflection coefficient.
3. What is interband transition. Distinguish between direct interbank transition and indirect band transition.
4. What is photoconductivity? Discuss a simple model of photoconductor. Show that sensitive photoconductors should have long response time.
5. What are traps? What is the effect of traps on photoconductivity?
6. Show the variation of photoconductive response versus illumination.
7. What is the distinction between fluorescence and phosphorescence?
8. What are colour centres. Discuss the model of F centres.
9. Describe the luminescence. What are the main groups of luminescent materials.
10. Discuss the decay mechanism of luminescence processes.
11. What are excitons? Explain. Distinguish between Wannier–Mott exciton and Frenkel exciton or tightly bound exciton.

Answers
Objective Type Questions

        1. (a)   2. (b)   3. (c)   4. (c)   5. (a)   6. (b)   7. (d)

Semiconductor Devices

17

17.1 Tunnel Diode

A tunnel diode is a type of semiconductor diode which is capable to very fast operation well into microwave frequency region, by using quantum mechanical tunnelling effect. In 1958, Leo Esaki, a Japanese scientist, discovered that if a semiconductor junction diode is heavily doped with impurities, it will have a region of negative resistance.

In the tunnel diode, the semiconductor materials used in forming a junction are doped to the extent of one-thousand impurity atoms for ten-million semiconductor atoms. This heavy doping produces a depletion region of very small thickness (~10 nm). In such a case, there is a finite probability that electron can tunnel from the conduction band of the n region to the valence band of p region. During the tunnelling the particle energy does not change.

The characteristic I-V curve for a tunnel diode is illustrated in Fig. 17.1.

The three most important aspects of this characteristic curve are:

1. the forward current increase to a peak with a small applied forward bias,
2. the decreasing forward current with an increasing forward bias to a minimum valley current,
3. and the normal increasing forward current with further increases in the bias voltage.

The portion of the characteristic curve between A and B is the region of negative resistance.

When the semiconductor is very highly doped, the Fermi level goes above the conduction band for n-type and below valence band for p-type materials. Figure 17.2 shows the equilibrium energy level diagram of a tunnel diode with no bias applied. The valence band of the *p* material overlaps the conduction band of the *n* material. The majority electrons

V. K. Jain, *Solid State Physics*, https://doi.org/10.1007/978-3-030-96017-9_17

Fig. 17.1 Characteristic I-V
curve of a tunnel diode

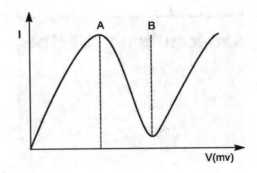

Fig. 17.2 Tunnel diode energy
diagram with no bias

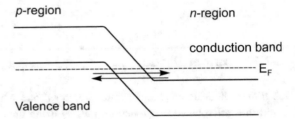

and holes are at the same energy level in the equilibrium state. If there is any movement
of current carriers across the depletion region due to thermal energy, the net current flow
will be zero because equal numbers of current carriers flow in opposite directions.

Figure 17.3 shows the energy diagram of a tunnel diode with a small forward bias
applied. The bias causes unequal energy levels between some of the majority carriers
at the energy band overlap point, but not enough of a potential difference to cause the
carriers to cross the forbidden gap in the normal manner. Since the valence band of the *p*
material and the conduction band of the *n* material still overlap, current carriers tunnel
across at the overlap and cause a substantial current flow. The amount of overlap between
the valence band and the conduction band decreased when forward bias was applied. With

Fig. 17.3 Tunnel diode energy
diagram with small forward
bias voltage

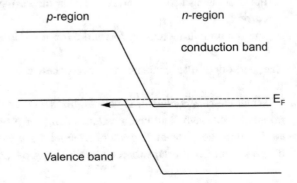

Fig. 17.4 Tunnel diode energy diagram with high forward bias voltage

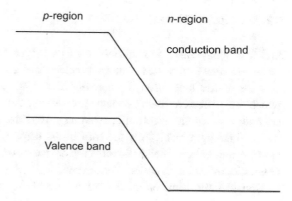

larger voltage the energy of the majority of electrons in the n region is equal to that of energy of the empty states (holes) in the valence band of the p region. This will produce the maximum tunnelling current.

As the forward bias voltage continue to increase the number of electrons in the n side that are directly opposite to the empty states in the valence band (in terms of energy) decreases. Therefore the decrease in the tunnelling current will start. As more forward bias voltage is applied the tunnel current drops to zero. But the regular forward current due to electron–hole injection increases due to lower potential barrier. Figure 17.4 is the energy diagram of a tunnel diode in which the forward bias has been large. The valence band and the conduction band no longer overlap at this point, and tunnelling can no longer occur. The portion of the curve from A to B in Fig. 17.1 shows the decreasing current that occurs as the bias is increased, and the area of overlap becomes smaller. As the overlap between the two energy bands becomes smaller, fewer and fewer electrons can tunnel across the junction. The portion of the curve between A and B in which current decreases as the voltage increases is the negative resistance region of the tunnel diode.

If the energy diagram of a tunnel diode in which the forward bias has been increased even further, then the energy bands no longer overlap and the diode operates in the same manner as a regular *p-n* junction.

The negative resistance region is the most important and most widely used characteristic of the tunnel diode. A tunnel diode biased to operate in the negative resistance region can be used as either an oscillator or an amplifier in a wide range of frequencies and applications. Very high frequency applications using the tunnel diode are possible because the tunnelling action occurs so rapidly that there is no transit time effect and therefore no signal distortion. Tunnel diodes are also used extensively in high-speed switching circuits because of the speed of the tunnelling action.

The advantage of Tunnel diode is: (i) High speed of operation, (ii) low cost, (iii) low noise, (iv) low power dissipation, (v) simplification in fabrication, (vi) longevity, and (vii) environmental immunity. The disadvantages are: (i) low output voltage, (ii) because it is a two terminal device, there is no isolation between input and output.

17.2 Gunn Diode

In some semiconductor materials such as n-type GaAs, the mobility of the electrons decreases above a certain voltage or electric field called threshold value E_{Th}. This is because as the field strength increases more and more electrons transfer to a state in which their effective mass becomes greater and thus decreasing their velocities. For field strength $E > E_{th}$ the electrons have a negative differential mobility that is an increase in the field strength result in a decrease in the drift velocity as shown in Fig. 17.5. This is called Gunn effect. The semiconductor device based on this effect is called Gunn diode. Gunn diodes are negative resistance device.

Consider the semiconductor like n-type GaAs or n-type InP or n-type CdTe which have closely spaced energy valley in the conduction band as shown in Fig. 17.6. The GaAs have an empty high energy conduction band separated from the lower energy filled (or partially filled) conduction band by a relatively narrow forbidden energy gap. The separation between valence band and conduction band Γ is 1.43 eV while the energy separation between Γ and L conduction band is only 0.36 eV. Under normal condition, the electron contribution to the current are in the high mobility partially filled energy band Γ. At a certain threshold voltage or electric field, the energy imparted to the electrons allow them to lifted to lower mobility or higher effective mass conduction band L, causing the drift velocity or current to decrease. Then as the voltage or electric field keeps increasing, electrons are removed from the low mobility to conduction band and current begins to increase again Fig. 17.6.

When the semiconductor material is not uniformly doped, there is region in the crystal where the concentration of electrons is relatively small and the conductivity is lower than in the rest of crystal. Because of this, the electric field in this region is stronger than in the rest of the crystal. Thus this region or domain is first area to transfer electrons into the higher energy conduction band when the voltage or electric field is increased. The electrons in this region are therefore slowed down as the electric field or voltage in

Fig. 17.5 Drift velocity v of the electron as a function of electric field strength

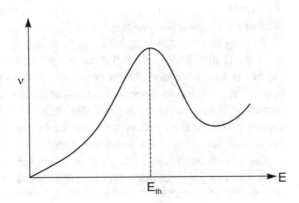

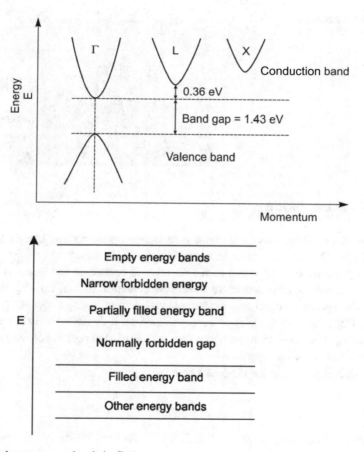

Fig. 17.6 Relevant energy levels in GaAs

increased. Therefore, this region or domain becomes a negative resistance region. Electrons in front of and behind this region are moving faster than the electrons within this
domain. Electrons from behind the region bunch up, decrease the gradient at the back of
negative resistance region. The electrons in front of this domain pull away from it, leaving
an area with a low concentration of electrons. In this way the negative resistance region
moves towards the anode at a speed approximately 10^5 m/s carrying along the bunch of
electrons. The arrival of the negative resistance region at the anode frees the electrons
and a new negative resistance region forms at the cathode. This domain begins its own
propagation towards anode. This creations and propagation of domain give rise to the
microwave oscillations. Figure 17.7 shows a schematic diagram of the Gunn oscillator.

Fig. 17.7 Construction of
Gunn diode

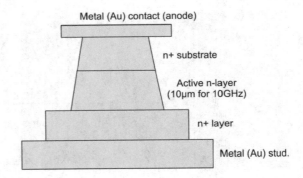

17.3   Varactor Diode

A junction diode which acts as a variable capacitor under reverse bias is known as a
varactor diode. The capacitance of varactor diode is in picofarad range.

When a p–n junction is formed depletion layer is created in the junction area. Since
there are no moveable charge carriers within the depletion region, this region acts as an
insulator. The p-type material and n-type material acts as positive and negative charged
plates, respectively. The diode may be considered as a capacitor with n region and p
region forming oppositely charged plates and with depletion region between them acting
as a dielectric as shown in Fig. 17.8

The capacitance of varactor diode is given by

$$C = \varepsilon \frac{A}{d}$$

where C is total capacitance of the junction, ε is the permittivity of the semiconductor
material, A is cross-sectional area of the junction and d is the width of the depletion
region.

Fig. 17.8 Varactor diode

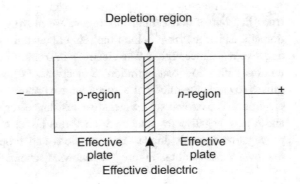

Fig. 17.9 Variation of
junction capacitance versus
reverse bias voltage

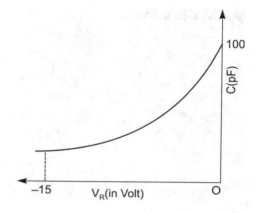

When the reverse bias voltage across the varactor diode is increased the width of depletion region increases. Therefore, the total junction capacitance C of the junction decreases. On the other hand if the reverse voltage across the diode decreased, the width of depletion layer decreases. Consequently the total junction capacitance C decreases. Figure 17.9 shows the variation of junction capacitance C versus reverse bias voltage V_R across the varactor diode. Thus C can be changed simply by changing the reverse bias voltage. A varactor diode is sometimes called voltage controlled capacitor.

Applications:

In circuits which requires voltage controlled tuning.

17.4 Photodiode

A photodiode is a reverse biased Si or Ge p–n junction in which reverse current increases when the junction is exposed to light. The reverse current in a photodiode is directly proportional to the intensity of light falling on its p–n junction.

When a photodiode is reverse biased it has a very small reverse current which is produced by thermally generated electron–hole pairs which are swept across the junction by the electric field created by the reverse voltage. When the p–n junction is illuminated with light, the reverse current increases with the increase in light intensity. This can be explained as follow:

When photons of energy greater or equal to band gap falls on the p–n junction it generates electron–hole pairs. These additional charge carriers will increase the reverse current. As the intensity of light incident on p–n junction increases the reverse current also increases. If the photons fall at some distance from the p–n junction, the free electrons and holes will recombine before they can join the flow of reverse current.

Fig. 17.10 Photodiode

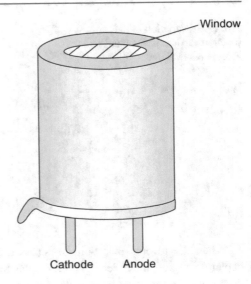

Window

Cathode   Anode

A photodiode consists of a p–n junction mounted on an insulated substrate and sealed inside a metal case as shown in Fig. 17.10. A glass window is mounted on the top of the case to allow light to enter and incident on the p–n junction.

When no light is incident on the p–n junction reverse current I_r is extremely small. This is called dark current. The resistance of photodiode with no incident light is called dark resistance R_R given by

$$R_R = \frac{V_R}{\text{Dark current}} = \frac{V_R}{I_R}$$

where V_R is the reverse voltage.

When light is incident on the p–n junction, there is a transfer of energy from the incident light to the atoms in the junction. This will creates more free electrons (and holes). These additional free electrons will increase the reverse current. As the intensity of light increases reverse current increases till it becomes maximum. This is called saturation current. There are two important characteristic of photodiode:

(i) Reverse current versus illumination curve: Fig. 17.11 shows the graph between reverse current I_R and illumination E of a photodiode. The graph is a straight line passing through the origin and is given by

$$I_R = mE$$

where m is the slope of the line and is called the sensitivity of the photodiode.

(ii) Reverse voltage versus reverse current: Fig. 17.12 shows the graph between reverse current and reverse voltage for various illumination levels. It is seen that for a given

Fig. 17.11 Reverse current versus illumination curve for photodiode

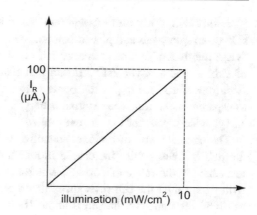

Fig. 17.12 Reverse current versus reverse voltage for different illumination level

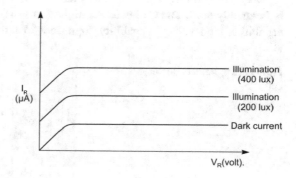

reverse biased voltage the reverse current increases as the illumination of the p–n junction of photodiode is increased.

Photodiodes are used in alarm circuits and counter circuits.

17.5 Photoconductive Cell

It is a semiconductor device in which resistance varies inversely with the intensity of light falling upon it. It is also known as photoresistive cell or photoresistor because it operates on the principle of photoresistivity.

The resistance of a semiconductor depends on the number of free charge carriers. In the absence of light illumination, the number of free charge carriers is small and hence the resistivity is large. When photons, having energy greater than the band gap of the semiconductor, strike the semiconductor, the free charge carriers, that is electrons and holes, are created. As a result of this the resistivity of the semiconductor decreases.

In this cell, a thin film of cadmium sulphide or cadmium selenide is deposited on one side of an iron plate and placed below a transparent foil of metal as shown in Fig. 17.13 When photons of sufficient energy fall on transparent metal foil the electrical resistance of CdS or CdSe layer gets reduced and hence its electrical conductance is increased. A current starts flowing in the battery circuit connected between the iron plate and the transparent metal foil. The external battery is included in the circuit to generate a direction and provide a path for the current to flow.

The characteristic of a photoconductive cell is shown in Fig. 17.14 The resistivity of the cell decreases with increasing illumination. When cell is not illuminated, it has a resistance in the range of 100 kΩ, which is known as dark resistance. When illuminated with strong light the cell resistance falls to only a few hundred ohms. The ratio of 'dark' to 'light' resistance of the cell is about 1000: 1.

Since spectral response of CdS is similar to that of human eye, a photoconductive cell is frequently used. The CdS cells are highly sensitive, low cost, long life, high dissipation capability, high voltage capability, high dark to light resistance. The disadvantages are

Fig. 17.13 Photoconductive cell

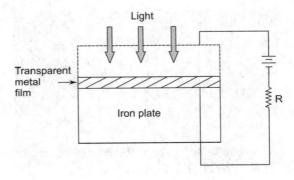

Fig. 17.14 Illumination characteristic of a typical photo conducting cell. The resistance is usually high, when the cell is dark and relatively low when illuminated

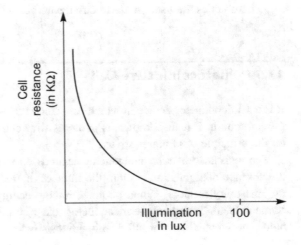

(i) the current changes with change in light intensity with a time lag and (ii) a relatively narrow response.

The photoconductive cell is widely used in OFF/ON circuits, light measurement and light detecting circuits.

17.6 Photovoltaic Effect

When a light of energy greater than the band gap energy of the semiconductor falls on an unbiased p–n junction, measurable current and voltage are developed in the circuit. This is known as photovoltaic effect. The word photovoltaic comes from light and electricity.

When light falls on the p–n junction, it may be reflected, transmitted or absorbed. The absorption of the photons causes creations of holes and electrons in the semiconducting material provided the energy of the photons is equal or greater than the band gap energy of the material. The electrons created in the p region (minority carriers) diffuse to the junction and because of this the potential barrier to the n side is decreased. Similarly, the holes created on the n side (minority carriers) diffuse to the junction and float up the barrier to the p side. The current I_L produced by the minority charge carriers due to photons thus increases. But since the diode is open circuited, the resultant current must remain zero. Therefore majority current should increase by the same amount as the minority carrier current. This increase in majority current is possible if the retarding electric field at the junction is reduced. This results in the reduction of potential barrier from V_0 to $V_0 - V$. A voltage equal to V therefore will appear as a measurable voltage in the external circuit. This is of the order of 0.1 V for the Ge cell and 0.5 V for Si cell.

The current I_L produced by the minority charge carriers due to photons increases are directed opposite to the forward biased current I and is independent of the built in voltage. In an open circuit, the two opposite current must be equal and hence

$$I_L = I_0 \left[\exp\left(\frac{eV}{\eta k_B T} \right) - 1 \right]$$

We get $I_L + I_0 \left[1 - \exp\left(\frac{eV}{\eta k_B T} \right) \right] = 0$

$$1 + \frac{I_L}{I_0} = \exp\left(\frac{eV}{\eta k_B T} \right)$$

$$\ln\left(1 + \frac{I_s}{I_0} \right) = \frac{eV}{\eta k_B T}$$

Therefore, photovoltaic e.m.f. V is

$$V = \frac{\eta k_B T}{e} \ln\left(1 + \frac{I_L}{I_0} \right)$$

Fig. 17.15 Plot of V versus I_L

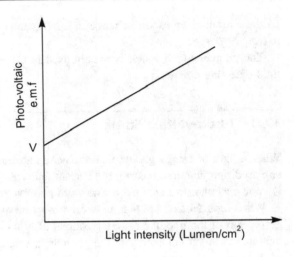

But $\frac{I_L}{I_0} \gg 1$, except for extremely small light intensities.

$$V = \frac{\eta k_B T}{e} ln\left(\frac{I_L}{I_0}\right)$$

This equation shows that the photovoltaic *e.m.f* V increases logarithmically with I_L and hence with illumination it has been shown in Fig. 17.15.

17.7   Solar Cell

A solar cell is a semiconductor device which directly converts sunlight into electricity. Light shining on the solar cell produces both a current and a voltage to generate electric power. Crystalline Si has been used as light absorbing semiconductor in most solar cells, even though it is relatively poor absorber of light and requires considerable thickness of material. Nevertheless, it has convenient because it yields stable solar cells with good efficiency (11–16%). The other crystalline semiconductor materials used for solar cells are cadmium telluride, copper indium gallium selenide, etc.

The basic operations of a solar cell are

(i)   The generation of charge carriers by light
(ii) The collection of light generated charge carriers
(iii) The generation of large voltage across the solar cell
(iv) The dissipation of power in the load.

When sun light falls on a material, the light may be reflected, absorbed or transmitted. The absorbed light, if its energy is greater than the band gap energy of the semiconductor material, creates electron–hole pairs. These charge carriers are metastable and will exist on the average for a length of time equal to minority carrier life time before they recombine. Under these conditions, the energy represented by the absorbed light is converted into heat in the semiconductor. The electrons produced in the p region and holes produced in the n region are minority charge carriers. Holes produced in p region (majority carriers) cannot diffuse from p to n side and electrons produced in the n side (majority carriers) cannot diffuse from n to p side because of the barrier voltage. However, the electrons from p region can diffuse to n region while holes from n region can diffuse to p region. All the minority charge carriers produced by the light may not cross over to the other side as it depends on the location where they are produced. If they are produced far away from the junction, then there is large probability that they will be lost by recombination before reaching the junction. The probability that the charge carriers will diffuse to the junction is large for charge carriers created near the junction. The recombination is prevented by the p–n junction which separates the electrons and holes. The carriers are separated by the action of electric field existing at the junction. If the minority charge carriers produced by the light reaches the p–n junction, they are swept across the junction by the electric field at the junction where it is now majority charge carrier. The presence of excess holes and electrons in their respective side leads to a reduction in the potential barrier. If the terminals of p–n junction are connected together (that is solar cell is short circuited) the light generated current flow through the external circuit. Figure 17.16 shows the I-V characteristic of a solar cell.

If an external resistance (load) R is connected across the solar cell, the current through it causes a potential drop across it. It makes the junction forward biased and results in a flow of current opposite the flow of photocurrent I_L. The load current is therefore

$$I = I_L - I_0\left[\exp\left(\frac{eV}{\eta k_B T}\right) - 1\right]$$

Under condition of open circuit voltage V_{OC} (it is the maximum voltage available from a solar cell and this occur at zero current) which correspond to the amount of forward bias on the solar cell junction with the light generated current, we have $I = 0$ and

$$I_L = I_0\left[\exp\left(\frac{eV_{OC}}{\eta k_B T}\right) - 1\right]$$

$$V_{OC} = \frac{\eta k_B T}{e}\ln\left(\frac{I_L}{I_0} + 1\right)$$

Under short circuit condition (when the voltage across the solar cell is zero), that is, $V = 0$

Fig. 17.16 **a** Short circuit
current and open circuit
voltage versus light intensity
for a Si solar cell. **b** Output
characteristic for Si solar cell
for various illumination levels

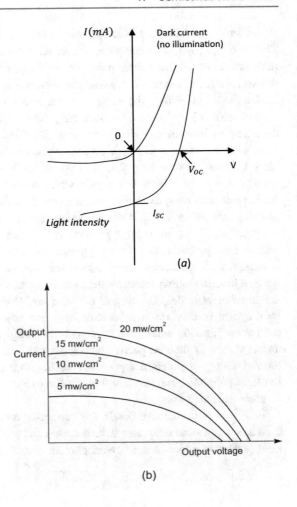

(a)

(b)

$$I = I_{SC} = I_L$$

where I_{SC} is short circuit current. It is due to generation and collection of light generated
charge carrier. The short circuit current is the largest current which may be drawn from
the solar cell. It depends on (i) area of the solar cell, (ii) the number of photons, (iii)
the spectrum of the incident light, (iv) the absorption and reflection properties of the
semiconducting material, and (v) the collection probability of the solar cell.

The power obtained from the ideal solar cell is given by

$$P = I^2 R = \left\{ I_L - I_0 \left[\exp\left(\frac{eV}{\eta k_B T} \right) - 1 \right] \right\}^2 R$$

Ideal power is

$$\text{Ideal power} = V_{OC} \times I_{SC}$$

It is convenient to define maximum load power

$$P_{max} = V_{mp} \times I_{mp} = V_{OC} \times I_{SC} \times F.F$$

where V_{mp} is the voltage at the maximum power, I_{mp} is the current at the maximum power, and F.F. is the fill factor. The F.F. is a parameter which in conjunction with V_{OC} and I_{SC} determines the maximum power from a solar cell. The F.F. is defined as the ratio of maximum power from the solar cell to the product of V_{OC} and I_{SC}

$$F.F. = \frac{V_{mp} \times I_{mp}}{V_{OC} \times I_{SC}}$$

The maximum conversion efficiency of a solar cell is defined as

$$\text{Efficiency} = \frac{\text{Output power}}{\text{Input power}} = \frac{P_{max}}{P_{input}} = \frac{V_{mp} \times I_{mp}}{I_T A} = \frac{V_{OC} \times I_{SC} \times F.F}{I_T A}$$

where I_T is the incident solar flux and A is the area of cell. For an efficient cell, it is desirable to have high value of fill factor, short circuit current and open circuit voltage.

Figure 17.17 shows a Si solar cell. A conventional silicon solar cell consists of a thin wafer of boron doped p-type silicon is laid down on a metallic layer. This metallic layer serves as an electrical contact and as a substrate to support the cell. The phosphorous is then diffused into the top surface forming a p–n junction near the top. With the p–n junction near the top, most of the sun light will easily reach the junction. An anti-reflecting coating is sprayed on the front surface to cut down on reflected light. The front surface contacts are made to collect the electrons arriving from the junction. The grid is constructed with few contacts as possible so that p–n junction is not shadowed too much.

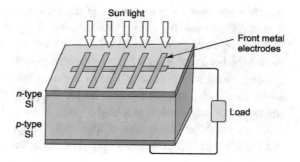

Fig. 17.17 Diagram of a typical silicon solar cell

The efficiency of solar cells is low because of the following:

(i)   Not all the photons which strike the cell penetrate it. Some are reflected by the cells surface and some strike the metal contact grid.
(ii) Not all the incident photons of sun light have sufficient energy to generate electron–hole pairs. On the other hand, photons which have too much energy generate electron–holes pairs, dissipating the surplus energy in the form of heat.
(iii) Not every electron–hole pair generated by the sun light is collected by the electric field of the junction and sent to external load, given that if charge carriers are generated far away from the junction, they may be lost by recombination.
(iv) The charge generated and collected in the depletion region must be sent to the outside. They are collected by metallic contact placed on the front and back of the cell. However, there remains a certain amount of resistance at the interface between Si and metal that provoke a dissipation that reduce the power transferred to the load.

Solar cells are used extensively in satellites and space vehicles to supply power to electronic and other equipments or to charge storage batteries they are receiving attentions even for terrestrial electric power generation. For it, it is planned to orbit big panels of solar cells outside the earth atmosphere for converting solar energy into electrical energy.

17.8   Light Emitting Diode (LED)

A light emitting diode or LED is a p–n junction, made from a very thin layer of heavily doped semiconductor material. It converts electrical energy into light energy.

When a p–n junction is biased in the forward direction, the resulting flow across the junction between the p and n regions has two components: holes are injected from the p region into the n region and electrons are injected from the n region into the p region. Thus minority carrier injection disturbs the charge carrier distribution from the equilibrium position. The injected minority carrier recombines with majority carriers until thermal equilibrium is re-established. As long as the current continue to flow minority injection continues. On both sides of the junction, a new steady state carrier distribution is established such that the recombination rate equals the injection rate.

Minority carrier recombination is not instantaneous. The injected minority carriers have to find the proper condition before the recombination process takes place. In the recombination process both the energy and momentum must be conserved. Energy conservation is easily met since a photon can take up the energy of electron hole recombination, but photon does not contribute much to the conservation of momentum. Therefore, an electron can only combine with a hole of identical and opposite momentum. Such conditions are not readily met resulting in a delay. Thus injected minority charge carriers have a finite life time called radiative life time before it recombines radiatively. The radiative recombination in not the only recombination path. There are also crystalline defects that can trap the injected minority carriers. This type of recombination process may or may not generate light. Energy and momentum conservation are met through the successive emission of phonons. The non-radiative recombination process is characterized by a life time called non-radiative life time. If radiative life time is larger than the non-radiative lifetime, as in the case of Si or indirect band gap semiconductor, electrons and holes recombine by non-radiative process and therefore no light is produced. If on the other hand, radiative life time is shorter, the light is emitted by recombination of electrons and holes for example in GaAs. Because of thin layer of the junction, a reasonable large number of photons can leave the junction and radiates away producing a coloured light output.

Assuming that photons are emitted, their energy should be equal to the band gap energy. However, as electrons in each energy band are not at a single discrete value of energy but rather may have a variety of energy values anywhere inside the bands. The most probable value as well as the distribution of electron in a band is given by Fermi–Dirac statistics which describe the probability of occupancy of an energy state. Statistics shows that the maximum population of electrons in the conduction band occurs at $\frac{1}{2}k_B T$ above the bottom of the conduction band, the energy of bottom of the conduction band is denoted by E_C. Similarly, the maximum population of holes in the valence band is $\frac{1}{2}k_B T$ below the top of valence band denoted by E_V. In such a situation the maximum spectral output is expected at a wavelength corresponding to band gap energy plus $k_B T$, the point where the concentration is maximum not at band gap energy. There will also be spread in photon energies since there are many possible energies in recombination as shown in Fig. 17.18.

Typical LED with wavelength emitted by them is given in Table 17.1.

Fig. 17.18 Output spectra of an LED

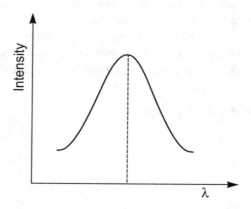

Table 17.1 LED and wavelength emitted by them

Semiconductor material	Wavelength (nm)
GaAs	850–940
GaAsP	630–660
AlInGaP	~540
GaAsP:N	585–595
AlGaP	550–570
SiC	430–505

17.9 Semiconductor Lasers

Lasing in a semiconductor laser is made possible by the existence of a gain mechanism plus a resonant cavity. In a semiconductor laser the gain mechanism is provided by the light generation from the recombination of holes and electrons. The wavelength of the light is governed by the band gap of the lasing semiconductor. In order for the light generation to be efficient enough to result in lasing, the active region of a semiconductor, where the carrier recombination occurs must be a direct band gap.

A p–n junction laser also known as diode laser, junction laser or injection laser, is the most widely used of all lasers. The first useful semiconductor laser was made of GaAs, which had been strongly doped. Laser action occurs in the transition region (~1 μm) between p and n doped material in a diode subject to a voltage applied in the forward direction. When the junction is forward biased (Fig. 17.19) electrons flow to p side and holes to n side. In the depletion region the injected holes and electrons appear in high concentration. A high hole concentration means a large number of empty sites are there into which electrons can fall. Thus a population inversion is created between the filled level near the bottom of conduction band and empty level near the top of the valence band. The recombination of electron and holes releases energy. If this energy is in the form of heat (as in the case of Si or Ge) the material is of no use for laser action. In GaAs most of the energy emitted appear as light of wavelength ~840 nm. If the photon so emitted is travelling exactly in the plane of the junction there may be amplification and if the photon continues to travel near the junction, the amplification grows. In order to make

Fig. 17.19 *p-n* junction

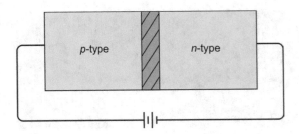

Fig. 17.20 Semiconductor
laser

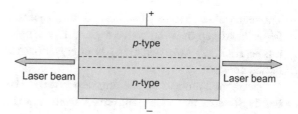

the amplification radiation travel to and fro as to increase the gain it is necessary to have
an optical resonator containing p–n junction. This is done by cutting the crystal so that the
two end faces are exactly perpendicular to junction and parallel to each other. Since most
of the semiconductors have a high refractive index, reflection at an air–semiconductor
interface is high and thus special coating is not necessary. The laser emission is restricted
to a very thin region ~1 μm, therefore, the beam emerges over a wide range of angles
(20°–30°). Because the laser transition occurs between the bands of energies, the emission
is not as monochromatic as from a gas laser.

A typical diode laser is physically small ~1 mm long and with an effective thickness
of ~1 μm. The lifetime of an electron–hole pair is ~1.0–10.0 ns. Thus, very high current
densities are required to obtain population inversion. The high current densities raise the
temperature of the diode, necessitating even higher current densities for laser action. As
diode heats up wavelength increases by 0.3 nm/°C because of change in band gap. Thus by
changing the temperature light can be tuned between 800 and 900 nm. Electrical pumping
to laser diode provided by current controlled DC power supply (a typical voltage ~2.4 V,
operating current 45 mA) is used. The threshold current is ~40 mA. Threshold current
density is ~500 A/mm². A schematic diagram of the diode laser is shown in Fig. 17.20.

Drawback of p–n junction laser
The lifetime of these lasers is short unless they are operated at low temperatures and in
pulsed mode. The optical transitions in the semiconductors are rather broad of the order
of 2 nm yielding a shape function with a width of the order of ~30 cm^{-1} for a wavelength
of 840 nm. Homojunction lasers (device use the same material for both p and n side of the
junction) are characterized by large threshold current. Such current prohibit continuous
operation at room temperature. Therefore continuous wave device require cooling at low
temperatures, making them impractical for many applications. The threshold condition for
laser current density can be reduced at room temperature by using a double semiconductor
heterostructure.

Improvement in the structure of the laser can be made by the use of hetrojunction, that
is, p–n junction fabricated from two different semiconductor materials having different
band gap. Device fabricated with hetrojunction is said to have hetrostructure. Such a structure requires two interfaces of different index of refraction one at the top and other below
the active region (double hetrostructure).The thickness of the active region is ~0.1 μm.

The material used for top and bottom of the active region is of lower refractive index and larger band gap than the active region. The material used for this is p-type AlGaAs and n-type AlGaAs. AlAs and GaAs both crystallizes in face centred cubic lattices and have lattice constant.

$a = 0.5662$ nm and 0.5652 nm, respectively. The active region is made of GaAs which has high refractive index and small band gap (Fig. 17.21). This difference in refractive index establishes an optical dielectric waveguide that confines photons to the active region. When a forward voltage corresponding to band gap of active layer is applied a large number of electrons from n side and holes from p side enter the active region. The injected carriers are confined to the active region by the energy barrier provided by the hetrojunction.

Recombination of electrons and holes in such a laser is shown in Fig. 17.22. For biases below threshold, a semiconductor laser emits a small amount of incoherent light spontaneously (Fig. 17.23). Above threshold, stimulated emission results in lasing. The relationship between lasing emission and the bias current of a semiconductor laser is linear. The threshold current of a laser this line is extrapolated to the point at which the

Fig. 17.21 Structure of hetrojunction laser

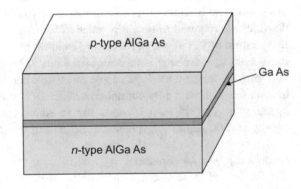

Fig. 17.22 Carrier confinement in the active region (-100 nm) which is provided by the energy barrier of the hetrojunction

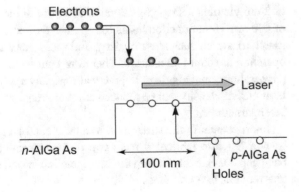

Fig. 17.23 Emission spectrum of semiconductor laser and LED

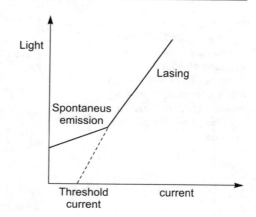

stimulated emission is zero (Fig. 17.23). Since the charge carriers in hetrojunction lasers are confined to a much smaller region than in homojunction laser, the heat deposition is much smaller while the threshold current densities are lower of the hetrojunction.

Applications.

1. In communication systems.
2. In compact disc (CD) players.
3. High speed printing.
4. Pump source of solid state lasers.
5. Laser pointers.
6. In medicine.

Emission Spectrum of a Semiconductor Versus Light Emitting Diode

When a p–n junction is forward biased, population inversion take place and, it emits light. This is spontaneous emission of radiation without reflector (Fig. 17.24).Such a junction operates as light emitting diode (*LED*). In order to make a *LED* do lasing it is necessary to increase the population inversion to the threshold value and reflectors are introduced to provide the feedback. As the refractive index of semiconducting material is high, the interface between the diode and air acts as a reflector with reflectivity ~0.3. The emission spectrum of a semiconductor laser as compared to that of *LED* is shown in Fig. 17.25.

Fig. 17.24 **a** In the absence of a photon, recombination of electron and hole produces spontaneous photon. **b** In the presence of a photon, electron hole recombination produces photons which are coherent with the previously existing photons

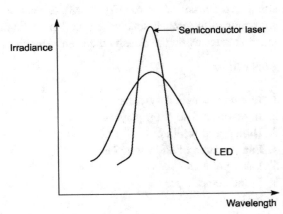

Fig. 17.25 Emission spectrum of a semiconductor laser and LED

17.10 Semiconductor Detector

The materials can be classified as conductors, semiconductors or insulator according to their conductivity or the magnitude of the forbidden energy gap (band gap) between valence and conduction band. The band gap for semiconductors is intermediate between conductors and insulators. The well-known examples of semiconductors are *Si* and *Ge*. The band gap for Si is 1.12 eV while for Ge it is 0.74 eV. When an energy equal or greater than the band gap is given to the semiconductor, the electrons jump from valence band to conduction band leaving a hole in the valence band. The average energy needed to create an electron–hole pair in Si at room temperature is 3.62 eV while in Ge it is 2.9 eV at 80 K. The average energy is larger than the band gap energy as some energy is lost as crystal excitation. The conductivity in the semiconductor is due to motion of electrons and holes. When pure semiconductor for example Si is doped with P(Vth group element) impurity, n-type semiconductor is formed and P donates electrons for conduction. The impurity is called donor. The energy level of P is in the forbidden gap just below the conduction band and a very little energy is needed to lift atoms from energy level of P to conduction band and conductivity is mainly due to electrons and such semiconductors are called n-type semiconductors. Similarly, a p-type semiconductor is formed by doping a third group element like boron (called acceptor) in Si. The energy level of B lies near the top of valence band of Si and requires a small amount of energy to lift electrons from valence band to impurity energy level and as a result holes are created in the valence band. Such a material is called p-type semiconductor and conductivity is mainly due to holes.

When p- and n-type of semiconductors are produced on the single piece of semiconductor then it is referred to as p–n junction. The p and n regions were originally neutral. As electrons from n-type migrate across the junction to p-type and holes from p-type diffuse to n-type, they uncover their respective donor and acceptor atoms. This results in positive charge on n side and a negative charge on p side. The migration or diffusion of electrons and holes across the junction leads to a region with a reduced concentration of mobile carriers. This region is called depletion region bounded by the conductive region, which are n and p-doped, respectively. Any voltage applied across the p–n junction will appear across the depletion region.

If a voltage is applied over the junction by connecting the negative terminal of the battery to p-type region and positive terminal to the n-type region, the junction is said to be reversed biased. The reverse biasing increases the width of the depletion region and therefore the sensitive detector volume. Simultaneously the capacitance decreases. Passage of ionizing radiation through the depletion region creates holes in valence band and electrons in conduction band. Electrons drift to the positive volt on the n side while holes drift to negative voltage on p side. This creates an electric output.

Ge detector which is usually used for γ-ray spectroscopy because of high Z value and density must be operated at liquid nitrogen temperature to minimize the thermal excitation of electrons into conduction band. Resolution of Ge detector is better than NaI(Tl) though the efficiency is smaller.

Surface Barrier Detectors

These kind of detectors are mainly used for α- and β-spectrum and for E and dE/dx measurements. It is a p-n type Si diode wafer characterized by a thin depletion region. It is made of n-type Si on which one surface has been etched prior to coating with a thin layer of gold and the other surface coated with a thin layer of aluminium to provide electrical contact. The depletion region is created by applying reverse biasing across the aluminium and gold contact, which provide a wide depletion region near the gold side and thin dead region near the aluminium side. The detector can be partially depleted (inactive entrance layer), totally depleted (no inactive layer) or over-depleted (higher applied potential than require for total depletion) depending on the applied potential. In totally depleted silicon surface barrier detector, the sensitive region extends through the whole thickness of the silicon.

A particle passing through the detector loses a small fraction of its energy dE/dx and may be completely stopped in a second having thicker depleted layer to rest of its energy. The A and Z of the particle can be determined from dE/dx and E with the use of proportionality

$$E \frac{\mathrm{d}E}{\mathrm{d}x} \propto AZ^2 \tag{17.1}$$

Lithium Drifted

In both Si and Ge, the material with highest available purity tends to be p-type. Donor atoms must therefore be added to the material to obtain the desired compensation. This can be achieved by using lithium atoms. Lithium does not occupy a substitutional site in Ge or Si. It enters in Ge or Si at interestial site. The ease of ionization of Li to Li$^+$ makes it a donor impurity. The Li is drifted from one side of the crystal by electric field. Its concentration at the entrance side becomes high and outnumbered the original acceptor and thus creating an n-type region near the entrance side, and then decreases towards the other end of the crystal. This is, however, not a stable situation when electric field is removed. Hence the final state is frozen by cooling of the drifted crystal to liquid nitrogen temperature.

When a potential is applied over such a crystal with positive terminal at the entrance side (high Li side), three volumes are created (Fig. 17.26), one of p-type, a middle intrinsic one and an n-type one (p-i-n detector). In the intrinsic region lithium donor electrons neutralizes the p-type acceptor. The intrinsic region becomes depleted and thus sensitive to nuclear radiation. The Li-drifted detectors are made either from Si known as [Si(Li) detector] or from Ge known as [Ge(Li) detector]. The Ge has high density and Z value and therefore useful for γ-ray spectroscopy in comparison with Si. The resolution of these detector is better than NaI(Tl) scintillation detector. However, NaI(Tl) has better efficiency.

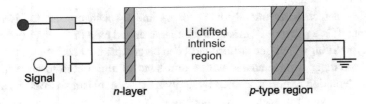

Fig. 17.26 *Li*-drifted semiconductor junction diode

Advantages

1. Superior in energy resolution.
2. Linear response over a wide range of energy.
3. Higher efficiency for a given size.
4. Fast pulse rise time.
5. Ability to operate in vacuum.
6. Insensitive to magnetic field.
7. Compact size.
8. Effective thickness that can be varied to match the requirement of the application.
9. Ionization energy is small ~3–4 eV.

Disadvantages

Limitation to small size and relatively high susceptibility of these devices to performance degradation with radiation induced damages.

Solved Examples

Example 1 For a photodiode, the reverse current with zero illumination is 50 μA. Determine dark resistance where reverse biased voltage is 10 V.

Solution

For zero illumination, the reverse current is dark current. The dark current is 50 μA and reverse bias voltage is 10 V. Therefore, dark resistance is

$$\text{Dark resistance} = \frac{\text{Reverse bias voltage}}{\text{Reverse current}} = \frac{10 \text{ V}}{50 \, \mu A} = 200 \, k\Omega$$

Example 2 A photodiode is exposed to light with an illumination of 2.5 mW/c m^2. If the sensitivity of the photodiode for the given condition is 37.4 μA/mW/cm^2, find the reverse current through the diode.

Solution

$$\text{Reverse current } I_R = \text{Sensitivity } m \times \text{Illumination } E$$

$$I_R = m \times E = 37.4 \times 2.5 = 93.5 \, \mu A$$

Example 3 For a solar cell, the open circuit voltage is 0.611 V, short circuit current is 2.75 A, voltage at maximum power is 0.5 V and current at maximum power is 2.59 A. determine the Fill Factor.

Solution

It is given that

$$V_{mp} = 0.5\,\text{V},\ I_{mp} = 2.59\,\text{A},\ V_{OC} = 0.611\,\text{V and } I_{SC} = 2.75\,\text{A}$$

The fill factor (F.F.) is given by

$$F.F. = \frac{V_{mp} \times I_{mp}}{V_{OC} \times I_{SC}} = \frac{0.5 \times 2.59}{0.611 \times 2.75} = 0.7707$$

Example 4 For a solar cell, the open circuit voltage is 0.611 V, short circuit current is 3.5 A, Fill factor is 0.7 and input power is 10 W. Determine the efficiency of the solar cell.

Solution

It is given that

$$F.F = 0.7,\ P_{in} = 10\,\text{W},\ V_{OC} = 0.611\,\text{V and } I_{SC} = 3.5\,\text{A}$$

$$\text{Efficiency} = \frac{V_{OC} \times I_{SC} \times F.F}{P_{in}} = \frac{0.611 \times 3.5 \times 0.7}{10} = 0.14945 = 14.945\%$$

Example 5 In a semiconductor laser the mode spacing is 1 nm for cavity of wavelength 250 μm. Now the length of the cavity is changed so that mode spacing is changed to 10 nm. What would be the length of the cavity?

Solution

$$v = \frac{c}{\lambda},\ dv = -\frac{c}{\lambda^2}d\lambda \text{ and } dv = \frac{v}{2L} = \frac{c}{2n_0 L}$$

$$d\lambda = \frac{\lambda^2}{2n_0 L}$$

$$\frac{\lambda^2}{n_0} = 2L\,d\lambda = 2 \times 250 \times 10^{-6}\text{m} \times 1 \times 10^{-9}\text{m} = 500 \times 10^{-15}\text{m}^2$$

$$L = \frac{\lambda^2}{2n_0 d\lambda} = \frac{500 \times 10^{-15}\text{m}^2}{2 \times 10 \times 10^{-9}\text{m}} = 2.5 \times 10^{-5}\text{m} = 25\,\mu m$$

Objective Type Questions

1. A p–n junction that radiates energy as light instead of heat is called
   (a) LED (b) Photodiode (c) Photocell (d) Zener diode
2. A photodiode is normally
   (a) Forward biased (b) Reverse biased (c) Neither forward nor reverse biased (d) Emits light
3. When the light increases the reverse current in a photodiode
   (a) Increases (b) Decreases (c) is unaffected (d) None of these
4. When the reverse voltage increases the junction capacitance
   (a) Decreases (b) Stays the same (c) Increases (d) Has more band width
5. The device associated with voltage controlled capacitance is
   (a) LED (b) Photodiode (c) Varactor diode (d) Zener diode
6. The varactor is usually
   (a) Forward biased (b) Reverse biased (c) Unbiased (d) In the breakdown region
7. LED are made out of
   (a) Silicon (b) Germanium (c) Gallium (d) Silicon and Germanium but not gallium
8. The process of emitting photons from semiconductor material is called
   (a) Photoluminescence (b) Gallium arsenide (c) Electroluminescence (d) Gallium phosphide
9. A tunnel diode is used
   (a) In high power circuits (b) In circuits requiring negative resistance (c) In very fast switching circuits (d) In power supply rectifiers
10. What types of diode is commonly used in electronic tuners in TVs?
    (a) Varactor (b) Schottky (c) LED (d) Gunn
11. A laser diode normally emits
    (a) Coherent light (b) Monochromatic light (c) Coherent and monochromatic light (d) Neither coherent nor monochromatic
12. In a photoconductive cell, the resistance of the semiconducting material varies with intensity of incident light
    (a) Directly (b) Inversely (c) Exponentially (d) Logarithmically
13. Which of the following parameters changes as an electron is transferred to upper conduction band
    (a) Rest mass (b) Total energy (c) Effective mass (d) Charge
14. The negative differential conductance is realized in Gunn diode because of
    (a) Reduced effective mass (b) Increase mobility (c) reduced mass, increased mobility (d) Increased effective mass and reduced mobility
15. The negative conductance region in Tunnel diode is the region
    (a) Between zero and peak value (b) Between valley voltage and peak voltage (c) Between negative voltage and peak voltage (d) Valley voltage onwards

16. A normal GaAs working as LED and GaAs working as Gunn diode differ from each other on the basis of
    (a) Oscillator frequency (b) Cross section (c) Effective mass same (d) Photon contribution

17. Solid state detectors are operated at
    (a) Forward biasing p–n junction
    (b) On reverse biasing of p–n junction
    (c) Both forward and reverse biasing of p–n junction
    (d) No biasing of p–n junction is required

18. In Solid state detectors a resistor of suitable value is connected with diode and battery in
    (a) Series
    (b) Parallel
    (c) Sometimes in series and sometimes in parallel
    (d) None of these

19. The word photovoltaic comes from
    (a) Wind energy (b) Brightness (c) Light and electricity (d) Video

20. A solar cell converts
    (a) Heat energy into electrical energy
    (b) Solar energy into electrical energy
    (c) Heat energy into light energy
    (d) Solar energy into light energy

21. Solar cells are made of
    (a) Al (b) Ge (c) Si (d) Cd

22. The efficiency of Si solar cell is about
    (a) 25% (b) 15% (c) 40% (d) 60%

23. The voltage of solar cell is about
    (a) 0.5–1 V (b) 1–2 V (c) 2–3 V (d) 4–5 V

24. The output of the solar cell is of the order of
    (a) 0.5 W (b) 1 W (c) 5 W (d) 10 W

25. The solid state detectors are operated at
    (a) Low voltage
    (b) High voltage
    (c) Very high voltage
    (d) None of these

26. Solid state detectors are based on
    (a) Transistors
    (b) Ionization of gases
    (c) p–n junction.
    (d) Triodes.

Problems

1. Explain the principle and working of tunnel diode.
2. What are the essential difference between a semiconductor junction diode and a tunnel diode.
3. Explain the principle and working of a Gunn diode.
4. What is the dark current of a photodiode?
5. Describe the basic operation of LED.
6. Discuss the working of varactor diodes.
7. Explain the salient features of photodiode and its characteristics.
8. Discuss the photovoltaic process. Show that the photovoltaic effect increases logarithmically with illumination.
9. State the principle of photoconductive cell. Describe its construction, working and uses.
10. What is photovoltaic effect? Describe the construction of solar cell and sketch its characteristic curve.
11. What is the basis of laser action in semiconductor diode lasers?
12. What confine carrier recombination to a narrow optical gain region in a hetrojunction semiconductor laser?
13. Is hole–electron recombination is instantaneous in a semiconductor laser? Define quantum efficiency for semiconductor laser.
14. The laser beam from semiconductor lasers emerges over a wide range of angles. Explain. Why resonators are not used in semiconductor lasers?
15. Explain construction and working of semiconductor laser. Why Si or Ge is not used as a laser material? Discuss the drawback of p–n junction laser. Why hetrostructure is used in semiconductor lasers?
16. Why emission from semiconductor lasers is not as monochromatic as from the gas lasers?
17. Discuss the working of a semiconductor detector. Explain the advantages and disadvantages of it.

Answers
Objective Type Questions

1.	2.	3.	4.	5.	6.	7.	8.	9.
(a)	(b)	(a)	(a)	(c)	(b)	(c)	(c)	(b)

10.	11.	12.	13.	14.	15.	16.	17.	18.
(a)	(c)	(b)	(c)	(d)	(b)	(a)	(b)	(a)

19.	20.	21.	22.	23.	24.	25.	26.
(c)	(b)	(c)	(b)	(a)	(b)	(a)	(c)

Selected Bibliography

1. Abragam A, Bleaney B (1970) Electron paramagnetic resonance of transition ions. Clarendon Press, Oxford
2. Akitt JW (1992) NMR and chemistry, 3rd edn. Chapman & Hall, London
3. Atherton NM (1973) Electron spin resonance. Wiley, New York
4. Ayscough PB (1967) Electron spin resonance in chemistry. Methuen & Co.Ltd., London
5. Azaroff LV (2004) Introduction to solids, Tata McGraw-Hill
6. Bednorz JG, Müller KA (1986) Z. Physik B 64:189
7. Bleaney BJ, Bleaney B (1970) Electricity and magnetism, 2nd edn. Oxford University
8. Blinder SM (2004) Introduction to quantum mechanics, Elsevier Amsterdam
9. Bransden BH, Joachain CJ (1999) Quantum mechanics, 2nd edn. Pearson Education 2000. Cambridge University Press
10. Cao G (2004) Nanostructure and nanomaterials, Imperial College Press
11. Dekker AJ (1957) Solid state physics, Englewood Cliffs. Prentice-Hall, N.J.
12. Drozdov AP, Eremets MI, Troyan LA, Kenofontov VK, Shylin SI (2015) Nature (London) 525:73
13. Drozdov AP, Kong PP, Minkov VS, Besedin SP, Kuzovnikov MA, Mozaffari S, Balicas L, Bulakirev FF, Graf DE, Prakapenka VB, Greenberg, E, Knyazev DA, Tracz M, Eremets M (2019) 569:528
14. Edelstein AS, Cammarata RC (1998) Nanomaterials, Institute of Physics
15. Epifanov GI (1979) Solid state physics. Mir Publishers, Moscow
16. Fitts DD Principles of quantum mechanics as applied to chemistry and chemical physics
17. Fox M (2007) Optical properties of solids, Oxford University Press
18. Gordy W (1980) Theory and applications of electron spin resonance, Wiley, Inc
19. Greenwood NN, Gibb TC (1971) Mössbauer Spectroscopy. Chapman & Hall Ltd., London
20. Griffiths DJ (2005) Introduction to quantum mechanics, 2nd edn. Pearson Education
21. Harrison WA (1970) Solid state theory. McGraw-Hill, New York
22. Kamihara Y, Hiramatsu H, Hirano M, Kawamura R, Yangi H, Kamiya T, Hosono H (2006) J Am Chem Soc 128:10012
23. Kittle C (2004) Introduction to solid state physics, VIIth edn. Wiley
24. Levi AFJ (2003) Quantum mechanics, Cambridge University Press
25. Levy RA (1968) Principles of solid state physics. Academic Press, New York
26. Liboff RL (1998) Introductory quantum mechanics, 3rd edn. Addison-Wesley
27. Marfunin AS (1979) Spectroscopy, luminescence and radiation centres in minerals, Springer, Berlin
28. Neamen DA (2003) Semiconductor physics and devices. McGraw Hill, New York

© The Author(s) 2022
V. K. Jain, *Solid State Physics*, https://doi.org/10.1007/978-3-030-96017-9

29. Omar MA (2005) Elementary solid state physics, Pearson Education Inc.
30. Orton JW (1968) Electron paramagnetic resonance. ILIFFE Books Ltd., London
31. Pauling L, Wilson EB (1935) Introduction to quantum mechanics with application to chemistry, McGraw-Hill
32. Peierls RE (1974) Quantum theory of solids. Clarendon Press, Oxford
33. Pilbrow JR (1990) Transition ion electron paramagnetic resonance. Clarendon Press, Oxford
34. Powell JL, Craseman B (1961) Quantum mechanics, Addison-Wesley
35. Schiff LI (1968) Quantum mechanics. McGraw-Hill, Kogakusha, Tokyo
36. Singh J (2003) Electronic and optoelectronic properties of semiconductor structure. Cambridge University Press, Cambridge
37. Wert CA (1970) Thomson Physics of solids, 2nd edn. McGraw-Hill, New York
38. Wertheim GK (1964) Mössbauer effect: principle and applications, Academic Press
39. Zettili N (2009) Quantum mechanics concepts and applications, Wiley

Index

© The Author(s) 2022
V. K. Jain, *Solid State Physics*, https://doi.org/10.1007/978-3-030-96017-9

Printed in the United States
by Baker & Taylor Publisher Services